Attention, Genes, and Developmental Disorders

Oxford Series in Developmental Cognitive Neuroscience

Series Editor
Mark H. Johnson, Centre for Brain and Cognitive Development,
Birkbeck College, University of London, UK

Attention, Genes, and Developmental Disorders

KIM CORNISH AND JOHN WILDING

UNIVERSITY PRESS

2010

Oxford University Press, Inc., publishes works that further
Oxford University's objective of excellence
in research, scholarship, and education.

Oxford New York
Auckland Cape Town Dar es Salaam Hong Kong Karachi
Kuala Lumpur Madrid Melbourne Mexico City Nairobi
New Delhi Shanghai Taipei Toronto

With offices in
Argentina Austria Brazil Chile Czech Republic France Greece
Guatemala Hungary Italy Japan Poland Portugal Singapore
South Korea Switzerland Thailand Turkey Ukraine Vietnam

Published by Oxford University Press, Inc.
198 Madison Avenue, New York, New York 10016

www.oup.com

Oxford is a registered trademark of Oxford University Press

Library of Congress Cataloging-in-Publication data

Cornish, Kim.
Attention, genes, and developmental disorders / Kim Cornish and John Wilding.
p. ; cm. — (Oxford series in developmental cognitive neuroscience)
Includes bibliographical references and index.
ISBN 978-0-19-517994-1
1. Developmental disabilities—Genetic aspects. 2. Attention. 3. Attention-deficit
hyperactivity disorder—Genetic aspects. I. Wilding, John M. II. Title.
III. Series: Oxford series in developmental cognitive neuroscience.
[DNLM: 1. Attention Deficit Disorder with Hyperactivity—genetics. 2. Adolescent
Development. 3. Attention. 4. Child Development. 5. Developmental
Disabilities—genetics. 6. Genetic Diseases,
Inborn—genetics. WS 350.8.A8 C818a 2010]
RJ506.D47C67 2010
618.92'8588042—dc22

9 8 7 6 5 4 3 2 1

Printed in the United States of America
on acid-free paper

To my husband John and my children Luke and Holly,

For your endless encouragement throughout the writing of this book—it truly could not have been done without your support and patience.
In memory of Walter Cornish for being such a wonderful and inspiring father
—Kim Cornish

To Nicolas, Davidek, Vincent, and Roberta,

Hoping that your generation will use expanding knowledge wisely.
—John Wilding

Preface

Interest in attention as a complex cognitive skill has undergone something of a renaissance in recent years among theoreticians, scientists, and clinicians alike. Collaboration across seemingly diverse disciplines, most excitingly between molecular genetics, developmental cognitive neuroscience, psychiatry, and education, has culminated in a wealth of new research discoveries that have the potential to elucidate the typical and atypical development of attention across multiple levels of analysis: the genetic, the brain, the cognitive, and the behavioral levels. This book places this new generation of research into a developmental context and urges the reader to recognize the importance of tracing trajectories of cognitive function and dysfunction from infancy onward rather than focusing on the mature end state.

Our collaboration began over a decade ago when we began to explore disorders of attention in children with known genetic causes. Using novel experimental paradigms developed by John Wilding, we found that similar degrees of behavioral symptoms of inattention across different neurodevelopmental disorders did not necessarily imply equivalent cognitive attention mechanisms, with disorders such as fragile X syndrome displaying a quite distinct "signature" profile of functional impairment compared to that of children with Down syndrome or those with symptoms of attention deficit/hyperactivity disorder (ADHD).

Since this initial observation, we have been privileged to be part of an exponential growth of research advances on several fronts. New methods are now available for viewing brain activity in real time, information on the complexities of the biochemistry of neural activity has expanded, and there is a greater appreciation of the complexity of the effects of single genes and their downstream targets. Analysis of the component processes included under the broad umbrella of "attention" has become increasingly sophisticated, and ingenious methods have been developed for measuring typical and atypical development of these processes from infancy into childhood and then into adulthood. With these advances, however, has also come the realization that neurodevelopmental disorders are extremely complex and dynamic in nature and are therefore unlikely to have simple causal mechanisms that explain putative gene–behavior relationships. So our journey will continue.

Our intention in writing this book was not to provide an exhaustive review of the current literature. Instead, we hope to provide the reader with a unique journey of exploration across an often overwhelming array of research studies and contradictory findings, with the ultimate goal of understanding the gene–brain–cognitive processes that drive atypical development, in our case, disorders of attention. We conclude with a push for systematic research that can bridge the gap between these exciting scientific research discoveries and the uptake of these discoveries by clinicians and educators. We strongly believe that research of this nature must have application in the wider community in order to enhance the potential of all individuals with neurodevelopmental disorders.

Even as we wrote each chapter, the ground was shifting beneath us with the advent of sophisticated imaging methods and more detailed genetic analysis, but the need for improved theory and experimental methodology remains pre-eminent. We hope that we have provided a solid foundation and structure within which new findings can be evaluated.

Inevitably many others have been involved, and we would like to thank all our collaborators over this period. We would especially like to highlight our gratitude to Professor Annette Karmiloff-Smith, an inspirational scientist and pioneer in the field of neurodevelopmental disorders whose work has inspired our own efforts to focus on the role of development in understanding attention dysfunction across differing genetic disorders.

We are indebted to the many scholars who have read individual chapters and provided invaluable comments: Dr. Gaia Scerif, Dr. Cary Kogan, Professor Rosemary Tannock, Dr. David Hessl, Dr. Peter Enticott, Dr. Darren Hocking. Dr. Debbie Mills, Dr. Lucy Cragg, Julie Hanck, Anna Tirovolas, Janice Kerfoot, Marina Ter-Stepanian, Britt Dash, and Jacalyn Guy, to name but a few. We are grateful to McGill University and the Canada Research Chairs Program that enabled Kim Cornish to take a sabbatical year to complete this project with John Wilding. The staff at OUP, in particular the executive editor, Catharine Carlin, and the series editor Professor Mark Johnson, have provided insightful comments and much needed encouragement during the writing of this book. And finally, above all, we thank the numerous children and young adults who have taken part in our many studies: they and their parents are the foundation of experimental research in this field.

Contents

I

Advances in Attention, Genes, and Developmental Disorders
An Introduction

> **CHAPTER SNAPSHOT**
>
> - Overview of chapter goals and structure
> - The critical importance of interdisciplinary research that recognizes the interplay between genes, brain development, behavior, and cognition
> - Introducing the dynamic role of development in attentional research

---

Imagine that you are in the supermarket searching for your favorite brand of yoghurt. It is, of course, sold in a tub with a distinctive size, shape, and color of label. No doubt you can easily picture it in your mind's eye. Sometimes your regular supermarket will rearrange its shelves, or you may go to an unknown supermarket and have to search for this yoghurt. First, you have to find the area where all the yoghurt is, so you will wander around in a reasonably systematic way searching for the general pattern of shapes and colors that is typical of yoghurt tubs. Then, with hundreds of varieties to choose from, you have to look for tubs with those special visual features that you have in your head. Sometimes you may think that you have found the right ones, but on closer inspection you find them to be merely similar and have to resume the search. Occasionally you may be distracted by seeing a friend, being asked a question by another shopper, or watching the antics of a toddler, then you have to recall what it was that you were looking for. More serious interruptions could happen, such as fire alarms and bomb scares, in which cases your current plan would be suspended for a longer time before being retrieved. Finally, unless the manufacturers have made the irritating choice to change their packaging, you do find your target. You pick it up, put it in the cart, and switch to the next item on your shopping list. Once you have all the items, you will complete your plan by going to the checkout, paying, and returning home. Many things can go wrong in the course of even a relatively simple sequence like this, and we have already noted some of them. Movements and other (sudden) changes in peripheral vision can distract us from the task, as can loud noises. Sudden changes may indicate danger in any context, and therefore, in the course of evolution, the ability to switch attention automatically and rapidly in order to focus on such events became a high priority. Switching of attention is rapidly followed by realignment of eyes and head to bring the events into central vision. Other features of stimuli have enormous power to attract, such as bright colors and sexual triggers. Advertisers continually make use of these forms of stimuli to persuade us to look at something other than what we are searching for. The ability to avoid distraction from such things is quite limited, since nature prefers to play safe rather than sorry, though a more demanding task that requires more "effort" will be less subject to distractions. However, preexisting concerns or demands (such as a dentist's appointment) will not only themselves impair concentration but also make it harder to suppress other distracters. And the longer the task goes on, the more likely it is that something will cause a distraction and impede the finding of the target.

Searching for an object in a supermarket is not just an exercise in simple problem solving; it is full of demands on attention in various different senses of that much-used term. Indeed, we may say that all successful behavior that is not instinctively preprogrammed or well learned requires selection and organization of a correct sequence of actions from among competing alternatives and hence involves selective attention, maintenance of attention (often referred to as sustained attention) and control of attention. Even instinctive behavior involves selection, due to triggering of preset patterns by stimuli that have a special favored status. In humans these instinctive responses are fairly simple, like those to a loud noise, but in other animals, whole sequences of behavior may be controlled by automatic responses to specific stimuli, which override other inputs.

The past decade has seen an explosion of research into attention. This has been driven by the realization that attention is not some semimystical concept that is an optional extra to hard scientific studies of how the brain controls behavior, but is essential to all efficient responding to the multiplicity of input from the environment that constantly bombards the senses. This expanding research has been facilitated by advances on several fronts. New methods are now available for viewing brain activity in real time, there is expanding information on the complexities of the biochemistry of neural activity, individual genes can be isolated and their functions identified, analysis of the component processes included under the broad umbrella of "attention" has become increasingly sophisticated, and ingenious methods have been developed for measuring typical and atypical development of these processes from infancy into childhood and then into adulthood.

This book is concerned with attention and its development, both typical and atypical, particularly with disorders known or assumed to depend on genetic abnormalities or variations. Major advances across seemingly diverse disciplines, including molecular genetics, pediatric neurology, psychiatry, cognitive psychology, and developmental neuroscience, have culminated in a wealth of new research discoveries that can elucidate disorders at multiple levels. The growing linkage and cooperation across different disciplines have developed most rapidly in research on attention, where connections are being mapped in increasing detail between genetic, brain, cognitive, and behavioral levels. Furthermore, it has become increasingly obvious that a full understanding of links between these levels requires study of development from birth onward, rather than focusing mainly on mature performance. To these ends it is vital to achieve clear definitions of the concept of attention, the component processes involved, and how they develop from infancy onward. Our coverage reflects this need in analyzing the wealth of behavioral data on all aspects of attentional performance. The available knowledge on the roles of genes and brain mechanisms in cognitive processes is as yet less extensive than the wealth of behavioral data. However, we stress that all these research areas are equally important to a full understanding, and we have examined in detail the current state of knowledge on genetic and brain mechanisms at the time of writing. This picture is likely to expand rapidly as intensive research provides new findings in these fields, and their contributions will become increasingly comprehensive and significant in the next decade.

We focus here, therefore, on three specific yet interlinking levels: the genetic blueprint (genotype), the developing brain, and the behavioral-cognitive outcomes (phenotype). Our aim is threefold: first, to demonstrate the interactive role of genes in development and to explore a common assumption that atypical attention is directly correlated with specific genes and their encoded proteins. Second, we aim to demonstrate the dynamic role of development itself in defining disorder-specific signatures of attention deficit. Here we will explore critically the common assumption that general cognitive "delay" can explain the increased prevalence of attention deficits observed in many neurodevelopmental disorders and that the nature of such deficits, irrespective of their etiology, will remain static across developmental time. Third, we hope to demonstrate that commonalities in behavioral symptomatology do not necessarily imply common cognitive mechanisms or pathways. In this context we explore critically the assumption that ADHD symptoms, notably inattentive and impulsive behaviors prevalent in many neurodevelopmental disorders, reflect the deficiency of a common attention mechanism that will result in equivalent deficits in cognitive processing irrespective of the specific attentional demands of a given task.

This book was not written with the intention of providing an exhaustive review of the current literature. Instead, we hope to provide the reader with a unique journey of exploration across an often overwhelming array of research studies and contradictory findings, with the ultimate goal of understanding the processes that drive atypical development: in our case, disorders of attention. We focus on six core neurodevelopmental disorders that have well-documented attention deficits: attention deficit/hyperactivity disorder (ADHD), autism, fragile X syndrome, Down syndrome, Williams syndrome, and 22q11 deletion syndrome. All of these disorders either have known genetic etiology or have a suspected genetic linkage. Collectively hundreds, if not thousands, of studies have been conducted on our targeted disorders. We aim to provide the most authoritative and extensive account to date of disorder-specific attention signatures and their development from infancy through to adolescence.

The book is divided into four sections: Section I provides the reader with a series of three "tutorials" that provide a framework for the later exploration of typical and atypical attention development. As such, they can be "dipped" into as needed. Chapter 2 provides a comprehensive account of what exactly we mean by the term "attention." Does it refer to a single process or a variety of different processes, and if the latter, are these independent or connected in some way? We also attempt to clarify the distinction between attention and executive functions, control processes that organize a wide range of aspects of behavior that we will define in detail in Chapter 2. Chapter 3 provides a comprehensive review of what has become one of the most extensively researched areas in developmental neuroscience: the relationship between specific genes and putative cognitive outcomes. In the domain of attention, we ask whether there are any critical "attention" genes, and if so, whether they are relevant only to a specific disorder or to multiple disorders. We also take this opportunity to provide detailed genetic backgrounds on our targeted disorders, and the reader will get the first taste of the complexities surrounding this research and, in particular, the search for attention susceptibility genes. Chapter 4 provides a comprehensive review of the brain functions associated with attention and also provides a snapshot of recent findings of the atypically developing brain as viewed with the remarkable techniques now available in brain imaging research. The main questions guiding this chapter are: To what extent do the known brain areas involved in attention differ both structurally and functionally across neurodevelopmental disorders, and also from the typical developing brain?

Section II provides a unique critique of how attention is currently measured at the behavioral and cognitive levels. Across both levels we provide a snapshot of the known atypical *signatures* and any developmental profiles. In Chapter 5, we highlight the most frequently used attention rating scales and checklists and ask whether attentive behaviors should be viewed as a "continuum" in which inattentive behaviors represent the end of a normal distribution rather than being qualitatively different from the normal profile? As it currently stands, most scales use a discrete cutoff criterion for defining abnormal attention that fails to capture the range of inattentive behaviors that may be impaired to different degrees across different neurodevelopmental disorders. In Chapter 6, we highlight the huge variety of tasks that have been deployed and developed to measure all aspects of attention, with critical evaluation of their usefulness in this regard. Specifically, we address the appropriateness of such tasks in assessing attention profiles in atypical populations, again using our target disorders as examples.

Section III comprises our three key chapters in which we examine the development of attention in typically and atypically developing infants, children, and adolescents. Ingenious new methods for examining attention in very young children and illuminating new findings are described. Some revolutionary ideas about development will be considered, which question the hitherto widely accepted view that the brain is organized at birth for the creation of "modules" that perform specific tasks (e.g., face processing) and that development of these modules proceeds more or less automatically. Instead, findings suggest that brain organization is a result of competitive interactions between different brain regions responding to environmental input that leads progressively to the type of specialization that we see in the adult. Similar competitive interactions also feature in influential theories of attention, so these ideas are particularly relevant to our main topic.

Chapters 8 and 9 examine the development of attention across our six neurodevelopmental disorders. In both chapters we highlight several current complex issues, most notably concerning methodology (for example, criteria for matching groups with disorders and typically developing groups), experimental design, sample size, and composition. Chapter 8 focuses on ADHD, and we provide one of the first examples to date of the trajectories of attention deficit (in selection, maintenance, and control) in toddlers, children, and adolescents with ADHD. We critique the various theories of ADHD and conclude that no one current theory can fully embrace the range of deficits that have been reported in ADHD. We also ask to what extent there is stability in ADHD profiles across development or whether different profiles change with age. In Chapter 9, profiles and trajectories of attention are described, analyzed, and compared in our remaining five neurodevelopmental disorders. A further unique component of this chapter is an analysis of current cross-syndrome studies that serve to differentiate one disorder from another and thus provide critical evidence for specific genetically driven attention signatures.

Section IV examines the implications of these research discoveries, both typical and atypical, for the wider community of educators, clinicians, and families. We propose that research plays a pivotal role in guiding clinical interventions, educational support strategies, and government policy. In Chapter 10, we begin by first evaluating the contribution of clinical interventions, specifically the impact of psychostimulants in improving attention outcomes in children with differing neurodevelopmental disorders, but with a specific focus on the ADHD research literature. In the second half, we highlight some very interesting innovations that are aimed at promoting research-led practice in classroom environments. We describe some examples of recently developed resources that help teachers and parents recognize the unique attention signatures associated with different neurodevelopmental disorders and the specific strategies that can be used to facilitate learning across a child's academic trajectory.

Our final chapter draws together the findings reviewed across our core chapters and provides some future avenues of research that will hopefully benefit from the knowledge gained over a decade of exciting, intensive, and illuminating research. Not unexpectedly, new research innovations can also learn from the many past mistakes that are inherent in research of this nature.

We have emphasized throughout this book the importance of development in determining later attention outcomes in both typical and atypical populations. We give particular emphasis to the critical importance of situating research within a developmental context and moving away from a reliance on "snapshots" in time that can never truly provide a complete developmental *signature*.

Chapter Summary

- The importance of research that is rooted in an interdisciplinary framework that can investigate attention across multiple levels: genetic, brain, cognition, and behavior
- Recognition that attention covers a range of cognitive processes that include selection, maintenance, and control, and methods of study must vary appropriately
- The importance of examining changes in attentional profiles over development rather than assuming that such profiles are relatively stable over time
- The need for research to maximize the potential of a new generation of sophisticated technologies that can help elucidate attentional trajectories and profiles in atypical development from infancy onward
- The need to establish whether commonalities in inattentive behaviors across six different neurodevelopmental disorders (ADHD, autism, fragile X syndrome, Down syndrome, Williams syndrome, and 22q11 deletion syndrome) imply similar cognitive pathways and mechanisms

---

Section I: Attention, Genes, and Brains

2

What Is Attention? Navigating Its Complex History and Facing the Challenges Ahead

---

CHAPTER SNAPSHOT

- Introducing attention
- Taxonomies of attention: examples from Posner and Mirsky
- Four traditions of attentional research—selection, divided attention, maintenance, and control
- The complex role of "executive functions"

In Chapter 1 we described how a fairly simple task of searching for an item on the supermarket shelves placed a variety of demands on processes that can all be regarded as requiring attention in some sense of that wide-ranging term.

One broad distinction runs right through the example. First, there is attention in the sense of the deliberate, controlled focusing on the task at hand, initiated in response to our intention to search for specific objects in a specific place. As already pointed out, a plan has to be constructed, search specifications set up, and so on. This planning and control requires what are known as *executive functions* (EFs) that are located predominantly in the frontal regions of the brain (for example, Berger & Posner, 2000; Robbins, 2007). Such functions are particularly prominent in human behavior and enable us to pursue complex and long-term goals, especially in unfamiliar situations, avoiding distracters that might break the required chain of actions if EFs were not operating to exclude their influence. Such executive functions control selection, maintain the goal, switch from one subgoal to another, and revise the plan in the face of difficulty. Given the central role of EFs in attention processing, it is not surprising that we, alongside researchers worldwide, have focused a considerable amount of our energies into understanding their function and trajectories across development.

Secondly, there are many specific operations or processes such as scanning the displays in the shop or moving on to the next item in a list or automatically switching attention from searching the displays to a noise on the other side of the aisle as someone knocks a stack of boxes off the shelf. Switching to a sudden event, however, is outside our control. Attention moves from one location to another, and the switch is accompanied by a sudden increase in alertness that is also automatic. See Box 2.1 for a simple example of this process.

What Is Attention?

Attention is a complex cognitive domain that has seen a resurgence of interest in the past two decades. Two important advances have moved this field forward. The first is recognition that behavioral aspects of inattention, for example, distractibility, impulsivity, and disorganization, may have distinct cognitive and biological pathways that need to be empirically investigated and isolated. This research plays a critical role in understanding one of the most intriguing of neurodevelopmental disorders, that of attention deficit/hyperactivity disorder (ADHD), a disorder initially defined at the behavioral level but whose genetic and cognitive "signature/s" are slowly being unraveled. We describe these in detail in Chapters 3 & 8. Our point here is to introduce the reader to the idea that attention encompasses both overt behavioral and underlying cognitive components. Although our focus in this chapter is on defining the cognitive constructs underlying attention, we describe elsewhere (Chapter 5) the behavioral constructs of attention and the current struggle to find a consensus as to whether behavioral manifestations of inattention constitute a unitary or multidimensional construct. In later chapters (8 & 9), we attempt to reconcile the often complex interrelationships that exist between behavior and cognitive constructs of attention and are most explicitly captured in the search for disorder-specific attention phenotypes.

In recent years, there has been a gradual moving away from a belief that attention comprises a single, unitary process, and instead, there is a growing acceptance that "attention" covers a multitude of processes.

Box 2.1 How Do I Get to the Zoo? Planning, Selection, and Maintaining Attention in Searching a Map

Here is another example of a task requiring lots of different sorts of attention, which is closer to our overall theme of attentional problems in children. A teacher might show a simple street plan to the class with various landmarks on it such as the station, schools, churches, the post office and so forth, and set a little problem: "I want you to find on the map how to get from the railway station to the zoo."

What are the necessary processes, and in what ways might they go wrong?

The child has to set up a plan for solving this task, a series of steps that will work out a route—and maybe turn the solution into words if that is what the teacher wants. First, the child must find the railway station on the map, which requires searching for a particular symbol (assuming that the children have learned how all the different landmarks are represented on maps). Then, a second search is needed to find the zoo, while holding the location of the station in working memory.

Then, the child must choose a series of steps from all the possible ways of moving away from the station, guided at each choice point by which move is best for getting closer to the zoo. Some of these may lead to dead ends because detours are required that at first sight seem to be going in the wrong direction. In such cases, the choice has to be cancelled, and the child must backtrack to an earlier choice. All the time the eventual goal must be retained in memory, and so must the successful steps. Above all, distractions must be resisted, both those on the map (like thinking about what there might be to see in the castle) and in the classroom (like a friend whispering a message).

Think of all the ways that the various processes of planning, selecting, and maintaining attention could go wrong. There are many clues in the supermarket example at the beginning of Chapter 1.

Cognitive neuroscientists have attempted with great success to distinguish the different aspects of attention in a variety of ways. We describe here two well-documented attempts to separate attention into varying components and highlight some of the difficulties inherent in this research.

Taxonomies of Attention: Posner and Mirksy

The most notable and perhaps the most widely accepted of all current theories of attention is the *"attention network"* theory proposed by Posner and colleagues (see Posner & Peterson, 1990, for a pioneering formulation, and Berger & Posner, 2000; Posner & Rothbart, 2007b, for later reviews). In this model, attention is broken down into three cognitive processes that operate through a series of neural networks: *alerting, orienting*, and *executive control*. All three networks are needed for most tasks, and the observant reader will note from our summary in Box 2.2 that all three processes are readily apparent in our supermarket example!

We will be considering Posner's proposals for three distinct attentional systems in more detail later in this chapter and throughout the book, but here we note that they are based on intense use of a single task (Attention Network Task), and, while they represent an impressive analysis of the mechanisms involved in this task, they cannot be regarded as a comprehensive analysis of all the processes that have been described in the course of extensive and varied research on attention.

Another analysis of the different aspects of attention was provided by Mirsky (1996) and illustrates the problems of obtaining unequivocal answers in attention research. Mirsky attempted to identify common clusters of attention measures from a battery of well-documented paradigms. Using a statistical technique known as "factor analysis" that groups together measures that are strongly related to each other and hence are

---

Box 2.2 Posner's Attentional Network Systems

- *Alerting*—our ability to achieve and maintain a high state of sensitivity to incoming information critical for optimal task performance
- *Orienting*—our ability to selectively attend to specific aspects of our environment and also to disengage from an existing focus and focus elsewhere
- *Executive Control*—our ability to monitor and regulate competing actions and thoughts

---

Box 2.3 Mirsky's Four Attention Factors

- *Focus/Execute*—to concentrate attentional resources on a specific task
- *Shifting of attention*—to shift attentional focus from one aspect of a stimulus complex to another in a flexible, efficient manner
- *Encode*—to hold information briefly in mind in a mnemonic capacity while performing some action or cognitive operation on it
- *Sustain*—to stay on task in a vigilant manner for an appreciable interval

---

assumed to reflect some common process, he was able to identify four core factors from a total of 10 measures comprising just six tasks. These are represented in Box 2.3.

Unfortunately, Mirsky's analysis of the data is not convincing. Only six tasks were included to extract four factors. Though 10 scores were available, in all cases where a task yielded multiple scores, those scores all loaded on the same factor. Therefore, the scores all reflected overall performance on a single task rather than distinct operations carried out in completing several different tasks. It is also debatable whether each task unequivocally measured the operations claimed for it; to identify the critical operation in each case would require two or more measures from different tasks to load on a common factor so that a unique effect could be isolated by deducing what the common component was in the different tasks. This was not achieved. For example, the Wisconsin Card Sorting Task (WCST, described in Chapter 6) is a complex attention-switching task, and there is no consensus on exactly what it is testing, so it is unclear what the factor represents that was deduced solely from this task. Other tasks that were related, such as the Arithmetic and Digit Span subtests from the Wechsler Intelligence Scale for Children (WISC), are not normally assumed to tap explicit attention components; both tested memory and both involved numbers, so either of these features could explain the association between them. It is, moreover, unclear in what sense they involved "encoding" operations that were absent in all the other tasks. And if the Continuous Performance Task (CPT) was intended to measure maintenance of attention (see our detailed description in

Chapter 6), some index of change in performance over the duration of the task would seem to be required, not simply a measure of overall performance, which might reflect any one of a number of cognitive functions.

Looking for Consensus on the Varieties of Attention

The agreement between Posner's and Mirsky's approaches is only partial, and if we were to consider also other less comprehensive proposals concerning the subdivisions of attention, the lack of general consensus would become still more obvious. There is no quick and easy way of achieving a tidy solution to this problem. Few or no tasks can be unequivocally said to engage, for example, selection or sustained attention, and opinions differ about what a particular task might test and what tasks to include in order to test particular functions. Different sets of tasks often produce different factors, and it should always be borne in mind that what comes out of a factor analysis is a function of what goes into it.

Though opinions may differ about the precise components to include in a comprehensive theory of attention, on a more general level a popular view has emerged that makes a threefold division between *selective attention*, *sustained attention*, and *control processes*. In addition, *divided attention*, when performing more than one task at a time, and *alertness* (or *arousal*) are also frequently included, the latter being particularly important when attempting to maintain attention on a specific task, especially if the task is boring. These distinctions, evolved from the different traditions of research discussed below, are obviously related to those suggested by Posner, though he also provides a more detailed analysis by dividing them into subcomponents. We will be developing this approach to attention as a complex hierarchy of functions, with control processes organizing lower, more basic processes in order to achieve selection, maintenance of attention, and complex sequences of behavior. Such control depends on manipulating excitation and inhibition to achieve switching of the focus of attention, to pursue a particular sequence of operations, to simultaneously carry out two tasks, to modulate arousal, and so on. Some of these component processes can also operate independently, as when sudden events cause a switch of attention and an automatic increase in arousal, or when the intrinsic interest of an object automatically engages attention.

There is currently a great deal of research attempting to identify precise subcomponents of attention, not only by the types of tasks that evoke each operation, but also by the areas of the brain that are active in carrying out the different processes. Before examining neurophysiological evidence, however, we will review research that has used a variety of tasks to test the different abilities, such as those involved in searching for the yoghurt in the example at the beginning of Chapter 1.

In reviewing the existing literature, we will generally use these distinctions between selective attention, sustained attention, and control processes, plus divided attention. Box 2.4 provides snapshot summaries of the four "traditions" of attentional research and highlights some of the questions that drive this field. Clearly weaknesses can occur in any of these. Furthermore, we have noted that each of these complex processes can be broken down into simpler components, such as disengaging and moving attention in the case of selective attention, so that impaired selective attention could result from a weakness in any of the component processes. Similarly, problems in maintaining attention could be caused by poor functioning in any of a number of component processes. Thus, with careful analysis and experimentation, it is possible in some cases to identify a very specific cause for a weakness that can have wide-ranging effects.

Box 2.4 Traditions of Research

- *Selective attention*—Where and how in the analysis of sensory information is some input processed in detail and other information suppressed?
- *Divided attention*—To what extent can two streams of information be processed simultaneously?
- *Maintenance of attention*—What factors affect the ability to continue attending to one aspect of the environment and to ignore others?
- *Attentional control*—What is the nature of the processes that organize selection, division of attention, and maintenance of attention and ensure that complex sequences of attentional switching are efficient?

Equally, and more disastrously, a weakness may be due to problems in the control processes that are essential for organizing attention in complex tasks. We need, therefore, to be able both to specify the different components of attention that are engaged by a task and the control and planning demands of the task. Then we can begin to identify more precisely the situations in which a child who demonstrates a problem in attending is at a disadvantage. Is it some basic operation that is faulty, or is it the more general ability to plan and control the execution of complex tasks that is inadequate?

There has been a huge increase in publications concerned with attention since the 1960s and 1970s (Cavanagh, 2004). For present purposes, the modern psychological study of attention has four main roots: (1) information processing models of selective attention initiated by Broadbent (1958) and subsequently pursued and developed in a flourishing but inconclusive literature (see Styles, 1997, for a review); (2) related investigations of divided attention and the extent to which humans can carry on more than two activities simultaneously, which were given a theoretical framework by the publication of Kahneman's (1973) book, *Attention and Effort*, proposing the key concept of capacity as a limitation on performance; (3) interest in sustained attention provoked by attempts to increase the reliability of radar operators during the Second World War who were keeping a lookout for rare but important events in an otherwise uninformative environment; tests such as the Mackworth Clock Test (Mackworth, 1948) were devised by researchers in Britain to measure vigilance performance, and a theoretical model of performance in detecting signals of low intensity was eventually developed in the USA, the theory of signal detection (Swets, Tanner, & Birdsall, 1961); (4) observations of impaired cognitive skills in neurological patients with brain damage, particularly in the frontal lobes, such as studies of wounded combatants from the Second World War by Luria and others, and parallel studies of the effects of specific brain lesions in animals, especially primates. This work led to the emergence of the now burgeoning subdiscipline of neuropsychology.

We will now describe the main traditions of the research in each of these areas in order to outline the concepts and findings that have emerged.

Traditions of Attentional Research

Selective Attention

Controlled Selection: Broadbent's Work on Auditory
Selective Attention

The experimental study of selective attention in ecologically important settings began in 1958, when Broadbent published *Perception and Communication*, summarizing a series of groundbreaking studies on auditory selective attention. He chose to focus on auditory attention partly because a practical problem presented itself relating to the ability of flight control operators to monitor messages from several sources, and partly because selection from auditory input cannot be achieved simply by pointing the ears in the right direction. Hence, the role of covert internal processes is more crucial, whereas with visual attention, selection is often due primarily to direction of gaze. Broadbent found that when two simple messages were presented overlapping in time, either from separate loudspeakers or dichotically (separately to the two ears), generally all the items from one speaker (or to one ear) were recalled first, followed by all the items from the other source, rather than recall occurring in the original temporal order of presentation. From this finding he argued that some simple features of all the input were analyzed first, and a physical feature, particularly the location of the source, was used to segregate the input into separate streams, defined by the sources. Broadbent further argued that, due to the limited capacity of the central processing system, complex analysis could only be performed on a limited amount of input at a time, so one of the "channels" would be selected for further processing, while all the other inputs were excluded by a filter (see Figure 2.1). Input from the excluded channels could be retained only briefly in a buffer store, and information retained in this way might be processed further if it were still available once the priority input had been dealt with; otherwise, it was lost completely. Many other researchers subsequently adopted this approach, but it is now widely questioned and this important issue is discussed later.

Thus, the proposed selective system consisted of three components: a buffer store holding input from which only a few physical features had

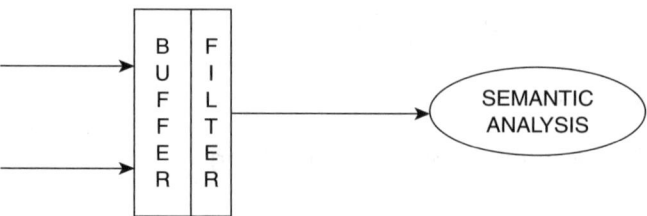

Figure 2-1. Broadbent's filter model.

been extracted, a filter that resolved competition by only allowing information with a specific physical characteristic to pass through (source on the left, right ear, male voice, etc.), and a central system that carried out further detailed analysis of meaning, with access to information in long-term memory. A similar visual system was proposed by Sperling (1960) following experiments on memory for letters and digits that had been briefly presented in three rows, each of four elements. About three or four items were typically recalled. However if, after presentation, recall was requested from only one of the three rows, recall of 3–4 items was again found, implying that this number of items was briefly available in every row, giving a total memory score of 9–12 items. Again Sperling suggested that a buffer store held all input, but only briefly. Once the usual 3–4 items had been recalled, the rest of the information in this buffer had decayed. So the results implied a short-lived buffer that holds the input, extracts simple physical features (the position in the display, in this case), then passes on a subset of this information, defined only by a physical feature, for further analysis and identification as individual letters or digits.

These early studies thus focused on the role of attention in stimulus processing. They envisaged a single, limited-capacity, all-purpose, inflexible system beginning with analysis of physical features of all input in parallel, followed by selection for more complex (semantic) analysis of a part of the input, and then leading to action, or at least awareness.

Problems for Early Selection Models

Some existing evidence already raised problems for this simple processing model. Cherry (1953) had developed an alternative method of testing auditory selection, the dichotic shadowing task. Separate passages of prose were delivered to each ear, and one of these had to be repeated back (shadowed), thus supposedly maintaining attention on only one ear. Requests for recall of any information from the other, unshadowed ear normally produced very low levels of performance, as the Broadbent model would predict, since unselected material would have only been analyzed for a few basic features. Cherry, however, found that if the listener's own name were included in the unshadowed passage, it would be noticed and recalled about a third of the time, suggesting that rejection was not complete. This is often called the "cocktail party phenomenon," as when you hear someone utter your name in a room full of chatter. It also became apparent in other tasks that inability to report an unattended item did not prove that it had not been processed at all. A number of studies demonstrated influence from unattended items in dichotic listening tasks on the responses to attended items (Corteen & Wood, 1972; Lewis, 1970; Mackay, 1973), and Tipper (1985, 2001) demonstrated the phenomenon of *negative priming*. An item that had to be ignored on one trial when identifying a specified target (e.g., the red letter when a red and a green letter were presented, and the green one had to be reported) took longer to process when it was presented as the target on the following trial, showing that it must have received some processing during the first exposure that inhibited further processing on the second exposure.

Treisman (Harvey & Treisman, 1973; Treisman, 1964, 1971) provided more evidence in support of this conclusion, also using the shadowing method. Though selection of one message against another was possible on the basis of a range of physical features (left vs. right ear, male vs. female, loud vs. soft voice, etc.), and differences in meaning or language were not helpful in enabling selection (as the filter theories predict), the meaning of information in the rejected ear was not completely unavailable. If the prose passages switched between ears in midsentence (see Figure 2.2), the listeners tended to switch their shadowing temporarily to follow the meaning, rather than continuing to shadow the same ear, as they had been instructed to do. This implied that meaning in the unshadowed message was at least potentially available and that absence of overt interference from that message did not prove that it was completely excluded. Treisman suggested that some elements (dictionary units) in long-term memory would respond to appropriate incoming information more readily than others. Such items as one's own name, important

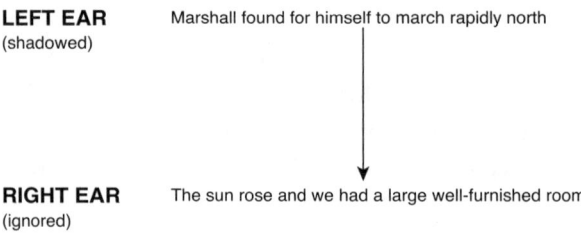

LEFT EAR   Marshall found for himself to march rapidly north
(shadowed)

RIGHT EAR   The sun rose and we had a large well-furnished room
(ignored)

Figure 2-2. Treisman's channel switching in the shadowing task.

elements with emotional connotations, and items primed by the meaning of preceding context, would have lower *thresholds* for excitation than other items and hence could be activated by a weaker input. Treisman suggested that, if Broadbent's filter weakened the unselected input, rather than excluding it completely, this theory could neatly explain the obtained results, because even weakened items in the unselected message could produce a response. Presumably this weakening occurred because fewer processing resources were devoted to the rejected input.

Deutsch and Deutsch: Late Selection

An alternative theoretical approach, pioneered by Deutsch and Deutsch (1963), argued against any early filtering system, claiming that such early exclusion is an unsound principle for survival. Ignoring brown animals is not a good rule if lions may appear. A sounder principle, they argued, would be to analyze all input to as great an extent as possible and select the most important of the competing channels as late in the sequence as possible. Clearly selection has to occur late if the meaning or importance of the input is the relevant factor for selection. They suggested that a threshold could be set so that important inputs exceeded it and were selected. This threshold would also vary depending on overall arousal, so that it would be exceeded more easily by the utterance of one's own name when awake than when asleep. Deutsch and Deutsch were unclear about how all inputs could somehow be graded on a single scale of importance before they were fully analyzed, and a simpler way of achieving this result would be through variation in the thresholds set for different items according to their current importance in the way suggested by Treisman. So, a person's name would be permanently set at a sensitive level, while the name of one's destination on a train journey would be temporarily adjusted to a sensitive level so that any announcements about it would attract attention. The history of research into selective attention has been dominated by this issue of the site at which selection occurs. A large number of supposedly decisive experimental tests between early and late selection have produced no clear resolution. As we shall see shortly, posing the question as an all-or-none issue was the main source of this stalemate.

Treisman's Feature Integration Theory

A type of selective attention task that is rather different from the shadowing and partial report tasks described above has become a major focus of more recent research. This is visual search, which requires participants to decide whether a specified object is present or not in a visual display. Use of this task has clearly indicated that in some situations there are strong effects from the nontarget elements, demonstrating that considerable processing of these distracters must occur.

Kahneman and Treisman (1984) distinguished between what they labeled *selective set tasks* like visual search, in which the target is specified beforehand by a single feature (or simple combination of features) and the response is simple (often Yes/No), and what they called *selective filtering tasks*, in which a complex input requires a complex response. The shadowing tasks and Sperling's partial report tasks, described above, are examples of the latter kind of task. Treisman's influence and the greater interest in and body of available research on visual processes have ensured that the majority of research in selective attention in recent years has focused on visual rather than auditory selection processes and on selective set tasks rather than selective filtering tasks.

Using visual search, Treisman and Gelade (1980) reported a fundamental difference between searching for a target distinguished from the distracters on a single feature (*feature search*, e.g., a green *H* among red *H*s or a black *H* among black letter *X*s) and searching for a combination of features among distracters that vary independently on both dimensions (*conjunction search*, e.g., a green *H* among red *H*s and green *X*s) (Figure 1.4 in Treisman & Gelade, 1980). In feature search, the number of distracters in the display had virtually no effect on the time it took to find the target, and in conjunction search, the decision time increased linearly as the number of distracters increased. In most real-life cases, conjunction search will be needed. It is unlikely that your tub of yoghurt has a single unique feature; it probably shares some features with one alternative brand and others with another alternative.

Treisman developed the feature integration theory (FIT) of visual selective attention to explain this difference between feature search and conjunction search. According to this theory, perceptual features are processed in parallel in different brain systems, which scan the whole display and respond to a feature such as color or shape. The system processing the target feature (color, in the example above) will automatically respond more strongly at the location where the target feature is present and activate a target-present response. This causes the target to stand out from the background (known as "pop-out"), irrespective of how many other items are present without that feature. This is compatible with what is known of the separate brain systems responsible for processing features such as color, shape, and movement. So if a green *H* in a background of red *H*s is present at the 11 o'clock position on the edge of the display, there would be strong activation in the green detector neurons that respond to that part of the display. Pop-out is independent of attentional focus elsewhere (Braun, 1998) and occurs more strongly when all distracters are identical to each other. It is likely, therefore, that it depends automatically on mutual inhibition between similar distracting inputs (all the red items interfere with each other), leaving the unique target with the strongest response.

However, where a combination of features defines the target, according to FIT, serial controlled processing is required to determine whether the two relevant features are at the same location. If a green *H* is presented among red *H*s and green *X*s, there will be many green responses in the color-detecting system and many *H* responses in the shape-detecting system, so a further process is needed to check whether green and *H* responses are occurring at the same location anywhere in the display. The additional process requires controlled matching (attention), according to Treisman and Gelade (1980). This distinction between *automatic processing* and *controlled processing* has become a major issue in attention research (see below for a more detailed discussion).

The FIT theory depends heavily on the assumption that coding of location is primary and preattentional in the visual system. The feature maps that encode shape, color, and other primary features are assumed to carry out such encoding across the whole visual field, and attention can be focused on a single spatial location when a unique color or shape is present; conjunctions of features are created by combining the different features that are present at the same physical location. This focus on location as a key feature in selection dates back to the demonstrations of Broadbent, Sperling, and others that location was an important cue for separating target information from irrelevant information, and their assumption that stimulus analysis involved a single series of steps from early physical to later semantic processing. Extraction of spatial position was assumed to occur early, prior to any selection process, and to be a major cue in subsequent selection.

However, it has become clear that this picture was incorrect on many counts. As we will describe in Chapter 4, location and object features are processed in two parallel and independent streams in the brain, location is not a single unambiguous feature but can be coded within several different frameworks, and links from input to output processes are apparent in the brain at a variety of levels, not solely at the end of a fixed sequence of analysis beginning with physical attributes and ending with semantic attributes interpreted through access to long-term memory. Consequently, many different selective attention processes can occur. Also, a variety of interactions can occur between the location and object-processing networks (see Allport, 1993, for a discussion of these issues, and Slagter et al., 2007, for an example of current evidence from brain imaging). As a result of these insights, a different theoretical tradition has emerged that places much less emphasis on location as a key factor in attention and suggests instead that it is already segmented *objects* like yoghurt tubs that are the focus of selective attention. (This theory necessarily also favors late rather than early selection.)

Duncan and Humphreys' Object-Based Selection

Duncan and Humphreys (1989) provided an alternative view to the approach that regards location as the critical variable in selection. They proposed that the visual field is first of all segmented into *objects*, and attention can only be focused subsequently on a selected object (see also Driver, 2001; Duncan, 2004; Scholl, 2001). For example, in an early demonstration, Rock and Gutman (1981) showed that attending to one of two overlapping pictures of objects produced no memory for the unattended object, even though it was at the same location. If attention is location based, it should process all items at the point of focus. Driver and Baylis (1998) cite other evidence indicating that segmentation and grouping in an array are more important in facilitating attention than focus on a location. If the location of distracters is held constant, similarities between distracting items, such as common movement or color (or a whole group of yoghurt tubs of a different make from the one being sought), enable them to be treated as a single group and excluded from attention all together much more easily than if they form a disparate collection of objects that cannot be combined into a single group. Also, judgments about whether two features are identical are more difficult if they are assigned to two different objects than if they occur in the same object (distance between the features being held constant in both cases), so separation into separate objects is the important factor (Duncan, 1980). In addition, Duncan has shown that it is possible to monitor more than one stream of visual input for targets and to detect the targets when they occur in any of the streams being monitored, provided that they do not occur simultaneously; it is, however, virtually impossible to detect two targets that occur simultaneously or to detect two objects that occur in rapid succession at the same location (Duncan, Martens, & Ward, 1997, did, however, show that it was possible to detect simultaneous targets in different sensory modalities.) This evidence all suggests that attention (and exclusion of other information) is restricted to a selected object rather than to a spatial location. Clearly, if attention focuses on objects, the input must have been separated into objects before this could occur. Therefore, segmentation into objects and into groups of similar objects must be preattentional, and preattentional processes that carry out this segmentation must involve more than simply parallel processing of features. They must also combine these features, so the FIT processing must precede selection.

Since Duncan and Humphreys (1989) argued that all selective processing requires the input to be already segmented into separate objects, which then become possible foci for selective attention, they could not accept that the difference between feature search (fast and parallel) and conjunction search (slow and serial) occurred because the former did not involve attention, while the latter required an attentional process to conjoin features and thus create an object description. For Duncan and Humphreys, attention is not a process that matches up features in order to create complex objects, but selection of already created objects. Why then should feature search and conjunction search differ so markedly?

Duncan and Humphreys suggested that the difference was due to the greater similarity between the target and the distracters in the latter case (because a green *H* among red *H*s and green *X*s shares one feature with all the distracters). This greater similarity would impede the segmentation of the target from the background objects. However, Treisman (1991) countered this argument by showing that conjunction search and feature search still differed in the same way when similarity was matched for both types of search. Furthermore, as will be shown in Chapter 4, conjunction search, unlike feature search, activates areas of the brain responsible for processing location, and damage to these same areas impairs conjunction search, offering further support to the Treisman position. Thus, FIT does appear to be correct in envisaging a key role for location in enabling the conjunction of visual features. The existence of a separate brain system for processing spatial position is in agreement with this view.

In conclusion, while Duncan and Humphreys demonstrate clearly that selection normally operates on already segmented objects, it is still possible that feature search involves a simpler process, whereby "pop-out" of the target occurs automatically due to mutual inhibition between identical distracters. Different levels of processing and different selective mechanisms are, on this view, operative in feature search and conjunction search.

Early and Late Selection in Visual Search:
Abandoning the Fixed Filter

Investigations by Lavie and colleagues (e.g., Lavie, 2001; Lavie, Hirst, de Fockert, & Viding, 2004), using visual search tasks, have suggested a resolution to the issue of where in the processing sequence selection occurs by demonstrating that the point of selection and the degree of processing

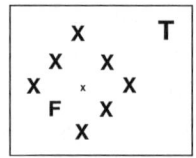

Figure 2-3. Lavie's singleton distracter task.

carried out on unattended inputs vary with the nature of the whole task context. They employed visual search for a target letter in which, in addition to the conventional nontarget distracters, a larger "singleton" distracter was sometimes presented, which would evoke the opposite response to that evoked by the target (see Figure 2.3). So if the task was to decide whether an *F* or a *T* was present in a background of *X*s, a single larger distracter (singleton distracter) was also present in the periphery, which might compete with the target (e.g., if the target were *T*, the peripheral distracter could be a competing *F* that had to be ignored, or a neutral letter). A singleton distracter in the form of the alternative target produced conflict and slowed responses to the target, but—and this was the key finding—this interference was reduced as the overall number of other "normal" distracters got larger. Hence, though the number of distracters normally has no effect on decision time in a feature search task of this kind, this factor does have an indirect effect on the influence of other distracting information. That is to say, the more demanding the task of segregating the target from the background letters, the less the capacity to process the distracter. Hence, when the task was more demanding, there was less interference from this distracter on the main task of identifying the target. These findings neatly fit the suggestions of Deutsch and Deutsch that as much of the input will be processed as possible, taking into account the overall load. When the main task is less demanding, the distracting letter will be processed and interfere with the main task. These variations in the degree of interference, and by inference, the degree to which distracters are processed, are an apparently automatic consequence of variation in perceptual load.

These findings clearly demonstrate that there is no fixed "filter" that excludes unwanted inputs completely, but that selection can occur at different points, and varying degrees of analysis of unattended material can occur. In Chapter 4, a variety of neurophysiological evidence will be cited that demonstrates that a great deal of analysis proceeds in parallel in the visual system, with selective procedures occurring at different sites and in a variety of ways, supporting the conclusion that selective influences of attention may occur earlier when task demands are higher. The findings undermine the assumption of theorists such as Broadbent that selection is always early and always all-or-none in order to protect from overload a limited-capacity central processor involved in analyzing all inputs. Rather, the degree of analysis, and therefore selection, depends on the processing resources available.

A further prediction was derived by Lavie et al. (2004) about factors affecting the degree of distraction that occurred. They argued that as well as the automatic effect of perceptual load, distracter effects depended on the efficiency of control processes. Therefore, if the control system were made less effective, by being already engaged on some other task beforehand, the distracter would have a greater effect. Note that this proposed effect of greater load on the control system is in the opposite direction from the effect of perceptual load at the time of target detection, where greater load on basic visual processing *reduced* the distracter effect. This new prediction was that the greater the load on the system that exercises control, the greater the effect of the distracter would be and the less efficient that system would become. So if in the supermarket you are thinking about a dentist's appointment or trying to remember what you had on the shopping list, you are more likely to mistake a different yoghurt tub for the one you are seeking. Lavie et al. identified working memory as a likely candidate for the exercise of these control processes and therefore assumed that by occupying this control system in a prior task, they would reduce its efficiency in directing the selective search process effectively (see Box 2.5 for a description of the Baddeley working memory model). Exactly this effect was found when working memory was engaged in a memory task presented just before the visual search task. Furthermore, to anticipate some of the discussion in the next chapter, when more activation of prefrontal areas of the brain (which support working memory) was observed, distracter interference increased. The inference is, of course, that such activity was due to the memory task rather than effective inhibition of interference, and the former interfered with the latter.

Box 2.5 Baddeley's Working Memory Model

Baddeley's model of working memory (Baddeley, 1986) envisages a central executive with two short-lived memory stores, the phonological loop holding sequences of auditory/articulatory items that can be refreshed by rehearsal within a period of about 2 seconds, and a visuospatial scratch pad that holds visual patterns briefly. The central executive has been less intensively investigated than the other two components but is now widely identified with the executive functions that we introduced earlier. It is assumed to be able to store and manipulate inputs from the senses or from long-term memory, and hence (among other things), it is responsible for setting up a series of instructions for performing a task that will activate the required procedures in the correct order. In 2000, Baddeley introduced a fourth component, the episodic buffer that provides temporary storage of information held in a multimodal code, that is capable of binding information into a single integrated structure or episode.

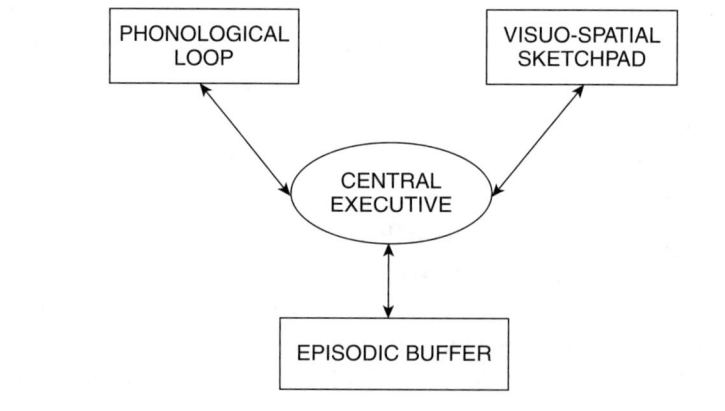

To sum up Lavie's conclusions, selection is not all–or–none but rather varies by degree, depending on other aspects of the task, in particular the perceptual load imposed by the display that needs selective processing and the current load on control processes. As much of the input as possible is analyzed to as great an extent as possible, irrespective of instructions about what to ignore, and final selection is left until late in the sequence. However, determining the appropriate balance between ignoring distracters altogether and permitting them to be analyzed, with the consequent risk of interfering with the main task, requires activity of the executive functions in the frontal lobes.

Why Is Selection Needed?

If attention is not automatically required to protect a limited-capacity processing system, why and when is it needed? One alternative approach to explaining the need for selection points to limitations on motor responses rather than limitations on input since only one response can normally be made at any one time, and selection of the most important input and response is necessary to prevent fatal immobility. The fact that, with practice, simpler responses can be combined into a complex response, or the fact that two well-practiced response sequences, such as driving and talking, can be performed approximately simultaneously does not fatally undermine this argument of the need for selective processes in many situations.

Another explanation of the need for selective processes was suggested by Desimone and Duncan (1995), who strongly reasserted the view that there are limitations on simultaneous processing of inputs and pointed to the necessity for selection due to the nature of the neural systems at higher levels. In the part of the perceptual system that processes information about objects, neurons at early stages in the processing sequence respond to a simple feature appearing in a small specific part of the visual field, while neurons at later stages in the sequence respond to complex inputs occurring anywhere over a wide area of the visual field. That is, they have *receptive fields* that are very large, covering an angle of some 25°, unlike neurons at earlier stages, which have fields of only 1° or so. The receptive field of a neuron is the area on the retina that, when stimulated, will evoke a response in the neuron. Thus, many objects in a scene could simultaneously stimulate any single neuron at the later stages of processing, so selection is necessary to ensure responses occur to the desired object or feature and not to irrelevant ones that also fall within the receptive field of the same neuron. In their *biased competition theory*, interactive competition between attributes serves to select the strongest candidate. Prior tuning from control systems (see below), as well as input features, will bias such competition according to current motives and goals, so that the required target evokes the strongest response. Attention therefore resolves competition between inputs or responses. We pointed out that it was difficult or impossible to perform two motor responses at once, so that one must be selected to avoid total inaction. Theorists such as Desimone and Duncan are suggesting that earlier processes in the sensory systems are also structured so that competition results in one input dominating and suppressing activation by other inputs. Without such selective processes, stalemate and inability to act might result. Thus, attention is necessary for any responses to be possible. Without it, you might remain in the supermarket forever with many types of yoghurt competing for your response!

New Issues: Capacity and Control

This more complex theoretical approach that we are now describing incorporates two important constructs in current attention research: processing "capacity" and control systems. We have demonstrated in the above discussion that, if there is flexibility in how much and what input is to be processed, and also in the degree of processing to be carried out, the assumption of a filter located at a fixed point in the processing sequence and working on rather simple criteria becomes untenable. Furthermore, questions as to how selection can be varied in different circumstances become critical. The issue of how (or even if) the filter was controlled was not discussed in the earlier studies, which barely considered such issues. Questions also arise about how much information can be processed simultaneously, to what extent processing can be carried out and responses evoked in parallel by two or more separate sources of information, what factors affect performance in such situations, and which systems organize these processes. These are the concerns of research on ability to divide attention between different tasks and acquire complex skills, to which we now turn.

Divided Attention and Attentional Capacity

The original theoretical underpinning of the need for selective processes was that the brain was a limited system that needed to avoid overload in order to function adequately, and that this had to be achieved by early selection systems that would prevent unnecessary processing of irrelevant inputs and hence ensure detailed processing of relevant inputs. The evidence discussed so far has shown the inadequacy of this position.

In his seminal book *Attention and Effort*, Kahneman (1973) developed a model of a flexible system that could both vary the available capacity, through deploying more or less "effort," and assign such capacity flexibly in accordance with the task in hand. Hence, more than one input channel might be processed if the total load were not too high.

While the demanding and unfamiliar tasks on which the original filter theories were based (such as dichotic listening, shadowing, and rapid presentation of large displays) did seem to restrict processing quite narrowly, this was not totally true even in these situations and may have been due to the shadowed message (for example) being, in effect, a continuous series of targets. Duncan's (1980) demonstration, described above, that only one target at a time can be processed, though more than one input stream can be monitored, would therefore explain restricted attention in the shadowing task. Other complex situations have, however, been studied that did enable two channels to be processed in parallel. The key factors appear to be the extent to which inputs and responses are naturally compatible, the degree to which different inputs (and different responses) overlap with each other in their demands on the processing system, and the degree to which the relevant skills have been practiced. Two distinct inputs, such as piano playing from sight by experienced pianists and shadowing (Allport, Antonis, & Reynolds, 1972), can be handled largely in parallel, and long practice with other pairs of tasks can produce similar results (see Spelke, Hirst, & Neisser, 1976, who trained two students to read stories and write down the categories of a series of acoustically presented words simultaneously). (See Azuma, Prinz, & Koch, 2004; Hazeltine, Ruthruff, & Remington, 2006; and Koch, 2008; Ruthruff, Van Selst, Johnston, & Remington, 2006 for more recent treatments of this issue). We may note at this point that the notion of attention has been extended in these studies to include the selection and control of sequences of input and action. This more comprehensive view of attention will also be apparent as we now consider control systems that modulate connections between inputs and responses as well as selection of inputs. This viewpoint enables us to consider the role of attention in a much wider range of situations.

Doing Two Things at Once: Controlled
and Automatic Processing

There is some doubt about whether pairs of tasks such as those considered so far really engage full capacity completely and continuously, or
whether rapid alternation between the two tasks might be possible.
Posner and Boies (1971) used a simpler task with better control of these
possibilities. The task required participants to decide whether two suc-
cessively presented letters were identical or not and respond by pressing
one of two keys with the right hand; they also had to press another key
with the left hand whenever a tone occurred (see Figure 2.4). The results
showed that responses to the tone were delayed when it coincided with
the presentation of either letter, or especially when it coincided with the
interval between the two letters. Clear evidence, one might suppose, for
a problem in processing two inputs at once, but when McLeod (1978)
substituted a spoken response to the tone instead of the key press, the
interference was eliminated. Evidently, interference depends on the
nature of the competing inputs and outputs, and on which responses are
required for which inputs. McLeod's results suggest that speech is an
automatic or *privileged* response to acoustic stimuli and is therefore produced easily, while a manual response is not. Consequently, the speech
response can be executed without interference from the visual judgment
process. Shaffer (1975) had already demonstrated that simultaneous copy
typing of a visually presented passage and repeating back another acoustically presented one was relatively easy, while the alternative combinations—typing to dictation and reading print aloud—produced a difficult
task. Likewise, for musicians, presumably playing an instrument has
become a well-practiced privileged response to a musical score, so that it
can be done at the same time as shadowing (Allport et al., 1972). To
anticipate, some unpracticed task combinations require new patterns of
stimulus–response connections to be set up by control systems; these can
be learned with practice and become automatic. At the unpracticed stage,
there are heavy processing demands that interfere with any concurrent
processes. Eventually the two sequences in the combination can proceed
independently, though there is likely to be some residual mutual interference, even when the combination is highly practiced. So you may be
able to do the supermarket shopping and dictate your memoirs on to a

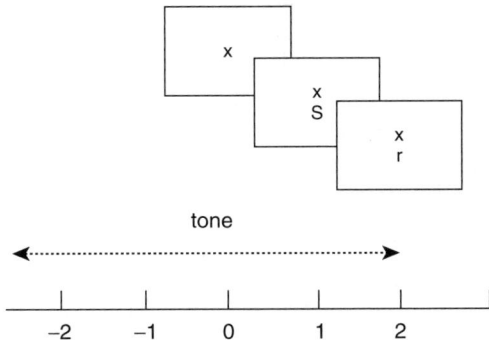

Figure 2-4. Posner and Boies' task.

pocket tape recorder at the same time, but only after considerable practice will you do this at all efficiently.

In two influential papers, Schneider and Shiffrin investigated the
ability to search for more than one target at a time (Schneider & Shiffrin,
1977; Shiffrin & Schneider, 1977). They, too, drew a distinction between
controlled processing and automatic processing. They constructed two
visual search conditions. In one, the targets were always letters and the
nontargets were always digits. There could be up to four possible target
letters and up to four items in the display, which might or might not
contain one of the target letters (see Figure 2.5). In this condition, the
size of the display to be searched (i.e., the number of items in the display
and hence the number of distracting items) did not affect decision time.
The letter target popped out automatically in a similar way to that
observed in feature search described above. In the other condition, both
letters and digits could be either targets or distracters, so the target set
might be *R 3 4 J* and the display set might be *T 5 P 2* (a no-target display). Thus, finding a letter in the display no longer guaranteed that it was
a target, and indeed, a letter that had previously been a target might later
appear as a distracter. In this condition, therefore, the control system had
to update the target constantly, and decision time increased linearly with
the display size.

However, Schneider and Shiffrin showed that this difference
was neither all-or-none nor due to anything intrinsic about the two
categories of written letters and digits (which are, of course, arbitrarily

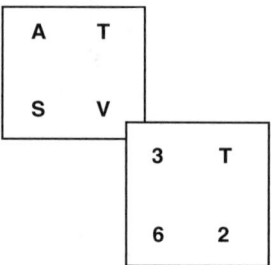

Figure 2-5. Shiffrin and Schneider's task.

constructed), but rather, it was due to long practice in learning to treat these two classes of items separately. They carried out a further study (Shiffrin and Schneider, 1977) in which they used the first half of the alphabet as targets and the second half as distracters, and after a very long period of practice, the search became automatic, and the number of distracters had no effect on time to find the target. Such training, therefore, can establish special networks for discriminating between two categories that are initially not distinct. The implication is that there is a continuum ranging from well-learned stimulus–response connections that are readily triggered by minimal input, through to unfamiliar stimulus–response associations that have to be constantly monitored and refreshed by priming from a control system (see also Cohen, Aston-Jones, & Gilzenrat, 2004, for a detailed discussion of control processes in attention and Schneider and Chein, 2003, for a review and development of the original Schneider and Shiffrin theory).

Duncan, Williams, Nimmo-Smith, and Brown (1993) have argued that attentional control may be closely related to general intelligence. Greater processing capacity (or intelligence) should be most apparent in difficult tasks, for example, when attempting to carry out two tasks simultaneously without prior practice. With practice and the development of skill, the programs for carrying out novel tasks become more autonomous and less dependent on total processing capacity. Hence, Duncan et al. argued that individuals with greater overall capacity (higher IQ) would show less interference than those with lower IQ between two simultaneous tasks that had not previously been practiced together.

To date, however, the evidence for this proposal is weak. Kane and Engle (2000) found that, in a task requiring divided attention, working memory capacity was unrelated to ability to exclude interference from earlier trials (proactive inhibition), and Oberauer, Lange, and Engle (2004) found little support for a relation between working memory capacity and ability to resist interference or coordinate two concurrent tasks. Whether working memory capacity, as defined by these authors, is equivalent to processing capacity in the sense of Duncan et al. (1993) is uncertain. The relations between these and related constructs, such as executive function, inhibition, and fluid intelligence, are unclear. These relations are discussed further below when we consider the effects of damage to the frontal lobes of the brain.

Summary of Selective and Divided Attention

Theories of attention have developed from a simple notion of fixed selection with a rigid sequence of processing to a more complex picture of variations in selective strategy dependent on overall task demands. The theoretical role of control processes has become more prominent, as selection among sensory inputs is seen as part of much wider system for selection between actions. Ability to cope with difficult tasks, due to their novelty and complexity, is dependent on such control processes, which coordinate selection of stimuli and responses and enable novel combinations to be carried out initially, then, with repetition, to become practiced and eventually autonomous. We return to these issues below in the discussion of attentional control, but before this we consider research on the maintenance of attention.

Maintenance of Attention and Arousal

The third tradition of attentional research involves investigations of ability to maintain attention. There are two aspects to this type of attention. One is concerned with short-lived switching and holding of attention and involves an increase in *arousal* (phasic changes) when a significant event or a warning signal occurs; the other involves the ability to maintain efficiency over a longer time span (tonic arousal). So in our supermarket example, the sound of falling boxes will cause our attention to switch, and we will focus on the source of the noise long enough to discover whether it is important to us and try to keep our mind on the main task, finding the yoghurt, and not to be distracted into something else. These two aspects of attention maintenance are related, respectively, to exogenous and endogenous control of attention and are encapsulated in research into the orienting reflex on the one hand and sustained attention on the other. Both phasic changes and tonic arousal involve concepts of activation, arousal, effort, alertness, awareness, and related ideas and engage functions of the reticular activating system of the brain stem and a number of other structures.

Maintaining Attention in Low-Input Situations

The first investigations of longer-term ability to maintain focus on a location or task looked at situations where little stimulation was arriving and targets were infrequent. This is usually referred to as vigilance. More recently the majority of research has been concerned with responding to occasional targets in a long sequence of events, most of which require no response.

In the 1940s, a practical problem presented itself because those watching radar screens for the occurrence of signals of approaching enemy planes showed a rapid decline in efficiency as the watch progressed. The period of efficient operation was surprisingly brief, as little as 20–30 minutes, after which the rate of missed targets increased steadily and responses became slower. The decline could be partially averted by a brief break or some arousing input such as a telephone call (Mackworth, 1948). The original explanation offered for this decline was that arousal declined as the task continued, and this produced increasing internal events like blinks of the visual system or a brief "nodding off," causing inputs to be missed so that no information was delivered to the neural decision system.

However, this simple explanation has turned out to be only partially true. Some 30 years later, Davies and Parasuraman (1977) found that when they measured evoked potentials in a vigilance task, the changes in brain activity that occurred over the course of the task suggested that it was not a decline in the efficiency of processes detecting the input that was responsible for the declining performance over time. Rather, it was brain activity associated with the decision-making aspects of the task that changed. Consequently, they argued that, while arousal was related to the overall level of performance (drowsy observers did worse than alert ones from the start), it was not a change in the level of arousal that was critical in determining decline, but changes in the observer's readiness to report a target (see Parasuraman, Warm, & See, 1998, for a more detailed discussion of the causes of the decline in performance over time).

But did observers get worse because they didn't see the targets or because they were less willing to report them? The key question was whether it was the ability to discriminate targets from nontargets, and targets from the background, that deteriorated in such situations (that is perceptual efficiency), or whether it was a change in the degree of caution (the "response criterion") being exercised by the observer. By "response criterion" we mean the amount of evidence required in order to make a specific response; a cautious criterion means that more evidence is required. We have already encountered an example of variation in caution or readiness to respond in Treisman's explanation of why some unattended items could be identified and others could not. She suggested that the threshold for activation was lower in the case of important elements such as one's own name, equivalent to responding to such elements on the basis of less evidence. In the 1950s signal detection theory (Swets, Tanner, & Birdsall, 1961) had provided a model and mathematical procedures for distinguishing between these two types of explanation. While a full exposition of this theory is unnecessary here, the issue will arise in due course as to whether impulsive responding (i.e., setting the response criterion at too low a level so that selection of a response occurs before adequate evidence has been obtained) is characteristic of children with attention deficit/hyperactivity disorder (ADHD) (see Chapter 8), or whether poor performance is due to an inefficient discrimination system. Hence, it is useful to expound SDT briefly here.

Signal Detection Theory (SDT) and Decline in
Attention Over Time

Returning to our original example of an observer watching a radar screen for signals of incoming aircraft, the observer has to decide whether a blip on the screen is random noise or a signal from a possible enemy aircraft (random interference, otherwise known as noise, can arise either in the radar system or in the nervous system). The quality of the observer's perceptual system will determine how accurately this distinction can be made. A better optical and neural visual system will discriminate between signals from real targets and those due to random noise more accurately; poor discrimination would lead to more errors in both directions (mistaking targets for nontargets and vice versa).

However, performance might vary between different observers or over time in a different way, depending on variation in preferences for reporting or not reporting possible signals as aircraft; such preferences will be determined by the potential consequences of different types of error, the expected probability of an aircraft appearing, the risk-taking propensities of the observer, and so forth. A cautious observer, one who expects only a few targets to appear, or one who is nervous about being reprimanded for making false reports, will set a conservative criterion for deciding that an event on the screen is due to an enemy aircraft and only make such a report when a large amount of evidence has been gathered in its favor. Such an observer will miss some targets but will also make fewer mistakes of reporting enemy aircraft when only random noise is present (*false alarms*). Conversely, an observer who is anxious to catch all possible intruders will report them on the basis of relatively little evidence and necessarily categorize some random noise as enemy aircraft, but only rarely *miss* real targets. Note these two different types of error that are possible. You might like to work out how they can occur when searching for the yoghurt.

The two sources of variation of individual variation in performance—precision in the detection system and caution in the criterion that is set—are independent and can be measured separately with the mathematical model established by SDT. Using this method, it turned out that with time, observers missed more targets, not so much because their detection system became less efficient but because they became more conservative in the amount of evidence they required before they would report a target (Stroh, 1971, p. 65). Though they missed more targets, they made fewer false alarms to nontargets. Also, their responses became slower. If their perceptual efficiency had declined, they would have made more mistakes of both kinds, misses and false alarms, but this did not occur. And, to complete the picture, if the task required a frequent nontarget event to be distinguished from a rare target one (flashes of different brightness), the "nontarget" responses became faster and the "target" responses slower, implying that the preference for reporting the former increased (Parasuraman & Davies, 1976). If decline in arousal had been the critical factor, all responses should have become slower, but the obtained result indicated that these nontargets were rejected more readily as time progressed, fitting the overall picture of a growing reluctance to report a target.

A popular explanation for these changes in response criterion is that observers begin with some expectations of how frequently targets will occur, but after they miss some or realize they initially overestimated the likelihood of such events, they become more cautious in identifying targets. This causes them to miss more, and so a downward spiral of detection frequency develops. Such a process is independent of any changes in sensory efficiency and must be ascribed to adaptive processes in some control system that evaluates success rates and modifies the criterion accordingly.

Maintaining Attention in High-Input Situations

There are some situations, however, in which perceptual processes do become less efficient over time on the task. The change in criterion described above is typical of vigilance situations where few target events occur and target events are defined simply in relation to the background (e.g., any blip on the screen). However, when the task presents inputs in rapid succession, and targets are defined in relation to other background events (e.g., a brighter spot on the screen among the occurrence of less bright spots, or a letter X following an O, as in the frequently used Continuous Performance Task, which we discuss in Chapter 6), ability to discriminate targets from nontargets does decline over the watch period. Searching for the yoghurt is another situation of this kind; as the search continues unsuccessfully, you are more likely to miss the target or to mistake some similar brand for the one you want. These situations impose a high perceptual load, and when the discrimination task is demanding, the decision mechanisms become overloaded and efficiency deteriorates. Such deterioration might be explicable in terms of reduced arousal or effort and concomitant reduced "capacity." Recent research on maintenance of attention has tended to use tasks belonging to the second type, such as the CPT, that are more common in ordinary life than watching blank screens for something to happen. The demands of such tasks are rather different from those of vigilance tasks, and hence, results should not be generalized from one type of task to the other.

Capacity and Arousal

We have presented evidence that declining arousal is not an adequate explanation for declining vigilance performance over time, but arousal remains a central construct in many attentional theories. We have already made brief mention of the proposed link between capacity and arousal when citing the theory of Kahneman (1973). He argued that the amount of capacity or effort deployed to cope with a task depended on assessment of task difficulty. Increased effort increased physiological arousal, which could be measured in several ways. Kahneman's favored method was to measure the degree of dilation of the pupil of the eye, with greater arousal producing a more dilated pupil. According to Kahneman, the task difficulty or task demands can be assessed by employing dual tasks. The greater the demands of a primary task, he argued, the less capacity is available for a subsidiary task, so performance on the latter provides an index of the demands of the primary task. However, we have seen already that the results of such dual tasks depend on the exact nature of the combination used, with some pairs of tasks easier to perform simultaneously than others. Hence, it is clearly an oversimplification to suggest that there is some fixed capacity available, and that we can measure how much of it a given task occupies by seeing how well a secondary task can be performed at the same time. In general, because no independent method of measuring capacity has evolved, this has not proved a particularly useful construct in theories of attention.

Pribram and McGuiness (1975) provided a more sophisticated elaboration of the construct of arousal and related concepts. They argued for three systems. One controls arousal and is based in the amygdala, an area in the midbrain involved in emotion that is responsible for phasic, short-term orienting responses to stimuli; a second system controls activation, is based in the basal ganglia of the forebrain, and produces tonic (longer-term) readiness to respond; the third system is based in the hippocampus and coordinates the other two systems, so it is equivalent to effort in Kahneman's sense. Sanders (1983) and his colleagues (for example, Sergeant, Oosterlaan, & van der Meere, 1999; Sergeant, 2005), have adopted and elaborated this model, postulating connections between the three energetic systems and different processing stages: arousal and encoding; effort and central processing; activation and response organization. They also followed Kahneman in linking effort to arousal (and also to their own construct of activation) and added an additional control system, termed *management* (see Figure 2.6). This addition seems somewhat redundant as effort appears to fulfill much the same function. The same research group has used this schema to investigate possible loci for the attention impairment in ADHD, arguing for an energetic rather than a cognitive deficit as the source of this disorder (e.g., Sergeant, 2005; see our critique in Chapter 8). However, the basis for these proposed connections between energetic and cognitive constructs is not clear and the linkages seem somewhat arbitrary.

What will emerge more clearly in Chapter 4 is that maintenance of attention is heavily dependent upon control processes in the frontal lobes of the brain, especially the right hemisphere, and the latter is also linked

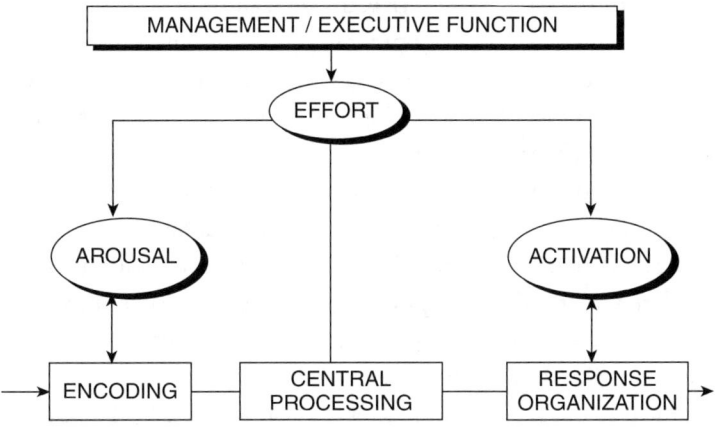

Figure 2-6. Sergeant's cognitive-energetic model.
From Sergeant, J. A. (2005). Reprinted by permission from *Biological Psychiatry 57*, 1249.

to the reticular activating system, which modulates arousal. The frontal lobes appear to exercise a controlling role in coordinating the necessary processes to maintain arousal and focus on a specific input or task.

Summary of Arousal and Maintenance of Attention

Short-term switches of attention involve the orienting reflex and arousal. Maintenance of attention requires interplay between control processes and arousal. Decline in ability to maintain performance over time may depend on changing criteria for responding when targets are rare and nothing else is happening, or to reduced perceptual efficiency following processing overload when input is continuous and targets are fairly frequent. These two types of test for maintenance of attention make different demands and need to be considered separately. Chapter 4 will show the importance of interactions between control processes in the frontal lobes and arousal systems in the brainstem in modulating arousal and maintaining efficient performance.

Attentional Control

We now turn to the fourth tradition of research, which developed out of different concerns from those that stimulated research into the other three areas. In the cases of selective, divided, and sustained attention, the most fruitful theoretical developments came through efforts to address practical questions: What improves selection or maintenance of attention, or to what extent can human beings carry on two or more tasks simultaneously? Tasks were devised to test the ability of interest and then varied systematically in order to reveal critical variables and enable models of the relevant processes to be developed. Specific tests of such models could then be undertaken. In recent years, with increasingly sophisticated methods of exploring brain activity, such models have helped to guide explorations of the relevant centers in the brain. Findings from these explorations will be described in Chapter 4.

In the case of executive functions (EFs), research has followed a rather different pattern. There has been a paucity of systematic attempts to analyze component processes in complex tasks and to develop standardized tests for these processes based on results from normal human beings.

In fact there is, as yet, no general consensus on whether a single complex control system or whether a variety of separate processes is to be assumed. Instead, the driving force for theory development has been observations of the breakdown of performance in varied situations following brain damage (principally to the frontal lobes) and the attempt to explain a diverse set of observations in terms of a unifying theory. For example, in a close parallel to our example of searching for the yoghurt, one real-life task sometimes employed in experimental studies has been to send participants out with a shopping list to test their ability to organize such an activity.

In consequence, though a number of researchers have associated damage to specific areas with specific operations, we do not have an agreed set of cognitive operations or of experimental procedures that can then be applied in cases of breakdown of functions to determine what specific control process is deficient. Still less do we have key manipulations of a given task that will implicate a specific process and reveal the critical cause of the impairment. As Denckla (1996) points out, few of the most frequently used tasks to test control functions include a matched condition that omits the assumed demands of the standard task on EFs but is equal in all other respects. Instead, a number of (mainly) rather complex tasks have been employed, the demands of which can be interpreted in several ways, leaving the exact implications of poor performance unclear (see Pennington, Bennetto, McAleer, & Roberts, 1996, for a detailed critique). These tasks have emerged not so much because it is clear what function they test, but largely because they tend to be performed poorly by patients with frontal lobe damage, and it is thereby inferred that they test some aspect of executive function. The danger of circular argument is obvious. Furthermore, no great consistency in results has been obtained with many of these tasks. It is not clear whether this is due to the constitution of the tasks, the fact that repetition of a novel complex task fundamentally changes the strategy employed, or the variety of functions dependent on the frontal lobes. The frontal lobes are a complex structure with complex links to the rest of the brain and support a wide range of functions. Uncontrolled damage is likely to affect a subset of these functions, different in different clinical populations. Moreover, several different neurotransmitters are involved in this area of the brain, presumably serving different functions. Genetic variation, in which we are particularly interested in the present context, can presumably affect any one or more of these transmitter systems, which would imply that results might be more consistent in these cases than in cases of physical damage through stroke or accident. These relations are explored in Chapters 3 and 4.

The Complex Role of Executive Functions

In 1922, the Italian neuropathologist Leonardo Bianchi noted that monkeys with damage to the frontal lobes of the brain would respond to objects in the same way as before their brain lesion but fail to carry through previously purposeful sequences of behavior. The animal would notice a handle previously used to open a door and grasp it but not follow through the sequence and open the door. Other observers have reported a variety of effects of damage to the frontal lobes of the brain. A well-established finding (Jacobsen, Wolfe, & Jackson, 1935) was the failure of such animals in delayed recall tasks. Two cups would be shown and food placed under one of them, but access would be delayed for a few seconds. Animals with frontal lesions appeared unable to retain the position of the food for more than a second or so. Originally interpreted as a problem in short-term memory, this weakness is now generally acknowledged to be part of a wider problem of behavioral organization. Malmo (1942) showed that reducing distractions helped and proposed that the animals were unable to continue focusing on the main task, being too readily distracted by other events in the vicinity.

Similar problems in purposive behavior were widely noted in soldiers with bullet wounds in the frontal lobes received during the First World War. Distractibility and difficulty in handling new situations were reported. One of the main pioneers in developing this new discipline of neuropsychology was Aleksandr Luria in Russia. His work was not well-known or available in translation in the West until some years later (see Luria, 1973, for a popular account). The earlier celebrated case of Phineas Gage (Harlow, 1868), who suffered a horrifying injury from an iron bar driven through the frontal part of his skull, also demonstrated a major personality change following the injury, though the accuracy of the reports has recently been questioned by Macmillan (2008). A previously organized competent worker became unruly, unstable, and unemployable, demonstrating that emotional aspects of behavior are also normally controlled from the frontal lobes.

So, we may well ask, how does this rather broad and varied picture fit into the concept of attention and illustrate deficiencies in specifically attentional aspects of behavior? Is it not just the ability to retain information and hence to learn that is impaired? The evidence for pure memory problems is not strong according to Pennington et al. (1996), who report that recognition, cued recall, and learning new information are often not impaired (see also Barkley, 1997). However, Turner, Cipolotti, Yousry, and Shallice (2007) have recently reported a pure memory problem associated with damage to medial areas of the frontal system and a retrieval problem associated with lateral damage.

Our task now, therefore, is to try to establish more precisely the nature of these control functions, the extent to which they are interconnected or independent of each other, and the experimental procedures that can be used to identify them. We will also need to address again the issue of the distinction between "controlled" processing that heavily involves such functions and "automatic" processing that involves them only minimally, and whether the difference between these two modes of operation is solely a matter of learning. In a later chapter we will discuss the typical development of the functions of the frontal lobe, such as planning ability, and the degree to which such functions depend on experience or only the maturity of the nervous system.

However, there is a need for caution. While the discussion to date may suggest that there is a one-to-one relation between frontal lobe operations and the cognitive construct of executive or control functions, such an assumption is overly simplistic. Some aspects of control lie outside the frontal lobes, and obviously the frontal lobes include operations other than executive control. Moreover, some supposedly key executive functions can apparently survive frontal damage.

Norman and Shallice's Model: Contention Scheduling
and the Supervisory Attentional System

For a discussion of control processes in attention, a useful framework is provided by Norman and Shallice's model of attention (1980, 1986, Figure 2.7). They argue that many aspects of routine behavior proceed

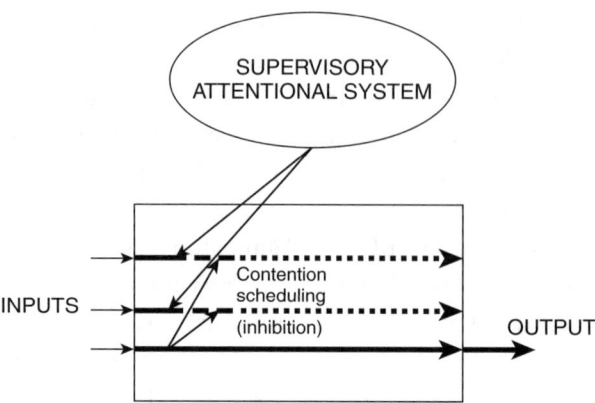

Figure 2-7. Norman and Shallice's SAS model.

without conscious control. "Decisions" about which one of several inputs to select and which one of several responses to make are often resolved by competition between possibilities. A set of such possibilities exists in the brain (as in Treisman's dictionary units) with thresholds set in accordance with long-standing importance and current motivations and priorities, and the first of these units to receive activation that raises the overall level above the threshold will fire and initiate behavior. This will, in turn, suppress other distractions until and unless they themselves create a high level of activation. So, the hand will pick up a cup if thirst plus the sight of the cup wins over hunger plus the sight of a bread roll, and the result will be to inhibit responses to the roll and activate those to the cup. Norman and Shallice called this system "contention scheduling." It has considerable resemblance to Deutsch and Deutsch's suggestion regarding selective attention (discussed earlier). Clearly, for all but very primitive segments of behavior, the routines between which selection occurs depend on a great deal of prior learning that has enabled them to become automated and to run off without thought. Automation means that the sequence of motor operations required is tightly linked together so that each one triggers the next appropriate action and not some irrelevant one. For complex behavior, such automated combinations are usually called *skills*.

However, Norman and Shallice also postulated an additional system to handle controlled behavior when a novel sequence of responses was required, perhaps to be learned for future automatic performance, as in learning to drive a car or to play a musical instrument. Sometimes this system is also needed to suppress previously automated sequences in favor of a novel sequence or at least to avoid responding to a potential distracter that might elicit an inappropriate response. For example, when proofreading the reader must focus on the orthographic patterns rather than the meaning of the words. Norman and Shallice proposed a supervisory attentional system (SAS) to fulfill this function, which resembles a flexible and more complex version of the original filter proposed by Broadbent. This system is attentional because it *selects*, not just individual stimuli, but a particular sequence of operations; a stimulus and the appropriate response may be selected, and the links between each such "unit" of behavior and the next. Execution and increased efficiency (i.e., learning) of complex sequences of behavior would be impossible without such initial selection, whether achieved by executing a series of instructions directly controlled "online" by the SAS or by setting up in advance excitatory and inhibitory links between components of the action to ensure that the chain of events occurs in the appropriate order. Hence, the SAS incorporates a set of executive functions. Baddeley (1993) equated the SAS with the central executive of his working memory model, which was outlined above in the discussion of Lavie's experiments, and this suggestion is supported by the study of Redick and Engle (2006), which will be described below.

Norman and Shallice proposed five general functions for the SAS: coping with novelty, planning or decision making, resisting temptation, trouble shooting, and dealing with dangerous situations. Basically, these reduce to two main functions, devising responses when a well-learned routine is not appropriate and suppressing habitual responses, or even to the single function of inhibiting the strongest response in order to enable alternatives. However, whether a single underlying mechanism is involved or whether several overlapping processes are to be distinguished is an empirical issue. The most plausible assumption is that the system is a hierarchical one where processes such as inhibition are activated in appropriate patterns to cope with different task demands.

Decomposing Executive Functions

Many other suggestions have been made, using various terminologies, for component processes in EFs that overlap with Norman and Shallice's list and with each other (e.g., Garon, Bryson, & Smith, 2008;

Miyake et al., 2000; Monsell, 1996; Pennington et al., 1996). These range from rather general constructs, like those given by Norman and Shallice, to the far more detailed and specific breakdown provided by Monsell. It is not possible to derive an unambiguous list of independent functions, but some of the main common themes are as follows:

1. Planning a strategy to cope with a novel task. These situations will include most complex novel tasks, including situations where dual tasks have to be handled. Situations requiring planning can be ana-lyzed to determine the specific task demands and the operations required to meet these demands, and these operations can also be divided into more specific subprocesses. For example:

   1a. One important aspect of planning is setting up and controlling the *execution of sequences* of selecting inputs and responses
      (i) These may in turn require *switching* between different inputs and responses and
      (ii) updating information as the task proceeds.
   1b. Many of the necessary operations for planning and sequence control involve both holding information in working memory and *inhibiting* inappropriate responses, particularly in situations where a well-learned response has to be replaced with a weaker, novel one, or has to be delayed or suppressed completely, or where switching from one input or response to another is required, or where interfering inputs have to be excluded. Whether or not these different forms of inhibition are distinct remains an open question.

2. The second major theme of executive function is *maintaining attention* to a plan, especially where there is strong competition or rel-evant inputs are intermittent, as in a vigilance situation. Such maintenance is also required when a response has to be delayed following a signal, or when a signal is awaited. Inhibition of responses to irrelevant inputs is clearly also involved in both these situations.

Various other functions have been cited less frequently: suspending action when an interruption occurs; holding the relevant instructions in a temporary store and resuming when appropriate (again, inhibition and storage are involved); or monitoring the results of actions and correcting errors. Careful analysis of impairments caused by damage to the frontal lobes (and other areas), and the increasing use of brain-imaging methods, will doubtless enable these constructs to be refined and clarified and their independence or nonindependence from each other to be defined.

While all these functions cannot be said unequivocally to serve *attention* in an agreed sense of the term, they are all crucial to coherent, organized, and goal-directed behavior that is to be maintained in the face of distractions. They are therefore relevant to any enquiry into genetic weaknesses that may affect such behavior. We need to define the possible sources of impairment as precisely as possible in order to understand the nature of such weaknesses and possible remediation. However, it is clear that overelaboration of potential components is likely to generate major problems in devising appropriate and unambiguous tests for each process. Moreover, the distinctions drawn are intuitive rather than empirical in most cases. A further implication is that at least some of the above processes are involved in all tasks of any degree of complexity, and this commits us to a view of attention as a feature of all organized behavior, rather than occurring in only certain types of tasks.

Decomposing EFs: The Analysis of Mikaye and Others

One attempt to derive more precise descriptions of EFs is the work of Miyake and colleagues (Friedman & Miyake, 2004; Miyake et al., 2000), who used structural equation modeling to test a model of three of the more specific EF components that have been suggested. In their second study, they tested the differences between three hypothesized variations of inhibition. They attempted to uncover the latent variables underlying a large number of tasks. Latent variables are the hypothesized common element underlying a group of tasks, and in structural equation modeling several tasks are administered that are assumed to fall into several different groups; each group of tasks reflects a different latent variable. The method tests how well the data match the initial assumptions about the task groupings. Miyake et al. administered nine tasks hypothesized to fall into three groups of EF components described in Box 2.6. Consider how these different operations are involved in searching the supermarket to fill your shopping basket.

Box 2.6 Miyake's Three EF Components

- Shifting between tasks or mental sets
- Updating working memory representations
- Inhibition of prepotent (i.e., dominant) responses

The proposed model of three EF components provided the best fit to the data, though intuitively it might seem that both set shifting and updating might require some inhibitory action. These three components were also differentially associated with the traditional tests of EFs. Set shifting was related most strongly to the Wisconsin Card Sorting Task (WCST), updating to Working Memory Span (WMS), inhibition to the Tower of Hanoi (TOH), and both inhibition and updating to Random Number Generation (RNG). This study by Miyake et al. was the first comprehensive attempt to elucidate the main component processes that together comprise EFs and supported the original assumption of three main components, Set Shifting, Updating Working Memory, and Inhibition (though they are unlikely to be completely separable). The study also supported assumptions that have been made about the key features of several popular tests of EFs, namely that WCST is testing ability to switch attention, that WMS depends heavily on ability continuously to update the content of working memory, that TOH requires ability to inhibit a strong but incorrect response, and that RNG requires both inhibition of overlearned number sequences and keeping a running record of recent responses. Lehto, Juujärvi, Kooistra, and Pulkkinen (2003) found a similar factor structure when testing children aged 8 to 13 years on a somewhat different range of tasks, but Huizinga, Dolan, and van der Molen (2006), while again identifying the shifting and working memory components, found no clear inhibition component.

In their further study, Friedman and Miyake (2004) tested a model of three possible different types of inhibition, using proposals by Nigg (2000) and others that distinguish between Prepotent Response Inhibition ("Behavioral Inhibition" in Nigg's terminology, e.g. Stroop task), Resistance to Distracter Interference (Interference Control, e.g., a word-naming task with a distracter word present), and Resistance to Proactive Inhibition (Cognitive Inhibition, e.g., working memory tasks where earlier lists may interfere with recall of later ones). Structural Equation Modeling demonstrated two rather than three distinct inhibition factors. The first factor included measures of both Prepotent Response Inhibition and Resistance to Distracter Interference. The second factor included only measures of Resistance to Proactive Inhibition, suggesting that the memory component was the distinctive feature of this factor, in which case the findings match the suggestions of Roberts and Pennington (1996) and Unsworth and Engle (2007, see the next section) that inhibition and working memory storage in combination are the key processes in EFs.

Despite the inevitable queries associated with the resulting models, the type of analysis carried out by Miyake et al. is superior to anything else available and makes a start in disentangling the complicated interrelations of the many functions proposed as components of EF. The main problem, of course, is that the factors that emerge are inevitably a function of the set of tasks employed, and different sets of tasks tend to produce different types of factor. We note that Miyake et al. did not include any tasks requiring maintenance of attention. It would be interesting to see whether inhibition (of any kind) is related to performance on such tasks or whether a separate maintenance factor might emerge. Furthermore, none of the tasks they used were matched with a control task lacking only the EF demands in order to test whether those demands were the critical feature in each case (cf. Denckla, 1996).

Working Memory and Attention

Unsworth and Engle (2007) have considered possible relations between working memory and inhibition in more detail. They have recently developed an analysis of one aspect of working memory operation that incorporates selective inhibitory processes closely resembling the two types of inhibition identified by Friedman and Miyake. This proposal (using different terminology) also develops the relation between the central executive of working memory and the SAS that was suggested by Baddeley. This work of Engle and others has focused on individual differences in working memory capacity (WMC).

Since the pioneering studies of Daneman and Carpenter (1980), working memory capacity has conventionally been measured using variations on a task that we have already mentioned, designed to measure working memory span. This type of task places requirements on both processing (e.g., evaluate the truth of a sentence or an arithmetical expression) and memory (e.g., after a series of such judgments, recall the words that accompanied the sentences or equations that had to be judged). The higher the word recall, the more spare capacity is assumed to have been available for processing and storing the words, hence the larger the WMC of the individual (however, we cast some doubt above on the validity of such a measure).

Redick and Engle (2006) found that groups differing in WMC also differed in their ability to resist misleading cues in Posner's Attentional Network Task, but not in the effects of alerting or orienting cues. This supports the suggestion of Baddeley and others that attentional control systems operate via working memory, and also indicates that some aspects of attention are independent of working memory.

Unsworth and Engle (2007) further analyzed the nature of WMC. They examined differences between individuals with high WMC and those with low WMC and proposed that the capacity difference depends on two processes. The span task requires maintenance of the words to be recalled in primary or immediate memory (PM) and/or their retrieval from secondary or long-term memory (SM). Maintenance in PM requires exclusion of irrelevant distracting information (from the senses or from memory) that might supplant the critical information. In the event of this maintenance failing, items have to be retrieved from a pool of possible items in SM, using contextual cues to narrow the possibilities and select the most plausible candidate(s). In particular, in the span task, intrusions from earlier lists have to be excluded by using cues specific to the most recently presented set of words. Hence, high ability both in maintaining information by inhibiting irrelevant interference and in inhibiting irrelevant contextual cues in retrieval will increase WMC. Unsworth and Engle (2007) present several lines of evidence in favor of this analysis, but we should also note the findings of Kane and Engle (2000) and Oberauer, Lange, and Engle (2004) quoted earlier, which showed no relation between WMC and dual-task performance. It would seem, therefore, that not all supposed EFs are closely related to WM. Dual-task performance was also unrelated to Friedman and Miyake's EF factors, further suggesting that it may have distinct properties.

Nevertheless, the linking of WM and EF processes through inhibitory constructs has considerable promise for clarifying the tangled concepts in this field.

Summary of Executive Functions

All new situations require sequences of responding to be planned to cope with them. Planning and execution of such sequences involves selection in working memory that enables switching between different inputs or responses, and updating of the representations of the current task situation. These operations require combinations of excitation and inhibition that are set up to achieve the current subgoals and longer-term goals. Thus, the complex construct of EFs can, in principle, be decomposed into a collection of operations, each achieved by the inter-play of excitation and inhibition and organized within working memory. A further necessary function is the maintenance and refreshing of goals and attentional sets in the face of distracting inputs, again involving the interplay of excitation and inhibition. As we have indicated frequently, the attention system is most plausibly conceived as a hierarchical organi-zation, in which basic processes are recruited to achieve selected aims as they arise.

The Varieties of Attention

At the beginning of this chapter, we identified three main aspects of attention that have been studied extensively across five decades: selection, maintenance, and control, together with ability to divide attention and arousal. Our analysis of the history of research in this field has attempted to develop a more integrated approach while considering these different areas of research.

At a number of levels, we have identified *selective* processes as essen-tial for adaptive behavior; these may involve selection of input or response from competing possibilities, selection of information in memory, or selection of a more complex sequence of actions. Some types of events produce automatic allocation of attention by evoking arousal and orient-ing to their location, but in many more cases selection depends on controlled allocation of attention in accordance with existing goals and motivation.

In all cases selection requires a combination of excitation and inhibition, whether achieved by already existing interconnections among elements that have been established during neural development or learning (as in some visual processes), or through a control pattern set up for a specific temporary purpose (such as "ignore the green items," "search for a specific design of yoghurt tub").

Complex chains of behavior require integrated sequences of selection through interconnections that trigger some sequences of responses and inhibit others. Initially, control of such complex sequences (which are rarely innate in human behavior) appears to depend on a "program" set up by the central executive of working memory, but once they are well practiced, such sequences become automatic and no longer require active control. When automaticity develops, it may become possible to carry out two or more sequences of behavior concurrently.

Attention is facilitated by arousal (at least up to some level), and arousal may also be stimulated by external cues that predict an important event (alerting) or attract attention (orienting), or by internal motivation. Maintenance of attentional focus on a specified stimulus source or course of action requires maintenance of arousal through brain centers that have reciprocal connections with the control system identified above. This system is therefore critical in the management of the overall organization of behavior, since it marshals both subsidiary processes and arousal in order to ensure integrated and effective action, free of competition from irrelevant distraction.

Though the degree of independence of the different aspects of attention is still under debate, it is clear that any comprehensive study of attentional development and disabilities in children requires testing of a wide range of functions ranging from simple processes like alerting and orienting to more complex control of behavior such as the ability to maintain focus over time on a stimulus source or to organize sequences of switching between different stimulus sources.

We turn now to examine in detail the link between attention and brain processes, and in Chapter 5 we review many of the tasks purporting to test the different functions that we have distinguished and assess their validity. In particular, we will attempt to highlight tasks that provide the best prospects for testing in children with attentional disorders across the different processes that we have identified.

CHAPTER SUMMARY AND KEY FINDINGS

- Selective processes are the core of attention; some are automatic but most require control processes to organize selection.
- Control processes (executive functions) marshal excitation and inhibition to select inputs and actions and organize complex sequences of behavior.
- Interactions between executive functions and arousal systems ensure that alertness and efficient functioning are maintained.
- Future research must continue to disentangle component processes involved in complex tasks and the hierarchical systems that enable selective behavior.

3

Genes and Atypical Attention

<table>
<tr><td>

CHAPTER SNAPSHOT

- Introducing our six neurodevelopmental disorders: autism, fragile X syndrome, Down syndrome, Williams syndrome, 22q11 deletion syndrome, and ADHD
- The complexities of linking genes to cognition—can developmental disorders lead the way?
- Searching for susceptibility genes in autism and ADHD
- Isolating "attention genes" in disorders of known genetic etiology

</td></tr>
</table>

Genes are located on the chromosomes and are specified by their sequence of bases in DNA. An estimated 20,000–25,000 genes comprise the human genome, representing a much smaller number than initially assumed. In recent years the quest has been to isolate gene function and specifically gene regulation or expression because genes play a critical role in influencing our characteristics by providing a hereditary "blueprint" with information needed to manufacture proteins. These encoded proteins provide the necessary building blocks for the development of functional skills and interactions of the organism with its environment.

We are witnessing one of the most exciting times in tracing the complex indirect route from genes to cognition. Huge advances in technologies have provided critical new insights into the world of genes, their function, and their impact on early brain maturation and cognitive development. New collaborations, hitherto unexplored, between the disciplines of molecular genetics, behavioral genetics, neurobiology, developmental neuroscience, and brain imaging have facilitated discoveries, alongside challenges, in our quest for understanding the dynamic pathways of gene–behavior interactions. This emerging interest in the complex mapping from genes to cognition represents one of the most innovative and potentially insightful avenues of research to date, but it also constitutes one of the most frustrating for researchers. As the reader will see throughout this chapter, and elsewhere in our book (see Chapters 8 & 9 specifically), it is rarely, if ever, the case that a specific gene or specific genes can be mapped in a linear fashion to an observable behavior. This quest is frequently referred to as the "genotype–phenotype relationship" ("*genotype*" is a term used to describe our genetic makeup, and "*phenotype*" is a term used to describe observable characteristics that are the joint product of both genotypic *and* environmental influence) and far from being simple is, in fact, extremely complex. To provide an example, gene expression (conversion of the information encoded in a gene first into messenger RNA and then to a protein, and the effects that behavior can have in return on gene expression), as we shall see later in this chapter, plays a fundamental role in predicting certain phenotypic traits. However, gene expression does not have a static impact on development but rather plays both a direct and indirect role on the developing brain, which in turn must also interplay with environmental influences and triggers. It should also be noted that behavior affects gene expression in return (as an example, see the excellent work of Meaney and colleagues, e.g., Diorio and Meaney, 2007), and so by ignoring the powerful and dynamic role of the unfolding environment in shaping our phenotypic end states, we limit the potential of future genetic-behavior studies to inform on two levels: first, to elucidate the multiple pathways that lead to a genetic variation (Altshuler, Daly, & Lander, 2008, provide an excellent critique of current state of play in genetic mapping of complex disorders), and second, to elucidate the critical role of development in investigating pathways leading from a specific genetic variation. We cannot assume that later end states directly reflect early infant start states. For the interested reader, Karmiloff-Smith (1998, 2007, 2009) provides an elegant critique of the narrow genetic mapping that some models of development advocate and of their limited application to understanding the complex genetic-environmental pathways that define many neurodevelopmental disorders.

Neurodevelopmental Disorders

The staggering advances made in the field of molecular genetics over the past 15 years have resulted in a wealth of new knowledge that has enabled many disorders to be more precisely defined at the genetic level, including those with already established etiologies, such as Down syndrome, first recognized in the 1860s, and those disorders that have been more recently characterized, such as Phelan–McDermid syndrome. There is also another group of disorders, such as autism and dyslexia, for which genetic etiology is not in doubt, but the genes involved are yet to be unambiguously determined. In these disorders, researchers have, however, begun to make substantial inroads into identifying possible "candidate" genes that contribute to susceptibility in certain families.

In this chapter we will focus on six well-documented neurodevelopmental disorders that have known attention impairments: autism, fragile X syndrome, Down syndrome, Williams syndrome, 22q11 deletion syndrome (formally known as velocardiofacial syndrome or DiGeorge syndrome), and attention deficit/hyperactivity disorder (ADHD). It is not our intention in this chapter to describe in detail the attention signatures that differentiate one disorder from another, but see Chapters 8 and 9 for detailed analyses of these issues. Instead, our objectives here are threefold: first, to provide the reader with an understanding of the underlying or putative causes of these neurodevelopmental disorders at the genetic and biological levels; second, to present the complexity facing researchers as they attempt to link genes to behavior and cognition; and third, to provide the reader with a snapshot of one of the most dynamic areas of genetic research—the search for so-called "attention genes," without, of course, denying the simplification inherent in this kind of shorthand of "gene(s) for X."

Table 3.1 provides a summary of the genes associated with each disorder.

Genetics of Autism

Autism is a complex neurodevelopmental disorder that has intrigued clinicians and researchers for decades. It was originally described by Kanner in 1943 and is characterized by a "triad of impairment" that includes a severe disruption in communication/language skills, impaired reciprocal social interactions, and restricted/repetitive patterns of behavior, interests, and routines. It is a lifelong condition with onset in the first 3 years of life. Autism is by no means a categorical disorder

Table 3-1. Summary of the Genes and Candidate Genes Associated With Autism, Fragile X Syndrome, Williams Syndrome, Down Syndrome, 22q11 Deletion Syndrome, and ADHD

Neurodevelopmental Disorder	Gene Name	Coding Protein(s)	Chromosome Location
Autism[a]	FMR1	Fragile X mental retardation protein (FMRP)	Xq27.3
	RELN	Reelin	7q22
	SHANK3	SHANK 3	22q13.3
Fragile X syndrome	FMR1	Fragile X mental retardation protein (FMRP)	Xq27.3
Williams syndrome[b]	ELN	Elastin	7q11.23
	LIMK1	LIMK1 protein	7q11.23
	GTF21	BTK-associated protein & TFII-I	7q11.23
	CYLN2	Cytoplasmic linker 2	7q11.23
Down syndrome[c]	–	–	21 (Trisomy)
22q11 deletion syndrome	Multiple	Multiple	22q11
ADHD	DAT1	Dopamine transporter	5p15.3
	DRD4	Dopamine receptor D4	11p15.5
	DRD5	Dopamine receptor D5	4p15.3
	5HHT	5-Hydroxytryptamine (serotonin) transporter	17q11.2
	5HHT1B	5-Hydroxytryptamine (serotonin) receptor 1B	6q13
	SNAP-25	Synaptosomal-associated protein, 25kDa	20p11.2
	COMT	Catechol-O-methyltransferase	22q11.21

[a] Selected candidate genes.
[b] Selected genes from more than 28 genes in the deleted region of chromosome 7.
[c] No specific gene name or protein associated with Down syndrome.

but instead represents one extreme of a broader phenotype known as "autistic spectrum disorder" (ASD). Within this broad category are housed a group of four disorders collectively known as *pervasive developmental disorders*: childhood autism, pervasive developmental disorder not otherwise specified (PDD-NOS), Rett syndrome, and Asperger syndrome. Recent epidemiological studies highlight a much greater prevalence of PDDs than was reported even a decade ago, with current rates for childhood autism at 20 per 10,000 children and that of all PDDs combined at 60–70 per 10,000 (see Fombonne, 2009). Although a variety of explanations have been put forward to explain this increase in prevalence, such as environmental causes (e.g., toxic exposure), it is likely that the broadening of diagnostic criteria to include children across a continuum of functioning may offer some explanation for this apparent increase.

Distinctive clinical features are not always a noted feature of autism, but recent developments in facial morphology techniques have identified subtle anomalies in facial morphology in children with autism. Using 3-D dense-surface models of face shape, Hammond and colleagues have identified a subset of boys with autism who show significantly different facial asymmetry from age and ethnicity matched comparison children, notably in the right depth dominance of the supraorbital region (Hammond et al., 2008). The authors speculate that these asymmetries may reflect underlying anomalies in brain maturation.

The causes of autism remain largely unknown, but compelling evidence from over a decade of research indicates a significant genetic component (for examples of recent reviews, see Losh, Sullivan, Trembath, & Piven, 2008; Steyaert & De la Marche, 2008). However, it is perhaps the clinical heterogeneity of this disorder that makes capturing the causal mechanism of autism so elusive. For example, there appear to be two causal pathways resulting in autistic behavior. The first has been labeled "idiopathic," in which autism presents independent of any other disorder. Idiopathic autism is found in familial clusters as identified by twin and family studies. The second is referred to as "syndromic autism" and occurs as part of a broader genetic phenotype such as fragile X syndrome or 22q11 DS (deletion syndrome). The idiopathic pathway has received considerable research interest, and we provide below a snapshot of some of the exciting discoveries that have begun to isolate putative susceptibility genes for autism. The second syndromic pathway has received less focus but is an emerging area that is likely to generate considerable interest in the next few years. What remains unknown is the extent to which the different autism *pathways*, presumably resulting from different etiologies, converge or diverge across development.

The Search for Autism Susceptibility Genes

It is unlikely that any one specific gene will produce the classic autism triad, although at least 10 susceptibility genes or regions of interest have been identified with a focus on four chromosomes: 7, 15, 16, and 22. In brief, the long arm ("q") of chromosome 7 at q21–q36 includes the candidate *Reelin* (*RELN*) gene at 7q22. Alterations to this gene are known to impact early cortical and cerebellar development, thus implicating a possible link between the reelin protein and autism pathology (Ashley-Koch et al., 2007; Persico et al., 2001; Skaar et al., 2005). The long arm of chromosome 15 at q11.2–q13 has also been indicated as a region of autism susceptibility. This region includes a gene cluster of $GABA_A$ receptor subunit genes ($\beta 3$, $\alpha 5$, and $\gamma 3$). During early brain development, GABA plays a critical role as the main inhibitory neurotransmitter, and when it is disrupted the maturation of neural circuits involved in complex cognitive behavior is affected. Recent findings from multiplex families, defined as those with multiple probands, implicate this region as housing one or more autism risk alleles (e.g., McCauley et al., 2004). However, findings are mixed. Most recently, Fatemi, Reutiman, Folsom, and Thuras (2009) report a reduction in $GABA_A$ subunit expression in the frontal cortex, parietal cortex, and cerebellum in individuals with autism. These areas have been frequently implicated in the pathogenesis of autism.

The short arm ("p") of chromosome 16 at p11.2 has been recently identified as a susceptibility region for autism if duplicated or deleted (Kumar et al., 2008; Weiss et al., 2008). On the long arm of chromosome 22 at q13, the *SHANK3* gene, whose protein product facilitates the expression of proteins necessary for the maturation and growth of dendritic spines and synapses in the developing brain, has been found deleted or mutated in some cases of autism (e.g., Durand et al., 2007; Moessner et al., 2007). The excitement generated by these new findings is in part due to major technological advances in genotyping; one of the most notable is *microarray technology*, which allows us to examine the expression profile of numerous genes within a given cell. In the field of autism, microarray profiling is beginning to shed light on the molecular pathways that might be dysregulated in this disorder. For example, Gregg et al. (2008) were the first to demonstrate gene expression differences in peripheral blood between children with autism and typically developing children matched in gender and age, and between subtypes of autism, in this case early onset autism compared to autism with regression. For an exceptional review of these recent technological advances in genetics and their application to autism, see Abrahams and Geschwind (2008). Although the above findings provide a glimpse of some very exciting work currently underway in the search for susceptibility genes linked to the autism phenotype, it has proven a difficult task to isolate any one gene that can be mapped directly to all features of the triad. Recent behavioral genetic findings may provide a clue as to why this search has so far remained elusive. Focusing on twin data, Ronald, Happé, and colleagues (Ronald et al., 2006) examined the strength of the relationship between the behavioral traits that comprise the three components of the autism triad in approximately 3,400 twin pairs aged between 7 and 9 years and found only modest correlations between the three areas. The authors suggest the tantalizing hypothesis that not all aspects of the triad are genetically related, and, instead, different impairments may relate to different risk genes, all of which need to be present to produce the autism spectrum disorder. It is therefore possible for some individuals to have one or two impairments within the triad but without the combination of all risk genes that are needed for a diagnosis of autism. See Happé and Ronald (2008) for a compelling theoretical account that argues for the "fractionable autism triad" approach to understanding the genetic variations and phenotypic outcomes in the broader autism spectrum.

Genetics of Fragile X Syndrome

Fragile X syndrome was first described in 1943 by Martin and Bell (originally labeled the "Martin-Bell" syndrome) and is the world's most common hereditary cause of developmental delay in males. Although still lacking consensus, recent estimates indicate a frequency of approximately 1 per 2,500 (Hagerman, 2008). The disorder is caused by the silencing of a single gene on the long arm of the X chromosome at q27.3. The gene, named the fragile X mental retardation gene 1 (*FMR1*), was identified in 1991 and is "turned off" in affected individuals. When this occurs there is an expansion of a trinucleotide (CGG) in the repeat region. In individuals unaffected by fragile X, there are between 7 and 55 CGG repeats, with 30 repeats the most common number. In clinically affected individuals (known as the *full-mutation status*), the CGG region expands to over 200 repeats, resulting in silencing of the gene and loss of the fragile mental retardation protein (FMRP). Due to X-linkage, almost all affected boys present with mental retardation compared to approximately a third of affected girls who, due to their second unaffected X chromosome, display less marked intellectual and cognitive impairment. See Chapter 9 for more details on the genetic variations in fragile X males and females and the impact on cognition.

In those individuals with CGG repeats between 55 and 200 (known as the *premutation "carrier" status*), a reduced level of FMRP sometimes occurs, but the majority have FMRP levels within normal limits. All carriers, however, produce increased levels of FMR1 messenger RNA (mRNA). These *FMR1* mRNA levels are increased from 2 to 8 times the normal levels with expanding CGG repeat size over the premutation range (Tassone, Hagerman, Chamberlain, & Hagerman, 2000; Tassone, Hagerman, Taylor et al., 2000). The carrier status is common, occurring in approximately 1 per 259 females and 1 per 813 males (but see Hagerman, 2008, for a revised estimate of 1 in 580 males). Premutations are unstable when transmitted from carrier mother to offspring with associated risk of giving rise to the fragile X full mutation phenotype should further expansion occur and reach a threshold of 200 repeats or above. The mechanism by which full expansion occurs transgenerationally is not yet understood.

The Importance of the Fragile X Mental Retardation Protein (FMRP)

FMRP is an RNA-binding protein that transports messages to the synapse and has a regulatory role in the translation of these messages into protein (see Garber, Visootsak, & Warren, 2008, for a review; Willemsen, Oostra, Bassell, & Dictenberg, 2004). FMRP typically inhibits the translation of messages, so that when it is absent or reduced there is an abnormal upregulation (this refers to changes in the levels of proteins) of a number of genes including *MAP1B*, which is important for synaptic plasticity and structure (Willemsen et al., 2004). In essence, without FMRP, dendritic spines, which are rich sites of synapses, do not develop normally. See Bassell and Warren (2008) for an exceptional review of recent findings. The extent to which these molecular discoveries explain the phenotypic outcomes in fragile X, notably at the cognitive and behavioral levels, is beginning to be revealed and will be discussed, in detail, in Chapter 9. See also Cornish, Turk, and Hagerman (2008) for a recent review of the genetic-cognitive correlates that comprise the fragile X continuum.

The clinical features that can characterize the syndrome include an elongated face, large prominent ears and forehead, and in males, postpubertal macroorchidism (Cornish, Levitas, & Sudhalter, 2007; Lachiewicz, Dawson, & Spiridigliozzi, 2000). See Figure 3.1 for an example of the fragile X facial phenotype. More subtle features can include narrow intereye distance, a highly arched palate of the mouth, and hyperextensible joints. However, the wide variability in manifestation in both boys and girls makes a diagnosis based on physical features alone almost impossible. It is precisely because of their relatively "normal" appearance that many affected children are not diagnosed with fragile X until relatively late in their development. Undoubtedly the most defining feature, especially in boys with the disorder, is mental retardation, and the resulting cognitive-behavioral phenotype, most notably the attentional control difficulties, spatial impairments, and autistic-like features that can accompany the syndrome from very early in development (Cornish, Scerif, & Karmiloff-Smith, 2007; Cornish, Turk, & Hagerman, 2008).

Fragile X and Autism: A Complex Relationship

There are currently very few single-gene disorders for which there is a certainty of the involvement of autism; fragile X is one. As a single-gene disorder, fragile X offers an interesting genetic model to explore the functions of FMRP regulation and the repercussions of its loss in early brain development. Commonalities across core social and language domains define the link between fragile X and autism, and it therefore seems highly plausible that similar neurobiological mechanisms are affected in both disorders. A recent study by Loesch and colleagues found that a common impairment in verbal skills best described the comorbidity of fragile X and autism at the cognitive level (Loesch et al., 2007). In an earlier study, Philofsky, Hepburn, Hayes, Hagerman, and Rogers (2004)

Figure 3-1. The fragile X facial phenotype.
Permission to reproduce kindly granted by the
photographer, Franklin L. Avery. (See color Figure 3-1)

reported a similar link in children with fragile X between exceptionally low verbal ability, in this case receptive language, and a dual diagnosis of autism, compared to children with fragile X alone in which verbal skills appeared to be a relative strength. Overall, children with a dual diagnosis tend to display more impaired cognitive performance than either children with autism alone or fragile X alone. See Cornish, Levitas, and Sudhalter (2007) for more detailed descriptions of the commonalities and differences between fragile X and autism across cognitive domains. See also the recent review by Moss and Howlin (2009) that highlights commonalities but also differences in autism profiles in genetically determined disorders.

The frequency of autism among fragile X individuals is still controversial, but approximately 2–6% of children with autism will have the

FMR1 mutation (Reddy, 2005; Wassink, Piven, & Patil, 2001), and between 33-67% of children with fragile X will fulfill the diagnostic criteria for autism (Clifford et al., 2007; Rogers, Wehner, & Hagerman, 2001) with more males than females reportedly meeting the autism cutoff. The relative increase in the incidence of dual diagnosis is likely due to the introduction of newer, "gold standard" diagnostic measures, such as the Autism Diagnostic Observation Schedule (ADOS-G; Lord, Rutter, DiLavore, & Risi, 1999) and the Autism Diagnostic Interview-Revised (ADI-R; Lord, Rutter, & Le Couteur, 1994). The recent development of a new instrument known as the Autism Observation Scale for Infants (Bryson, Zwaigenbaum, McDermott, Rombough, & Brian, 2008) shows promise for clinicians and researchers to extend their explorations of the early signs and symptoms of autism in toddlers under the age of 3.

Alongside the findings of an autism–fragile X link associated with the full mutation, recent preliminary studies indicate that the premutation (carrier) status may also confer an increased risk of autism compared to the general population (Aziz et al., 2003; Loesch et al., 2007), suggestive of a continuum of clinical involvement in fragile X that includes a vulnerability to autism.

From a *biological perspective*, a number of studies have shown a correlation between FMRP status and IQ (e.g., Loesch et al., 2002), but the level of FMRP is not a predictor of the presence of autism (Bailey, Hatton, Skinner, & Mesibov, 2001). This suggests that the genotype in children with fragile X who also have autism must involve additional genes that are unique to this subgroup. Because FMRP is absent in the full mutation of fragile X, this syndrome provides a unique window to assess how the multiple genes that bind to FMRP are dysregulated in fragile X and may contribute to the autistic phenotype. For recent mouse models of autism see Moy and Nadler, (2008), for an excellent review of the advances and challenges of this research, and also Martin, Goldowitz & Mittleman, 2010, for the role of cerebellar pathology in autism.

Mouse models of fragile X (e.g., Zhang, Shen, Ma, Ke, & El Idrissi, 2009) also offer promising insights and directions towards understanding what will undoubtedly be complex gene–gene interactions alongside the modifying effects of environment and development on the fragile X–autism phenotype.

However, the development of novel technologies, for example *microarray profiling*, has the exciting potential to locate gene expression differences across autism subtypes including the fragile X–autism subtype, and between differing genetic disorders. In the next decade, the application of these newer technologies to an understanding of the different cellular pathways that are disrupted both within a given disorder and between disorders will provide critical information that will help to guide clinical diagnosis and prognosis, and promote future strategies for therapeutic interventions. In the case of autism, as alluded to above, gene expression studies have the added potential of locating possible genetic markers that will help identify "at risk" individuals in early infancy.

Genetics of Down Syndrome

Down syndrome, or trisomy 21, was first described nearly 150 years ago by Down (and reported by Lejeune, Gautier & Turpin in 1957) and is the most common genetic cause of mental retardation. The syndrome has an estimated frequency of 1 in 730 (Canfield et al., 2006) and in most cases results from an additional copy of chromosome 21. Advanced maternal age is seen as a main risk factor for Down syndrome. Possible environmental risk factors have also been identified in recent years, for example, maternal smoking, but with mixed results (see Sherman, Allen, Bean, & Freeman, 2007, for a comprehensive critique). The clinical phenotype of Down syndrome is well documented and includes oblique eye fissures with epicanthic skin folds (a skin fold of the upper eyelid covering the inner corner of the eye) that give an "almond" shape to the eyes, poor muscle tone, a flat nasal bridge, and a single palmar fold (a single rather than a double crease across one or both palms). The disorder is also linked with an increased risk for Alzheimer's disease (AD) that is clearly associated with increasing age (>50 years) but has a variable prevalence in that not all adults with Down syndrome develop AD (see Coppus et al., 2006; Zigman & Lott, 2007).

Perhaps more than any other neurodevelopmental disorder described in this book, Down syndrome is often viewed as comprising a more "global" cognitive delay compared to the well-documented signatures of autism or fragile X syndrome. However, closer inspection reveals a characteristic profile of weaknesses in language comprehension, verbal short-term memory, especially for auditory sounds and selective attention, but also relative strengths in receptive language, long-term memory for verbal information, visuoconstructive skills, and inhibitory control, although all are below chronological age levels (e.g., Abbeduto, Warren, & Conners, 2007; Cornish, Scerif, & Karmiloff-Smith, 2007; Jarrold, Baddeley, & Phillips, 2007; Pennington, Moon, Edgin, Stedron, & Nadel, 2003).

Candidate Genes for Down Syndrome

Since 2000, when the full DNA sequence of chromosome 21 was revealed (Hattori et al., 2000), at least 400 genes have been identified. However, not all of these genes appear to be directly involved in causing Down syndrome. The field is still in its infancy, but there is general agreement that elevated expression of certain protein-coding genes on chromosome 21 must produce a deleterious effect on early brain development that causes the Down syndrome phenotype. Isolating those genes, however, has proven to be especially complex. One of the few ways in which specific genes can be isolated is through analysis of the rare cases that have a partial trisomy, in which not every cell has three copies of chromosome 21. In these individuals it may be possible to isolate a specific region that contains dosage-sensitive genes that contribute to the Down syndrome phenotype. In recent years, progress in this field has identified a specific duplicated region on chromosome 21 referred to as the "Down syndrome chromosomal region-1" (DSCR), which is hypothesized to contain specific genes that when triplicated are linked to the cognitive characteristics associated with the syndrome (see Rachidi and Lopes, 2008, for a review of recent developments pertaining to this model). An alternative to the critical region hypothesis proposes that the Down syndrome phenotype results from the cumulative nonspecific effects of the many hundreds of genes on chromosome 21. The resulting global disruption to gene expression will have widespread repercussions on the developing cortex. The "developmental instability" model therefore proposes that it is the number rather than the function of genes that serves as the most critical factor in producing the Down syndrome phenotype (see Olson, Richtsmeier, Leszi, & Reeves, 2004). However, there is increasing acknowledgment that both models may be valid.

One of the most powerful tools to elucidate the genotype–phenotype relationship in Down syndrome has been through the generation of mouse models. This research has produced some exciting discoveries that have begun to link specific genes on chromosome 21 to key brain regions known to contribute to specific cognitive functions. Experimental over-expression of these genes in mice appears to cause significant impairment in early brain development. Recently, a number of critical genes have been proposed, one of the most intriguing being the *DYRK1A* gene that is located in the Down syndrome critical region and expressed in cortex, hippocampus, and the cerebellum. Findings from studies employing transgenic mice (i.e., those that have been experimentally altered such that a gene of interest is either overexpressed or prevented from being expressed) show that overexpression of this gene can cause difficulties in spatial learning and cognitive flexibility (Altafaj et al., 2001; 2008), both well-documented deficits in Down syndrome. However, the reader must be aware of the complexity of gene expression in general and in particu-lar when examining the atypical developing brain. In the case of Down syndrome, understanding of gene dosage effects and the possible interac-tive role in brain development is only just beginning to emerge.

Genetics of Williams Syndrome

Williams syndrome, also known as Williams-Beuren syndrome, was initially identified independently by Williams and Beuren in 1961. Until quite recently, it was assumed to be a rare disorder—1 in 15,000–20,000, but one recent estimate indicates the frequency at approximately 1 in 7,500 (Stromme, Bjornstad, & Ramstad, 2002). The disorder is not usually hereditary and results from the deletion of approximately 25–30 genes on the long arm of chromosome 7 at q11.23. Physical characteris-tics of Williams syndrome include a distinct facial phenotype (full cheeks, small and widely spaced teeth, long philtrum, and a short, upturned nose), short stature, cardiovascular abnormalities, connective tissue abnor-malities, and increased calcium in the blood (hypercalcemia, although this is quite rare) (Donnai & Karmiloff-Smith, 2000; Morris, Demsey, Leonard, Dilts, & Blackburn, 1988; Tassabehji & Donnai, 2006).

Mental retardation, although variable in severity, is a core character-istic of almost all children with Williams syndrome, but perhaps most salient is their distinctive cognitive phenotype that distinguishes toddlers and children with Williams syndrome from those with other types of neurodevelopmental disorders, as well as extremely friendly social interaction. At first glance the cognitive phenotype appears to be very straightforward, with poor spatial and numerical skills alongside relatively proficient face-processing skills and language abilities (see Mervis & Becerra, 2007, for a comprehensive review). However, a decade of study by Karmiloff-Smith and colleagues has elucidated a more subtle yet "atypical" profile in so-called domains of "proficiencies" (see Karmiloff-Smith, 1998, 2007, and Karmiloff-Smith et al., 2004, for examples of how so-called *proficiency* in face-processing skills is far from equivalent to their typically developing peers even though their scores often fall within the "normal" range). These findings question the feasibility of using an adult so-called "modular" approach to understanding atypical cognitive performance and instead underscore the importance of recognizing that development itself is a key to understanding cognitive end states in disorders such as Williams syndrome (see Karmiloff-Smith, 2009, for a critique of this developmental approach).

Linking Genes to the Williams Syndrome Phenotype

The first gene to be linked to the phenotype was the *elastin (ELN)* gene, which encodes for a protein that provides strength and elasticity to vessel walls. This gene was found to be disrupted in a family with dominant cardiovascular problems (Ewart et al., 1993), and subsequent studies have identified haploinsufficiency (i.e., the presence of only one instead of the two copies of a gene) of the *ELN* gene to be the single most important genetic mutation responsible for the cardiovascular problems inherent in Williams syndrome (see Pober, Johnson, & Urban, 2008, for a review of the genetic mechanisms underlying elastin expression and supravalvular aortic stenosis (SVAS) in Williams syndrome). For some time, *ELN* was argued to contribute also to the facial phenotype, but more recent studies of partial deletion patients have ruled out *ELN* as the major contributor to facial dysmorphology and identified, through human and animal studies, the *GTF21RD1* gene located toward the telomeric end of the Williams syndrome critical region (Karmiloff-Smith et al., 2003; Tassabehji et al., 1999, 2005). Indeed, one of the most fruitful approaches to elucidating the role of the deleted genes and their correlation with the Williams syndrome phenotype is to provide detailed characterizations of individuals with partial deletions. Although these individuals are uncommon, their phenotypes provide important new insights into the genetic pathways of the "typical" Williams syndrome phenotype. A number of genes have already been explored and linked to the cognitive profile, for example, *LIMK1*, *GTF21*, *CYLN2*, and *GFF21RD1*. The findings of these studies will be described briefly, but the reader is cautioned against drawing definitive conclusions from what are often preliminary data.

The Lim-Kinase1 (*LIMK1*) gene, as it applies to the Williams syndrome cognitive phenotype, is controversial. The gene encodes a protein kinase that is strongly expressed in the developing brain and has been postulated to play a significant role in the development of normal visuospatial cognition (e.g., Morris et al., 2003). Its deletion in the Williams syndrome critical region (WSCR) has led some researchers to argue that the absence of the *LIMK1* gene is associated with abnormal brain function and the resulting visuospatial processing deficits reported in Williams syndrome (Frangiskakis et al.,1996; see Meyer-Lindenberg, Mervis, & Berman, 2006, for review). In contrast, other studies have not replicated this association (e.g., Tassabehji et al., 1999), and some very interesting recent data by Gray, Karmiloff-Smith, Funnell, & Tassabehji (2006, see also Smith Gilchrist, Hood, Tassasbehji, & Karmiloff-Smith, 2009) challenge the assumption that spatial deficits even characterize individuals with partial deletions of LIMK1. They found performance on a wide range of measures that tap varying aspects of spatial cognition to be comparable to a normative sample. The authors conclude that if LIMK1 plays a contributory role in visuospatial functioning, it must be in combination with other proteins that act together to impair spatial functioning.

Two other well-documented genes are the *GTF21* gene and the *CYLN2* gene, and both have been postulated to contribute to the cognitive phenotype of Williams syndrome. In the case of GTF21, the protein product is implicated in gene regulation through interactions with different tissue-specific transcription factors and chromatin-remodeling complexes (Makeyev et al., 2004; Roy, 2001), and it is part of the TFII-1 family that also includes the *GTF21RD1* gene. Recent findings implicate *GTF21* and its family member *GTF21RD1* in craniofacial development and cognitive development (Tassabehji et al., 2005), and the *GTF21* gene in particular, as contributing to the mental retardation that is associated with the disorder (Morris et al., 2003). The cytoplasmatic linker protein 2 (*CYLN2*) gene has also been implicated in the Williams syndrome phenotype, although its role is less understood at the cognitive level. Preliminary data gathered from animal models and individuals with partial deletions including CYLN2 indicate a possible contributory, but not exclusive, role in hippocampal dysfunction and particular deficits in motor coordination (Hoogenraad et al., 2002; van Hagen et al., 2007). See also Li et al. (2009) for gene networks involved in hypersocialiability and deficits in motor coordination in partial deletion mice.

Thus, the picture that emerges, at least in the three genes we have briefly examined, is the absence of any clear, linear link between a Williams syndrome candidate "gene" and specific cognitive impairment. The lack of consensus across research studies also adds to the confusion and is likely due, in part, to differing cognitive methodologies employed by differing studies. One of the most powerful approaches to unraveling the effects of one or more genes on phenotypic outcomes in Williams syndrome is the partial deletion approach. We have described some of these case studies above, and their findings add a new richness to our current understanding of how individuals with different combinations of genes deleted in the WSCR (whether it be 1, 2, 3, 10 or more from the 28 genes that comprise the critical region) differ from each other across multiple aspects of cognitive functioning.

Genetics of 22q11 Deletion Syndrome

22q11 DS (also referred to as DiGeorge syndrome, Velo-Cardio-Facial syndrome, and Shprintzen's syndrome) is the most common microdeletion in humans. Initially delineated in 1978, the disorder has an estimated frequency of 1 in 4,000 (Óskarsdóttir, Vujic, & Fasth, 2004) and 1 in 6,000 (Botto et al., 2003). 22q11 DS results from a microdeletion of chromosome 22q11.2, and in the majority of cases the deletion is de novo (i.e., new to the family). The clinical features are well documented but extremely varied in both nature and degree of severity. Core physical characteristics include cardiac deficits, cleft palate, and a distinct, but subtle, facial phenotype (long face, hooded, swollen eyelids, malar flattening, broad nasal ridge, narrow mouth, and slightly low-set ears) (Hayes, 2007; Óskarsdóttir, Holmberg, Fasth, & Strömland, 2008). Similar to fragile X syndrome, reliance on facial characteristics alone can rarely serve as a diagnostic tool. 22q11 DS is also strongly associated with a high prevalence of psychiatric disorders including schizophrenia (see Gothelf, Schaer, & Eliez, 2008, for a review) and ADHD, with an estimated risk of 25–30 times that of the general population.

In contrast to the well-documented physical and clinical features of 22q11 DS, the cognitive phenotype has only recently been identified and research is still ongoing. Almost all affected children will display some developmental delay. To date, studies report a characteristic profile of weaknesses in visual-spatial memory, numerical cognition, executive functions, and visual attention, with relative strengths in word knowledge, auditory memory, and factual information (see van Amelsvoort, & Zinkstok, 2005, for a review; Bearden et al., 2001; Simon, Bearden, Mc-Ginn, & Zackai, 2005; Swillen et al., 1999).

Linking Genes to the 22q11.2 DS Phenotype

The strong association with schizophrenia alongside neurocognitive deficits that appear to "mirror" the well-documented profile reported in individuals with schizophrenia (Lewandowski, Shashi, Berry, & Kwapil, 2007) has led to an intensive search for candidate genes that may shed light on this association. Of the genes studied to date, the *Catechol-O-methyl transferase gene (COMT)* mapped to 22q11.2 has created the most research interest primarily because of its link with schizophrenia, but also, as we shall see later in this chapter, because it is also widely cited as a risk factor for ADHD.

COMT GENE

Converging evidence suggests that the gene that codes for COMT may play a significant role in the pathogenesis of the cognitive processes found to be disrupted in 22q11 DS. The keen interest surrounding this gene is its connection to dopamine. Dopamine is a neurotransmitter that has been fundamentally implicated in prefrontal cortical information processing.

It is known to modulate complex, higher-order cognitive functions such as working memory and cognitive flexibility, including attentional set shifting (e.g., Abi-Dargham et al., 2002; Robbins & Roberts, 2007). We will visit the dopamine family of genes in the next section when we focus on ADHD.

The COMT protein is a major enzyme in dopamine catabolism. It influences cortical dopamine flux, notably in prefrontal cortex, and is responsible for the degradation of dopamine in this region (see Karoum, Chrapusta, & Egan, 1994; Meyer-Lindenberg et al., 2005). The *COMT* gene confirms a polymorphism, which means the gene can take different forms, that determines high or low activity of the COMT enzyme. The *Met* allele is associated with very low COMT activity, resulting in increased dopamine levels in the prefrontal cortex and so is hypothesized to confer a cognitive advantage on skills that tap the prefrontal cortex (e.g., executive functions). In contrast, the *Val* allele is associated with high COMT activity, resulting in increased prefrontal dopamine catabolism and so is hypothesized to confer a cognitive vulnerability. For example, a recent study of individuals homozygous for the Val/Val genotype showed poorer performance on tests of executive control, along with inefficient prefrontal cortical activity when compared with *Met* carriers (Blasi et al., 2005).

A series of recent studies by Goldberg and colleagues examining healthy individuals and patients with schizophrenia unrelated to 22q11 DS found that individuals with the Val/Val genotype showed a reduction in performance on a range of executive function tasks including those that tapped cognitive flexibility (e.g., Wisconsin Card Sorting Task; WCST) and working memory (Diaz-Asper et al., 2008; Egan et al., 2001; Goldberg et al., 2003, but also see Gothelf et al., 2005, for an alternative finding). However, despite these promising findings, a recent meta-analysis of reported correlations between COMT genotype and cognitive functions found little or no significant association between the COMT genotype and executive function or working memory performance (Barnett, Scoriels, & Munafò, 2008). The relationship between COMT and ADHD is also explored later in this chapter and briefly in Chapter 8.

Other genes of interest located on 22q11 include the *PRODH* gene, which has been linked with poor prepulse inhibition (PPI), a marker of poor sensorimotor gating (inhibition of motor responses to irrelevant stimuli), and is also deficient in schizophrenia (see Paylor et al., 2001). Other genes have also been identified that may play specific roles in producing the 22q11 DS phenotype, such as the *GNB1L* gene, the armadillo repeat gene, and the *Tbx1* gene (see Prasad, Howley, & Murphy, 2008, for an excellent comprehensive review of gene findings linked to 22q11 DS).

Genetics of ADHD

Attention deficit/hyperactivity disorder (ADHD) is one of the most common childhood psychiatric disorders with recent population-based studies reporting rates of diagnosed ADHD at approximately 8–12% worldwide (Schneider & Eisenberg, 2006; Visser, Lesesne, & Perou, 2007). Although classified as a childhood disorder, accumulating research now suggests that ADHD is a lifelong condition that persists into adulthood (Kessler et al., 2006). There is also a well-documented gender skew with a higher prevalence in boys than girls (see Rucklidge, 2008).

ADHD is characterized by a triad of clinical features, inattention, impulsivity, and hyperactivity (DSM-IV; American Psychiatric Association, 2000; see Box 5.1, Chapter 5 for an overview of the ADHD diagnostic criteria), and is frequently accompanied by comorbid disorders that include reading disability, motor coordination difficulties, and oppositional defiant problems. Consequently, children with ADHD are at greater risk for low academic achievement, substance abuse, and antisocial behavior/delinquency in adolescence, alongside an increased risk of psychopathology in adulthood (Elkins, McGue, & Iacono, 2007; Ruchkin, Lorberg, Koposov, Schwab-Stone, & Kratochvil, 2008). Although initial ADHD symptoms may be noticed as early as the preschool years, beginning as young as 3 years (see Greenhill, Posner, Vaughan, & Kratochvil, 2008, for an excellent review), full ADHD diagnosis does not usually occur before 7 years of age. However, although ADHD symptoms or "subtypes" can present quite strongly in the early years, notably hyperactivity, there is some promising longitudinal research that highlights instability of symptoms from preschool through to midchildhood (see Lahey, Pelham, Loney, Lee, & Willcutt, 2005).

In recent years, there has been extensive discourse regarding possible ADHD subtypes, and although this is still a matter of considerable debate, the DSM-IV does identify two dimensions of symptoms: inattention and hyperactivity-impulsivity. Based on these dimensions, three subtypes have been delineated: the combined type (CT), the predominantly inattentive type (IT), and the predominantly hyperactive-impulsive type (HT). The cognitive signatures that distinguish these three ADHD subtypes will be discussed at length in a later chapter (Chapter 8).

Although there is no single known cause for ADHD, the findings from a substantive research pool of family and twin data indicate a strong genetic influence. A recent review of 20 twin studies by Faraone and colleagues found the heritability of ADHD to be 76% (Faraone et al., 2005), making it among the most heritable of psychiatric disorders. Research in this field is dynamic, and it would be an almost impossible job and, indeed, would not do justice to the tremendous work in this field, to attempt to condense a decade of intensive genetic work into a relatively brief summary. As essential reading, we recommend reviews by Waldman and Gizer (2006), Kieling, Goncalves, Tannock, and Castellanos (2008), and Sharp, McQuillin and Gurling (2009).

The search for susceptibility genes that might contribute to the ADHD phenotype has been the subject of intense investigation, and it is now widely accepted that several genes, each contributing a small amount to the total genetic variance, are implicated in ADHD (Fisher et al., 2002). Furthermore, converging evidence suggests that ADHD is highly likely to represent the extreme of a trait distributed continuously throughout the general population (Cornish et al., 2005; Hay, Bennett, Levy, Sergeant, & Swanson, 2007; Levy, Hay, McStephen, Wood, & Waldman, 1997). Using rating systems such as the newly developed SWAN Scale (reviewed in Chapter 5) that allow for a distribution of attention behaviors rather than a scale that restricts responses to fixed categories (and often results in skewed data), we can measure wider population variation by extending the frequently used 4-point scale to 7 points. This extended scale has the benefit of accruing additional responses from non-ADHD individuals, thus allowing a broader range of attention/inattention behaviors to be recorded from the general population. We will describe some of the main genetic findings that have resulted from viewing ADHD as a continuous, rather than discrete or categorical, variable in the population.

The Search for ADHD "Endophenotypes"

ADHD, like many other psychiatric disorders, has a complex hereditary pattern such that its genotype remains elusive. This is quite different from other disorders that have a direct genetic origin such as fragile X syndrome or Down syndrome. In both these cases, genes that influence the phenotype are directly linked to genes on a given chromosome, be it the X chromosome in fragile X or chromosome 21 in Down syndrome. In neurodevelopmental disorders such as ADHD, and to some extent autism, research has to rely on identifying measurable indices of risk, termed *endophenotypes*, that represent simple but important clues to the genetic underpinnings of a disorder. Castellanos and Tannock (2002) define an endophenotype as "heritable, quantitative traits that index an individual's liability to develop or manifest a given disease." As such, endophenotypes can be measured across multiple levels including the neural, anatomical, and cognitive levels. Given that these traits can be manifested quite subtly in affected individuals, it is important that tools are sufficiently sensitive to capture these signatures. Endophenotypes are also most likely to be continuous in nature rather than dichotomous and should demonstrate stability when assessed by different measurement techniques. See Tannock et al. (2009) for a detailed critique of the critical features of endophenotypes. The current literature, however, as it pertains to putative endophenotypes for ADHD, is surprisingly limited, although an emerging number of studies have highlighted candidate cognitive endophenotypes in response inhibition, working memory, and moment-by-moment variability in task performance. The findings from these new data will be discussed in detail in Chapter 8 when we explicitly focus on ADHD across multiple levels of analysis.

Candidate Genes

In the following section, we summarize the candidate genes for ADHD that have received the most attention in recent years, but as the reader will quickly ascertain, there is no clear consensus across studies, with some finding a positive relationship and others finding no relationship. See also Posner, Rothbart, and Sheese (2007) for a description of recent findings relating genes to attentional networks and Kebir, Tabbane,

Sengupta, and Joober (2009) for a more general discussion of candidate genes and neuropsychological phenotypes in ADHD.

Seven genes will be explored: dopamine transporter gene (*DAT1*), dopamine D4 receptor gene (*DRD4*), dopamine D5 receptor gene (*DRD5*), serotonin transporter gene (*5HTT*), serotonin receptor gene (*HTR1B*), synaptosomal-associated protein (*SNAP-25*), and the catechol-O-methyltransferase gene (*COMT*).

THE DOPAMINE FAMILY

Numerous findings now attest to the critical involvement of the dopaminergic system in the pathogenesis of ADHD (e.g., DiMaio, Grizenko, & Joober, 2003; Krause, Dresel, Krause, la Fourgere, & Ackenheil, 2003). The efficiency of pharmacological agents that inhibit the dopamine transporter, for example, methylphenidate (MPH) (Ritalin©), is well documented (Iversen & Iversen, 2007; Joober et al., 2007; Stein et al., 2005), together with supporting findings from neuroimaging and animal studies (Gilbert et al., 2006; Volkow et al., 2007). Accordingly, most molecular studies have therefore focused on testing candidate genes that are linked to the dopaminergic pathways.

Dopamine Transporter Gene (DAT1)   *DAT1* has been studied extensively as a candidate gene for ADHD but with very mixed results. The gene is located on the short arm of chromosome 5 at p15.3, which is also one of the genes deleted in the Cri-du-Chat syndrome or 5p-deletion syndrome, a rare genetic disorder that also includes excessive hyperactivity (see Cornish & Bramble, 2002, for a review). The gene has a variable number tandem repeat (VNTR), and the VNTR polymorphism in the gene is not silent but has been shown to affect the expression of the transporter, resulting in increased levels of messenger RNA (mRNA). The two most common alleles are 9 and 10, and it is the *DAT1* 10-repeat allele (10/10) that is known as the "high risk" allele in ADHD. Other combinations such as the heterozygous 9/10 alleles or the homozygous 9/9 alleles do not appear to confer the same risk. Based on knockout mouse research (a genetically altered mouse that has one or more candidate genes switched off), *DAT1* has been shown to be important for the reuptake of dopamine. Knocking out the *DAT1* gene leads to two core ADHD-like features: hyperactivity and reduced inhibitory behaviors (Gainetdinov et al., 1999). The critical role of dopamine in the expression of ADHD symptoms is also evident when stimulant medication such as methylphenidate is used in the treatment of ADHD. Methylphenidate, for example, is a dopamine reuptake *inhibitor* that acts to block the dopamine transporter as one mechanism for alleviating ADHD symptoms. Numerous studies have been conducted to test the efficacy of stimulant medication in ADHD children and adults, and there is widespread consensus that it is indeed beneficial for individuals with clinically diagnosed ADHD and appears to have no long-lasting effects on brain development (see Grund et al., 2006, for a recent review, and Chapter 10).

Cook et al. (1995) were the first to report an association between the 10-repeat, "high risk" *DAT1* allele and ADHD. Since then, this finding has been replicated in some studies (e.g., Chen et al., 2003; Friedel et al., 2007; Mill et al., 2005; Simsek, Al-Sharbati, Al-Adawi, Ganguly, & Lawatia, 2005; Todd et al., 2005; Waldman et al., 1998) but not in others (Cheuk, Li, & Wong, 2006; Holmes et al., 2000; Roman et al., 2001). This heterogeneity in data sets may be due to a number of factors, most notably differences in sample selection criteria or assessment and/or simply chance variance. A recent study by Cornish and colleagues was the first to show an association between ADHD symptoms measured as a behavioral *dimension* in the general population and the 10/10 genotype (Cornish et al., 2005). Specifically, the authors found that by isolating the extreme ends of the distribution (children scoring >90th and <10th percentile on the *SWAN* Scale), there was a significant overall association between DAT1 genotype and high and low ADHD sub-groups. In the sub-group defined by the high *SWAN* cutoff, a greater percentage of children had the 10/10 genotype (see Figure 3.2). Furthermore, the findings suggested that one pathway from DAT1 to ADHD behaviors may operate through variation in response inhibition such that children with the 10/10 genotype were more likely to struggle on tasks that required the withholding of a prepotent reponse. Extending these findings, Bellgrove, Hawi, Kirley, Gill, & Robertson (2005) found children with ADHD with the DAT1 10/10 genotype to display more response variability on a task requiring focused attention relative to ADHD children with a low DAT risk and typically developing controls.

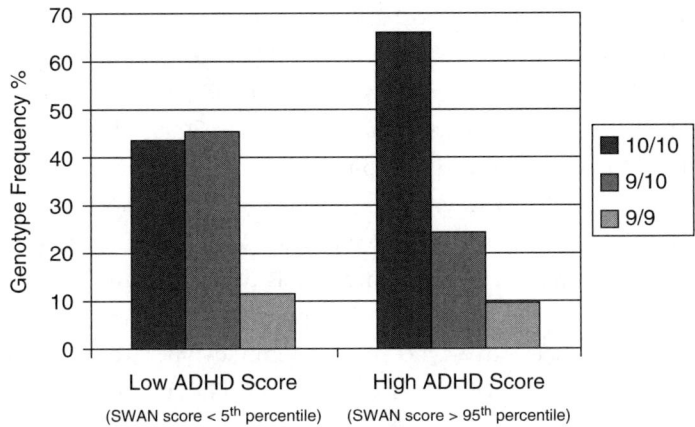

Figure 3-2. DAT1 genotype by SWAN high- and low-attention groups. From Cornish et al. (2005). Reprinted by permission from *Molecular Psychiatry* *10*, 690.

Most recently, Bellgrove et al (2009) have extended their findings to show a link between selective attention and DAT1.

Dopamine D4 Receptor Gene (DRD4)   The *DRD4* gene has been studied less intensively than the *DAT1* but shows more consistency in its findings. The gene is located on the short arm of chromosome 11 at p15.5. Ten alleles have been found in the population (2–11), and studies have consistently replicated an association between the 7-repeat allele and increased risk of ADHD (e.g., Gornick et al., 2007; Li, Sham, Owen, & He, 2006; see Faraone et al., 2005, for a meta-analysis of existing studies pre-2005), but paradoxically at the cognitive level, the *presence* of the 7-repeat allele confers superior performance by children with ADHD on sustained attention tasks (Bellgrove, Hawi, Lowe et al., 2005; Johnson, Kelly, Robertson et al., 2008; Manor et al., 2002). In contrast, the *absence* of the 7-repeat allele appears to be associated with drifting sustained attention in ADHD possibly linked, as suggested by Johnson et al., to diminishing arousal that gives rise to inconsistent performance.

Dopamine D5 Receptor Gene (DRD5)   The *DRD5* gene is the least studied of all the potential dopamine "candidate" genes. The gene is located on the short arm of chromosome 4 at p15.3. A tentative association between the *DRD5* and ADHD came from an early family study by Daly, Hawi, Fitzgerald, and Gill (1999) and has since been reviewed in two meta-analysis studies (Li et al., 2006; Maher, Marazita, Ferrill, & Vanyukov, 2002) and replicated in a joint study of 14 groups (Lowe et al., 2004). However, two recent studies have failed to find an association (Bakker et al., 2005; Mill et al., 2005). Interestingly, in the Lowe et al. study, the authors found the association with *DRD5* to be confined to particular subtypes or symptom traits, notably the ADHD predominantly-inattentive-type and combined-type. To the authors' knowledge, there are currently no published studies that have linked any cognitive attention profiles to this association.

Catechol-O–Methyltransferase Gene (COMT)   The *COMT* gene has already been discussed briefly above because of its association with 22q11 DS. *COMT* is highly expressed in prefrontal cortex and plays a critical role in regulating synaptic dopamine levels in this brain region. Most molecular genetic interest in this gene has focused on *Val/Met* polymorphisms that determine high or low activity of the COMT enzyme. The *Val* allele is associated with high COMT activity, resulting in increased prefrontal dopamine catabolism, and is associated with cognitive vulnerability especially on tasks that require prefrontal flexibility (e.g., perseverative errors on the Wisconsin Card Sorting Task, reviewed in Chapter 6) (see the landmark study of Egan et al., 2001; also Barnett, Jones, Robbins, & Muller, 2007). In contrast, the Met allele is associated with very low COMT activity, which results in slower dopamine catabolism and is associated with better function of prefrontal cortex in adults (Bilder et al., 2002).

Given the importance of the dopaminergic system in ADHD and the associations between the neurocognitive deficits associated with the *Met/Val* genotype, it is understandable why *COMT* was initially viewed as a candidate gene for ADHD. Unfortunately, the enthusiasm has not translated into positive correlations. Eisenberg et al. (1999) were the first to report an association between the ADHD impulsive–hyperactive type and the Val genotype, but subsequent studies have failed to replicate this positive association (Bellgrove, Domschke, Hawi, et al., 2005; Hawi, Millar, Daly, Fitzgerald, & Gill, 2000; Mills et al., 2004; Tahir et al., 2000).

Moreover, the same group of investigators failed to replicate their original finding (Manor et al., 2000). The findings from a recent meta-analysis of 11 family-based and 2 case-control studies also confirmed the lack of association between the COMT Val/Met polymorphism and ADHD (Cheuk & Wong, 2006).

The pattern of these findings suggests that the *COMT* gene plays a targeted role in some aspects of prefrontal cognition but has limited, if any, association with ADHD per se.

SEROTONERGIC SYSTEM

There is accumulating evidence of a possible but complex link between ADHD and two serotonergic genes: serotonin transporter gene (*5HHT*) and the serotonin 1B receptor (*HTR1B*). Abnormalities in the functioning of the central serotonergic system have also been linked to the pathogenesis of a range of other psychiatric disorders including suicidal behavior, depression, and alcoholism. The interest in ADHD is due to the growing number of findings that suggest that serotonergic inputs may moderate dopamine's effects on attention. Indeed, the interconnections between dopamine and serotonin are well documented. For example, animal studies have shown an interaction between the dopamine system and the serotonin transporter gene, such that *5HHT* has an inhibitory effect on the dopamine reward system. Other studies have demonstrated that serotonin can regulate dopamine release in the striatum midbrain and prefrontal cortex, mediated in part by the serotonin 1B receptor (Sershen, Hashim, & Lajtha, 2000). Additional studies in knockout mice have shown a critical role of serotonin in the neurobiology of ADHD-like symptoms. Specifically, mice lacking the 5HT1B receptor gene are more likely to exhibit impulsive behavior and hyperactivity (e.g., Bouwknecht et al., 2001). Together, these findings suggest that mutations in the serotonin genes may be important risk factors for the development of ADHD.

Serotonin Transporter Gene (5HHT)   The *5HHT* gene is one of the most widely studied genes in psychiatric research. Like the *DAT1* gene, the serotonin transporter is a protein responsible for the reuptake of neurotransmitters, in this case, serotonin. The gene plays a key role in regulating synapse levels of available serotonin, and serotonin turnover and is strongly associated with impulsive behavior. *5HHT* has also been hypothesized as a causal factor in ADHD (see the excellent review by Waldman & Gizer, 2006). The gene is located on the long arm of chromosome 17 at q11.2 and has two alleles—the short (S) allele and the long (L) allele. The S allelic variant has been shown to reduce 5-HHT transcription, resulting in diminished serotonin levels. In contrast, the L 5-HTTLPR allele and the long/long (LL)-genotype appears to be associated with rapid serotonin reuptake (Greenberg et al., 1999). An association between ADHD and *5HHT* genotype has been demonstrated by different groups in population- and family-based case control studies. Seeger, Schloss, & Schmidt (2001), for example, found an enhanced expression of the L/L genotype in children with severe hyperkinetic disorder (similar to ADHD combined-type) compared to other subtypes or controls. Similar findings of an L/L–ADHD association have also been reported (e.g., Zoroglu et al., 2002), but others have found no association across a wide range of family studies worldwide (e.g., Langley et al., 2003; Wigg et al., 2006; Zhao et al., 2005). Indeed, one study even found the S/S genotype to be *over-* rather than underrepresented in ADHD (Li et al., 2007). Part of the discrepancies in the research may be due to the close association between conduct disorders and ADHD combined-type, such as aggression, that are known to be independently affected by the *5HHT* gene. There is a need to disentangle these two conditions to determine whether the 5HHT effect is associated solely with a diagnosis of ADHD alone or with a comorbid diagnosis of ADHD + conduct disorder. More extensive studies are needed that tease apart the impact of ADHD comorbidities at the genetic and cognitive levels. We return to this issue in Chapter 8.

Serotonin Receptor Gene (5HTR1B)   The serotonin receptor gene is a relatively understudied candidate gene for ADHD, but as described above, serotonin may play a critical role in producing the impulsive behaviors associated with ADHD. The gene is located on the long arm of chromosome 6 at q13. Of the known polymorphisms, the most widely studied to date has been the relatively common G861C polymorphism, referred to as the "G" allele. To date, three genetic studies have reported an association between the *HTR1B* gene and ADHD (Hawi et al., 2002; Quist et al., 2003; Smoller et al., 2006), but one other found no association (Mill et al., 2005). Interestingly, the Smoller et al. study found the strongest link between the HTR1B gene and ADHD inattentive–type, and they suggest that this gene in particular is a susceptibility locus for the inattentive subtype.

SYNAPTOSOMAL–ASSOCIATED PROTEIN (SNAP-25)

Researchers have also begun to explore candidate genes outside of the major neurotransmitter systems. These genes are typically chosen because of their function and possible etiological role in ADHD. The *SNAP-25* is one of these genes and is located on the short arm of chromosome 20 at p11.2. The gene encodes a protein essential for synaptic plasticity and neurotransmitter release (Sollner et al., 1993). Two polymorphisms mapped to SNAP-25, 1065 and 1069, were identified initially by Barr and colleagues as possible sites for association between ADHD and *SNAP-25* gene (Barr et al., 2000). Recent studies have struggled to replicate this finding in its entirety (see Brophy, Hawi, Kirley, Fitzgerald, & Gill, 2002), and there have been some interesting findings that have linked the *SNAP-25* gene to ADHD via a *parent-of-origin* effect. In brief, we each have two copies of each gene that are inherited from our parents, and in the case of *SNAP-25*, recent findings suggest that the association of *SNAP-25* variants with ADHD is largely due to transmission of alleles from paternal (inherited from an individual's father) chromosomes, suggesting that genomic imprinting may be operating at this locus (Kustanovich et al., 2003; Mill et al., 2004). This promising line of research is ongoing, but as with many other genes associated with ADHD, the precise causal variant has yet to be established.

Overall, the findings from a decade of intensive research resulting in numerous published reports (only a percentage of which were reviewed here) demonstrate just how complex ADHD is as a genetic disorder. All seven of the genes reviewed here showed significant evidence of an association, but findings were as divergent as they were convergent. Individual variations such as gender and ethnicity alongside methodological and sample differences across studies may contribute to the heterogeneity of findings. It is also possible that genetic vulnerability in ADHD leads to the interaction of one or more candidate genes, each having a small effect, but substantial enough to modify early brain development. Add to this model additional modifiers such as environmental factors and the process of development itself, and ADHD phenotypes emerge but with no one specific pathogenic pathway.

Candidate "Attention" Genes and Other Neurodevelopmental Disorders

As we shall see in later chapters, ADHD symptoms and cognitive inattention are prevalent across a range of neurodevelopmental disorders. Given the intense interest in isolating putative candidate genes that might give rise to the development of ADHD, it is not surprising that a growing number of studies have sought to examine relations between candidate genes for ADHD and known neurodevelopmental disorders that have attention as a core feature. However, we offer the reader a word of caution here because it can be tempting to make the assumption that similar inattentive behaviors across disparate disorders imply similar cognitive mechanisms or etiologies to ADHD. As the reader will see throughout the course of this book, it is unwise to make a priori assumptions that commonalities at the behavioral level across disorders, X, Y, and Z (with their varying combinations of gene–gene interactions), can inform us about either the typical development of attention or ADHD.

Autism

As we have described already, autism is a severe neurodevelopmental disorder that impacts a range of cognitive and behavioral domains. Children with autism also demonstrate high levels of ADHD symptoms above what would be expected in the general population. We have also indicated that autism is a disorder with strong genetic susceptibility, but the genes have yet to be fully mapped and identified, and there is evidence of genetic heterogeneity that possibly reflects the heterogeneity observed in the autism phenotype. The serotonin genes, in particular the serotonin transporter gene (5HHT), have been linked to autism notably because the *5HHT* gene has been associated with specific behaviors that mirror some of those seen in the autism phenotype (e.g., impulsivity, ADHD, mood disorders). There is also tentative evidence that autism is associated with an elevation of serotonin levels (e.g., Anderson et al., 2002) and that judicious clinical use of serotonin-reuptake inhibitors may be useful drugs for treatment of a range of autistic symptoms including inattention and hyperactivity (Hollander, Kaplan, Cartwright, & Reichman, 2000; Hollander, Phillips, & Yeh, 2003). A number of studies have examined the relationship between autism and the transmission of the 5HHT long/short allele variants, with the S/S genotype being the most vulnerable for ADHD, although the reader will recall that the findings in the ADHD literature are themselves far from conclusive. A recent meta-analysis of the available data from simplex and multiplex families found no evidence of an association between either of the 5HHT L/S genotype and autism (Huang & Santangelo, 2008, but see Conroy et al., 2004). However, the authors did find a complex profile of the allele transmission related to ethnicity with an overrepresentation of the S allele in some ethnic groups but not others.

Only two studies to date have looked at other ADHD candidate genes and their relation to autism. One examined possible linkage between the COMT and DRD4 polymorphisms and found no relationship (Yirmiya et al., 2001), and the other looked at a possible link between DRD4 and autism, but although they found individuals with the "high-risk" 7-repeat allele, it was not beyond what was expected by chance (Grady et al., 2005).

Fragile X Syndrome

One study to date has examined the association between the serotonin transporter gene (*5HHT*) and fragile X syndrome with somewhat puzzling findings (Hessl et al., 2008). Specifically, the L/L genotype (rather than the S/S genotype) appeared to be overrepresented in fragile X males and was linked to high levels of aggression, self-injurious behavior, and stereotypic behavior. Unfortunately, the authors did not use a measure of ADHD behaviors, which is a core behavioral concern in fragile X syndrome and ADHD and is especially linked to 5HHT function.

No studies have been published looking at possible links between candidate dopamine genes (e.g., *DAT1*) and ADHD in fragile X. The current dearth of studies is disappointing given that anecdotal evidence suggests that not all children with fragile X display the same degree of cognitive impairment. It might well be possible that a subsample of affected children actually have symptoms that correlate with, for example, the *DAT1* 10/10 genotype or the *COMT* Val/Val genotype, placing them at greater risk of developing ADHD in addition to their diagnosis of fragile X syndrome. An examination of the potential associations between candidate genes, ADHD behaviors, and cognitive inattention is currently in progress in a sample of young males as part of a longitudinal study of attention and inhibitory deficits (Cornish & Scerif, 2009, personal communication).

Down Syndrome

Although not substantive, promising early research attempted to link increased COMT activity with Down syndrome, but findings were inconclusive (Brahe, Bannetta, Serra, Opitz, & Arwert, 1986; Gustavson, Floderus, Jaqell, Werrerberg, & Ross, 1982). The *COMT* gene has recently been revisited in Down syndrome and no association was found (Bhowmik, Dutta, Sinha, Chattopadhyay, & Mukhopadhyay, 2008). A similar absence of association was reported in the same study for the high-risk 7-repeat allele of the *DRD4* gene. Together, these findings indicate that ADHD candidate genes are not linked to the inattentive behaviors found in children with Down syndrome. On an interesting note, recent studies of adults with Down syndrome have observed elevated mRNA in the *SNAP-25* gene, but as yet its role in the behavioral phenotype of Down syndrome is unclear (Greber-Platzer, Fleischmann, Nussbaumer, Cairns, & Lubec, 2003).

Williams Syndrome

To date, we can find no published studies that have examined an association between Williams syndrome and ADHD candidate genes, yet the syndrome has a relatively well-documented ADHD profile. The paucity of research is puzzling, although it could simply be that ADHD symptoms really do not characterize the syndrome, making such research less focused and thus less promising from a genetic viewpoint.

22q11 Deletion Syndrome

The *COMT* gene is of obvious interest to this disorder given its linkage to 22q11. Although this research is still in its infancy, there is an emerging but somewhat contradictory literature. As the reader will recall from our description of the *COMT* gene above, the gene is important as it encodes the enzyme that is critical for dopamine metabolism in the prefrontal cortex and has been a popular gene candidate because of the long hypothesized role of dopamine in schizophrenia. However, given the lack of an association between *COMT* and ADHD, it is not surprising that few studies have pursued this link in children with 22q11 DS. However, the findings from a recent study by Gothelf and colleagues suggest an intriguing relationship, with the Met genotype significantly more prevalent in 22q11 DS with ADHD than in those without ADHD (73.9% vs. 33.3%) (Gothelf, Michaelovsky, Frisch, et al., 2007). Further research is needed to clarify this association and to establish whether there is a cognitive cost related to the Met genotype and ADHD in children with 22q11 DS.

Conclusions

In the present chapter we have attempted to capture the staggering advances in technology and knowledge that have facilitated our understanding of how genes interact with brain development to produce cognitive phenotypes. We have focused on six neurodevelopmental disorders that are associated with specific attention difficulties at the behavioral and cognitive levels. Four of these disorders have a known genetic etiology (fragile X syndrome, Williams syndrome, Down syndrome, and 22q11 DS), and two have strong genetic influences, but their causal mechanisms are, as yet, not determined (autism and ADHD). From just a brief snapshot of the genetic findings, the reader will appreciate the depth and complexity of the task we face in tracing the pathways from genes to behavior. Although major inroads have been achieved and we highlight many of them throughout the course of this book, it is clear that research is only beginning to unravel these complex relationships. One of the key areas of concern is the variability of findings, perhaps most notable in the ADHD literature, which clearly indicate that one-to-one mapping of a gene to a specific behavioral trait or cognitive function is extremely difficult if not impossible.

In addition, environmental factors will undoubtedly play a role in phenotypic outcomes in many neurodevelopmental disorders, as will biological factors such as gender, genomic imprinting, and ethnicity. We refer the reader to Chapters 8 and 9 where we describe how these factors impact on specific disorders. The interested reader may also want to look closely at the findings of Meaney and colleagues, whose elegant research on primates clearly demonstrates how behavior and environmental triggers can shape genetic and epigenetic mechanisms to produce phenotypic outcomes and diversity (e.g., Kaffman & Meaney, 2007; McGowan, Meany, & Szyf, 2008; Szyf, McGowan, & Meaney, 2008). These findings together with those in human populations (see Karmiloff-Smith, 2007, 2009, but also Pennington et al., 2009, who provide a conceptual framework for understanding environment × gene interactions in relation to attention and reading) illustrate that gene–brain–cognition relations are not unidirectional but multidirectional, requiring careful investigation from infancy onward.

The real advances in gene profiling are opening up new and exciting avenues of research. One avenue in particular that shows real promise is the concept of linking "attention" genes to neural substrates of attention, for example, locating specific allelic variations for sustained attention and then linking these to known disorders of attention such as ADHD. Bellgrove and his colleagues are at the forefront of this endeavor, and we discuss their work further in the context of ADHD later in this book (Chapter 8; see also Bellgrove & Mattingley, 2008, and Chambers, Garavan, & Bellgrove, 2008, for excellent critiques of the current research evidence). When we place this research in the context of advances in neuroimaging techniques of brain and gene expression (see Chapter 4), then the reader can appreciate the potential of this new strategy to advance our understanding of the interplay between genes, neural systems, cognition, and pathology. Go one step further to include a developmental framework, and we can explore the exciting potential of this approach to the clinical and educational domains. See Hariri and colleagues for a seminal discussion on this issue (e.g., Viding, Williamson, and Hariri, 2006).

CHAPTER SUMMARY AND KEY FINDINGS

- Despite numerous studies there is still no conclusive evidence linking specific genes to specific behaviors even in single gene disorders such as fragile X syndrome.
- Searching for "attention genes" in disorders of known genetic origin requires careful investigation across multiple levels of analysis: genetic, brain, and cognitive levels. The feasibility of this research is yet to be determined, but it certainly represents an exciting new direction in teasing out the genetic contribution to the attention problems that characterize many neurodevelopmental disorders.
- Inconsistencies in findings across ADHD studies may in part be due to differing methodologies that include variations in sample sizes and composition, ADHD diagnosis criteria, and a failure to control for the possible influence of comorbid disorders on phenotypic outcome.
- Future research must consider the critical roles of gene–gene and gene–environment interactions in determining disorder-specific attention profiles.

4

Brains and Atypical Attention

CHAPTER SNAPSHOT

- Recent advances in understanding the developing brain—innovative methodologies opening up new gateways of discovery
- The feasibility and pitfalls of using imaging technologies in neurodevelopmental disorders
- What can imaging studies of the typical development of attentional processing tell us about atypical brain maturation?
- Do advances in imaging technologies offer clues to underlying causes of attention deficits?

In this chapter we focus on a whirlwind of new discoveries that have helped shape our current understanding of the developing brain and how cognitive functions such as attention can be captured and observed in real time in typical and atypical development. In order to situate the reader in the context of these new developments, we first describe the basic anatomy and functions of the human brain and then go on to provide details of how brain imaging technologies can be applied to the study of attention and developmental disorders. This review is by no means a critique of cognitive neuroscience methodologies but instead aims to provide the interested reader with a glimpse of how advances in brain technologies can inform the study of atypical development of attention. Where possible we provide signposts to more detailed investigations and critiques for those wanting to pursue developments in more detail. But let us begin with an update of attention from Chapter 2.

We examined cognitive processes that are generally acknowledged to involve attention in one of the wide range of meanings of the term and outlined the main theoretical constructs of attention. We presented some of the current debates and experimental designs that have been conceived to investigate and refine these theoretical constructs.

In brief, the findings demonstrated that the process of selection is not the rigid filter at a fixed stage of processing envisaged in the original Broadbent model but is flexible and dependent on a variety of factors relevant to the situation. Hence, various control processes operate to ensure selection of relevant inputs or responses. The ability to divide attention between two or more stimuli or response sequences also depends on the nature of the stimuli or responses and the level of practice and, as with selection, is mediated through the control systems. These control systems are often referred to as the executive functions and are primarily dependent on the frontal lobes of the brain. The ability to maintain attention is also dependent on this control system working in conjunction with the midbrain and reticular activating systems of the brain stem that control arousal. Finally, we examined the complexities of the hypothesized executive system in an attempt to define more precisely which functions might be involved in this highly developed control system on which so much behavior depends.

In the present chapter we will be looking at studies of the organization of the brain that underlie these different constructs and attentional processes. Additionally, we will examine how different areas of the brain are active during different processes, and how they are interrelated during the performance of complex tasks. Our focus is on atypical development and, in particular, how advances in imaging techniques can inform attention development in different neurodevelopmental disorders. This is a relatively new field, and so we highlight some of the complexities and challenges that face researchers as they embark on this journey.

Originally, knowledge of brain function was obtained by looking at the effects of damage to various areas of the brain, such as acquired brain insults or strokes. Study of the effects of lesions in different parts of the human brain is, however, dependent on accidental events such as blows to the head, carbon monoxide poisoning, disease, and strokes, which do not affect a neatly circumscribed area, and until recently the type of damage could only be localized accurately postmortem. Recovery or partial recovery can also occur over time, suggesting either that the originally observed consequences were due to side effects of the trauma on surviving brain areas or that considerable plasticity of function is possible.

More recently, *in vivo* observation employed through a variety of methods has increasingly allowed neuroscientists to obtain high-resolution images of the structure and activity of the brain. These brain images can capture "snapshots" of brain activity in order to determine which areas are active during task performance. This allows us a glimpse of the complexities of the interconnections across brain regions that facilitate accurate and fast responding. In the domain of attention, research has provided clues to the complex nature of the brain networks and neurotransmitters involved in the selection, maintenance, and control of information. Undoubtedly, imaging technology is far superior to earlier brain damage investigations, but we hope that the reader will come to appreciate that each method has its unique limitations as well as strengths.

The human brain comprises many components that include the cerebrum, cerebellum, and brainstem. The cerebral cortex can be broadly subdivided into four lobes that comprise a range of cognitive functions including those uniquely involved in attention. Box 4.1 provides a summary of brain functions and lobes, but the reader needs to remember that the brain functions as a whole by its interconnected networks and pathways. For example, the task of attending to a specific sound in the distance and trying to locate its source will incorporate a number of brain regions working together as whole. We recommend the excellent text by Kandel, Schwartz, and Jessell (2008) for a detailed guide to brain structure, function, and control of action.

Pathways in the Human Brain

Figure 4.1 illustrates the main visual pathways from the primary visual cortex or striate area to the prestriate areas and beyond. There are two major pathways, or visual information streams, known as the ventral and dorsal streams (Merigan & Maunsell, 1993; Ungerleider & Mishkin, 1982). Cortical mechanisms operating within the *ventral visual stream* receive substantial input from parvocellular (P) layers of the lateral geniculate nucleus (LGN) that are tuned to high spatial / low temporal frequencies and are wavelength (or color) sensitive. The ventral stream passes via the visual area V4 in the prestriate cortex (extrastriate ventral

Box 4.1 Brain Function in the Different Lobes of the Brain

Frontal lobe—Important for performing complex actions that include planning, remaining "on-task," inhibition, and motor execution. The prefrontal cortex plays a critical role in controlling attention.

Parietal lobe—Important for shifting or disengaging attention from its primary focus, and also the integration of visual, spatial and somatosensory information

Temporal lobe—Important for perceptual attention, language, auditory perception, memory and object perception

Occipital lobe—Important for visual perception

Cerebellum—Important for the coordination of voluntary motor movements and balance

area that is selective to form and color information) to the inferotemporal cortex and from there to the ventrolateral prefrontal cortex, where it plays a critical role in the processing of object features. Conversely, mechanisms operating within the *dorsal visual stream* receive substantial input from magnocellular (M) layers of the LGN that are tuned to low spatial/high temporal frequencies that are not wavelength selective (i.e., Merigan & Maunsell, 1993; Schiller, Logothetis, & Charles, 1990;

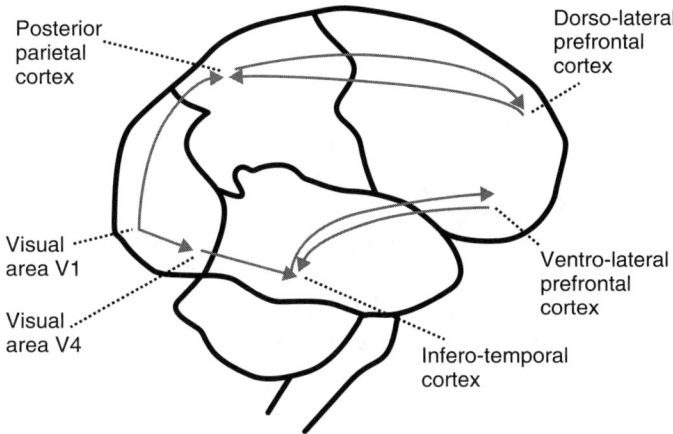

Figure 4-1. Dorsal and ventral processing streams for visual input.

Shapley, 1990; Tolhurst, 1975). Neurons located in extrastriate dorsal areas (i.e., visual area V5 or the medial temporal [MT] area) are highly selective for motion information and are involved in processing information concerning position via the posterior parietal cortex to the frontal eye fields, and dorsolateral areas of the prefrontal cortex. Gaze direction, position of objects relative to the body, and visual information relevant to motor activity, such as grasping, are computed in the parietal areas to prepare and control actions toward objects (Goodale & Milner, 1992; Milner & Goodale, 1995).

Effects of attention (via cueing, instructions, or past training) on the responses of single cells or brain areas to subsequent input have been observed throughout these pathways. Generally, findings suggest that attentional focus becomes more critical in determining cell responses as activation is passed up the system from the primary visual cortex (areas 17 and 18) to the prefrontal areas, which work to retain an "attentional template" of the position or nature of the target objects and feed such information back to the parietal and temporal areas that select targets and initiate action. The prefrontal cortex also affects endogenous arousal (arousal that arises from internal processes rather than external inputs) via its connections with the reticular activating system.

Assisting the brain in its integration of attentional information is a network of biological neurotransmitters. We highlight three—dopamine,

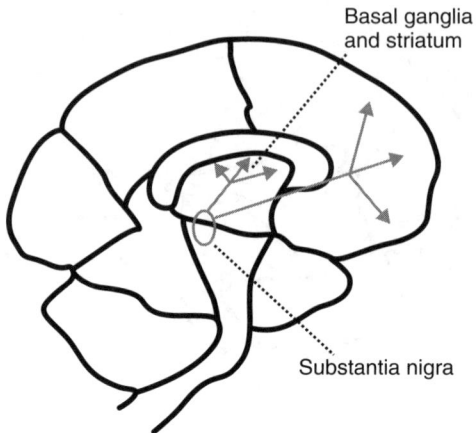

Figure 4-2. Dopamine circuits.

serotonin, and noradrenaline/norepinephrine—because of their role in modulating attentional control. The dopamine-based cortical pathways originate principally in the substantia nigra. This nigrostriatal system connects to the basal ganglia and striatum. A second main pathway connects to the frontal and prefrontal cortex. See Figure 4.2. Dopamine system dysfunctions have been associated with a variety of effects such as impulsive responding, problems in switching attention, and generally reduced effectiveness of frontal control systems (e.g., Mehta, Manes, Magnolfi, Sahakian, & Robbins, 2004; Robbins, Milstein, & Dalley, 2004; see also the excellent critique by Diamond, 2007).

The main serotonin pathway originates in the rostral raphe nuclei and connects to the cerebellum and has a sequence of connections to the thalamus in the midbrain, frontal, motor, parietal, and occipital cortex. See Figure 4.3. Recent findings suggest that low serotonin may impair the ability to maintain attention through modulation of selected brain areas, including the prefrontal areas and basal ganglia (Wingen, Kuypers, van de Ven, Formisano, & Ramaekers, 2008).

The main noradrenergic network connects from the locus coeruleus to the same cortical areas as the serotonin circuit and, like the latter, also has connections to the cerebellum. See Figure 4.4. Norepinephrine is also involved in arousal, and its overproduction during high arousal may

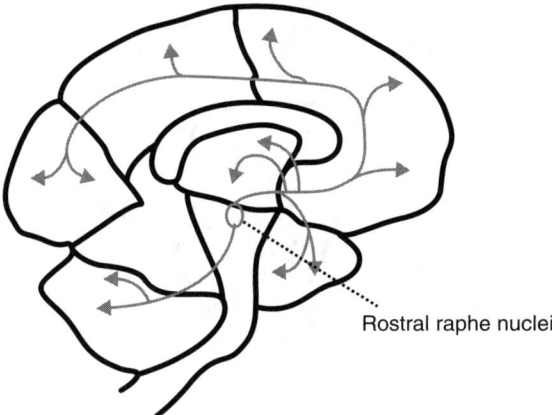

Rostral raphe nuclei

Figure 4-3. Serotonin circuits.

impair EF functions as well as impair selective attention (Robbins et al., 2004).

All three neurotransmitter functions are referred to and discussed throughout the book and therefore only briefly summarized here, but we recommend the reader to look closely at Chapter 3 for a detailed account of their role in the functioning of so-called "attention genes."

Brain Development and Attention Performance

Having reviewed a basic primer of adult brain structure or function, let us now turn to its development. By the age of 6 years, the brain has achieved 95% of its adult volume. In general, regions serving primary functions such as sensory and motor processes mature earlier than those in the temporal and parietal lobes that are active in language and attentional functions. The last areas to mature are in the prefrontal and lateral temporal lobes. Indeed, the effects of the late maturation of these brain areas can be readily observed in performance on tasks requiring attentional control (for example, well-known tasks of inhibition such as the Stroop task and Go–NoGo tasks, all reviewed in Chapter 6), and changes continue to occur at least up to adolescence (see Casey, Getz, & Galvan, 2008, for a comprehensive review of current literature). Correlations between development of the prefrontal areas and memory have been

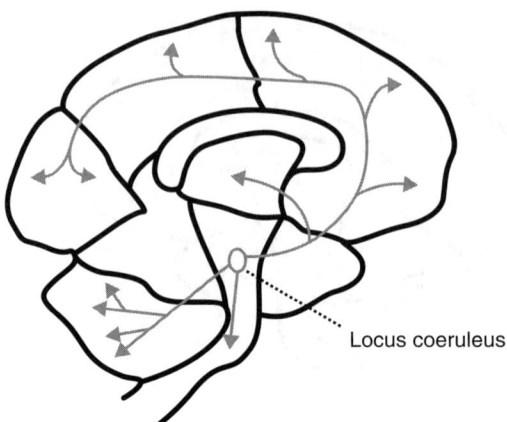

Locus coeruleus

Figure 4-4. Noradrenaline circuits.

demonstrated by Sowell et al. (2001), and between development of these areas and prepotent response inhibition by Casey et al. (1997). Monk et al. (2003) found that emotional information produced more activity in the amygdala in adolescents than in adults, whereas adults showed more activity in the ventral prefrontal cortex, implying greater inhibitory control of the emotional response. Mabbott, Noseworthy, Bouffet, Laughlin, and Rockel (2006) found that increases in white matter from 6 to 17 years were related to speed of visual search, particularly in frontal-parietal areas, indicating maturing integration of the control and spatial attentional systems.

An important question is the following: Are processes of brain development "prewired" so that we are seeing the development of preprogrammed, relatively self-sufficient, processing modules devoted to specific tasks? Or is development due to a wider process of interaction between different brain regions in response to inputs from the environment, leading to specialization of function within a given area and the creation of links to other areas that carry out different functions? Many theorists have advocated the preprogrammed modular alternative, but more recently arguments have been raised in favor of the second option. For example, Johnson, Halit, Grice, and Karmiloff-Smith (2002) point out that, if modules simply mature as a child gets older, more areas of activation should become apparent during the performance of a given task with increasing age, whereas the reverse is actually the case. For example, the immature brain processes faces bilaterally whereas in adults this function is localized in the right hemisphere. Likewise with word recognition: there is widespread activation early in development, but a focus in the left temporal lobe later.

Attention in the Brain

Posner's Attentional Networks Model

What do we know more specifically about neural systems involved in adult attention? Posner and colleagues (Posner & Dehaene, 1994; Posner & Petersen, 1990; Posner & Rothbart, 2005) have developed the most detailed integrated model of the operations of different areas in controlling attention. Figure 4.5 provides a summary of this model and its related brain regions. They distinguish three main attentional systems: first, the posterior system, which responds to an event by disengaging from the current focus of attention (this operation being controlled by the parietal lobe, damage to which can produce neglect of the contralateral hemisphere) and moves to the location of the event (controlled by the superior colliculus, damage to which slows shifts of attention) and engages attention on the new location (pulvinar lobe of the thalamus in the midbrain). Second, a vigilance network (right parietal and frontal lobes and the brain stem) maintains and adjusts alertness. This network increases the efficiency of the posterior system, while reducing ongoing activity in the anterior system (see below), thus suppressing competing instructions to the posterior system from the anterior system. The third system identified by Posner is the anterior system (anterior cingulate gyrus, supplementary motor areas—Brodmann area 6, basal ganglia) and acts as a more general executive system to resolve conflicts and direct the actions of the other systems to meet current goals. For example, blood flow increases in this area as the number of targets to be detected increases.

There has been particular interest in the role of the *anterior cingulate gyrus* in attention and cognitive control. It is connected to both the

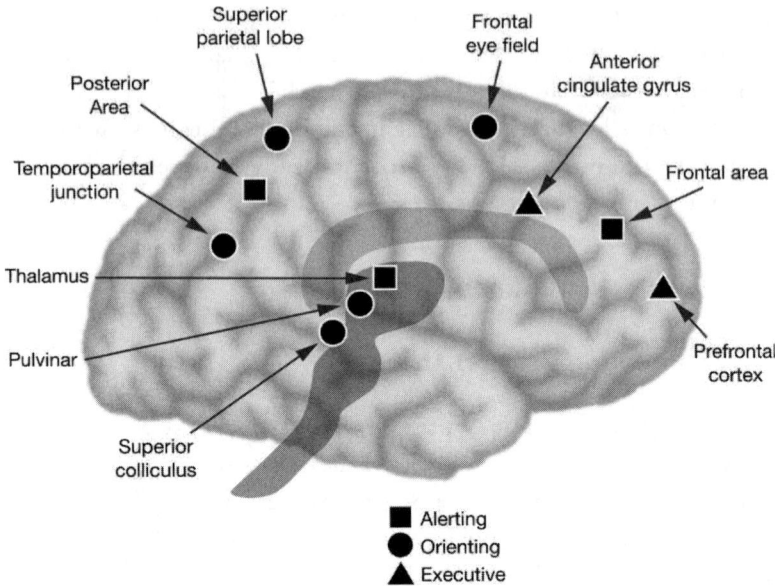

Figure 4-5. Posner's attention networks model.
From Posner and Rothbart (2007b). Adapted with permission from the *Annual Review of Psychology 58*, 6.

posterior parietal lobe and the dorsolateral prefrontal cortex, so it is part of the dorsal stream. However, it has more complex functions than simply visual attention and is also involved in what is termed "self-regulation" (see Posner, Rothbart, & Sheese, 2007). A good example of a task that involves self-regulation is the Stroop task (reviewed in Chapter 6), which requires the participant to read and ignore an overlearned word like "blue" and instead respond with the name of a conflicting ink color. Imaging studies have found the left dorsal anterior cingulate and prefrontal cortex to be most active in aspects of the Stroop that require most effort in handling conflict (e.g., Fan, Flombaum, McCandliss, Thomas, & Posner, 2003). Moreover, Botvinick and colleagues (Botvinick, Braver, Barch, Carter, & Cohen, 2001; also Botvinick, 2007; Botvinick, Cohen, & Carter, 2004) have suggested that a major function of the anterior cingulate is to monitor for difficulty induced by conflict such as that between possible responses, especially when a prepotent response has to be overridden, or when a current strategy is producing errors.

The authors present evidence that activation of the anterior cingulate cortex (ACC) accompanies conflict rather than situations requiring strong control.

We turn now to a review of brain regions associated with specific aspects of attention: selection, maintenance, and control.

Selective Attention

The ability to select relevant information from the myriad of distracters that bombard us is a fundamental attention skill. Directing attention primes relevant processing systems in the cortex. Activation in response to a stimulus is higher in visual centers when attention is directed to visual stimuli than when it is directed to auditory stimuli and vice versa (Shomstein & Yantis, 2004). Moving colored shapes evoke the highest activation in areas processing movement, color, or shape, depending on which feature is the focus of attention (see Nobre, Rao, & Chelazzi, 2006). Furthermore, when superimposed pictures of faces and houses are presented and attention is directed to one of the two objects then switched to the other, the main activated brain area changes accordingly (Serences, Schwarzbach, Courtney, Golay, & Yantis, 2004). Many studies have shown that directing attention to a specified location results in higher activation to a stimulus presented at that location than when attention is directed to a different location (Desimone & Duncan, 1995; Kastner & Ungerleider, 2000; Reynolds & Chelazzi, 2004). Hence, selective attention of various types consistently facilitates brain activation in the appropriate processing centers.

More complex tasks that demand voluntary control of selective attention recruit control systems in the frontal lobes of the brain. Switching or focusing of attention in response to instructions or cues produces transient activity in prefrontal and postparietal areas of the cortex (Loose, Kaufmann, Tucha, Auer, & Lange, 2006; Serences et al., 2004; Shomstein & Yantis, 2004). Likewise, demands such as dividing attention (Johnson & Zatorre, 2006) or carrying out a difficult selection (Nebel et al., 2005) engage frontal systems. Evidence on the role of prefrontal cortex in controlling spatial attention can be derived from studies of attentional hemineglect. This occurs when brain damage leads to a problem in selecting information from one side of the visual field.

Typically, damage to the right prefrontal cortex leads to neglect of the left visual field, whereas damage to the left prefrontal cortex has much smaller effects. A number of types of hemineglect occur, depending on damage to different sites (Allport, 1993; Colby & Goldberg, 1999), details of which will not be covered here. The difficulty appears to be due to a control weakness in disengaging from the right visual field rather than a perceptual inability to "see" the neglected side (Posner, Walker, Friedrich, & Rafal, 1984; Loetscher & Brugger, 2007), since instructions or other manipulations to aid disengagement have some compensatory effect.

Maintenance of Attention

For maintaining attention, lesion and imaging studies show that the right prefrontal cortex is crucial in both sustained and phasic (short-term) alertness and attention to events, as well as the maintenance of information in working memory (e.g., D'Esposito, Ballard, Zarahn, & Aguirre, 2000; Swick & Knight, 1998). A recent study by Shallice, Stuss, Alexander, Picton, and Derkzen (2008) found that sustained counting judgments (slow vs. fast rates) were impaired following lesions to the superior medial (SM) and right lateral (RL) frontal cortex, confirming the critical role of frontal cortex in attention. However, unlike patients with lesions in the SM regions who displayed impairment across both the slow and fast conditions, patients with lesions to the RL region displayed impairment on the fast condition only. The slow (vigilance) condition was not impaired. These findings failed to confirm the Posner and Petersen (1990) proposal that the RL frontal cortex is critical for vigilance but conflict with many other findings.

Rueckert and Grafman (1996) found a greater decline in performance on a Continuous Performance Task (CPT, reviewed in Chapter 6) that presented frequent stimuli at regular intervals, following right hemisphere lesions. Koski and Petrides (2001) also found that in a group with right hemisphere lesions, reaction times declined in a lengthy task requiring regular responses to cued targets at regular intervals. Imaging studies provide more precise information. Lawrence, Ross, Hoffmann, Garavan, and Stein (2003) identified two circuits associated with good performance when detecting targets during a rapid presentation of visual stimuli; one was a right frontal-parietal circuit that was activated, and the other circuit was deactivated and included temporal lobe and cingulate cortex activity together with the limbic system of the midbrain.

Attentional Control

A variety of evidence on the brain areas involved in attentional control has already been cited in Chapter 2 while clarifying the notion of executive function. Dorsolateral and ventrolateral prefrontal cortices are involved in planning, response organization, and decision making. Frontal systems and particularly dorsolateral prefrontal cortex are also involved in control of selection, activated when changes in focus or division of attention are required. Damage to areas of prefrontal cortex demonstrates their importance in a number of functions, such as the ability to inhibit prepotent responses or previous responses when a switch is required (see Robbins, 2007, for a review), alongside other executive functions (e.g., Funahashi, 2001; Raye, Johnson, Mitchell, Greene, & Johnson, 2007; Tanji & Hoshi, 2008). The complexities of brain activation in specific operations, especially those involving executive functions, are highlighted in recent functional imaging studies. Using functional magnetic resonance imaging (fMRI, see our description below), Stevens, Pearlson, and Kiehl (2007) investigated the key circuits involved in inhibiting responses on "NoGo" trials in the Go–NoGo task, one of the most widely used tasks of response inhibition. In this task, occasional rare trials indicate that no response is required; this tests inhibition of the strong positive response that has been induced by the majority of the trials. They identified three interacting circuits, each involving a number of brain areas. One circuit combined frontal, striatal, and thalamic centers and normally inhibited the other two circuits, but inhibition was changed to activation on trials requiring responses to be withheld. The other two circuits involved parietal and premotor areas in one case and a frontal-parietal combination in the other. These two circuits acted reciprocally to inhibit or reactivate motor responses. This important study illustrates the complexity of the interactions across the brain that are involved in achieving a single inhibitory function.

Images of the Brain: Methodology and Use in Developmental Studies

The past decade has seen unparalleled advances in our ability to view the living brain and its activity while undertaking different tasks. These techniques allow a more detailed mapping of cognitive pathways and networks and their intercorrelations across development. The current literature is extensive, and it is not our aim in this section to provide the reader with a definitive critique of all existing studies, but rather to highlight within the domain of attention some of the innovative findings that have begun to enhance our understanding of the typical and atypical developing brain and the complex networks that drive brain maturation across the lifespan. For outstanding examples of progress in this field, see the work and critiques of Casey and her colleagues (Casey, Jones, & Hare, 2008; Casey, Tottenham, Liston, & Durston, 2005; Casey et al., 2004; Durston & Casey, 2006). We now briefly describe each imaging technique in turn, concluding with the most recently developed ones. Our main aim is by no means to provide a complete overview of neuroimaging methods, but instead to discuss, where possible, recent examples of typical and atypical developing studies that highlight the feasibility of each technique. We focus specifically on our six neurodevelopmental disorders described in detail in Chapter 3: ADHD, autism, fragile X syndrome, Down syndrome, Williams syndrome, and 22q11 DS.

Positron Emission Computed Tomography (PET) and Single Photon Emission Computed Tomography (SPECT)

PET is among the earliest functional imaging techniques and one that provides noninvasive, three-dimensional, quantitative images of the distribution of radioactivity in the body. This method can be used to trace blood flow, regional glucose metabolism, or neurotransmitter content, for example, dopamine transporter levels. There are, however, potential risks associated with using PET imaging with children, notably ethical concerns about pharmacological effects of the PET radiopharmaceuticals, the use of sedation in some instances, and exposure to ionizing radiation. Sundaram, Chugani, and Chugani (2005) provide an extensive review of the main ethical concerns associated with PET and also address the applications of this technique to disorders of mental retardation and brain pathology.

SPECT works similarly to PET, involving the injection of radiop-harmaceuticals, which then distribute throughout the brain, emitting single photon radiation, such as gamma rays, as they decay. The more active the brain area, the greater the blood flow, resulting in more concentrated amounts of the radioactive tracers that can be picked up by the SPECT camera. Compared to PET imaging, SPECT is relatively inexpensive, but both techniques have relatively poor spatial resolution, and their use in tracing typical developmental changes is extremely difficult due to ethical constraints.

Neurodevelopmental Disorders

PET and SPECT have been used quite extensively as structural imaging tools across a range of neurodevelopmental disorders. In ADHD, PET studies have found that dopamine signaling is altered specifically in the midbrain, demonstrated by a decreased binding potential (a measure of the density of available neuroreceptors) of dopamine transporter (DAT), which may be linked to the hyperactivity seen in ADHD (Jucaite, Fernell, Halldin, Forssberg, & Farde, 2005). A recent SPECT study examining the effects of methylphenidate (a stimulant medication also known as "Ritalin") on the resting regional cerebral blood flow (rCBF) in children with ADHD found that, after treatment, normalization of the rCBF was observed in the somatosensory area, right striatum, ventral higher visual areas, and the superior prefrontal area (Lee et al., 2005). In children with autism, results from PET studies demonstrate an asymmetry in cortical serotonin synthesis with specific impairment associated with different hemispheres. For example, severe language impairment was associated with serotonin decreases in the left hemisphere (e.g., Chandana et al., 2005). Specific decreases in rCBF in the temporal, parietal, and frontal cortices and the caudate nucleus have also been reported, suggesting that several brain areas may be involved (Degirmenci et al., 2008). However, this profile does not appear to be specific to autism, with similar abnormalities in the fronto-parieto-temporal region reported in fragile X syndrome (Kabakus et al., 2006) and Down syndrome (Aydin, Kabakus, Balci, & Ayar, 2007). Together, these findings suggest that a common profile of brain abnormalities may underlie the attention difficulties that represent these genetically disparate disorders. In turn, the complex role of fronto-parietal-temporal circuits in the development of attention more generally clearly highlights the need to understand further: (1) how distinct attentional profiles emerge from these seemingly common brain abnormalities, (2) and how we can provide an account for the extreme vulnerability of frontoparietal and temporal circuits across disorders of distinct etiology. For example, there is potential involvement of both parietal and frontal systems in Williams syndrome, but a relatively greater involvement of frontostriatal pathways in fragile X.

To date, and not unsurprisingly given the constraints of this technology, no studies have used PET or SPECT to assess developmental changes in the atypical brain.

Magnetic Resonance Imaging (MRI)

Magnetic resonance imaging has been used extensively in typically developing children since the late 1980s to provide structural images of the brain. The technique is noninvasive, so it involves no injections, and it does not use ionizing radiation, so it is safe for infants and children. Moreover, this technique can be used repeatedly on the same participant, thus facilitating longitudinal, developmental studies. MRI uses a powerful magnetic field and radio waves to create cross-sectional brain images and has excellent spatial resolution, allowing for optimal discrimination between tissue types (white matter, gray matter, and cerebrospinal fluid), specific brain structures, and regions of interest (ROI).

MRI has been an extremely valuable tool in tracing developmental changes in typically developing infants and children with noted longitudinal changes in gray and white matter (e.g., Gogtay et al., 2004; Wilke, Krägeloh-Mann, & Holland, 2007), cortical thickness (Shaw et al., 2008; Sowell et al., 2004), and brain-cognitive correlations (Sowell et al., 2004). For a comprehensive review of recent MRI developmental studies, see Lenroot and Giedd (2006) and Giedd (2008). The interested reader might also want to explore the recent findings of studies using a technique known as diffusion tensor imaging, which specializes in imaging white matter of the brain and tracking connectivity and pathways.

Neurodevelopmental Disorders

A substantial body of research now attests to the feasibility of using MRI in atypically developing children, and findings point to important differences in both brain volume and structure that distinguish children with differing neurodevelopmental disorders from each other and from their typically developing peers. For example, a recent study of children with ADHD examined the developmental trajectories of cortical maturation and found a marked delay (a difference of 3 years), compared to controls, in prefrontal regions important for attention control (Shaw et al., 2007). In autism, a recent systematic review and meta-analysis of 46 published MRI studies found a generalized brain enlargement with increases to the cerebral hemispheres, cerebellum, and caudate nucleus. In contrast, there is a reported decrease in the size of the corpus callosum and tentatively the midbrain and vermal lobules VI–VII and VIII–X (Stanfield et al., 2008). Some of these abnormalities can be quite clearly associated with the attentional difficulties experienced by individuals with autism (as, for example, the involvement of the caudate nucleus in frontostriatal circuits), but others are less explicitly linked to attention and control difficulties. In addition, while the reported structural abnormalities seem reliable and replicable, what is lacking is an understanding of their precise relationship to the attentional profiles of individuals with autism and of their specificity to attention in autism compared to other disorders.

In fragile X syndrome, one of the most consistent structural abnormalities is that of increased caudate volume (e.g., Gothelf et al., 2008—see Figure 4.6; Hoeft et al., 2008), which, alongside a recently reported anomaly of the lateral prefrontal cortex in toddlers with fragile X (mean age 35 months) (Hoeft et al., 2008), has been suggested as a possible precursor to the well-documented impairments in attentional control associated with impaired response inhibition. Indeed, inhibitory deficits are now seen as part of the signature cognitive profile in fragile X and are observable from infancy onward (e.g., Cornish, Scerif, & Karmiloff-Smith, 2007; Munir, Cornish, & Wilding, 2000b; Scerif, Cornish, Wilding, Driver, & Karmiloff-Smith, 2004, 2007). In Williams syndrome, there are presently no published studies that specifically link brain abnormalities with attention dysfunction, as clearly witnessed at the behavioral and cognitive levels. However, this may just be a matter of time as an

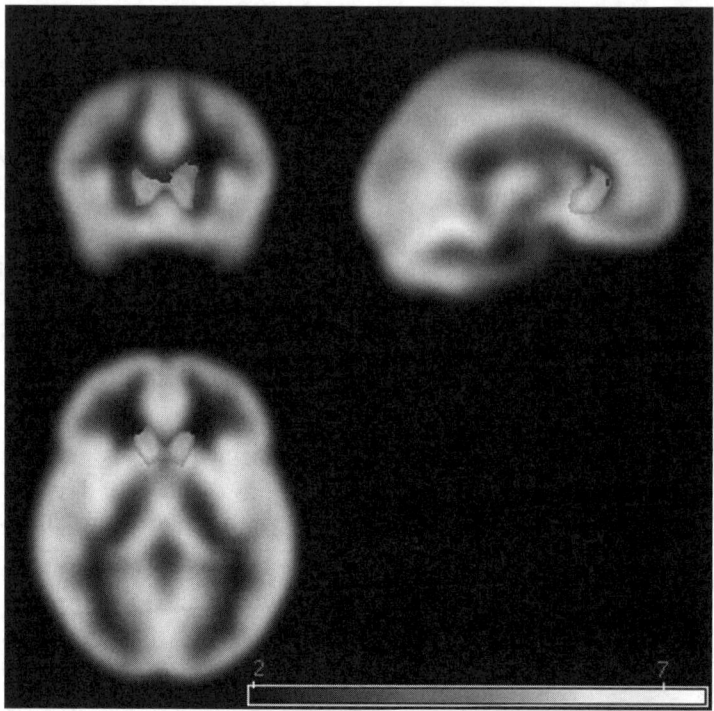

Figure 4-6. MRI images of regions of increased gray matter volumes in individuals with fragile X syndrome compared to typically developing controls.
From Gothelf et al. (2008). Reprinted by permission from *Annals of Neurology 63*, 40–51. (See color Figure 4-6.)

emerging literature is clearly demonstrating neuroanatomic correlates with specific cognitive functions known to be especially impaired in Williams syndrome, such as visuospatial and numerical abilities (see Martens, Wilson, & Reutens, 2008, for an excellent review of recent findings; also Reiss et al., 2004, for an earlier example).

In children with 22q11 DS, a recent study found significant reduction in cerebellar gray matter, and white matter reductions in the frontal lobe and cerebellum compared to typically developing children (Campbell et al., 2006). In children with Down syndrome, frontal lobe volumes were found to be reduced alongside disproportionately smaller volumes of the cerebellum (Pinter, Eliez, Schmitt, Capone, & Reiss, 2001). Both 22q11 DS and Down syndrome are associated with significant problems in executive functions and attention, functions known to be associated with the frontal lobes. See Schaer and Eliez (2007) for an excellent review of how structural imaging research has furthered our knowledge of the pathogenic processes underlying atypical brain development.

Functional Magnetic Resonance Imaging (fMRI)

Functional MRI is one of the newest and most widely used techniques for measuring brain functioning. It is noninvasive and in its basic format measures increases in blood oxygenation (blood flow represents delivery of oxygen to recently used cells) related to neural activity in specific brain regions, providing a map of activity across brain areas as specific tasks are performed. Limitations include poor to moderate temporal resolution and restrictions in the populations with which it can be used because the machine can be considered noisy when the images are being taken in, the tunnel in which the participant has to lie is constricted and movement can distort the signals. However, more refined MRI technology has made these issues less of a concern in recent years.

A growing literature has begun simultaneously to chart early cognitive and neural development in typically developing children. The general consensus is that cortical development and cognitive development occur concurrently across childhood through to adolescence. For example, Crone and colleagues, using a cross-sectional design to examine age-related changes in task-switching—a core attentional control function (see Chapter 7 for more detailed descriptions of these functions)—found an age improvement that persisted into adolescence (Crone, Donohue, Honomichl, Wendelken, & Bunge, 2006). This improvement, however, developed along two different trajectories depending on the task component being traced. Although cross-sectional studies such as these can provide important information about changes in brain structure and function at distinct age time points, one of the most critical uses of fMRI is its potential to isolate developmental progressions, both typical and atypical. See Kotsoni, Byrd, and Casey (2006) for an excellent review of the basic principles and applications of fMRI to developmental research, and Casey, Galvan, and Hare (2005) and Durston et al. (2006) for specific examples.

Neurodevelopmental Disorders

The use of fMRI in children with neurodevelopmental disorders, although not as frequent as with typically developing children, is nonetheless impressive. The feasibility of this technique in detecting the functional organization of the atypical brain is now well established, with the greatest intensity of research activity focused on the ADHD population. A recent study by Booth et al. (2005) demonstrated just how impressive this research can be. The authors compared performance across attention subcomponents (selection vs. response inhibition) in children with ADHD (aged between 9 and 12 years) and a typically developing control group matched on chronological age. The findings indicated minimal group differences in brain activation on a selective attention measure (a visual search paradigm) but significant differences on a response inhibition measure (a Go–NoGo paradigm). In the latter, there was widespread hypoactivity in the ADHD group in the frontostriatal regions during task performance. The pattern of these findings is commensurate with the numerous cognitive findings of impaired inhibitory control in ADHD compared to control children (see Chapter 8 for a review of these findings). See also Suskauer et al. (2008), for an extension of these findings, and Kelly, Margulies, and Castellanos (2007) for a recent review of functional brain imaging studies in ADHD.

Across other neurodevelopmental disorders, the most successful application of fMRI has been in subsamples of children and adolescents with higher levels of cognitive functioning. For example, a recent study by Gomot, Belmonte, Bullmore, Bernard, and Baron-Cohen (2008) explored the brain regions involved in the detection of novel auditory stimuli using a task that required focused attention on selected inputs while ignoring distractions (a hallmark of selective attention) in a sample of children and adolescents with high-functioning autism or Asperger syndrome (aged between 10 and 15 years). Findings indicated greater brain activation in the right prefrontal-premotor area and left inferior parietal regions in the autism group compared to controls. The authors argue that individuals with autism may overfocus on specific events when their attention is actively drawn toward target stimuli. There are other numerous examples of fMRI studies in autism and a growing number that have begun to identify brain–cognitive correlates in attention processing, but most, unfortunately, are conducted on adult populations (for an interesting example, see Gilbert, Bird, Brindley, Frith, & Burgess, 2008).

In fragile X syndrome, although there is a growing literature that has successfully used fMRI, the majority of published research has tended to focus on females with a skew toward later adolescence/early adulthood. The reason for this is that fragile X females are much less intellectually impaired than their male counterparts and are therefore more able to perform a broader range of cognitive tasks using this technique. We recommend the reader turn to Chapters 3 and 9 for more detailed information about the genetic variations in fragile X and their impact at the brain and cognitive levels. To date, Reiss and colleagues provide the only examples of fMRI studies conducted in young males and females with fragile X with a specific focus on attention control. Using a traditional Go–Nogo paradigm, requiring participants to view a series of letters and respond with a key press to every letter except the letter X, for which they had to withhold a response, findings indicate that frontostriatal regions, known to be involved in response inhibition, are especially affected in fragile X, irrespective of gender. For example, Hoeft et al. (2007) compared performance in fragile X male adolescents (mean age 15.4 years) and two control groups: an IQ-matched developmentally delayed group and a typically developing group matched on chronological age. Their findings are noteworthy in two respects: first, Go–NoGo performance by fragile X males, unlike that of control males, was not associated with increased activation in the *right* frontostriatal regions. Second, successful performance was instead associated with increased activation levels in the *left* frontostriatal network. See Figure 4.7 for an example of brain activation during this task as recorded by fMRI.

The pattern of these findings led the authors to make the tantalizing conclusion that response inhibition in fragile X may be guided by compensatory processes brought about by a complex interaction between the effects of genetic factors (in the case of fragile X—the *FMR1* gene and its encoded protein—the FMRP, see Chapter 3) on early brain

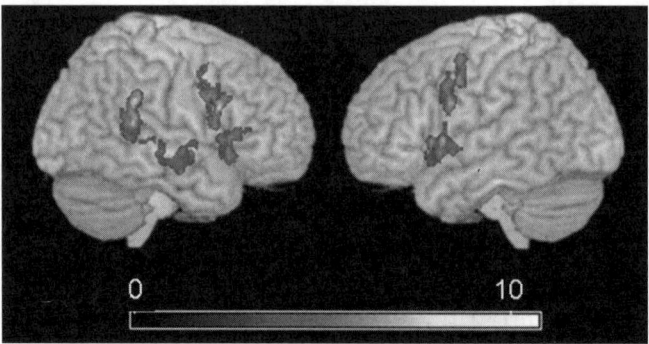

Figure 4-7. fMRI images of the brain activation during a response inhibition task: Go–NoGo paradigm.
From Hoeft et al. (2007). Reprinted by permission from *Human Brain Mapping 28*, 543–554. (See color Figure 4-7.)

maturation, with particular vulnerability in the frontostriatal network. A similar prefrontal dysfunction was also observed in a study of females with fragile X (mean age 15.9 years) using the same Go–NoGo paradigm (Menon, Leroux, White, and Reiss, 2004). Undoubtedly, these intriguing findings await further exploration, but they clearly demonstrate that at least by adolescence, fragile X is associated with anomalous brain development in regions that are involved control of attentional responses.

In the first fMRI study to examine response inhibition performance in adolescents with 22q11 deletion syndrome (mean age 17.8 years), Reiss and colleagues (using the No–GoNo paradigm described above) reported abnormal brain activation in the parietal regions with greater activation in the left parietal network during successful task performance, a pattern that was not observed in controls (Gothelf, Hoeft, Hinard, et al., 2007).

The authors suggest the possibility that anomalous development of the parietal region and the subsequent impact on developing attention processes (see Bish, Ferrante, McDonald-McGinn, Zackai, & Simon, 2005) may have resulted in compensatory strategies that involve the recruitment of the parietal networks to perform a simple response inhibition task. While tantalizing, such a suggestion would, of course, need to be substantiated by further studies, and in particular, ones with a developmental or longitudinal component. These are necessary to tease apart processes that are atypical from the outset, from those that may be the result of dynamic compensatory changes in the developing brain of individuals with 22q11 deletion syndrome. Of note, these compensatory changes would be distinct from those displayed by adolescent boys with fragile X syndrome described above, highlighting again the need for direct cross-syndrome comparisons in understanding atypical patterns of brain activation.

To date, we can locate no published studies using fMRI in children or adolescents with Down syndrome or Williams syndrome that have focused on attention processing. However, there is an emerging literature in Williams syndrome that demonstrates the feasibility of this technique, at least in adults, to discriminate changes in parietal and visual cortical pathways (dorsal and ventral streams). For example, Mobbs, Eckert, Menon et al. (2007) recently demonstrated a distinct profile of reduced visual cortical activation in specific dorsal stream structures during performance of a processing task that required identifying the global components of a shape stimulus (e.g., a big triangle or a big square). The authors conclude that Williams syndrome, unlike controls, may be associated with an early and specific disruption of the dorsal stream pathway, leading to the well-documented cognitive difficulties in visual-spatial processing. The relative sparing of the ventral stream may explain the juxtaposition between their poor visual spatial abilities and their proficient visual perceptual abilities.

Transcranial Magnetic Stimulation (TMS)

TMS is a noninvasive method of brain stimulation in which magnetic fields are used to induce electric currents not only in focal areas of the motor cortex but also in many other cortical regions. A small coil of wires is held against the scalp, and a powerful electric current is passed through, which produces a brief electrical pulse. The pulse travels unimpeded to the cortex, where it generates an electrical current that causes activation of neurons in that region. TMS has a very high level of spatial resolution, so it is possible to pinpoint very specific brain areas for stimulation. TMS can be used to temporarily disrupt cortical activity, a technique, often referred to as "virtual lesioning." There are a number of modes of TMS—for example, paired-pulse TMS, which is used to probe GABAergic systems, and theta burst, which can be used to probe neural plasticity. A *single-pulse* mode that delivers a single stimulation to the target area allows the researcher to determine when a given brain area is required to perform a cognitive task.

TMS is a sophisticated tool and as a complementary approach to functional neuroimaging, it has the potential to provide a detailed understanding of brain–behavior correlates by mapping the functional connectivity between brain regions. Additionally, TMS goes one step further than fMRI by being able to directly infer causality rather than simply state an "association." There is now a well-established literature that demonstrates clear functional changes in cortical maturation using this technique. One of the key findings is of developmental changes in the inhibitory mechanisms within the motor cortex, specifically in intracortical inhibition, which appears to increase significantly from childhood (post–10 years) into adulthood (e.g., Mall et al., 2004). The authors suggest a tentative link between earlier studies that had suggested a relationship between decreased levels of intracortical inhibition and increased practice-dependent neuronal plasticity. See Garvey and Mall (2008) for their exceptional review of recent developmental studies using TMS, including the Mall et al. study.

Neurodevelopmental Disorders

The potential use of TMS as a tool to map atypical brain functioning and behavior in neurodevelopmental disorders is just emerging. Focusing primarily on disorders with known motor abnormalities, TMS has been used most recently to track abnormal intracortical inhibition in children with ADHD (e.g., Buchmann et al., 2007; Gilbert, Sallee, Zhang, Lipps, & Wassermann, 2005). Findings indicate a strong association between severity of hyperactive symptoms and intracortical inhibition that reduces in intensity when on stimulant medication such as methylphenidate (MPH) (see Chapter 10 for a review of ADHD and MPH). In the Buchmann et al. study, for example, children with ADHD (aged 7 to 15 years) were assessed twice, once at baseline without stimulant medication (e.g., methylphenidate [MPH]; see Chapter 10 for a review of studies investigating the efficacy of MPH in reducing ADHD symptoms) and again, when on medication for at least 10 days. Their findings offer some interesting insights into the neurophysiological mechanisms underlying the motor deficits in ADHD by demonstrating that a reduction in ADHD symptoms was specifically correlated with MPH-induced intracortical inhibition, resulting in clinical improvement in hyperactive behaviors.

Although studies outside of ADHD are limited, recent findings suggest the feasibility of TMS in autism (e.g., Sokhadze et al., 2009), and we would argue that it has a strong potential utility in fragile X syndrome where motor problems including hyperactivity and clumsiness are core features of the phenotype.

Event-Related Potentials (ERPs)

ERP electrophysiological techniques record brain activity following sensory or cognitive stimulation. This technique provides direct, high-temporal resolution of neural activity in the brain following an input. Typically, a positive component occurring about 90 milliseconds after stimulus onset (known as P90 or P1) is followed by a negative component after 150–200 msec (N1), then a sequence of other components (P2, N2, etc). The principal components known to be affected by attention, for example, are P1, N1, and especially P3. These are generated by successive processing stages, P1 and N1 in the sensory cortex and the complex P3 component by multiple sources, including parietal and frontal systems (Anllo-Vento, Schoenfeld, & Hillyard, 2004; Herrmann & Knight, 2001). ERPs are excellent tools to monitor early brain development, and there is now a well-established literature demonstrating developmental trajectories from infancy through to adolescence and adulthood. The reader has a choice of excellent reviews to consult for comprehensive updates and critiques of ERP methodology across typical and atypical development (Banaschewski & Brandeis, 2007; Nelson & McCleery, 2008; Picton, & Taylor, 2007).

Neurodevelopmental Disorders

There is now a growing literature documenting abnormalities in ERPs in children and adolescents across a range of neurodevelopmental disorders. The increasing sensitivity of this technology to detect changes in brain activity makes this a popular alternative or, in some cases, a complement to fMRI. The ERP has been studied extensively in ADHD populations (see for example, the recent studies by Gumenyuk et al. 2005 and van Mourik, Oosterlaan, Heslenfeld, Konig, & Sergeant 2007) that have used ERPs to measure distraction (selective attention). In autism, ERPs have been used to examine self-monitoring in error processing (Vlamings, Jonkman, Hoeksma, van Engeland, & Kemner, 2008) and orienting to speech sound changes (Lepistö et al., 2005). See Jeste and Nelson (2009) for a recent comprehensive review of ERP findings in autism.

In one of the few studies to use a cross-syndrome comparison, Groen, Wijers, Mulder, et al. (2008) examined response monitoring in four groups of children aged between 10 and 12 years; children with autism, children with ADHD on stimulant medication, children with ADHD without medication, and typically developing control children matched on chronological age and IQ. ERP data revealed a response monitoring performance impairment (error detection and later error awareness) that differentiated children with ADHD from children with autism and typically developing children. These findings concur with the well-documented neuroimaging and cognitive findings of impairment in frontostriatal pathways in ADHD.

Less well studied, but with an emerging literature, are fragile X syndrome (Castrén, Pääkkönen, Tarkka, Ryynänen, & Partanen, 2003), Williams syndrome (Grice et al., 2003), and Down syndrome (Karrer, Karrer, Bloom, Chaney, & Davis, 1998). Across these studies the feasibility of using ERP has been demonstrated, but the lack of more recent research using this technique is surprising and may relate more to the difficulties in obtaining reliable data than to a lack of research effort. The results nonetheless are very interesting and warrant further investigation. For example, Castrén and colleagues (Castrén et al., 2003) identified abnormal processing in the auditory afferent pathways in boys with fragile X (mean age 11.6 years) with more N1 frontal distributions than control boys. The authors link these findings to the possible atypical development of frontal temporal pathways in fragile X. In Williams syndrome, Grice and colleagues (Grice et al., 2003) found atypicalities in the neural mechanisms that support early perceptual processing and suggest that disruption to these early visual pathways will have long-term detrimental effects on later visual processing, resulting in the well-documented visuospatial deficits associated with Williams syndrome.

Magnetoencephalography (MEG)

This electrophysiological technique utilizes magnetic and electrical currents generated by neurons, allowing for these magnetic fields to be measured. MEG detects the activity of the magnetic fields and is reliable at localizing and mapping *sources* of activity in the brain. As a technique, MEG is more child friendly than fMRI, with less scanner noise and not as much physical restraint needed of the participant. A core advantage is that it can allow the fast tracking of brain activations in the order of milliseconds alongside relatively good spatial accuracy. MEG offers an exciting approach to profiling patterns of neural activity during the processing of sensory inputs and may help us understand how these basic inputs influence more complex neural networks involved in the development of higher-order processing such as face processing (e.g., Kylliäinen, Braeutigam, Hietanen, Swithenby, & Bailey, 2006) and auditory skills (Huotilainen, Shestakova, & Hukki, 2008).

Neurodevelopmental Disorders

MEG studies that include children with neurodevelopmental disorders are limited. Of the few published studies to date, almost all have focused on ADHD populations in which differences related to the limbic regions appear to differentiate children with ADHD from typically developing control children. Notably, Mulas et al. (2006) found that group differences emerged in performance on a modified version of the Wisconsin Card Sorting Task (a measure of cognitive flexibility, reviewed in Chapter 6) specifically in the very earliest preparatory stages of processing. The authors suggest that the well-documented impairments in higher-order processing such as executive functions may in part be due to a disruption of early sensory processes. See also the recent findings of Tannock and her colleagues (Dockstader et al., 2008), who have used MEG to explore somatosensory processing in adults with ADHD. Unfortunately, outside of ADHD, few studies have used this technique to explore differences, but its feasibility has recently been explored in children with autism (Gage, Siegel, Callen, & Roberts, 2003; Wilson, Rojas, Reite, Teale, & Rogers, 2007), where findings indicate atypical maturation of the auditory cortical systems. In fragile X syndrome, MEG has been tentatively explored in adults with an early study by Rojas et al. (2001) reporting large auditory evoked response differences in fragile X adults compared to typically developing controls. Regrettably, this appears to be the only published study available focusing on fragile X.

In adults with Down syndrome, two recent studies have demonstrated the feasibility of MEG to discriminate atypical cortical activation in the sensorimotor regions (Virji-Babul et al., 2008; Virji-Babul, Cheung, Weeks, Herdman, & Cheyne, 2007). To the best of our knowledge, there appear to be no published studies exploring the feasibility of MEG in Williams syndrome or 22q11 DS.

Dynamic Imaging Approaches

Of growing popularity is the recognition that a combined approach using differing, complementary brain imaging techniques will allow us to gain deeper insight into the brain networks and interconnections that guide brain function, in essence, to begin to move a step closer to making causal statements of brain functions. Examples are the use of ERP with fMRI and most recently a combined approach using TMS (a stimulation technique) with fMRI. Together, these methods allow us to not only see how different neural areas interact across a given cognitive function or task but also how and when activity of one area impacts on another area. TMS and EEG have also been used to look at the brain at rest and when engaged in an activity, to create a "TMS-evoked potential." See Taylor, Walsh, and Eimer (2008) for an exploration of the use of TMS and EEG to explore the functional dynamics of brain connectivity and cognitive function.

In summary, the findings from this wealth of new data, made possible by newer imaging technologies, demonstrate the feasibility of imaging the typical and atypical brain in childhood. However, there is still some way to go in providing comparable data of atypical brain development in children with significant cognitive impairment. The most successful studies have tended to use participants with IQs within the normal range, such as children with ADHD or high-functioning autism. There is also a tendency for studies to incorporate older childhood and adolescent samples rather than younger age groups. There are obvious reasons for this, but by focusing solely on later outcomes, we may be tapping only the end state rather than exploring age changes in brain maturation over the course of development. Future studies may also need to consider including larger sample sizes, especially in disorders that show wide behavioral and cognitive variability such as ADHD and fragile X. Increasing the sample size will also allow for observations of any age-related changes from a cross-sectional perspective. This will provide important data, but the most dynamic approach of all will involve longitudinal observations of typical and atypical development. This field is emerging, however, and as techniques become more sophisticated and accessible, we will be able to witness the dynamic differences in brain networks and functions that reflect the differing developmental pathways that characterize different neurodevelopmental disorders.

Recognizing Potential Challenges

Alongside the significant advantages gained in knowledge from a decade of intensive neuroimaging research, there are also potential concerns, namely in the interpretation of data. For example, in functional imaging, while it is generally assumed that greater activation indicates that a given area is involved in carrying out processes necessary to a particular neural computation, it is also possible that such activation may reflect cognitive processes that are preventing interference with processing occurring elsewhere in the brain, or it may simply indicate greater difficulty in carrying out the required task, rather than more effective processing (Durston & Casey, 2006). A second example relates to identifying the brain location responsible for a specific computation. The conventional technique is the subtraction paradigm; two tasks are devised, one of which requires the computation and the other of which does not. It is then assumed that differences in brain activity between the two tasks reflect the additional computation required in the first case. A simple example would be a comparison of switching overt attention with eye movement and switching attention covertly to a given location without eye movement. Brain activity in the second case is subtracted from that obtained in the first case to identify areas specifically recruited in order to implement the eye movements. Clearly this procedure implies that the control of eye movements is a straightforward addition that would not alter the other computations in the brain and ignores the possibility that, for example, specific processes are necessary to prevent eye movements in the covert case. Such assumptions are questionable even in this relatively simple example.

Another example concerns the interpretation of differences that may be found in data derived from different selected populations (e.g., Down syndrome vs. typically developing children). Such differences can be interpreted either as indications of the cause of the behavioral differences, or as modifications in the brain structure or function that have occurred as a consequence of the basic abnormality, perhaps due to modifications that attempt to compensate for deficient processing. Many studies have assumed a static, modular organization in the brain where specific sites carry out specific operations in virtually all typically developing individuals. As a result they tend to conclude that a location in the brain producing more or less activity in the atypical group than in the control group is the site of the processes that are behaviorally deficient in the atypical group. This approach has been dubbed "modern, albeit high-tech phrenology" by Peterson (2003, p. 826). However, it is also possible that the changes in activity levels might be due to atypical processing of the task in an attempt to circumvent a weakness in some other area of the brain.

Consequently, while there is enormous promise in these technologies for advancing our understanding of fundamental processes underlying performance in both typical and atypical development, research enquiry needs to be driven by sound experimental designs and careful interpretation of data. The two must complement one another.

Recognizing the Importance of Development in Understanding Typical and Atypical Brain Functions: Current Issues and Future Directions

One of the prevailing theoretical orientations about cognitive development (and we have alluded to this already) in the past couple of decades has been that the brain is "prewired" in a modular organization, with specific modules that perform specific functions, and process proprietary inputs (e.g., Fodor, 1983), and develop naturally as the brain matures, with input merely triggering the prespecified modular function. The infant, child, and adolescent brain is thus viewed as an immature, developing form of the adult brain, with certain functions inactive or inadequate at different stages. Thus, disorders of infancy and childhood are viewed as involving losses of or damage to domain-specific functions; there is, as it were, a hole in the overall normal pattern. Alternatively, development is assumed to proceed normally but at a slower pace than for typically developing children. These two views are known, respectively, as the difference and delay interpretations of neurodevelopmental disorders and have dominated theories about the nature of such disorders. A very different position is that of neuroconstructivism (Johnson, 2001; Karmiloff-Smith, 1998) in which domain-specific functions are viewed as the *outcome* of a dynamic interactive process between soft initial constraints and the structure of an input. Initially, several cortical areas attempt to process all inputs, but with time the one that is more domain relevant wins out and becomes domain specific after repeated processing of input. In other words, specialization, localization, and modularization of function are the emergent product of development, and in many neurodevelopmental disorders, even where proficient overt behavior is observed, this process of specialization does not occur (for a seminal critique, see Karmiloff-Smith, 2009).

We argue that neurodevelopmental disorders should not be viewed as parallel to adult disorders that follow brain damage through stroke or other physical insult. The latter can damage or destroy an established processing system developed for a specific purpose (inhibition, for example), giving the impression that a highly segregated modular system is potentially available from birth and only requires normal environmental input in order to mature into the standard finished article. However, a system that is imperfect from the start will develop compensatory strategies for coping with problems set by the environment. Weak or nonfunctional processes are likely to have pervasive effects on how the whole system develops. Development is likely to follow a different path from that observed in typically developing cases and may differ in subtle ways from child to child, due to differences in the degree or nature of the initial impairment and the input from the environment.

One of the most exciting research developments in the next decade will be the emerging field of "developmental imaging genetics." This initiative brings together molecular genetics, developmental cognitive neuroscience, and neuroimaging to explore the biological pathways and mechanisms that drive typical and atypical development at the neural, environmental, and behavioral levels. Using state-of-the-art imaging technology and genetic data, research can begin to investigate the impact of specific genes on cognitive function by isolating developmental changes at the brain level. See Viding, Williamson, and Hariri (2006) and Fisher, Munoz, and Hariri (2008) for seminal accounts of this strategy and its potential role in identifying early indicators of later adult psychopathology. In terms of understanding the interplay between genes and brain maturation in the developing attention system, the reader might also find the recent eloquent review by Brocki, Clerkin, Guise, Fan, and Fossella (2009) to be especially insightful. As a complement to this review Casey, Soliman, Bath and Glatt (2010) highlight and address specific challenges and promises of imaging genetics to understanding gene-brain-behaviour interactions in typical and clinical populations.

The applicability of this approach to neurodevelopmental disorders has been recently reviewed by Meyer-Lindenberg and Zink (2007) in the context of Williams syndrome, in which the authors propose a convincing rationale for the use of functional imaging technologies to specify and dissect neural mechanisms that underlie the phenotype and then map these neural dissociations onto the known genetic variants that comprise the Williams syndrome genotype. However, for this approach to provide maximal clinical benefit, rather than be of purely theoretical interest, research must be situated within a longitudinal framework with appropriate evaluations of interventions and their impact across the developmental trajectory. With this research agenda, the field of developmental cognitive genetics undoubtedly holds considerable promise for the identification and remediation of early cognitive and behavioral impairments that define later phenotypic end states.

CHAPTER SUMMARY AND KEY FINDINGS

- There is an emerging literature that demonstrates the feasibility of using a variety of new imaging technologies to understand the typically developing brain, but currently these technologies have a more limited use with atypical populations.
- The most widely documented neurodevelopmental disorder to undergo imaging investigation is ADHD, and this research offers promising insights into how, in this disorder, the atypical brain processes information in the attention domain.
- Issues such as degree of intellectual impairment, anxiety levels, and attention difficulties pose significant problems in obtaining reliable imaging data in some neurodevelopmental disorders. Small sample sizes and skewed sample compositions further limit the generalizability of findings.
- The dynamic role of development itself is now being acknowledged as a critical future focus for brain research. This approach moves away from the more traditional view of the brain as a "prewired" modular organization.

Section II: Measuring Attention

5

Measuring Attention at the Behavioral Level
Rating Scales and Checklists

CHAPTER SNAPSHOT
• The importance of examining attention at both the behavioral and cognitive levels • Ratings of inattentive behavior: reliability and consistency of different sources • The feasibility of current rating scales in capturing the range of inattentive behaviors across differing neurodevelopmental disorders • The use of categorical (diagnostic) and dimensional (quantitative) approaches to charting inattentive behaviors

---

Inattentive behavior, as observed in the classroom, has a detrimental and long-lasting impact on academic attainment and developmental outcomes from preschool onward (Loe & Feldman, 2007; Smallwood, Fishman, & Schooler, 2007). But what do we mean by "inattentive" behavior? Difficulties in concentration and focus, distractibility, impulsivity, and disorganization represent a constellation of behaviors that come under the rubric of inattention and are included in both clinical assessments, for example, the Diagnostic and Statistical Manual of Mental Disorders, 4th Edition (DSM-IV, see Box 5.1 for a summary of criteria), and in parent/teacher-based assessments. The severity and impact of these behaviors varies across development such that, in some cases, early problems dissipate as the child matures and is able to focus on higher-level cognitive demands necessary for classroom learning. Other children, however, remain inattentive throughout their childhood, making them vulnerable to the development of more serious clinical disorders such as attention deficit/hyperactivity disorder (ADHD), either the

Box 5.1 DSM-IV Diagnostic Criteria for Attention
Deficit/ Hyperactivity Disorder

I. Either (A) or (B)
   A. Six (or more) of the following symptoms of **inattention** have
      persisted for at least 6 months to a degree that is maladaptive
      and inconsistent with developmental level:

Inattention

   a. Often fails to give close attention to details or makes care-
      less mistakes in schoolwork, work, or other activities
   b. Often has difficulty sustaining attention in tasks or play
      activities
   c. Often does not seem to listen when spoken to directly
   d. Often does not follow through on instructions and fails
      to finish schoolwork, chores, or duties in the workplace
      (not due to oppositional behavior or failure to understand
      instructions)
   e. Often has difficulty organizing tasks and activities
   f. Often avoids, dislikes, or is reluctant to engage in tasks that
      require sustained mental effort (such as schoolwork or
      homework)
   g. Often loses things necessary for tasks or activities (e.g.,
      toys, school assignments, pencils, books, or tools)
   h. Is often easily distracted by extraneous stimuli
   i. Is often forgetful in daily activities
  B. Six (or more) of the following symptoms of **hyperactivity-
    impulsivity** have persisted for at least 6 months to a degree
    that is maladaptive and inconsistent with developmental
    level

Hyperactivity

   a. Often fidgets with hands or feet or squirms in seat
   b. Often leaves seat in classroom or in other situations in
      which remaining seated is expected

Box 5.1 DSM-IV Diagnostic Criteria for Attention
Deficit/ Hyperactivity Disorder *(continued)*

---

    c. Often runs about or climbs excessively in situations in
       which it is inappropriate (in adolescents or adults, may be
       limited to subjective feelings of restlessness)
    d. Often has difficulty playing or engaging in leisure activities
       quietly
    e. Is often "on the go" or often acts as if "driven by a motor"
    f. Often talks excessively

Impulsivity

    g. Often blurts out answers before questions have been
       completed
    h. Often has difficulty awaiting turn
    i. Often interrupts or intrudes on others (e.g., butts into
       conversation or games)

 II. Some hyperactive-impulsive or inattentive symptoms that caused
   impairment were present before age 7 years
III. Some impairment from the symptoms is present in two or more
   settings.
IV. Clear evidence of clinically significant impairment in social,
   academic, or occupational functioning
 V. The symptoms do not occur exclusively during the course of
   PDD or schizophrenia or are not better accounted for by other
   mental disorder.

**Based on These Criteria, Three Types of ADHD Are
Identified:**

1. ADHD, *combined type*: if both criteria IA and IB are met for the
   past 6 months
2. ADHD, *predominantly inattentive type*: if criterion IA is met but
   criterion IB is not met for the past 6 months
3. ADHD, *predominantly hyperactive-impulsive type*: if criterion IB is
   met but criterion IA is not met for the past 6 months.

combined-type (CT) comprising both inattention and hyperactivity-inattentive behaviors or the inattentive-type (IT) comprising predominantly inattentive behaviors (approximately 25% of ADHD cases, but note that there is still a lack of consensus as to the validity of ADHD subtypes; see Chapter 8 for further discussion).

The cost of unrecognized inattention on a child's academic and social development is now well documented (see Spira & Fischel, 2005, for a review). For example, inattentive behavior in kindergarten children, but not hyperactive behavior, predicts poor reading outcomes in Grade 1 and also in Grade 5, independent of kindergarten reading-related skills and concurrent levels of hyperactivity (Dally, 2006; Rabiner & Coie, 2000). Inattentive behavior in the classroom is also strongly associated with mathematics difficulties in elementary school children (Dobbs, Doctoroff, Fisher, & Arnold, 2006; Fuchs et al., 2005) and later executive functioning difficulties in adolescence (Friedman et al., 2007). Moreover, inattentive behavior has been found to predict poor response to evidence-based reading and math instruction (Fuchs et al., 2005; Rabiner, Malone, & Conduct Problems Prevention Research Group, 2004). Accordingly, the presence of some persistent inattentive behavior (e.g., distractibility, poor concentration, and organization) in childhood is considered a developmental risk factor for poor academic outcomes (Warner-Rogers, Taylor, Taylor, & Sandberg, 2000). Other research has also demonstrated a link between cognitive function, namely, working memory, and academic outcomes in reading and math (e.g., Gathercole & Pickering, 2000). Thus, current theoretical and empirical work suggests a possible *triad of impairment* that encompasses inattentive behavior, impaired cognitive attention, and poor academic functioning (Tannock & Martinussen, 2007).

The concept of "attention" encompasses both behavioral and cognitive components, which has given rise to confusion because their interrelationship is complex (for reviews of this issue, see Cornish, Wilding, & Grant, 2006; Tannock, 2003). Although the focus of this chapter is on behavioral measures of inattention, the overall aim of the book is to provide the reader with a clear understanding of both the behavioral and cognitive nature of attention and their interplay across development.

From a *developmental perspective*, there are relatively few studies of inattentive behaviors that have been conducted in preschool children (2 to 5 years) and fewer still that have charted the trajectories of *atypical* behaviors along a longitudinal time course. Yet, clearly inattention difficulties have their origins early in development and do not suddenly "emerge" in midchildhood. Possible explanations for the paucity of current knowledge are linked to the somewhat "transient" nature of any early inattention behaviors and their natural decline with age from toddlerhood to childhood. As we shall see in Chapter 7, the development of attention is intrinsically linked with an emerging reliance on endogenous factors that allow for behaviors to be guided by effortful, voluntary control. There is also the diagnostic challenge in differentiating the "typical" range of inattentive behaviors from symptoms that indicate an actual preschool disorder with clinical significance. The option of simply modifying existing, clinical diagnostic tools such as the DSM-IV for use with preschoolers is not an optimal choice, given that such tools were not developed for children under the age of 5 years and would therefore not be based on an age-appropriate, standardized population. However, given the recent findings of a general (if somewhat weak) stability between a diagnosis of ADHD inattentive-type in young preschool children and its trajectory and cognitive outcomes into later childhood (e.g., Lahey et al., 2006; Lahey, Pelham, Loney, Lee, & Willcutt, 2005), there is a critical necessity to develop age-specific screening tools that can differentiate the range of inattentive behaviors in very young children and chart their developmental time course. With this knowledge comes the potential to predict pathways of early inattentive behaviors that are likely, in later childhood years, to have significant clinical and academic importance.

Attention Rating Scales

The past two decades have seen an unprecedented rise in the development of behavioral rating scales specifically focused on elucidating disorders of attention either at the *symptom* level (e.g., rated as a "poor attenders" by their teachers or parents) or at the *diagnostic* level (e.g., has met the clinical criteria for a diagnosis of ADHD). Such scales are frequently used in clinical and research contexts and most are available commercially. A critical component of a good rating scale is that it is both *reliable* and *valid*. In essence, does it measure what it claims to measure? By *reliability*, we mean that the scale must measure a particular construct in the same way across multiple observations, environments, and time frames. This is especially important for scales that record frequency and duration of inattentive behaviors in different settings, such as the school or home. By *validity*, we mean: Does the scale measure what it was designed to measure, in our case, inattentive behaviors? In order to obtain good validity, scales need to ensure *"content validity,"* whereby items are derived from well-established clinical criteria such as the DSM-IV, and *"construct validity,"* which assess the extent to which a rating scale correlates with other scales with established validity measuring the same or a similar construct. The higher the correlation, the more likely the scale is to measure what it proposes to measure. Without strong reliability and validity, rating scales leave themselves open to random errors and misinterpretations in terms of under- and overrepresentation of behavioral problems.

Undoubtedly, all rating scales have their strengths and weaknesses. In terms of *strengths*, rating scales have the potential to enhance the ability to screen for inattentive behavior across varying, everyday settings such as school or the home, thus facilitating the viewpoint of multiple informants (teacher, parent). If sufficiently sensitive, rating scales can identify early precursors of "high-risk" behaviors that may later evolve into a clinical problem. This information informs early interventions that target, and therefore reduce, the impact of inattentive behaviors. In terms of *weaknesses*, rating scales are by their very nature only recorded observations of behavior across snapshots of time, be it in a given day, week, or month. They are also reliant on the accuracy of information from a variety of sources that include trained professionals as well as from parents and teachers. Moreover, rating scales often have minimal value outside their normative age bracket, which in some cases can be quite restrictive. For example, and as we have addressed already, a rating scale designed to assess the presence of inattentive behaviors in children aged 5 to 12 years may not be sufficiently sensitive to pick up subtle, atypical behaviors in children aged 4 years or under, and may be under- or oversensitive for children aged 13 years or over. Also, rating scales designed to tap quite "significant" inattentive behaviors at the clinical level are often unable to chart more subtle profiles of difficulty that require a general population distribution of inattentive scores. Instead, many scales lend themselves only to categorical placement (e.g., severe, moderate, mild) based on clinical cutoff scores. However, see the growing enthusiasm for the SWAN scale developed by Swanson and colleagues (reviewed below) that allows inattentive behaviors to be viewed from a dimensional perspective.

Who Assesses Inattentive Behaviors?

A core question that is still unresolved is: To what extent do parent or teacher ratings, or even combined responses, make a stronger (or equal) contribution to a diagnosis of attention impairment than a clinical interview alone? The findings are mixed. On the one hand, studies have reported low rater agreements between parents and teachers, indicating a lack of agreement across scale items (e.g., Mitsis, McKay, Schulz, Newcorn, & Halperin, 2000; Power et al., 1998; Wolraich, Lambert, Bickman, et al., 2004). Other studies, however, report that teacher ratings as compared to parent ratings may make a stronger contribution to the prediction of attentional impairments (Power et al., 1998). Conversely, recent clinical studies indicate that parent reports are as informative as teacher reports in accurately and robustly identifying behavior outcomes in medication trials of children with ADHD (Biederman, Gao, Rogers, & Spencer, 2006; Biederman, Faraone, Monuteaux, & Grossbard, 2004). It would seem sensible, therefore, to advise the reader to bear in mind that some lack of agreement is inevitable, especially when different informants are responding to behaviors in very different situations, time frames, and environments. For example, inattentive behaviors presented in the context of the home, during the course of an evening or weekend, are going to be somewhat different from those behaviors presented within the context of a school environment during the daytime. What is important to note here is that all respondents who are directly observing and interacting with the child on a daily basis will provide valuable information that will help guide diagnosis and treatment. However, the ability of rating scales alone to predict whether children reach DSM-IV criteria for ADHD is still an issue of debate.

Measurement of Inattentive Behaviors in Neurodevelopmental Disorders

A core concern, and an issue that remains unresolved, is the lack of appropriate rating scales that are sufficiently sensitive to capture attention signatures in children with differing neurodevelopmental disorders. Current research has tended to focus almost exclusively on behavioral descriptions of inattention in children with ADHD. Indeed, this extensive literature attests to the importance of charting attention difficulties at the behavioral as well as cognitive levels. In contrast, the attention phenotype of children with other neurodevelopmental disorders has not received comparable attention. We suggest two core reasons for the paucity of available information. The first is an assumption that clinicians, parents, and teachers cannot dissociate behavioral concerns from cognitive disability; and the second, is an assumption that all neurodevelopmental disorders, irrespective of etiology, will present with the same degree of severity in behavioral inattention due to their additional mental retardation status. However, perusals through those studies that have assessed inattention profiles in disorders, such as autism, Down syndrome, fragile X syndrome, and others, highlight the importance of careful consideration of disorder-specific constellations of inattentive behaviors. For example, a series of recent cross-syndrome studies by Cornish and colleagues compared children with fragile X syndrome (a disorder caused from the silencing of a single gene on the X chromosome) and Down syndrome (caused by a trisomy on chromosome 21) and found group differences in the severity of inattentive behaviors and levels of hyperactivity, which, upon further investigation, corresponded to distinct "cognitive attention" signatures (Cornish, Munir, & Wilding, 2001; Cornish, Munir, & Wilding, 1997; Munir, Cornish, & Wilding, 2000b). We therefore argue that it is essential that both behavioral and cognitive descriptions of attention are incorporated within a single design, irrespective of degree of intellectual or cognitive impairment. This will help clarify, for example, how the two levels map onto one another, and whether behavioral and cognitive inattention differentially map onto or predict later academic or cognitive outcomes.

The aims of the present chapter are threefold. First, to provide a comprehensive, yet not exhaustive review of existing attention rating scales with a focus specifically on subscales that measure inattentive behaviors. We have selected those scales that have been used extensively in studies to screen for inattentive behaviors in population-based samples and in atypical samples. Accordingly, 11 scales will be reviewed in their most recent formats. Of these 11 scales, 6 specifically measure inattentive behaviors per se: Conners Parent and Teacher Rating Scale (CPRS; CTRS), Child Behavior Checklist (CBCL), ADD-H Comprehensive Teacher Rating Scale (ACTeRS), Swanson, Nolan, and Pelham Checklist (SNAP-IV), Strengths and Weaknesses of ADHD Symptoms and Normal Behavior Scale (SWAN), and ADHD Rating Scale-IV (ADHD RS-IV). A further five scales measure inattentive behaviors within a broader context, for example, mental health problems: Developmental Behavior Checklist (DBC) and Strengths and Difficulties Questionnaire (SDQ); executive functioning: Behavior Rating Inventory of Executive Function (BRIEF); and temperament: Child Behavior Questionnaire (CBQ) and the Early Childhood Behavior Questionnaire (ECBQ). Table 5.1 provides a summary of each of the rating scales in terms of their structure, age range, normative data, and strengths and weaknesses. See also Collett, Ohan, and Myers (2003) for a comprehensive 10-year review of many of attention rating scales cited above.

The second aim is to provide a summary of findings from a developmental perspective across each of the scales, and the third aim is to highlight the feasibility, alongside concerns, of utilizing rating scales in our six neurodevelopmental disorders: autism, fragile X syndrome, Down syndrome, Williams syndrome, 22q11 DS, and ADHD. Detailed summaries of each of these disorders are provided in Chapters 3, 8, and 9 and so will not be discussed in detail here.

Conners Rating Scales-R (CRS-R)

The Conners Rating Scales-Revised (Conners, 1997) assess the presence and severity of behavior problems reflecting the ADHD inattentive type and hyperactive-impulsive type from the DSM-IV (APA, 1994, see Box 5.1) and is the result of 30 years of work (Kollins, Epstein, & Conners, 2004). The Conners Teacher (59-item long-form version, and 28-item short-form version) and Parent (80-item long-form, and 27-item short-form) Rating Scales (CTRS-R, CPRS-R) provide substantive normative data on children aged 3 to 17 years and have been used extensively worldwide across diverse cultures including Asia, India, Spain, and Turkey (Dereboy, Senol, Sener, & Dereboy, 2007; Farre-Riba & Narbona, 1997; Luk & Leung, 1989; Pal, Chaudhury, Das, & Sengupta, 1999; Shur-Fen Gau, Soong, Chiu, & Tsai, 2006) for the assessment and treatment

Table 5-1. Attention Rating Scales

Scale Reference	Forms/Items	Relevant Attention Subscales	Normative Age Range	Strengths	Weaknesses
ADD-H Comprehensive Teacher's Rating Scale (ACTeRS) (Ullmann, Sleator, & Sprague, 2000)	Teacher (24 items) Parent (25 items)	Inattention Hyperactivity	6–14 years but poorly described	Strong focus on inattentive behaviors; short to complete, making it suitable for repeated applications	Normative data is unclear, especially for the parents form; paucity of studies testifying to its effectiveness as a screening tool
ADHD Rating Scale-IV (ADHD RS-IV) (DuPaul, Power, Anastopoulos, & Reid, 1998)	Teacher (18 items) Parent (18 items)	Inattention Hyperactive/ Impulsivity	5–18 years and stratified by age	Closely based on diagnostic criteria from DSM-IV; discriminates children with inattentive behaviors from nonclinical peers	
Brief Rating Inventory of Executive Function (BRIEF) (Gioia, Espy & Isquith, 2003; Gioia, Isquith, Guy, & Kenworthy & Brown, 2000)	*Preschool version:* Teacher & Parent form (63 items) *School age version:* Teacher & Parent form (86 items)	Inhibit, Working Memory, Plan and Organize	2–5 years and 5–18 years and stratified by age	Covers a range of executive functions that have strong clinical utility; some sensitivity in detecting the range of inhibitory behaviors in developmental disorders	Time consuming to complete; limited range of inattentive behaviors

Measure	Version/Items	Scale/Subscale	Age Range	Strengths	Limitations
Child Behavior Checklist (CBCL) (Achenbach, 1991; Achenbach & Rescorla, 2000)	*Preschool version:* Teacher & Parent form (99 items) *School age version:* Teacher & Parent form (112 items)	*Syndrome scales:* Attention Problems *DSM-Oriented Scales:* Attention Deficit/Hyperactivity Problems	1.5–5 years 6–18 years and stratified by age	Extensively researched and used worldwide; significant congruence between the Attention Problem Scale and a clinical diagnosis of ADHD; used to detect the range of attention problems across different developmental disorders	Does not distinguish between inattentive and hyperactive behaviors
Child Behavior Questionnaire (CBQ) (Rothbart, Ahadi, Hershey, & Fisher, 2001)	*Parent form only:* Standard version (195 items) Short version (94 items) Very short version (36 items)	Attentional Focusing Impulsivity, Inhibitory Control	3–8 years of age and stratified by age	Normative data includes longitudinal analysis; assesses wide range of inattentive behaviors; translated into numerous languages	Paucity of studies assessing children with neurodevelopmental disorders
Conners Rating Scales–Revised (CRS–R) (Conners, 1997)	*Teacher form:* Long (59 items) Short (28 items) *Parent form:* Long (80 items) Short (27 items)	Cognitive Problems/ Inattention	3–17 years and stratified by age	Clear clinical and research applications; discriminates children with inattentive behaviors from nonclinical peers; can monitor the effects of treatment on range and intensity of inattentive behaviors; translated and standardized across diverse cultures	Few studies assessing children with differing neurodevelopmental disorders outside of ADHD

(continued)

Table 5-1. Attention Rating Scales (*continued*)

Scale Reference	Forms/Items	Relevant Attention Subscales	Normative Age Range	Strengths	Weaknesses
Developmental Behavior Checklist (DBC) (Einfeld & Tonge, 2002)	*Parent form* (96 items) *Teacher form* (93 items) *Short version* (24 items) *Early Screen* (17 items)	No specific subscale	4–18 years	Only behavioral measure to be directly applicable to children and adolescents with intellectual disabilities; translated and standardized across diverse cultures; quick to complete	Few inattention/ hyperactivity items make it difficult to identify ADHD symptom clusters
Early Childhood Behavior Questionnaire (ECBQ) (Putman, Garstein & Rothbart, 2006)	*Parent form only:* Standard version (201 items)	Attentional Focusing Impulsivity, Inhibitory Control	1.5–3 years	Normative data includes longitudinal analysis; assesses wide range of inattentive behaviors	Extremely time consuming to complete; paucity of studies assessing children with neurodevelopmental disorders

Measure	Informant (items)	Domains	Age range	Description	Notes
Strengths and Weaknesses of ADHD Symptoms and Normal Behavior (SWAN) (Swanson, Schuck, et al., 2005)	Teacher and Parent (18 items)	Inattention Hyperactivity/ Impulsivity	Unspecified but broadly 5–12 years	Examines the continuum of attention behaviors in a general population; sufficiently sensitive to discriminate between children with and without ADHD; used as primary outcome measure in clinical efficacy studies and genetic studies (e.g., DAT1 polymorphism)	No normative data
Swanson, Nolan, and Pelham IV Questionnaire (SNAP–IV) (Swanson, 1992)	*Teacher and Parent:* Standard form (90 items) Short form (18 items)	Inattention Hyperactivity/ Impulsivity	Unspecified broadly 5–12 years	Used extensively in ADHD research and appears sufficiently sensitive to discriminate between children with and without ADHD; used as primary outcome measure in clinical efficacy studies	No normative data
Strengths and Difficulties Questionnaire (SDQ) (Goodman, 1997)	Teacher and Parents (25 items)	Hyperactivity/ Inattention	3–16 years	Focuses on strengths as well as difficulties across a range of problem behaviors; some sensitivity in screening children with ADHD symptoms in genetic studies; translated and standardized across diverse cultures	Relatively few inattention items

of ADHD.[1] In the *long form*, six subscales comprise the Teacher form and seven comprise the Parent form. Subscales that are the core of both forms are: Cognitive Problems/Inattention, Hyperactivity, Oppositional, Anxious-Shy, Perfectionism and Social Problems, Conners' Global Index, and the DSM-IV Symptom subscales (Inattention, Hyperactivity/ Impulsivity, Total). In the *short form*, four subscales comprise the Teacher form and Parent form: Oppositional, Cognitive Problems/Inattention, Hyperactivity, and the ADHD Total. Normative data were based on a large ethnically heterogeneous sample of children (Teacher long form, n = 1973; Parent long form, n = 2482; Teacher short form, n = 1897; Parent short form, n = 2426) across the United States and Canada. Internal consistency reliabilities for the various subscales were moderate to excellent across all Teacher forms (long version, .77 to .96; short version, .88 to .95) and Parent forms (long version, .73 to .94; short version, .86 to .94). Test–retest reliability was also moderate to excellent for the Teacher forms (long version, .47 to .88; short version, .72 to .92) and Parent forms (long version, .47 to .85; short version, .62 to .85). There is also a Self-Report form (long and short versions) for adolescents.

Focusing on the Cognitive Problems/Inattention subscale, internal consistency and test–retest reliability were moderate to excellent for the Teacher form (long version, .91, .47; short version, .89, .92) and Parent form (long version, .93, .69; short version, .92, .73). On the DSM-IV Symptoms-Inattentive subscale (long version only), there was also excellent internal consistency and test–retest reliability for both the Teacher form (.95, .70) and Parent form (.92, .67). In terms of construct validity, the Cognitive Problems/Inattention subscale correlated satisfactorily with other similar rating subscales including the Attention Problem subscale of the Child Behavior Checklist (CBCL, reviewed next in this section).

SCORING

Responses are based on a 4-point Likert scale in which behavior is rated as occurring 0 (*Never, Seldom*), 1 (*Occasionally*), 2 (*Often, Quite a bit*), and 3 (*Very Often, Frequent.*) Subscale raw scores can be summed and

---

[1] A Connors Third Edition came in to publication at the time of writing—Conners 3 (Conners, 2008)

Question Number	Never, Seldom	Occasionally	Often, Quite a Bit	Very Often, Frequent
Q 4. Forgets things he/she has already learned (Teacher form-Short)	0	1	2	3
Q 17. Avoids, expresses reluctance about, or has difficulties engaging in tasks that require sustained mental effort (such as school work or homework) (Parent form-Short)	0	1	2	3
Q 19. Has trouble concentrating in class (Parent form-Long)	0	1	2	3
Q 26. Inattentive, easily distracted (Teacher form-Long)	0	1	2	3

Inset 5.1

converted into T-scores (standardized scores with a mean of 50 and a standard deviation of 10). A T-score above 65 is considered to be in the clinical range. Examples of the items that comprise the Cognitive Problems/Inattention subscale are provided in Inset 5.1. Behavior is rated within the last month (all forms).

From a *developmental perspective*, data are available by age group and gender for ages 3 to 5, 6 to 8, 9 to 11, 12 to 14, and 15 to 17. For the Cognitive Problems/Inattention subscale, the youngest children scored much lower (i.e., had fewer reported attention problems) than all other age groups (both Teacher and Parent forms). There was a significant gender by age interaction (Parent only), revealing that differences between boys and girls increased with age.

NEURODEVELOPMENTAL DISORDERS

The majority of studies using the Conners Rating Scale have tended to include children already clinically diagnosed with ADHD or with a suspected diagnosis. It is now well established that the scale represents a valuable clinical and research tool for differentiating young children and adolescents with inattentive behaviors and ADHD from nonclinical peers (e.g., Conners, Sitarenios, Parker, & Epstein, 1998a, 1998b; Hale, How, Dewitt, & Coury, 2001; Kumar & Steer, 2003). However, it may be less sensitive at discriminating children with ADHD from children with other clinical diagnoses (Sullivan & Riccio, 2007), and at discriminating ADHD as a distinct disorder in children with differing degrees of intellectual impairment or specific developmental disorders (Deb, Dhaliwal, & Roy, 2008).

Child Behavior Checklist (CBCL)

The CBCL (Achenbach, 1991; Achenbach & Rescorla, 2000) is one of the most widely used checklists for assessing problem behaviors in pre-school and school children as reported by parents, teachers, and young people themselves. It has strong cultural validity and is available in over 79 languages. The checklist covers two broad age ranges: 1½ to 5 years (99 items, Parent form and Teacher form) and 6 to 18 years (112 items, Parent form and Teacher form). There is also a Self-Report form (112 items) for adolescents.

A core advantage of this CBCL is the diversity of the scales and clusters representing a wide range of behavior problems. There are two main types of scales:

- *"Syndrome" Scales*, which comprise the following subscales: Attention Problems, Emotional Reactive (preschool version only), Anxious/Depressed, Somatic Complaints, Withdrawn, Sleep Problems (parent versions only), Aggressive Problems, Social Problems, Rule-Breaking Behavior, and then two general categories, namely, Internalizing Behaviors (relating to problems that are mainly within the individual or self), and Externalizing Behaviors (relating to problems that result in conflict with others).
- *DSM-Oriented Scales* are based on the DSM-IV criteria and comprise the following subscales: Affective Problems, Anxiety Problems, Pervasive Developmental Problems (preschool version only), Attention Deficit/Hyperactivity Problems, Oppositional Defiant Problems, Somatic Problems (school-age version only), and Conduct Problems (school-age version only).

The normative data for the Preschool version of the CBCL were based on a moderate to large heterogeneous sample of children (Parent form, n = 700; Teacher form, n = 1192). Internal consistency for the various subscales was moderate to excellent for the Teacher form (.52 to .97) and Parent form (.66 to .95). Test–retest reliability was moderate to excellent on the Teacher form (.68 to .91) and Parent form (.68 to .92). For the DSM-Orientated Scale, internal consistency was moderate to excellent for the Teacher form (.68 to .93) and Parent form (.63 to .86). Test–retest reliability followed a similar profile and was moderate to excellent on the Teacher form (.57 to. 87) and Parent form (.74 to .87). Focusing on the Attention Problem subscale, internal consistency and test–retest reliability were excellent for the Teacher form (.89, .84) but only moderate for the Parent form (.68, .78). On the Attention Deficit/Hyperactivity subscale of the DSM-Oriented Scale, there was moderate to excellent internal consistency and test–retest reliability for the Teacher form (.92, .79), and only moderate for the Parent form (.78, .74).

The normative data for the school-age version of the CBCL were based on a large heterogeneous sample of children (Parents form, n = 1753; Teacher form, n = 2319). Internal consistency was moderate to excellent on the Teacher form (.72 to .97) and Parent form (.78 to .97). Test–retest reliability was also moderate to excellent on the Teacher form (.60 to .96) and Parent form (.82 to .92). Focusing on the Attention Problem subscale, internal consistency and test–retest reliability were excellent for the Teacher form (.95, .95) and for the Parent form (.86, .92). On the Attention Deficit/Hyperactivity subscale of the DSM-Oriented Scale, there was also excellent internal consistency and test–retest reliability for both the Teacher form (.94, .95) and Parent form (.84, .93). A similar profile of reliabilities has also been reported in a sample of children diagnosed with *mental retardation* aged between 5 to 12 years (Miller, Fee, and Netterville, 2004). As noted above, construct validity is satisfactory when compared with the Conners Rating Scale.

SCORING

Responses are based on a 3-point Likert scale in which behavior is rated as 0 (*Not True*), 1 (*Somewhat* or *Sometimes True*), and 3 (*Very True* or *Often True*). Subscale raw scores can be summed and converted in to T-scores. A T-score above 69 is considered to be in the clinical range.

Examples of the items that comprise the Attention Problem subscale, preschool and school-age versions, are provided below in Insets 5.2 and 5.3.

Preschool Version   Behavior is rated now or within the past 2 months (all forms).

Question Number	Not True	Somewhat or Sometimes True	Very True or Often True
Q 5. Can't concentrate, can't pay attention for too long (Parent form)	0	1	2
Q 24. Difficulty following directions (Teacher form)	0	1	2
Q 59. Quickly shifts from one activity to another (Parent form)	0	1	2
Q 64. Inattentive, easily distracted (Teacher form)	0	1	2

Inset 5.2

School Version   Behavior is rated as occurring now or within the past 2 months in the Teacher form, and in the past 6 months in the Parent form.

From a *developmental perspective*, normative data are available by age group and gender for ages 1½ to 3 (preschool version), 6 to 7, 8 to 9, 10 to 11, 12 to 13, 14 to 15, and 16 to 18 years old (school-age version). Analysis reveals that scores on the Attention Problem subscale were much higher for clinically referred children than for nonreferred children across

Question Number	Not True	Somewhat or Sometimes True	Very True or Often True
Q 4. Fails to finish things he/she starts (Parent form)	0	1	2
Q 8. Can't concentrate, can't pay attention for too long (Teacher form)	0	1	2
Q 41. Impulsive or acts without thinking (Teacher form)	0	1	2
Q 78. Inattentive or easily distracted (Parent form)	0	1	2

Inset 5.3

each age group. In the typically developing preschool children, scores seemed to decline slightly with age but remained stable thereafter. The profile of scores on the Attention Deficit/Hyperactivity subscale also revealed a similar age profile with scores declining in the preschool years and then stabilizing across childhood into adolescence. With regard to gender, there was a general trend for girls to receive lower scores than boys on both subscales.

NEURODEVELOPMENTAL DISORDERS

In terms of *inattentive behaviors*, there is significant congruence between the Attention Problem subscale and a clinical diagnosis of ADHD (Steingard, Biederman, Doyle, & Sprich-Buckminster, 1992; Steinhausen, Drechsler, Foldenyi, Imhof, & Brandeis, 2003) but less congruence between parents' and teachers' responses (Derks et al., 2008; Tripp, Schaughency, & Clarke, 2006). Findings also indicate developmental stability, with overall prevalence of CBCL Attention Problems not dramatically changing from midchildhood onward (7 to 12 years) (Barnow, Schuckit, Smith, & Freyberger, 2006; Rietveld, Hudziak, Bartels, van Beijsterveldt, & Boomsma, 2004) or during adolescence (13 to 17 years). This pattern contrasts with the more transient nature of inattentive behaviors frequently reported in the preschool years (see also an earlier study of the long-term stability of the CBCL by Biederman et al., 2001). As with other rating scales to be discussed in this chapter, the CBCL appears to have considerable difficulty distinguishing between clinically diagnosed ADHD "inattentive" versus "combined" subtypes (Forbes, 2001) but is excellent in discriminating ADHD, on the basis of Attention Problem scores, from non-ADHD samples (e.g., Kalff et al., 2002; Mahone et al., 2002).

The CBCL has also been used quite extensively on other types of neurodevelopmental disorders and appears to demonstrate consistency in its ability to identify high levels of attention problems in a variety of developmental disorders including autism and pervasive developmental disorder (PDD) (Eisenhower, Baker, & Blacher, 2005; Holtmann, Bolte, & Poustka, 2007; Pandolfi, Magyar, & Dill, 2009), fragile X syndrome (Cornish, Munir, & Wilding, 2001; Hatton et al., 2002; Kau et al., 2004), and 22q11 DS (Jansen et al., 2007). One of the only studies to perform a cross-syndrome comparison of the CBCL Attention Problem subscale reported comparably high levels of attention problem scores in children and adults (6 to 31 years) with Williams syndrome, but significantly lower scores in children with Down syndrome (Graham, Rosner, Dykens, & Visootsak, 2005). A similar finding of lower CBCL Attention Problem scores in children with Down syndrome compared to children with fragile X was also reported by Cornish, Munir, and Wilding (2001). Together, these findings provide some evidence that not all neurodevelopmental disorders show equivalent levels of inattentive behaviors.

ADD-H Comprehensive Teacher Rating Scale (ACTeRS)

The ACTeRS (Ullmann, Sleator, & Sprague, 1991, 2000) is a somewhat lesser-known scale than the more popular Conners and CBCL. The emphasis here is on *inattentive and hyperactive behaviors* as rated by teachers (11 items) and parents (10 items) for children aged 5 to 14 years (K to Grade 8). Items are generally comparable to DSM-IV descriptions. Additional subscales include Social Skills, Oppositional Behaviors, and Early Childhood (Parents only). There is also a Self-Report form (35 items) for adolescents and adults.

The normative data for the teacher version of the ACTeRS are based on a large sample of ethnically heterogeneous children (n = 2362). In contrast, the normative data for the Parent form are not clearly specified and so cannot be reported. Internal consistency was excellent for the Teacher form (.92 to .97) and Parent form (.81 to .89). Test–retest reliability was moderate for the Teacher form (.78 to .82), but no data are provided for the Parent form. Focusing on the Attention subscale, internal consistency was excellent for the Teacher form (.97) and Parent form (.89). Test–retest reliability was available for the Teacher form only and was moderate (.78). In terms of construct validity, correlations with the Conners Cognitive Problems/Inattention subscale (.48) and the ADHD Index (.49) reached moderate agreement. A somewhat higher correlation of .64 was reached between the Attention subscale of the ACTeRS Teacher form and the Inattentive/Cognitive Problem subscale of the Conners (Forbes, 2001), suggesting that the two scales, while not in perfect unison, do measure similar inattentive constructs. See also Erford and Hase (2006) for a recent comparison of the Conners and the

Teachers' ACTeRS in a large sample of children from 5 to 11 years of age.

SCORING

Responses are based on a 5-point Likert scale in which behavior is rated from 1 (*Almost Never*) to 5 (*Almost Always*). Subscale raw scores can be summed and converted into percentile scores. A percentile score ranging from 25 to 40 suggests a moderate problem. Examples of the items that comprise the Attention subscale are provided below in Inset 5.4. The time frame is not specified, but the rater is asked to compare the child's behavior with that of his or her peers.

From a *developmental perspective*, the normative data are inconclusive as only summary tables are presented. These figures indicate some variations in attention stability across age and also by gender, with girls obtaining higher scores from Grade 3 onward, but no explanations of the profiles are provided by the authors, who simply state, "although there are some statistically significant differences across grades, graphic representations of these data in Figures 2.1 and 2.2 show few clear trends" (p .3). See also Erford and Hase (2006), who express similar reservations. In a series of studies by Wilding and Cornish using the ACTeRS Scale to differentiate "good attenders" from "poor attenders" as rated by their teachers, very low correlations were recorded between age and inattention (.07 and .12) and between age and hyperactivity (-.11 and -.03)

Question Number	Almost Never				Almost Always
Q 1. Works well independently (Teacher form)	1	2	3	4	5
Q 3. Completes assigned tasks satisfactorily with little additional assistance from parents, brother or sister (works steadily; doesn't continually ask for help or express frustration) (Parent form)	1	2	3	4	5
Q 4. Follows simple directions accurately (Parent form)	1	2	3	4	5
Q 6. Functions well in the classroom (Teacher form)	1	2	3	4	5

Inset 5.4

(calculated from data available from Munir, Cornish, & Wilding 2000b; Wilding, 2003).

NEURODEVELOPMENTAL DISORDERS

Although the authors of the ACTeRS provide data showing the scale's ability to distinguish between children who have been clinically diagnosed with ADHD (n = 356) from children without ADHD (n = 221), there are surprisingly few recent studies that can substantiate these findings. In terms of other neurodevelopmental disorders, Cornish, Munir, and Wilding (2000b) have successfully used the ACTeRS to distinguish inattention profiles in two syndrome groups: fragile X syndrome versus Down syndrome, and between these groups and typically developing children.

Swanson, Nolan, and Pelham-IV Questionnaire (SNAP-IV) and Strengths and Weaknesses of ADHD Symptoms and Normal Behavior Scale (SWAN)

The SNAP-IV (Swanson, 1992) and the SWAN (Swanson et al., 2004) are two of the few rating scales to provide a dimensional scaling (as opposed to the often used categorical scaling) of the DSM-IV items for inattention, impulsivity, and hyperactivity. The benefits of using a dimensional approach is that it provides a broader range of scores thus allowing for normal variations in the ADHD phenotype that are often masked in more categorical scoring systems, which assume that inattentive behavior is qualitative (either present or absent). Swanson and colleagues, in the past two decades, first developed the SNAP Rating Scales (Versions I–IV) and more recently the SWAN Scale, to allow a normal distribution range of attention and activity levels in a given population to be rated, rather than just the presence of inattentive behavior.

The SNAP-IV can be completed by both parents and teachers, and although originally designed for use with the DSM-III classification, it has been updated for use with DSM-IV. The scale comprises both a short form (18 items) and a long form (90 items), with the latter including DSM-IV criteria for ADHD (Inattentive and Hyperactivity/Impulsivity) and Oppositional Defiant Disorder. Despite several large-scale studies (e.g., Newcorn et al., 2001) using the SNAP-IV, there are no representative normative data available. Internal consistency is moderate to excellent (Stevens, Quittner, & Abikoff, 1998), and although validity is not well established, it is supported by the fact that the SNAP-IV is based on items included in other ADHD scales and the DSM-IV. Interestingly, a Chinese version of the SNAP-IV has recently been standardized on a large sample of children (n = 3534) from Grades 1 to 8, reporting moderate to excellent internal consistency and test–retest reliabilities (Shur-Fen Gau et al., 2008).

SCORING

Responses on the SNAP-IV are based on a 4-point Likert scale in which behavior is rated as 0 (*Not at All*), 1 (*Just a Little*), 2 (*Quite a Bit*), and 3 (*Very Much*). However, instead of summing items into a total scale score or a total problem score, item averages are calculated for each scale, with scores tentatively above the 95th percentile considered clinically significant. Detailed scoring information is provided on a specialist website (www. ADHD.net). Examples of the items that comprise the Inattention subscale are provided in Inset 5.5. The time frame is not specified, but the rater is asked to compare the child's behavior with that of his or her peers.

From a *developmental perspective*, few studies have examined age-related changes, using the SWAN. However, the findings from a recent cross-sectional study of elementary-aged school children (with an age range of 5 to 11 years) did not find developmental amelioration of inattentive behaviors during this age range, nor did they report large gender effects (Bussing et al., 2008). The scale authors therefore support the unstratified cutoff approach offered by the SNAP-IV but caution in

Question Number	Never, Seldom	Occasionally	Often, Quite a Bit	Very Often, Frequent
Q 1. Often fails to give close attention to details or makes careless mistakes in schoolwork or tasks	0	1	2	3
Q 3. Often does not seem to listen when spoken to directly	0	1	2	3
Q 4. Often does not follow through on instructions and fails to finish schoolwork, chores, or duties	0	1	2	3
Q 8. Often is distracted by extraneous stimuli	0	1	2	3

Inset 5.5

using the teacher SNAP ratings, in particular, as a diagnostic tool for ADHD.

SWAN

The SWAN is an 18-item scale that measures inattentive, hyperactive, and impulsive behaviors. It is an adaptation of the SNAP-IV in which different wording is adopted alongside a new 7-point scale. The first nine items correspond to the Attention Deficit scale and the last nine items to the Hyperactivity/Impulsivity scale. Although reliability and validity data are not available (not unexpected given this scale is in its early development), there have been at least three recent large-scale studies that have demonstrated the normal distribution of the SWAN, covering strengths as well as weaknesses in inattentive behavior in samples from Australia (Hay, Bennett, Levy, Sergeant, & Swanson, 2007, n = 528 twin pairs, age range 6 to 9 years; n = 488 pairs, age range 12 to 20 years), the United Kingdom (Cornish & Hollis, unpublished data, n = 1684, age range 6–11 years), and the Netherlands (Polderman et al., 2007, n = 561 twin pairs, age 12 years).

SCORING

The SWAN scoring requires each item to be rated across a 7-point Likert scale, which goes from a low level of problems (3, 2, 1) through average (0) to high level (-1, -2, -3). Summed scores range from a minimum

Question Number	High level Problems			Average level of Problems			Low level of Problems
Q 1. Gives close attention to details and avoids careless mistakes	–3	–2	–1	0	1	2	3
Q 2. Sustains attention on tasks or play activities	–3	–2	–1	0	1	2	3
Q 3. Listens when spoken to directly	–3	–2	–1	0	1	2	3
Q 4. Follows through on instructions and finishes school work/chores	–3	–2	–1	0	1	2	3

Inset 5.6

of -27 to a maximum of 27 for each scale: Inattention and Hyperactivity/ Impulsivity. This approach produces a normal distribution of the data, with scores in the 5% range considered as clinically significant. Examples of the items that comprise the Inattention subscale are provided in Inset 5.6. The time frame is not specified, but the rater is asked to compare the child's behavior with that of his or her peers.

From a *developmental perspective*, one of the unique contributions of the SWAN is its ability to provide a continuum of attentive behaviors within a given population. Thus, by its very nature it cannot provide age trajectories. However, recent studies have highlighted the potential of the SWAN to identify cognitive variability, such as reading or inhibitory performance, in children at both extremes of the SWAN continuum (e.g., top and bottom 10%) and across a broad age spectrum (5–11 years) (Savage, Cornish, Manly, & Hollis, 2006; Wilding & Burke, 2006).

NEURODEVELOPMENTAL DISORDERS

In terms of inattentive behaviors, there is considerable evidence that both the SNAP-IV and SWAN are sufficiently sensitive to discriminate between children with and without ADHD (e.g., Lubke et al., 2007; Newcorn et al., 2001). The SNAP-IV in particular has been used extensively in ADHD research, most notably as a primary treatment outcome in the Multimodal Treatment Study of Children with ADHD (MTA) (Jensen et al., 2007; Swanson et al., 2001) and on clinical trials of stimulant medication for children and preschoolers with ADHD (e.g., Abikoff et al., 2007; Steele et al., 2006; Vitiello et al., 2007; Wigal et al., 2004).

One of the most important and unique strengths of the SWAN scale is that it provides a broader, phenotypic description of ADHD symptoms in a general population and can therefore facilitate not just the selection of children for clinical efficacy studies but also for genetic studies of ADHD. As we described in Chapters 3 and 8, there is now overwhelming evidence that variations in ADHD have some genetic component (see Mick & Faraone, 2008, and Stevenson et al., 2005, for excellent reviews). Moreover, we know that heritability appears to be the same for extreme cases of ADHD, as it does for individual differences in ADHD behaviors in the general population (Gillis, Gilger, Pennington, & DeFries, 1992; Levy, Hay, McStephen, Wood, & Waldman, 1997), suggesting that the categorical diagnosis of ADHD represents the extreme of a geneti-cally influenced continuous trait in the population. The implication is that both categorical (diagnostic) and dimensional (quantitative trait) approaches are valid molecular genetic strategies when studying ADHD. In a series of recent studies Cornish and colleagues, using the SWAN to screen an epidemiological sample of 6- to 11-year-old children from Central England (UK), found a significant association between the high-risk DAT1 genotype and high SWAN scores (e.g., scores above the 90th percentile) (Cornish et al., 2005). Furthermore, these same children demonstrated specific impairments on measures that tapped response inhibition and selective attention (Cornish et al., 2005), but not sustained attention (Cornish, Wilding, & Hollis, 2008, but see also Bellgrove, Hawi, Kirley, Gill, & Robertson, 2005).

Although to date there have been no studies that utilized the SWAN Scale on children with other types of neurodevelopmental disorders, a recent clinical study by Posey et al. (2007) used the SNAP-IV to assess ADHD symptoms in children diagnosed with either autism, Asperger syndrome, or pervasive developmental disorder (with an age range of 5 to 14 years). The authors found a decrease in the severity of inattentive symp-toms reported by both teachers and parents from baseline to the conclu-sion of the treatment (Week 8). These findings, although preliminary, suggest some sensitivity of the SNAP-IV to detect changes in inattentive behaviors in children with significant developmental impairments.

ADHD-Rating Scale-IV (ADHD RS-IV)

The ADHD Rating Scale-IV (DuPaul et al., 1998) is an 18-item scale linked directly to the DSM-IV diagnostic criteria for ADHD and com-prises two subscales: Inattentive and Hyperactive/Impulsive. The scale has been used extensively as a clinical tool to gather parent and teacher ratings regarding the frequency of each of the ADHD symptoms and is based on substantive normative data on individuals aged from 4 to 20 years (Teacher form, n = 2000; Parent form, n = 2000). Internal con-sistency and test–retest reliability were excellent for the Teacher form (.96, .89 for Inattention and .88, .88 for Hyperactivity/Impulsivity) and the Parent form (.86, .78 for Inattention and .88, .86 for Hyperactivity/Impulsivity). In terms of construct validity, DSM-IV RS is moderately correlated with other similar rating scales such as the Conners and the CBCL. Although the scale was originally standardized on a US sample, it has also received validation for use in Europe, but with some restrictions (see Magnusson, Smari, Gretarsdottir, & Prandardottir, 1999; Servera & Cardo, 2007; Zhang, Faries, Vowles, & Michelson, 2005).

SCORING

Responses are based on a 4-point Likert-type format: 0 (*Never* or *Rarely*), 1 (*Sometimes*), 2 (*Often*), and 3 (*Very Often*). Scores for each subscale are summed and placed within a percentile. The percentile for predicting the presence of the ADHD inattentive-type and the ADHD combined-type are Teacher ratings at or above the 90th percentile and Parent ratings at or above the 93rd percentile. Examples of the items that comprise the Inattention subscale are provided in Inset 5.7. Behavior is rated over the past 6 months or the current school year.

From a *developmental perspective*, normative data are available for four age groups: 4 to 7, 8 to 10, 11 to 13, and 14 to 18 years old. Analysis revealed significant age effects for both subscales, with Hyperactivity/ Impulsivity showing most variation, with the youngest group displaying the highest scores followed by the 8- to 10-year-olds. The two oldest groups both received lower ratings than the younger groups. There were fewer age differences reported for ratings of Inattention symptoms, the 14- to 18-year age group being rated as less inattentive than the younger groups. In terms of gender, boys received higher ratings than girls on both the Inattention and Hyperactivity/Impulsivity subscales.

Question Number	Never or Rarely	Sometimes	Often	Very Often
Q 3. Has difficulty sustaining attention in tasks or play activities	0	1	2	3
Q 5. Does not follow through on instructions and fails to finish work	0	1	2	3
Q 11. Avoids task (e.g. schoolwork, homework) that require sustained mental effort	0	1	2	3
Q 15. Is easily distracted	0	1	2	3

Inset 5.7

NEURODEVELOPMENTAL DISORDERS

A major strength of the ADHD RS-IV is its ability to discriminate young children with ADHD, combined- and inattentive-types, from their non-clinical peers (e.g., Hattori et al., 2006; Kadesjo, Hagglof, Kadesjo, & Gillberg, 2003; Power, Costigan, Leff, Eiraldi, & Landau, 2001). A comparison study using the ADHD-RS-IV and the CBCL assessed a large sample of children (with an age range 3 to 7 years) comprising three groups: ADHD and a dual diagnosis of oppositional defiant disorder (ADHD + ODD), children with ADHD alone (ADHD), and nonclinical peers (Kadesjo et al., 2003). Both scales were able to discriminate children with ADHD, irrespective of an ODD diagnosis, from nonclinical comparison children. However, only the ADHD-RS-IV was able to discriminate the three groups separately with significantly higher ratings given to the ADHD + ODD group, followed by the ADHD group and then the nonclinical comparison group. To date there are no published studies that have utilized the ADHD-RS-IV across children with other types of neurodevelopmental disorders.

Developmental Behavior Checklist (DBC)

The DBC (Einfeld and Tonge, 1995, 2002) is one of the only behavioral measures to be directly relevant to individuals with significant intellectual impairments. Unlike any of the other measures reviewed in this chapter, the DBC items were derived from over 7,000 clinical files of children and adolescents with mental retardation and have so far been translated into Arabic, Chinese, Croatian, Dutch, Finnish, French, German, Greek, Hindi, Italian, Japanese, Norwegian, Portuguese, Spanish, Swedish, Turkish, and Vietnamese). The checklist has been extensively used as a research and screening tool in epidemiological studies (e.g., Cormack, Brown, & Hastings, 2000) and in neurodevelopmental disorders, notably autism (Gray & Tonge, 2005; Gray, Tonge, Sweeney, & Einfeld, 2008). Ninety-six items comprise the long version of DBC and are divided between five core subscales: disruptive/antisocial behavior, self-absorbed behavior, communications disturbance, anxiety problems, and social-relating problems. See Bontempo et al. (2008) for the most recent factor analysis of these subscales. Normative data are derived from a large multicenter epidemiological study in New South Wales and

Victoria, Australia (see Einfeld & Tonge 1996a, 1996b), and data are also available for boys and girls and for the mild, moderate, and severe mental retardation groups (Einfeld & Tonge, 2002). The DBC Parent form has moderate to excellent internal consistency (.61 to .91) and test–retest reliabilities (.51 to .87). No current data are available for the Teacher form. See Hastings, Brown, Mount, and Cormack (2001) for a critique of the psychometric properties of the DBC. More recently, two new versions have been developed: a short form (DBC-P24; Taffe et al., 2007) and an autism-screening form (DBC-ES; Gray & Tonge, 2005). Although there is no explicit subscale related to inattention, there are six items that reflect inattentive and hyperactive behaviors (Overexcited, Overactive, Poor Attention Span, Impatient, Impulsive, and Noisy). Hastings, Beck, Daley, and Hill (2005) report that a total score for these items showed good internal consistency and was able to differentiate children with and without a clinical diagnosis of ADHD (see also Einfeld & Tonge, 2002).

SCORING

Responses are based on a 3-point Likert scale in which behavior is rated as 0 (*Not True as Far as You Know*), 1 (*Somewhat or Sometimes True*), and 2 (*Very True or Often True*). Scores can be derived from each subscale alongside a total score that can be used as an overall measure of clinically significant problems using a cutoff score of 46+ (see Inset 5.8). Behavior is rated over the past 6 months.

From a *developmental perspective*, Clarke, Tonge, Einfeld, and Mackinnon (2003) found that the DBC was sensitive to change in

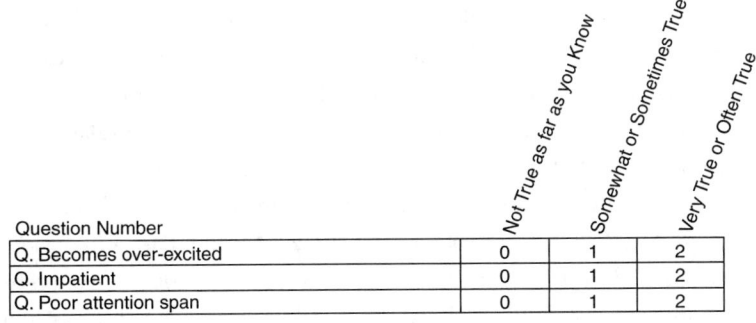

Question Number	Not True as far as you Know	Somewhat or Sometimes True	Very True or Often True
Q. Becomes over-excited	0	1	2
Q. Impatient	0	1	2
Q. Poor attention span	0	1	2

Inset 5.8

behavior with age. Although their sample size was quite small (n = 37), it was robustly significant. To date, there are no age trajectories on the DBC from either a longitudinal or cross-sectional perspective in non-clinical populations.

Focusing specifically on ADHD behaviors, Hastings et al. (2005) assessed a combined sample of children and adolescents, aged 4 to 18 years, with an intellectual disability, some with a known etiology (e.g., Down syndrome, autism) and some with unknown etiology for their intellectual disability. Using a subset of questions from the DBC–Parent version, the most interesting finding to emerge was a gender difference, with more boys than girls showing hyperactivity/inattentive symptoms. However, it was unclear from this study whether the DBC was sufficiently sensitive to capture ADHD clinical profiles. Of note are other published studies, not especially focused on attention, but which have demonstrated the utility of this measure across a range of neurodevelopmental disorders including longitudinal analyses of changes in behavioral patterns in fragile X syndrome (Einfeld, Tonge, & Turner, 1999) and Williams syndrome (Einfeld, Tonge, & Florio, 1997; Einfeld, Tonge, & Rees, 2001). The recent development of the DBC as a screen for autism is further testament to its unique versatility in the field of neurodevelopmental disorders (Brereton, Tonge, & Einfeld, 2006; Gray et al., 2008).

Strengths and Difficulties Questionnaire (SDQ)

The SDQ (Goodman, 1997) is used worldwide as a clinical screening instrument to estimate prevalence rates of mental health problems and to evaluate the efficacy of intervention outcomes. It is also used as a research tool to evaluate associations between problem behaviors including inattentive behavior and academic and social risk factors. Similar to the SWAN Scale, the SDQ also focuses on strengths as well as difficulties and has been translated into over 50 languages and used across diverse cultures including Europe, North America, Brazil, Middle East, and Asia (see Marzocchi et al., 2004; Woerner, Fleitlich-Bilyk et al., 2004). Twenty-five items comprise the SDQ and are divided between five subscales, with five items in each: Hyperactivity/Inattention, Emotional Symptoms,

Conduct Problems, Peer Relationships, and Prosocial Behavior. The SDQ can be completed by both parents and teachers. There is also a Self-Report form (25 items) for adolescents.

The SDQ is standardized on large, ethnically heterogeneous samples including children from the USA (n = 10,367; Bourdon, Goodman, Rae, Simpson, & Koretz, 2005), Netherlands (n = 562; Muris, Meesters, & van den Berg, 2003), Germany (n = 930; Klasen, Woerner, Rothenberger, & Goodman, 2003), the United Kingdom (n =10,000; Goodman, 2001), Australia (n = 1,359; Hawes & Dadds, 2004), and China (n = 2,128; Du, Kou, & Coghill, 2008). For each of the countries cited, reliabilities, including internal consistency and test–retest reliabilities, were satisfactory. For example, Goodman (2001), reporting on a large epidemiology sample of British children, found internal consistency to be moderate on both the Teacher form (.70 to .87) and the Parent form (.57 to .82). Test–retest reliabilities were also moderate for the Teacher form (.65. to .82), and Parent form (.57 to .72). Focusing on the *Hyperactivity/Inattentive* subscale, internal consistency and test–retest reliabilities were excellent for the Teacher form (.88, .82) and moderate for the Parent form (.77, .72). The relatively good reliabilities obtained on the Hyperactivity/Inattention subscale have been replicated extensively and across different cultures (e.g., Mellor, 2004; Woerner, Becker, & Rothenberger, 2004). In terms of construct validity, the SDQ also appears to correlate well with other attention rating scales, including the longer established CBCL (e.g., Becker, Woerner, Hasselhorn, Banaschewski, & Rothenberger, 2004; Goodman & Scott, 1999).

SCORING

Responses are based on a 3-point Likert scale in which behavior is rated as 0 (*Not True*), 1 (*Somewhat True*), and 2 (*Certainly True*). The scoring can be quite complex, so we advise the reader to go to the SDQ website for the variations on the scoring criteria. See www.sdqinfo.com.

Examples of the items that comprise the Hyperactivity/Inattention subscale are provided in Inset 5.9. Behavior is rated over the past 6 months or the current school year.

From a *developmental perspective*, there are surprisingly few studies that have assessed age trajectories on the SDQ from either a longitudinal

Question Number	Not True	Somewhat True	Certainly True
Q 13. Easily distracted, concentration wanders	0	1	2
Q 14. Thinks things out before acting	0	1	2
Q 15. Good attention span, sees things through to the end	0	1	2

Inset 5.9

or cross-sectional perspective in nonclinical populations. Although a glimpse of stability in profiles is gleaned from the many studies that have provided extensive normative data on the SDQ, few of these studies provide age-specific or gender norms that allow for the charting of trajectories across the subscales. Two recent studies assessed age-related changes, and both suggest stability with age on the Hyperactivity/ Inattention subscale. The first is a recent large population-based study of Australian children aged 4 to 9 years (Hawes & Dadds, 2004), and the second is a longitudinal study of children at ages 5 and 6 years (Perren, Stadelmann, von Wyl, & von Klitzing, 2007).

NEURODEVELOPMENTAL DISORDERS

Although not specifically designed to discriminate ADHD from non-ADHD symptoms, as the Conners Rating Scale does, for example, recent findings do suggest that the Hyperactivity/Inattention subscale demonstrates some sensitivity in screening children with ADHD symptoms in genetic studies (Curran et al., 2001) and appears to correspond positively with clinical diagnoses of ADHD (e.g., Becker et al., 2004; Hawes & Dadds, 2004; Muris, Meesters, Eijkelenboom, & Vincken, 2004). Moreover, there is emerging evidence to suggest that the SDQ is sufficiently sensitive to provide predictive information on inattentive behaviors in the classroom (Hayes, 2007). See Vostanis (2006) for a comprehensive review of the clinical and research applications of the SDQ.

To date, there are relatively few studies that have assessed inattentive behaviors in other neurodevelopmental disorders. One noteworthy exception examined ratings of hyperactivity/inattention in children with autism and children with ADHD compared to typically developing controls and found that the SDQ suggested greater overall impairment in the ADHD group, followed by the autism group and then the typical controls (Happé, Booth, Charlton, & Hughes, 2006). Although not focused on specific disorders, Simonoff, Pickles, Wood, Gringras, and Chadwick (2007) examined a large general population sample of adolescents aged between 12 and 15 years to ascertain whether individuals with mild intellectual disabilities exhibited more ADHD behaviors than those without an intellectual disability. Findings confirm expectations, with lower IQ scores associated with greater ADHD symptoms using the SDQ.

Behavior Rating Inventory of Executive Function (BRIEF)

The BRIEF (Gioia, Espy, & Isquith, 2003; Gioia, Isquith, Guy, Kenworthy, & Baron, 2000) is not strictly a rating scale of inattentive behaviors as defined by the ratings scales already reviewed in this chapter. Instead, the BRIEF focuses on ratings of executive function behaviors in the everyday context as defined by parents and teachers. The BRIEF provides normative data for two core age ranges: 2 to 5 years (63 items, identical Teacher and Parent forms) and 5 to 18 years (86 items; separate Teacher and Parent forms). There are five clinical subscales in the preschool version: Inhibit, Shift, Emotional Control, Working Memory, and Plan/Organize; and eight subscales in the school-age version: Inhibit, Shift, Emotional Control, Initiate, Working Memory, Plan and Organize, Organization of Materials, and Monitor. Both versions also include a Global Executive Composite Index. A core advantage of this rating scale is that it covers a wide range of executive functions that have strong clinical utility.

The normative data for the *preschool version* are based on moderate, ethnically heterogeneous samples of children (Teacher form, n = 302; Parent form, n = 460). There was excellent internal consistency across the subscales for both the Teacher form (.90 to .97) and Parent form (.80 to .95). Test–retest reliability was also moderate to excellent on both the Teacher form (.65 to .94) and Parent form (.78 to .90). Focusing on three subscales, two of which—Inhibit and Working Memory—are presumed to overlap with diagnostic criteria for ADHD inattentive-type and ADHD combined-type (Gioia et al., 2000), and the third of which—Plan/Organize—is widely regarded as a core executive function skill (see Chapter 2 for a detailed discussion of executive functions). Internal consistency and test–retest reliability were moderate to excellent across all three subscales: Inhibit (Teacher form = .94, .94; Parent form = .90, .90), Working Memory (Teacher form = .94, .88; Parent form = .88, .85), and Plan/Organize (Teacher form = .97, .85; Parent form = .80, .78).

The normative data for the school-age version are based on moderate, ethnically heterogeneous samples of children (Teacher form, n = 702, Parent form, n = 1419). There was excellent internal consistency across the subscales for both the Teacher form (.90 to .97) and Parent form (.80 to .97). Test–retest reliability was also excellent on both the Teacher form (.83 to .92) and Parent form (.76 to .86). Internal consistency and test–retest reliability was also excellent across the three subscales of Inhibit (Teacher form = .96, .91; Parent form = .91, .84), Working Memory (Teacher form = .93, .86; Parent form = .89, .85), and Plan/Organize (Teacher form = .91, .88; Parent form = .90, .85). In terms of construct validity, the BRIEF teacher and parent versions appear to correlate moderately well with other rating scales such as the CBCL and the ADHD-RS-IV (see Mahone et al., 2002), suggesting some similarities in construct.

SCORING

Responses are based on a 3-point Likert-type format: 1 (*Never a Problem*), 2 (*Sometimes a Problem*), and 3 (*Often a Problem*). Scores for each subscale are summed and converted to T-scores and percentile ranks. A T-score of 65 or above is considered to be in the clinical range. Examples of the items that comprise the Inhibit subscale only (preschool and school-age versions) are provided in Insets 5.10 and 5.11.

Preschool Version   Behavior is rated over the past 6 months.

Question Number	Never a Problem	Sometimes a Problem	Often a Problem
Q 23. Is fidgety, restless or squirmy	1	2	3
Q 28. Is impulsive	1	2	3
Q 54. Has trouble putting the brakes on his/her actions after being asked	1	2	3
Q 56. Completes task or activities too quickly	1	2	3

Inset 5.10

Question Number	Never a Problem	Sometimes a Problem	Often a Problem
Q 38. Does not think before doing (Teacher form)	1	2	3
Q 42. Interrupts others (Teacher form)	1	2	3
Q 49. Blurts things out (Patent form)	1	2	3
Q 65. Talks at the wrong time (Parent form)	1	2	3

Inset 5.11

School–Age Version   Behavior is rated over the past 6 months.

From a *developmental perspective*, normative data are available by age group and gender: 2 to 3, and 4 to 5 years (preschool version), and 5 to 6, 7 to 8, 9 to 13, and 14 to 18 years (school-age version). Small age effects were observed, most notably on the Inhibit subscale, whereby ratings improved with age. Also on the Inhibit subscale, boys generally showed elevated scores across age ranges. The only age by gender interaction was observed in the school-age version (Parent and Teacher forms), again specifically on the Inhibit subscale, with boys demonstrating a decrease in reported problems with increasing age.

NEURODEVELOPMENTAL DISORDERS

The BRIEF has been used extensively to assess atypical variations in executive behavior across a range of neurodevelopmental disorders including ADHD (Gioia, Isquith, Kenworthy, & Barton, 2002; Mahone & Hoffman, 2007; Mahone et al., 2002; Sullivan & Riccio, 2007; Toplak, Bucciarelli, Jain, & Tannock, 2009), autism (Gioia et al., 2002), and 22q11 deletion syndrome (Kiley-Brabeck & Sobin, 2006). In one of the few studies to date to provide a cross-syndrome comparison of BRIEF ratings on the Inhibit, Working Memory, and Plan/Organize subscales, Gioia et al. examined profiles in children with ADHD inattentive-type, ADHD combined-type, and autism. Additional comparison groups included Reading Disability and Traumatic Brain Injury (TBI). On the Inhibit subscale, the most impaired group were children in the ADHD combined-type group, followed by children in the autism group and then children in the ADHD inattentive-type group. On the Working

Memory subscale, both the inattentive-type group and combined-type group displayed comparable ratings. On the Plan/Organize subscale, the inattentive-type group scores were elevated above those of the combined-type group and the autism group.

Children's Behavior Questionnaire (CBQ) and Early Childhood Behavior Questionnaire (ECBQ)

The CBQ (Putnam & Rothbart, 2006; Rothbart, Ahadi, Hershey, & Fisher, 2001) and the newly developed ECBQ (Putnam, Garstein, & Rothbart, 2006) questionnaires were designed for parents to assess childhood temperament (CBQ; age range of 3 to 8 years; 195 items in the standard version, 94 items in the short version, and 36 items in the very short version) and toddler temperament (ECBQ: age range 1 1/2 to 3 years; 201 items in the standard version). Both questionnaires include subscales that tap a range of inattentive behaviors: Attentional Focusing, Impulsivity, and Inhibitory Control.

There are extensive normative data across different sources for the CBQ, standard and short form (e.g., Rothbart et al., 2001, n = 411) and the ECBQ (Putman et al., 2006, n = 317). Both questionnaires have also been translated into a variety of languages for use across diverse cultures: Swedish, Spanish, Italian, Russian, Romanian, Hebrew, and Chinese. Internal consistency is moderate to excellent for the CBQ scales (.64 to .94) and ESBQ scales (.62 to .90). To our knowledge, no studies of test–retest reliabilities have yet been conducted on the CBQ and ECBQ scales.

SCORING

Responses are based on a 7-point Likert scale: 1 (*Extremely Untrue of Your Child*) to 7 (*Extremely True of Your Child*). The authors recommend the use of continuous scale scores (rather than categories) in the interpretation of results. Examples of the items that comprise the Attentional Focusing subscale only, ECBQ and CBQ versions, are provided in Insets 5.12 and 5.13.

Early Child Behavior Questionnaire (ECBQ).   Behavior is rated over the past 6 months.

Question Number	Extremely Untrue of your Child						Extremely True of your Child
Q 126. When playing alone, how often did your child becomes easily distracted?	1	2	3	4	5	6	7
Q 196. When playing alone, how often did your child move from one task or activity to another without completing any?	1	2	3	4	5	6	7
Q 169. While looking at picture books on his/her own, become easily distracted?	1	2	3	4	5	6	7

Inset 5.12

Question Number	Extremely Untrue of your Child						Extremely True of your Child
Q 16. When practicing an activity, has a hard time keeping her/his mind on it	1	2	3	4	5	6	7
Q 21. Will move from one task to another without completing any of them	1	2	3	4	5	6	7
Q 84. Is easily distracted when listening to a story	1	2	3	4	5	6	7

Inset 5.13

Attentional Focusing Subscale

Child Behavior Questionnaire (CBQ): Behavior is rated over the past 6 months.

Attentional Focusing Subscale
From a *developmental perspective*, the normative data do include longitudinal stability estimates that appear relatively stable across a 2-year interval

of the CBQ (5 to 7 years) and 6 and 12 month spans of the ECBQ (from 18 to 36 months). A recent study by Spinrad et al. (2007) has also reported longitudinal stability over a 1-year period, from 18 months onward, on the Attention Focusing subscale. As far as we are aware, there are no published normative data examining gender by age interactions.

NEURODEVELOPMENTAL DISORDERS

Validation of the CBQ as a temperament scale is well documented (see Putnam & Rothbart, 2006), but its use outside of that domain, especially in relation to inattentive behaviors, is only in its infancy. However, two studies to date provide a glimpse of the potential of this questionnaire to capture differences in attention in children with Williams syndrome (Klein-Tasman & Mervis, 2003) and children with autism (Konstantareas & Stewart, 2006). Both of these syndromes are discussed in detail in Chapter 9, but the findings from these two studies suggest that ratings on the attention subscales (e.g., Inhibitory Control, Attentional Focusing) can differentiate children with autism, for example, from their typical peers, but have less sensitivity in differentiating between differing atypical groups, for example, between Williams syndrome and a mixed etiology group. Further research is clearly warranted across differing developmental disorders including ADHD populations.

Conclusions

We have not attempted to cover the entire spectrum of behavior-rating scales but rather limit our review to those that have been used extensively on children and adolescents with attention problems. As can be observed from Table 5.1, each rating scale has its own profile of strengths, but a consistent weakness in almost every scale reviewed was the lack of available age-related trajectories, both from teachers and parents, on typically developing children as well as those with specific neurodevelopmental disorders. Yet with a greater understanding of how attention develops at the cognitive level, it becomes important to elucidate the interrelationships between inattentive behavior and cognitive inattention, and to define how this relationship changes developmentally from preschool to adolescence. There is also a pressing need to develop standardized attention scales that can be used to explore the range and severity of inattention behaviors across differing neurodevelopmental disorders, with and without significant intellectual impairment. Albeit not surprisingly, current research is almost exclusively focused on discriminating children with ADHD subtypes from their nonclinical peers. Comparatively few studies have explored inattention profiles in other neurodevelopmental disorders, and there are virtually no studies that have explored atypical trajectories across different disorders.

CHAPTER SUMMARY AND KEY FINDINGS

- Attention rating scales are an important tool in capturing the range of severity of inattentive behaviors in ADHD but are less effective in discriminating inattentive behaviors in disorders that also present with significant cognitive impairment.
- This conclusion prompts the development of more sensitive rating tools to explore the range and severity of attention signatures at the behavioral level across different neurodevelopmental disorders and across development. This will complement the current cognitive research on disorders of attention.
- Future research needs to embrace the concept of attention as a continuum of involvement, thus allowing a broader understanding of attention phenotypes.

6

Measuring Attention at the Cognitive Level
Batteries and Tasks

CHAPTER SNAPSHOT
• Choosing a task: assessing the maze of possibilities
• Issues of complexity: Standardized versus experimental paradigms. What works and what doesn't in neurodevelopmental disorders.
• The importance of incorporating age- and ability-appropriate tasks to assess subtle attention trajectories across age in typical and atypically developing populations

Attention, as we have discussed in previous chapters, is a multidimensional construct with both behavioral and cognitive components. In Chapter 5, we described in detail the behavioral characteristics of "inattention" and its measurement. In brief, behavioral descriptions of inattention include disorganization, forgetfulness, distractibility, and poor concentration. At one level, these descriptions are useful because they reliably document the manifestations of inattention in everyday contexts such as the classroom or home. Moreover, these behavioral ratings have identified children at risk for poor developmental outcomes. However, when used alone these ratings are inadequate to guide educational or clinical approaches because inattentive behavior must arise from a range of underlying cognitive problems that may need different educational or clinical interventions in different cases. For example, telling a student to "pay attention" will not be effective if the behavior is attributable to cognitive problems in differentiating relevant from irrelevant information but may help if the student has a problem in switching focus to a new source of information.

In Chapter 2, three major attentional components were specifically addressed: *selection* of relevant input from the huge amount of information arriving at the different senses; *maintenance* of attention on a specific input or task; and *control* processes that organize sequencing of responses, switching of attention and retrieval of relevant information from memory. In Chapter 4, the relations between different attentional processes and activity in different areas of the brain were surveyed, and a broad pattern of agreement was found with the conclusions drawn in Chapter 2. The picture that emerges is one of a complex system that operates in different ways in many subsystems across the whole brain and is designed to select for action. This may require specific inputs to be favored, specific responses to be connected to such inputs, instructions to be held temporarily until triggered by a prior event in the sequence, sequences to be organized that switch attentional focus from one source to another, and so forth. Clearly, these attention functions combine in real-life activities. For example, even many "simple" tasks, such as searching for a specific symbol on a road map or finding your favorite dessert in a supermarket, require a combination of processes that tap selection, maintenance, and attentional control. Given these complexities, the reader can appreciate the challenge of establishing a set of cognitive measures that will adequately test the full range of attention functions that contribute to efficient performance. The focus of the present survey is to consider the available tasks and attempt to evaluate their adequacy for this ambitious goal.

From a *developmental perspective*, as we shall see in the detailed discussions in Chapter 7, identifying tasks that are sensitive enough to produce age-related changes across a wide age span, say, from infancy to midchildhood, is extremely difficult, and such studies are vulnerable to floor and ceiling effects. As a result, the majority of research has tended to focus on quite restricted age ranges, for example, focusing on toddlers between the ages of 2 and 4 years, or on children aged between 6 and 10 years. Although not ideal, this approach is perhaps the most sensible because it allows for finer-tuned attention profiles to emerge within a given, albeit restricted, age range.

Attention Batteries and Tasks

One glance at the current literature reveals an extensive range of tasks purporting to measure one or more components of attention. Such tasks, as with rating scales, can be used both as research and clinical tools to

Box 6.1 Criteria for Task Selection

- Validity and reliability
- Test–retest reliability
- Age appropriateness
- Purpose
- Sensitivity and specificity

assess attention performance in typically and atypically developing children. But what is the criterion for choosing the most appropriate task? We will address five essential criteria outlined in Box 6.1

A critical component of any task choice is whether the task measures the attention processes it claims to engage. We have already come across the terms "validity" and "reliability" in Chapter 5 in our discussion of attention rating scales and so will not replicate that information here. However, there are specific issues that directly pertain to choosing an attention task or battery that are not as essential when choosing a rating scale.

The first issue is that in order to ensure that the task provides convincing evidence of *validity* and *reliability*, we must identify which feature or features of a particular task are critical to its validity. As we have seen, considerable uncertainty surrounds the different functions included under the domain of *attention* and particularly under executive function. Many of the popular tasks employed are so complex that they inevitably engage a variety of processes rather than any single component of attention. So in some cases, these tasks may be valid measures of attention, but it is unclear exactly what aspect ensures this validity. Ideally, prior exploration should establish, by varying a number of parameters, which features of the task are critical to the effect under investigation. For example, which variation of the task produces the clearest difference in performance between a group of children with impaired attention and typically developing children? Unfortunately, many tasks are widely employed without any systematic investigation of this nature.

The second issue is one of *test–retest reliability*. For in order to establish the reliability of a task, one of the basic prerequisites is that the same task, or two different variants of it, is given on two different occasions in order to discover whether the result is similar on both occasions. This is not especially straightforward because whenever a task is novel there will always be some practice effect on the second attempt at the task. As already noted, this is a particular problem with EF tasks, since EF weaknesses are often manifest most clearly with novel tasks where old knowledge has to be applied in a new way; hence, the pattern of results may be very different at the second time of testing. Since it is often not appropriate to repeat the administration of such a task, we may be unsure whether an individual with specified attention impairment will perform poorly on repeated presentations of the task or only on the first occasion that the task is given. One solution to this problem is to compare performance on two halves of a single administration of the task (*split-half reliability*), either comparing the first and second halves of the test or comparing odd-numbered trials with even-numbered trials (where the task consists of a series of test trials). This method, however, is also sometimes inappropriate, for example, where a problem has to be solved and the measure taken is the total time or the number of moves to achieve success.

Age appropriateness of a given task is also a critical concern and one that is especially relevant to this chapter. All too frequently, tasks initially developed to assess adult performance are "modified" for use with children without any consideration given to whether such a modification reduces or even nullifies the sensitivity of the original task demands. Some basic questions that need to be addressed include the following: Are the requirements of the task likely to be comprehensible to the participant group, and is the task sufficiently motivating to make it likely that they will complete it? Is the duration reasonable for children of that age (especially if several tasks are likely to be administered in one session), and is it sensitive enough to capture developmental changes? Simply reducing the duration of an adult task does not address the above concerns.

A further concern relates to the *purpose* for which the test is to be used. Established task batteries, even if their validity and reliability are well established, only measure what they have been designed to measure. They assess whether an individual or a group, compared with norms established from a suitably representative sample of the population, differs significantly on some aspect of behavior. But batteries are in the main fairly blunt instruments and usually offer few or no means of pursuing the fine details of differences that they may detect. To do this requires additional manipulations of the parameters of the tasks, in which case existing materials and norms will be inappropriate. Standardized tasks, at best, offer a limited number of variations that may not include the one required for a given investigation. For example, we may want to discover whether the number of nontargets in a visual search task or their similarity to each other (or both) cause particular difficulty for children with a specified condition, but one or both of these variations may be unavailable in the test. Hence, the ease of creating a purpose-made modification of the task becomes a relevant concern. This is particularly problematic if a task is part of a computerized battery that cannot readily be amended or extended to pursue further questions. Consequently, for research purposes, it is usually necessary to take information about useful tasks as a guide to designing one's own versions rather than simply using the tasks in their original form.

Finally, once a measure is finally adopted as a valid index of the ability under investigation, its sensitivity and specificity also need to be considered. *Sensitivity* refers to the ability to identify individuals with impairment in that ability without missing a large proportion of cases, while *specificity* refers to the ability to avoid misclassifying as impaired those individuals who are in fact unimpaired. Both these types of error need to be minimized in a satisfactory task.

Though we would like to have adequate information on all these issues when comparing different tests and deciding which are the most appropriate for testing children (and often very young children) with various impairments, in practice we will often have to make informed guesses.

Many batteries of cognitive tests include one or two token tests claiming to measure attention, but there are few systematic attempts in such batteries to define or measure distinct components. Attempts by cognitive psychologists to identify such components and create appropriate batteries are surprisingly rare; consequently, it is only recently that test batteries have been created purporting to define and measure independent aspects of attention that can then be used to probe the precise nature of attentional disorders. The goal of the present chapter is threefold. First, to further clarify the different functions involved in attention by examining ways of testing them, and to provide descriptions of tasks that have been employed in testing attention in children and adolescents. We will first describe and evaluate five batteries that aim to test a variety of aspects of attention, and then other individual tasks designed to probe specific aspects. The second goal of this chapter is to provide a summary of findings from a developmental perspective across each of the batteries and tasks, and the third goal is to highlight the feasibility, alongside concerns, of using attention tasks with our six different neurodevelopmental disorders: autism, fragile X syndrome, Down syndrome, Williams syndrome, 22q11 DS, and ADHD. Detailed summaries (genetic, behavioral, and cognitive) of each of these disorders are provided in Chapters 3, 8, and 9 and so will not be discussed in detail here.

Attention Batteries[1]

A number of task batteries have been devised to test different aspects of attention, often in the wider context of testing general cognitive performance. Such batteries provide standardized norms derived from typically developing populations. Consequently, there are several problems in using such batteries when investigating attention in atypical populations, particularly in cases where general cognitive abilities may be impaired. As indicated, batteries are developed and normative performance established using populations with typically developing abilities. Hence, tasks may be too difficult for populations with genetic or other neurodevelopmental impairments, and testing may produce scores at or close to the floor level for the test, thus precluding any possibility of differentiating levels of performance within or between such groups. The results merely show that individuals from these populations cannot carry out the task, and the reason for failure is unclear. Other assumptions of the task, such as the need to understand instructions and task requirements, may not be fulfilled. A related problem is that a given set of tasks is commonly only

---

[1] We have selected batteries that have been used most extensively. There are other batteries from which detailed data are lacking, such as the Amsterdam Neuropsychological Tasks (de Sonneville, 1999), the KITAP (Zimmerman & Fimm, 1993), and the Delis-Kaplan Executive Function System (Delis, Kaplan, & Kramer, 2001).

suitable over a limited age range, and some batteries offer different sets of tasks to test the same functions at different ages. The equivalence of such alternative test sets is not always convincing. It is therefore difficult to find or devise tasks that permit a sufficiently wide range of performance to be tested with both typical and atypical populations over a wide age range.

A second major problem with task batteries, particularly commercially available batteries with a fixed format, is their rigidity. As already indicated there is little or no scope for varying parameters of the tasks to explore details of a revealed weakness or to test hypotheses relating to it. In other words, batteries can only play a limited role in investigative research. The need is to develop flexible tasks in which the difficulty and task requirements can be readily varied in order to establish the sources of difficulty and be able to isolate any differences within and between different neurodevelopmental disorders. In the following chapters we outline some of the innovative tasks that have been developed to address this issue both in the assessment of attention in typically developing children (Chapter 7) and in children with different neurodevelopmental disorders (Chapter 8 and 9). See Table 6.1 for a summary of the attentional functions that are assumed to be tested by the five batteries that we are to consider. This box lists other relevant information such as the age range over which the battery is appropriate, whether norms are available, and the main strengths and weaknesses of each task battery.

The Attention Network Test

The Attention Network Test (ANT; Fan, McCandliss, Sommer, Raz, & Posner, 2002; Fan & Posner, 2004) is one of the most widely employed standardized tasks of attention designed to engage the three independent attention systems (alerting, orienting, and control) identified by Posner and his colleagues (Posner & Petersen, 1990). It is not strictly a battery, but rather a tool for experimental investigation, consisting of a single task with variations that enable specific weaknesses to be identified. The ANT is a modified version of the original task used by Posner and colleagues.

In the adult version of the ANT, cues indicate whether a specific target will appear above or below a fixation point, and a symbolic stimulus (an arrow) indicates whether a left or right response is required.

Table 6-1. Summary of Task Batteries Assessing Attention

Battery Reference	Tasks/ Measures	Age Range	Administration Time and Method	Normative Data	Primary Usage	Strengths	Weaknesses
Attention Network Test (ANT) (Child version) (Rueda et al., 2004)	Alerting Orienting Control of conflict	4–12 years	25 min; computerized	Yes	Research investigation	Fair reliability	Lengthy for younger groups; unrealistic time constraint on responses; few data from atypically developing groups
NEPSY: A Developmental Neuropsychological Assessment (Korkman, Kirk, & Kemp, 1998)	Maintenance Planning Switching Inhibition	3–4 years 5–12 years (only some tests common to both age groups)	Time not precisely specified; apparatus and instructions	Yes	Clinical diagnosis	Automated scoring of various kinds; good reliability; designed for children; some usage with atypically developing groups	Validity of tasks unclear; no norms for atypically developing groups

Measure	Attention components	Age	Time/format	Publicly available	Purpose	Design/reliability	Validity/notes
Cambridge Neuropsychological Automated Testing Battery (CANTAB) (Fray, Robbins, & Sahakian, 1996)	Selection Maintenance Planning/control Switching Inhibition	4 years plus (some tasks only)	Different tasks last 5–10 min; computerized	No	Research investigation	Automated	Not designed for or meaningful to children; no norms for typically or atypically developing groups; reliability weak on some tasks
Test of Everyday Attention in Children (TEA-Ch) (Manly et al., 2001)	Selection Maintenance Control	6–16 years	40–45 min (complete battery); apparatus and instructions	Yes	Research investigation	Designed to interest children, but some tasks tedious; good reliability	Validity unclear (see text); data available from ADHD studies but not other neurodevelopment groups
Wilding Attention Tests (WATT) (Wilding, Munir, & Cornish, 2001)	Selection Maintenance Switching	6–17 years (also versions for younger children)	15–20 min (complete battery); 10 minutes (Selection and Maintenance tasks); computerized	Yes	Research investigation	Successful in engaging children's interest; variety of scores available; successfully used with atypically developing groups; Fair reliability	Not yet publicly available; scoring needs streamlining.

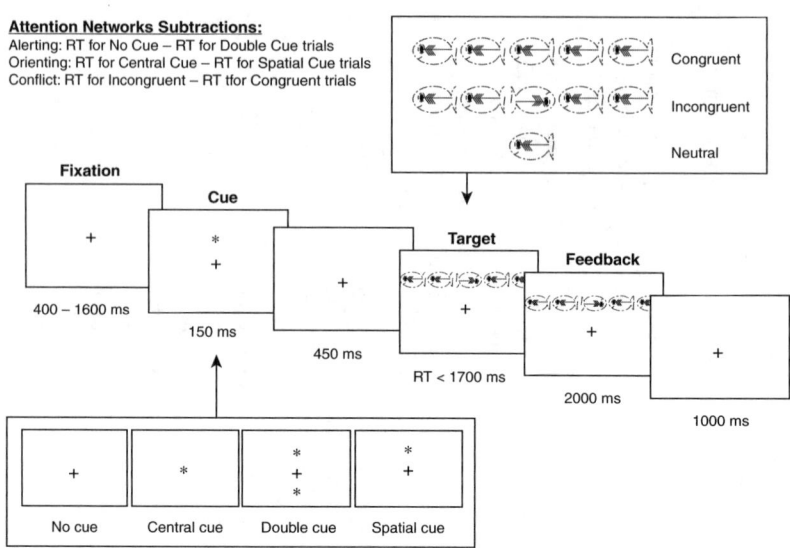

Figure 6-1. Schematic of the child version of ANT.
From Rueda et al. (2004). Adapted by permission from *Neuropsychologia 42*, 1071.

First, the fixation point appears in the center of the screen, and after a variable interval (400–1600 msec) a cue is given lasting for 150 msec, indicating the position of the forthcoming target. The cue may be neutral (in the same position as the fixation point) or double, appearing both above and below fixation. See Figure 6.1. In both these cases, it only serves an *alerting* function, giving a warning of stimulus onset and providing no information concerning the target position. Alternatively, an *orienting* cue in the form of an arrow above or below the fixation point may indicate target position as well as alerting to stimulus onset. In the fourth condition, no cue is presented.

Following the cue there is a delay of 450 msec, with the fixation point present, followed by the target stimulus, which remains present until response, or for 1,700 msec if no response occurs. There are three display conditions. In the *neutral* display condition, the target consists of a single central arrow pointing right or left and requiring the appropriate response; in the *congruent* condition, there are two arrows on each side of the central target arrow pointing in the same direction as the target. In the *incongruent* condition, these arrows point in the opposite direction to the target. These flanking arrows are to be ignored.

Facilitation due to alerting is measured by subtracting the median latency or the error rate for the double cue condition from the corresponding measure for the no-cue condition, thus giving an index of the effect of a warning cue before stimulus onset. Note that there is no test of the effect of different intervals between cue and target that would provide information on the time required to alert attention and the duration of alerted attention. *Facilitation due to orienting* is measured by subtracting the median latency or error rate for the congruent cued condition from the corresponding measures for the neutral condition. *Effects of conflict* are measured by the difference between congruent and incongruent trials. Note that when calculating the median latency for each condition, all other variations in factor are ignored; for example, latencies for double-cue trials are calculated across all flanker conditions and vice versa. These procedures assume that the cueing and interference effects are independent and do not vary according to the combination of the two factors. However, Fan et al. (2002), Callejas, Lupianez, and Tudela (2004), and Redick and Engle (2006) all report some interactions between these two effects, implying that interference effects vary depending on the presence or absence of an alerting cue. Fan et al. (2002) argue that, because the two effects are not significantly correlated, independence can be assumed, but Redick and Engle (2006) point to the low reliability that is typical of scores derived by calculating the difference between two measures and the possibility that the low correlations reflect this unreliability rather than true independence.

Evaluation of the ANT

The ANT therefore incorporates 12 conditions (4 cue conditions by 3 flanker conditions). It does not include conditions where some spatial cues falsely indicate the position of the ensuing target, as in the original Posner task. Instead, ability to handle conflict (the control system) is tested by the incongruent flanking condition. Since the latter requires inhibition of interference and the former requires overcoming a prepared response, these two procedures might not be equivalent. However, Friedman and Miyake, (2004) found that these two types of inhibition loaded on the same factor, implying that they do engage the same processes, so this criticism is not a major problem.

The adult version of the ANT is a reliable measure. Performance is stable in normal adult samples, similar in males and females, and the attentional measures are not affected significantly by practice. Test–retest correlations, a critical indication of reliability, are moderately high, ranging from .52 to .77, implying that the test tends to produce similar results when carried out more than once. The relations between the measures and specific attentional processes are also supported by brain imaging studies indicating the operation of separate brain systems that involve different neurotransmitters (see Chapter 4).

While this task clearly represents a useful tool for investigating some basic key processes in attention, we would reiterate the point made earlier that the task is limited in its demands and does not tap many of the varieties of attention that we have discussed in Chapter 2. For example, it assumes that control functions can be assessed by the single task of excluding distraction and provides no means of assessing planning, switching, sequencing, or maintenance of attention. The task does have the virtue that each effect is assessed by comparing the critical condition against a baseline in which the key factor is absent. Its relation to other tasks needs to be explored to see if efficiency of these simple operations has consequences for more complex attentional processes (see the discussion in Chapter 2 of Redick & Engle, 2006, for one example of this).

From a *developmental perspective*, surprisingly few studies have modified the adult ANT to make it suitable for use with children. The only studies to date that the authors are aware of are by Rueda, Posner, and colleagues (e.g., Rueda et al., 2004; Rueda, Posner, & Rothbart, 2005; Rueda, Posner, Rothbart, & Davis-Stober, 2004; and see Fan & Posner, 2004, for a review). In brief, as the task is described in detail in Chapter 6, the arrows of the adult version are replaced with pictures of yellow fish on a blue-green background that face left or right, and the child is asked to feed the central fish by pressing the appropriate button depending on the direction in which the fish is facing. Although Rueda et al. had some degree of success in modifying the adult version, a concern is that the task duration (approximately 20 min) and the cognitive requirements, such as continued fixation, might be overdemanding for younger children. Even in the 6- to 10-year-old range, low reliability was found over a 6-month delay. More studies to complement those of Rueda et al.

are needed to determine the applicability of this task across a developmental timeline.

Neurodevelopmental Disorders

Given the advanced level of attentional capacity required for the successful completion of the child ANT, it is not surprising that relatively few studies have examined performance in children with neurodevelopmental disorders. Of the few studies that have, one has focused on children with ADHD subtypes (combined-type, inattentive-type) (Booth, Carlson, & Tucker, 2007), and one on children with 22q11 DS (Sobin et al., 2004). Findings from both studies (to be discussed in later chapters) suggest that the ANT task is sufficiently sensitive to discriminate executive network, and alerting networks of attention in neurodevelopmental disorders. However, the main constraint of the ANT, alluded to in both studies, is the reaction time limit. One improvement would be for researchers to modify the task to allow for slower processing time.

The NEPSY

The NEPSY-I (Korkman, Kirk, & Kemp, 1998) is a developmental neuropsychological assessment tool, originally devised in Finland by Korkman and colleagues and subsequently modified and expanded in the USA. It comprises 27 subtests and assesses five cognitive domains in children aged 3 to 12 years old: attention and executive functioning, language, sensorimotor, visuospatial ability, memory, and learning.

The reader should note that a second edition of the NEPSY (NEPSY-II; Korkman, Kirk, & Kemp, 2007) is now available in which the Tower of London task (planning), Visual Attention, and Knock and Tap have been replaced by three new EF tasks (Animal Sorting, Clocks, Inhibition), and the age range has been extended to 16 years. At the time of writing, there are few published studies using or evaluating the NEPSY-II available for review, so our discussion considers only the first edition.

We focus here on *attention and executive function*, which comprises six "core" tests and for which there has been extensive research. See Box 6.2 for a summary description of subtests.

Box 6.2 The NEPSY Subtests (Attention and Executive Function)

- *The Tower* (age 5 to 12 years only): In this task, colored balls with a hole through the center are placed on a display of three vertical pegs and have to be moved one at a time between the pegs from an initial position to end up matching a target position; only the top ball on a peg may be moved. This transformation is to be achieved in a prescribed number of moves (which in some cases exceeds the minimum possible and thus requires suppression of a move toward the goal in favor of a "detour"). However, a time limit of 30 seconds is imposed on the first four problems and 45 seconds per item on the remainder, and the test closes if failure occurs on four trials in succession. This task is generally regarded as a test of planning, though it has been argued that an important factor is ability to inhibit moves that make the display more similar to the target display but are not the correct moves that will eventually attain the goal.

- *Auditory Attention and Response Set* (AARS) (age 5–12 years only): The auditory attention part of this task requires detection of the word "red" in a rapid sequence of words and response by selecting a red token. The Response Set task requires selection of a yellow token in response to "red," a red token in response to "yellow," and a blue token in response to "blue."

- *Visual Attention* (all ages): In this task, specific target pictures have to be found in a randomly organized array, and in the 5–12 year age group only, examples of two different faces have to be found in a linearly organized array of faces that share features with the targets. These are all continuous visual-search tasks of varying degrees of difficulty. The last task, the most difficult, requires two target descriptions to be maintained in working memory in a conjunction search.

- *Statue* (age 3–4 years only): In this task, a motor position has to be adopted following instructions (such as standing still while pretending to be holding a flag), then maintained for 75 seconds. This task assesses control and maintenance of attention.

- *Design Fluency* (age 5–12 years only): In this task, dot patterns are presented and have to be connected by straight lines to make a

Box 6.2 The NEPSY Subtests (Attention and Executive Function) *(continued)*

new design in each case. A structured and a random array are given, each for 60 seconds. This task assesses various aspects of attentional control, including planning and checking.

- *Knock and Tap* (age 5–12 years): This task requires children to knock with their knuckles on the table when the examiner taps with a flat palm and do nothing when the examiner knocks with their knuckles. This pattern is then reversed. This is a Go–NoGo task requiring the inhibition of the prepotent direct imitation, followed by a reversal of the newly learned response pattern.

Evaluation of the NEPSY

The authors of the NEPSY (Kemp, Kirk, & Korkman, 2001) write:

> The NEPSY was based on theory and traditions of neuropsychological assessment rather than Factor Analysis. Factor structure inherently suggests that all of the subtests contributing to a particular factor are highly correlated. Therefore they measure one construct. Neuropsychological functions are not that simple. Complex cognitive functions require contributions from different domains to a greater or lesser degree. (p. 21)

The authors go on to argue that tests were chosen to permit brief screening for different disorders, acquired or developmental, on the basis of current research and that the subtests included within a given domain (for example, attention/executive function) were not primarily intended to give a combined picture of a single function, but to pick up dysfunction in different components of such a function. Individual differences, they claim, are best defined by examining the pattern of performance in the different subtests, combined with a variety of other evidence on educational and clinical history, and so forth. The validity of the tasks for testing the claimed functions is unsupported by any formal data, as is the independence of the different tasks in their cognitive requirements. However, Kemp et al. (2001, p. 16) do report respectable reliabilities for these tasks, and the tasks have, of course, been used successfully with typical and atypically developing children, as our more detailed discussion below will demonstrate. Having said that, a cautionary note to the reader is to remember that the NEPSY is primarily designed for clinical evaluation, rather than as a research tool where variations can be devised for research purposes.

From a *developmental perspective*, in terms of attention functioning, the NEPSY provides only a small snapshot of development from 3 to 4 years old, with a more comprehensive snapshot for 5- to 12-year-olds, covering selective attention (AARS and Visual Attention), divided attention (two faces visual search in Visual Attention), maintenance of attention (AARS), planning (possibly Tower, Design Fluency), and some aspects of inhibitory control (Tower, AARS Response Set, Knock and Tap). In terms of developmental trends, Korkman, Kemp, & Kirk (2001), in a sample of 800 children aged 5–12 years, found significant age effects across all attention subtests. For the four tasks used over the whole age range, the greatest improvement for Tower, AARS, and Visual Attention was in the younger age groups (5 to 7 years), with no significant improvement after 10 years. The Design Fluency task showed a different profile, with most improvement from 9 to 10 years and no further improvement after 11 years.

Neurodevelopmental Disorders

NEPSY group differences between typically developing children and differing neurodevelopmental disorders have been quite extensively reported. For example, children with high-functioning autism display significantly poorer performance on the Tower and AARS subtests compared to typically developing controls, even when IQ is accounted for, but show comparable performance on the Visual Attention subtest (Hooper, Poon, Marcus, & Fine, 2006). On the Knock and Tap subtest, children with autism performed much worse than typically developing children matched on age and IQ (Joseph, McGrath, & Tager-Flusberg, 2005). On the Statue subtest, children with ADHD (Mahone et al., 2006) and children with comorbid hyperactivity–oppositional defiance behaviors (Youngwirth, Harvey, Gates, Hashim, & Friedman-Weieneth, 2007) performed at a significantly lower level than typically developing controls. In a population of children with 22q11 DS, impaired performance was especially noted on the visual attention subtest. Further qualitative observations indicated that children's scanning of the targets lacked strategy and focus and did not seem to be attributable to being overly impulsive (Sobin et al., 2004).

The CANTAB

The Cambridge Neuropsychological Testing Automated Battery (CANTAB) was originally designed to diagnose dementia in elderly individuals (Fray, Robbins, & Sahakian, 1996) but has been used increasingly to assess the development of attention, planning, and spatial working memory in children and adolescents (Luciana & Nelson, 2002, and see Luciana, 2003, for a review). It is a computerized battery of tests covering visual memory, visual attention/working memory, and planning. We will look at the last two categories. See Box 6.3 for a summary description of subtests.

Evaluation of the CANTAB

The CANTAB therefore covers the following aspects of attention by providing computerized versions of a variety of existing tasks: visual selection (Match to Sample), maintenance (Rapid Visual Information

---

Box 6.3 The CANTAB Subtests (Visual Attention, Working Memory, and Planning)

- *Intradimensional/Extradimensional Shift*: In this task, two forms of switching of attention are required: intradimensional (reversing stimulus response pairings), and extradimensional (changing the relevant feature of the display). The method used is similar to that of the Wisconsin Card Sorting Task, discussed below.
- *Matching to Sample*: In this task, a target pattern is shown in the middle of the screen, then between one and eight similar choices are shown around it. The exact match has to be found. The task is similar to the Matching Familiar Figures Task (Kagan, Rosman, Day, Albert, & Phillips, 1964), which is commonly assumed to test ability to inhibit premature responses, but can also be regarded as a visual-search task.

(continued)

Box 6.3 The CANTAB Subtests (Visual Attention, Working
Memory, and Planning) *(continued)*

- *Delayed Match to Sample* uses a similar format, but the target does
  not remain when the search display is presented, and there are only
  four items in the latter, so the task taps predominantly visual recognition memory.
- *Rapid Visual Information Processing*: This is a Continuous Performance Test (CPT) in which responses have to be made to targets
  in a rapid sequence of mainly nontargets presented at a rate of
  100 items per minute. The targets are specified sequences of three
  digits (e.g., 2-4-6). Measures calculated include the number of hits
  and misses, and latency. The task assesses maintenance of attention
  with a high input rate and takes approximately 7 minutes. A simpler version with 1-2-3 as the target is used with children aged
  4 to 8 years.
- *Spatial Memory Span*: In this task, an array of squares is shown and
  a sequence of these lights up; the sequence then has to be recalled
  by touching the squares in order. The task begins with a sequence
  of two items and progresses to longer sequences (maximum nine
  items). This is the commonly used Corsi blocks test, originally
  designed to measure spatial working memory span (Milner,
  1971).
- *Spatial Working Memory*: In this task, an array of squares is presented,
  and squares have to be probed for the presence of (initially) four
  targets until one is found; erroneous repetitions on empty squares
  are recorded (within trial errors). Once a target has been found,
  search begins for a new target that will be in a different square; repetitions on squares already found to contain targets are also recorded
  (between trial errors). The task requires memory for squares already
  probed and development of a strategy to avoid repetitions.
- *Stockings of Cambridge (SOC)*: This is the Tower of London
  (described below) presented as colored balls in stockings instead of
  on pegs. Measures include times and the number of moves. The
  task takes approximately ten minutes. It is typically viewed as a
  measure of planning.

Processing), planning (possibly SOC, Spatial Working Memory), and set shifting (ID/ED shift). The Spatial Memory Span task is classified as a Planning/EF test but involves basically spatial memory. Likewise, two of the other tasks, Delayed Match to Sample and Spatial Working Memory, have major memory components. Information is given for most tasks about which brain areas they engage, but the independence of the functions tested by each task is not discussed nor obvious. The battery was designed primarily for neurological examination and hence focuses on EF functioning far more than on other aspects of attention. Though it does provide a wealth of measures for performance on each task, interpretation of many of these must await further research to determine what distinct abilities, if any, they might reflect. There is no comprehensive set of norms, but instead, normative data are compiled from a wide variety of published studies and unpublished data sets.

From a *developmental perspective*, in one of the most comprehensive studies to assess developmental trends in performance on the CANTAB, Luciana and Nelson (2002) examined children aged 4 to 10 years old and an adult comparison group (18–30 years) on six subtests that included the set-shifting task (ID/ED). Unfortunately, no other attention measures were included in this study, although the Tower of London was included as a planning measure. The rationale for the choice of subtests was their validation in previous adult studies, task duration, and task component.

Performance across all six subtests improved with age but did so differentially across subtests. Adult performance on the ID/ED set-shifting task was reached by age 7 years, with children at this age easily completing all nine stages of the task. A similar age trajectory on this task has also recently been reported by Waber et al. (2007) on a large sample of children aged 6 to 17 years, assessed on a broad range of cognitive tasks that included three of the CANTAB subtests (ID/ED, Spatial Working Memory, and Spatial Span).

Neurodevelopmental Disorders

Performance on the CANTAB (or at least some of its subtests) has been documented across a variety of neurodevelopmental disorders, most notably those with known attentional and EF impairments such as ADHD (e.g., Coghill, Rhodes, & Mathews, 2007; Rhodes, Coghill, & Mathews, 2004) and autism (e.g., Hughes, Russell, & Robbins, 1994; Landa & Goldberg, 2005; Ozonoff et al., 2004). All studies find impairments in performance relative to typically developing controls, but these studies and others tend to focus only on a small number of EF/attention subtests, for example, the ID/ED set-shifting task and the SOC task. Of note, three recent studies, again using these same subtests, have compared performance in children with autism to that of children with ADHD with very consistent findings.

The first study by Goldberg et al. (2005) assessed performance on three samples of children aged 8 to 12 years: children with high-functioning autism, children with a diagnosis of ADHD, and typically developing children, all with IQs above 75. The authors found no group differences on either the ID/ED task or the SOC task. The second study by Happé, Booth, Charlton, and Hughes (2006) also comprised three samples of children aged 8 to 12 years: children with high-functioning autism, children with a diagnosis of ADHD, and typically developing children, but in this case with a slightly lower IQ cutoff of 69 or above. Both the ID/ED task and the SOC subtests were used, and both failed to elicit any group differences. The third and most recent study is by Sinzig, Morsch, Bruning, Schmidt, and Lehmkuhl (2008), which comprised four samples of children aged 6 to 18 years: children with dual diagnosis of high-functioning autism and ADHD, children with a diagnosis of ADHD, children with a diagnosis of high-functioning autism, and typically developing children. The IQ cutoff was 80 or above. Although there were no overall group differences in performance on the ID/ED task and the SOC task (in accordance with the Goldberg et al. and Happé et al. studies), differences emerged on the basis of effect sizes, with the dual diagnosis (Autism + ADHD) group producing more errors and requiring more time to complete the task compared to the group of children with a diagnosis of autism alone ($d = 0.6$) and the typically developing controls ($d = 0.6$). On the SOC task, there was also a medium effect size between children with autism alone and children with ADHD alone ($d = 0.6$), with poorer performance in the autism group compared to the ADHD group.

To date, performance on the CANTAB attention/EF tasks (at least on the two subtests documented with high-functioning autism and ADHD samples) appears to be restricted to relatively "high-functioning" atypical populations, namely, those with IQs within the low-average to normal range.

The TEA-Ch

The Test of Everyday Attention (TEA; Robertson, Ward, Ridgeway, & Nimmo-Smith, 1994) and, more recently, the Test of Everyday Attention in Children (TEA-Ch; Manly et al., 2001) are comprehensive batteries of attention tasks designed to test three different aspects of attention. We will concentrate here on the TEA-Ch. The tasks that compose the TEA-Ch are expressly tests of *everyday* attention; they aim to employ tasks with some face validity that demand similar abilities to those required in daily life and cover the three most widely accepted aspects of attention: selection, maintenance, and control (see Chapter 2 for a detailed description of these components). Consequently, the tasks used here are very different from those used in the highly constrained tasks of the Posner ANT; the latter employs displays requiring rapid decisions and timed responses in order to test operations like alerting and attention. The TEA-Ch tasks require continuous responding over a considerable time period, but the authors do not draw overly ambitious conclusions about highly specific processes that an individual task may engage. The battery was designed primarily as a clinical tool to identify children with possible attention impairments.

The TEA-Ch is based on a threefold division derived from earlier factor analysis of tasks comprising the Adult-TEA, which the authors adapted when creating the TEA-Ch battery. Confirmatory Factor Analysis (CFA) on data from 293 children aged 6 to 16 years was used to test the hypothesis that the TEA-Ch tasks engaged three separate attentional processes as specified above. Factor analysis was explained briefly in Chapter 2 when discussing Mirsky's research. It is a method of deducing the underlying processes involved in several tasks by examining the degree to which individuals who perform well or badly on one task will show a similar pattern on other tasks. In CFA, instead of simply waiting to see what the data reveal, hypotheses are made about which tasks test

Box 6.4 The TEA-Ch Subtests (Selective Attention, Maintenance of Attention, and Control)

Selective attention is measured by two tasks requiring the ability to detect targets from distractors.

- *Sky Search*: This task requires the search for "target" spaceships placed in pairs on a sheet; targets are defined as pairs where the two spaceships are identical and distracter pairs differ. The measure taken is the mean time to find a target pair, with motor time removed (this was measured by the children marking targets on a sheet displaying only targets; thus, the specific effect of the selective requirement could be isolated).
- *Map Mission*: This task records the number of targets (knife and fork signs) found on a map in a fixed short time span (1 minute), again a measure of speed; motor time is not measured separately.

Sustained attention is measured by four tasks requiring the ability to maintain attention over a period of time.

- *Score*: This task requires children to keep count of a number of "scoring" sounds on a tape over a period of 10 trials comprising between 9 and 15 tones at variable intervals (500 to 5,000 msec).
- *Score DT*: This is another listening task, but this time, as the children count, they also have to listen out for an animal name read as part of a news bulletin. After each trial they name the animal and the number of target tones.
- *Code Transmission*: This task requires children to listen to monotonous sequences of digits, one every 2 seconds, listening for two 5s in a row. When they hear the specified sequence, they have to recall the preceding digit. The task is a form of Continuous Performance Task (see below) and takes approximately 12 minutes.
- *Walk/Don't Walk*: This task requires children to "walk" along a paper pathway, one step at a time, each time they hear a specific tone. Unpredictably, they hear a tone that is different from the rest, and they must inhibit the next step and not continue walking until the target tone resumes. In essence, this requires them to listen carefully to the tones, to maintain attention, and not get carried away.

Box 6.4 The TEA-Ch Subtests (Selective Attention, Maintenance
of Attention, and Control) *(continued)*

- *Sky Search DT*: This task was originally designed to test a supposedly distinct ability to carry out a dual task but was subsequently
  included with the Sustained Attention tasks after the factor analysis. This is a particularly complex test that requires children to
  combine the two previous tasks: finding the space ships (Sky
  Search) and keeping a count of scoring sounds (Score!). Measures
  include the number of correct responses to the sounds, the number
  of correctly identified ship pairs, and the time taken.

Control of Attention is measured by two tasks requiring switching of
attention or response.

- *Creature Counting*: In this task, a picture is shown of "aliens" in their
  burrows, and children have to scan along counting the aliens; however, at intervals, an arrow is inserted pointing upward, in which
  case the incremental counting has to continue, or if the arrow
  points downward, decremental counting is required. Time and
  errors are recorded, and if the child is correct on three or more
  trials, time per switch is calculated for correct trials only; there was
  no control for baseline speed or adjustment for overall accuracy (in
  fact, very inaccurate participants are excluded by the above
  criterion).
- *Opposite Worlds*: This task measures the ability to reverse an established habit. In the "same-world" condition, children are required
  to follow a path scattered with the digits 1 or 2 and to name each
  one in the order they are named on the pathway. In the "opposite-
  world" condition, they do the same task, but this time when they
  see 1, they have to say "two," and if they see 2 they have to say
  "one." Time taken to complete each condition is recorded.
  Unfortunately, since errors have to be corrected before proceeding
  to the next stimulus, the main contribution to this score is probably time taken to appreciate and correct a mistake, rather than
  interference due to response conflict; incidence of errors is not
  recorded.

similar abilities, and the match between these predictions and the results is tested. The CFA carried out on the tasks in the TEA-Ch confirmed the threefold division, and the authors went on to test differences between an ADHD group and a comparison group of children. These findings are discussed in detail in Chapter 8. Is the conclusion that the TEA-Ch distinguishes three distinct aspects of attention justified? This question will require a detailed examination of the structure of the TEA-Ch. See Box 6.4 for a summary description of subtests.

Evaluation of the TEA-Ch

Validity of the TEA-Ch was assessed to some extent by correlations with four other generally accepted tests of various aspects of attention. Accepting a (liberal) significance level of .01 (40 correlations were computed), the following significant relations were obtained. The search measures were correlated with the Stroop and Trails tests (see below for a description of these), both of which employ measures of speed in selective tasks (though Manly et al. argue that the correlations are due to the common demands of the tasks rather than the common measure). Score!, Score DT, Code Transmission, and Creature Counting were significantly related to the Matching Familiar Figures Task (MFFT), often regarded as a measure of ability to inhibit impulsive responding. However, MFFT was unrelated to the Walk and Opposites tasks, which are likely to require inhibition, and Code Transmission was also correlated with Stroop and Trails. Thus, no comprehensive or clear-cut data are provided on the validity of the separate TEA-Ch tasks. The authors place the most emphasis on the results of the CFA, with its three separate factors. Reliability of the TEA-Ch is good, with test–retest correlations between scores obtained 5 to 20 days apart being over .7 on most of the measures, with age effects removed.

Wilding (2005) has criticized the factors derived from these tasks on a number of grounds, arguing that the differences between the selection and sustained factors obtained in the factor analysis may have resulted from confounding measurement differences with task differences. The three factors that were extracted differed not only in the assumed task demands (selective, sustained, or control), but also in the nature of the scores employed. Both of the selective attention scores were speed measures, virtually all the sustained attention scores were accuracy measures, and the attentional control scores were again speed measures (with some uncertainties attached to them, as indicated above). Accuracy and speed are commonly closely related, but the correlation may either be negative, where faster responding is at the expense of lower accuracy, known as *speed accuracy trade-off*, or positive when difficult trials yield slower and less accurate responses. Indeed, the two measures may also be unrelated. Such relations may occur across different participants or within the results from a single participant: for example, slower responders might be more accurate if speed–accuracy trade-off is operating, or an individual might make fewer errors when responding slowly. Which type of relation was present in these search tasks is an empirical issue that the authors did not address. There is, therefore, no a priori justification for assuming that these two types of measure were equivalent and interchangeable, and the fit of the data to the model tested may have reflected these confounded differences of measurement between the sets of tasks, rather than the difference in the type of attention purportedly being measured. Whether or not this was the case cannot be decided on the basis of the available data from the TEA-Ch, but it is at least possible that the distinction drawn between the three types of attention might not depend on the claimed difference in task demands. Other work by Wilding and colleagues (described later in Chapter 7) has, in fact, suggested that speed measures and accuracy measures, at least in this type of task, may reflect distinct aspects of cognitive function. Hence, claims by the TEA-Ch authors that their tasks measure different attentional factors may need reassessing.

From a *developmental perspective*, Manly et al. (2001) have provided normative data across nine subtests of the TEA-Ch in children aged 6 to 16 years. The subtests all showed age-related improvements. Inspection of Table 2 in Manly et al., (2001, p. 1072) suggests some interesting differences in the age at which performance stops improving. Measures of accuracy (Score!, Code Transmission, Walk, Creature Counting) show a consistent tendency for performance to improve up to the 9- to 11-year-old age group and show little further improvement thereafter, whereas speed of performance was still improving at age 15 to 16 years (Skysearch, Mapsearch, Creature Counting time).

Neurodevelopmental Disorders

In contrast to the developmental research, the TEA-Ch subtests have been used extensively to document the range of attentional impairments in atypical populations, notably children with ADHD, but also in children with other neurodevelopmental disorders such as autism, fragile X syndrome, and Down syndrome (the findings of these studies are reviewed in Chapters 8 and 9). Rarely in the research literature is the TEA-Ch battery given as a whole. Instead, it is more common to see findings that report an individual subtest measuring a particular aspect of attention or a cluster of subtests measuring a subdomain such as attentional control.

Of those few studies that have assessed attention using the full TEA-Ch battery, some interesting findings have emerged. For example, Heaton et al. (2001) assessed children with a diagnosis of ADHD and non-ADHD comparison children aged between 6 and 15 years. Performance was most significantly impaired in the ADHD group on the Score! subtest, a measure of sustained attention. Hood, Baird, Rankin, and Isaacs (2005) assessed whether performance on the TEA-Ch had an immediate improvement with stimulant medication (methylphenidate [Ritalin©], which is reviewed in Chapters 8 & 10) in children diagnosed with ADHD, aged 7 to 11 years. Performance was compared to typically developing children across two time points: Time 1 (baseline: ADHD group on no medication) and Time 2 (ADHD group on medication). Both testing sessions occurred in the same day. Performance at Time 1 revealed a significant impairment in the ADHD group versus comparison children on two sustained subtests: the Score! task and Sky Search DT task. The authors suggest that these two measures, in particular, may be sensitive enough to differentiate children with ADHD from typically developing children. When the groups were retested at Time 2, performance on all TEA-Ch subtests was improved and was at a comparable level to the comparison children. See Chapters 2 and 10 for a more in-depth discussion of these findings and others that have used the TEA-Ch to assess the efficacy of medication in ADHD.

The WATT

Wilding designed a battery of computerized tasks (Wilding, Munir, & Cornish, 2001), which have been subsequently modified over the years to test different aspects of attention (selective, divided, sustained, and control) across a broad age range (4 to 16 years) of typically and atypically developing children The tasks are user friendly, relatively brief, and involve "scenarios" that a young child can relate to and participate in.

WATT Subtests

Visearch is a *visual search* task that comprises both single- and dual-target search conditions and tests selective attention and attentional switching. Single-target search tasks are commonly employed as tests of selective visual attention, and the dual-target search task tests attentional switching, widely regarded as requiring control or executive functions.

The Visearch task presents a display of a river, trees, and various "holes" of differing colors, shapes, and sizes, all on a green background. Children have to search for a specific type of hole in which monsters are hiding. They are asked to find the "king of the monsters," by clicking on the targets with the computer mouse or pointing at a touch screen, but (unknown to the child) only small monsters appear until the 20th target is found. The task terminates after 50 responses if the child makes a large number of errors. Two easy single-target conditions (one with vertical black ellipses as targets and another with horizontal brown ellipses, both of which share features with nontargets) are followed by a dual-target condition in which the child is required to alternate between these two types of target. There are 25 targets randomly positioned among 100 holes in all in the single-target tasks and 15 of each type of target in the dual-target task. Subsequently, a further difficult single-target condition has been added with additional distracters that closely resemble the targets. See Figure 6.2 for an example of the Visearch display.

Before each task there is a demonstration and practice session until it is clear that the children understand which of the stimuli represent the targets and what they have to do. Each run takes about a minute to complete. The task ends if the king is not found after 50 clicks. Mean time per hit is calculated, and the total number of errors (false alarms, i.e., clicks on nontargets or background, and repetitions on already located targets, plus failures to switch target in the alternating task). Mean time per hit is measured from the previous response in all cases, whether that previous response was a hit or an error; hence, time spent on errors is removed from this measure. Mean distance per hit was not considered

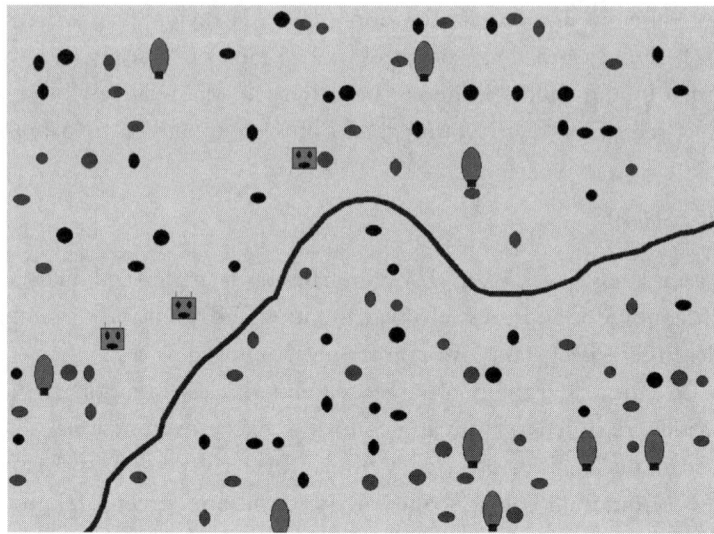

Figure 6-2. The display from the Wilding Visearch task. (See color Figure 6-2.)

after the initial study (Wilding et al., 2001), since Wilding (2003) suggested that time and the number of false alarms were the key dependent variables.

Vigilan is a sustained attention test using the same type of visual display as Visearch and specifically measures *vigilance*, the ability to maintain attention while awaiting the appearance of a target. In the Vigilan task, a target appears at irregular intervals, ranging from 4 to 14 sec. When a yellow line appears around one of the black vertical ellipses, this indicates that a monster is "at home," and if the child clicks on the shape within 7 seconds, the monster will appear. The king appears on the 16th target displayed after about 4 minutes on task. The measures calculated are the number of hits, mean time per hit, the number of false alarms, and the distance wandered while awaiting a target.

Wilding Monster Card Sorting Task (WMST) is a child friendly, computerized version of the Wisconsin Card Sorting Task (WCST), which is described below and is frequently used with adult patients with brain injury to test aspects of attentional control, specifically, ability to switch the focus of attention when the current focus proves ineffective. In the WMST version, four monster kings are shown on the screen, varying

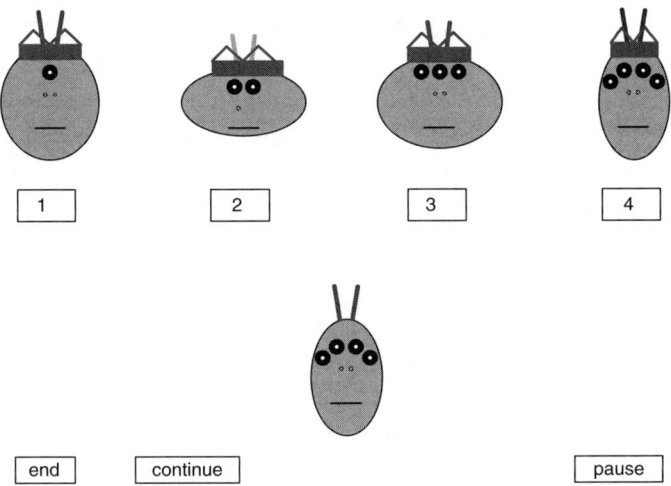

Figure 6-3. The display from the Wilding Monster Sorting Task.
(See color Figure 6-3.)

across three categories: color of their antennae (red, blue, green, or
purple), head shape (four shapes, ranging from a wide horizontal ellipse
to a tall vertical ellipse), and number of eyes (one to four). See Figure 6.3
for an example of the display.

Children are asked to help the kings choose teams for the monster
sports and so should allocate each monster in turn to one of the teams.
If they are correct, the king will be happy and smile, but if they are not
correct, he will be upset and make a sad face. As with the WCST, if the
child makes a series of 10 correct choices, the sorting criterion changes
without any signal being given. The program calculates a variety of measures (similar to the WCST) including the total number of errors made
on 64 trials, the number of perseverative responses (i.e., continuing to
make choices on the basis of a feature despite negative feedback), and the
mean time to respond on the trial immediately following an unambiguously correct response (designed to reflect whether a correct inference
had been made concerning the critical feature in the preceding trial).

Two studies have employed the WMST task (Wilding et al., 2001;
Zhan et al., 2010). The first of these found that the number of correct
responses and categories solved were the measures that best discriminated between good and poor attention groups. Zhan et al. (2010),

who did not include ratings of attentional ability, found that several of the measures commonly derived from the original WCST were highly correlated (number of correct responses, number of categories solved, number of conceptual level responses). All these measures, together with the number of perseverative errors, were still improving at age 17, while the speed of responses reached a maximum at about age 13.

Evaluation of WATT

The WATT visual-search tasks are similar in their main features to the visual-search tasks of the TEA-Ch described above, with the important difference being that the computerized presentation enables a variety of error measures to be recorded (the TEA-Ch search tasks record only speed of performance). This difference has proved crucial in the findings derived from the search and vigilance tasks. As discussed in detail in Chapter 7, differences between good and poor attention groups (differentiated by teacher ratings) occurred in the number of false alarms to nontargets, especially in the more difficult single-target search and the dual-target search, but not in the speed of performance. Wilding (2005) demonstrated that when both types of measure (speed and accuracy) were employed in the WATT tasks, they were not highly correlated within tasks, but the speed measures were correlated across both selective and sustained tasks, as were the accuracy measures. The speed measures were also correlated with the parallel measures from the Skysearch and Mapsearch tasks in the TEA-Ch. Thus, factor analysis detected two factors, one based on speed and the other on false-alarm rates (Wilding & Cornish, 2007). Moreover, speed measures were related to chronological age, and accuracy measures were related to ratings of attentional ability (Cornish, Wilding, & Hollis, 2008; Wilding, & Cornish, 2007). Hence, the supposedly separate selective and maintained attention factors did not emerge in these data; instead, separate factors emerged depending on type of measure, only one of which was related to attention as reflected in the general ratings by teachers.

Further analysis of the data from Wilding (2003) has shown that the measures of false alarms from the two difficult search tasks (difficult target–foil discrimination and switching between two targets) were significantly but not highly correlated ($r = .29$, $p < .01$), implying that they

Table 6-2. Correlations for Mean Correct Response Time and Log False
Alarms Between Two Runs on the Four Visual-Search Tasks and the
Vigilance Task Using Different Displays

	Condition 1	Condition 2	Condition 3	Condition 4	Vigilance
Mean Correct Response Time	.75	.75	.62	.81	.78
Log False Alarms	.52	.59	.62	.56	.53

engage some common process but also incorporate substantially different demands upon attention. Multiple regression demonstrated that they
independently and significantly predicted the attention ratings ($p < .01$
in both cases). These findings support the view that these more difficult
search tasks engage EF/control processes that have both general aspects
and specific components required for each task.

The reliability of these tasks was assessed by correlations of performance on two different displays of each task carried out in succession.
Table 6.2 gives a summary of results and demonstrates that the mean
correct response times were more reliable than the false-alarm rates.

The WATT tasks were designed specifically for children and have
proved motivating, congenial, and easy to administer. The tasks can be
adapted for use to answer research questions, but of necessity the available software does not enable purpose-built modifications to be made,
other than those changes that can be achieved through instructions,
practice, and so forth.

From a *developmental perspective*, as indicated above, the Visearch and
Vigilan tasks appear to be especially sensitive in distinguishing, at a cognitive level, children with good and poor attention as recognized by teachers
who evaluated their behaviors in the classroom. This consistent finding has
been reported across a series of studies of preschool children (Wilding &
Burke, 2006) and elementary-aged children (Cornish, Wilding, & Hollis,
2008; Rezazadeh, Wilding, & Cornish, paper submitted, 2010; Wilding
et al., 2001; Wilding, 2003; Wilding, Pankhania, & Williams, 2007). More
recently, a study of Chinese children aged 16 to 17 years successfully
extended the range over which these tasks could be used (Zhan et al.,
2010). A recent study (Rezazadeh et al., paper submitted), using a simpler

form of the original tasks, with fewer and larger stimuli designed to be suitable for younger children, found that the accuracy scores from these tasks were the best predictors of attentional ratings, and that no improvement in prediction was achieved by including a Continuous Performance Task and the Day–Night task that requires reversal of the standard naming responses (see later in the chapter for full descriptions of these tasks), plus chronological age (CA) and IQ. Together, these findings demonstrate that the tasks are valid measures of the variable that they were designed to assess.

A version of the visual search task designed for use with very young children, requiring responses by means of a touch screen, has been successfully used with 2- to 4-year-olds (Scerif, Cornish, Wilding, Driver, & Karmiloff-Smith, 2004, 2007; Wilding & Burke, 2006). Younger children were, in general, slower, less systematic, and less accurate than older toddlers.

Zhan et al. (2010) successfully used the WMST in a Chinese population of typically developing children and adolescents ranging in age from 6 to 17 years and demonstrated different age trajectories for different measures, as described in Chapter 7.

Neurodevelopmental Disorders

The WATT has now been used quite extensively to study inattention across a range of neurodevelopmental disorders (fragile X syndrome, Down syndrome, and Williams syndrome). Although these findings are described at length in Chapter 9, of interest here is that age-related changes in performance across the Visearch and Vigilan tasks appear to differentiate disorders from infancy onward. For example, toddlers with fragile X, Down syndrome, and Williams syndrome do not differ from controls on measures such as search speed and path, but they do differ in their pattern of errors (Cornish, Scerif, & Karmiloff-Smith, 2007; Scerif et al., 2004). Both Williams syndrome and fragile X infants display atypical processing of target–distracter similarity, but with only fragile X infants showing difficulties in inhibiting previously successful responses. In toddlerhood, Down syndrome children did not differ from younger typically developing controls. By childhood, there are again markedly different error patterns between fragile X, Down syndrome, and typically developing children (Munir et al., 2000b; Cornish, Munir, & Wilding, 2001; Wilding, Cornish, & Munir, 2002).

Summary of Task Batteries

See Table 6.1 for a summary of attention batteries. A general assessment will be given at the end of this chapter. These batteries do not, of course, exhaust the tasks that have been used to test the different attention functions, and we therefore go on to describe and assess other tasks that have been devised and employed for these purposes.

Individual Attention Tasks

A variety of other tasks have been used to test the range of attentional functions that we have identified. The reader is cautioned, however, to look closely at the issues of validity, reliability, and appropriateness when selecting a given task that purports to measure a specified attentional component. One of the key difficulties is that not all tasks are developed in a standard format and so vary from study to study, each with slightly different parameters. Consequently, it is unlikely that comprehensive norms will be available. In this next section, we describe a wide range of the more commonly used attention tasks, but our coverage of them will necessarily be much briefer than that for the batteries. In many cases, detailed description has been withheld until Chapter 7, where developmental data derived from some of the most frequently used tasks are discussed in detail. In addition, there are a variety of other tasks that we will not include, due to paucity of description, usage, and data.

The majority of the tasks are assumed to have some degree of validity, and in addition, especially in the case of the tasks testing EF functions, evidence is often available from patients with brain damage or brain scanning that the task engages frontal lobe operations. Data on reliability are rarely available, and in many cases tasks designed for adults have been used with children without direct evidence of their suitability for this purpose. See Table 6.3 for a summary of core attentional tasks and their relevant information such as the age range over which the task is appropriate,

Table 6–3. Summary of Individual Tasks for Testing Attention

Task/Measure	Age Range	Administration Time and Method	Normative Data	Primary Usage	Strengths	Weaknesses
Tests of Selection						
Comprehensive Trail-Making Task Version A (Reynolds, 2002)	11+	5–12 min Paper and pencil	Yes	Clinical, Research	Good reliability; some data from atypically developing groups Does not require literacy; some data from neurodevelopmental groups; correlated with other versions	Speed measure may be affected by correction of errors
Color Trail Making Task (Williams et al., 1995)	6–16	5 min Paper and pencil	Some	Clinical, Research	–	–
Negative priming (colors) (Pritchard & Neumann, 2004)	5–12	15 min Cards	Some	Research	Clear effects with young children	Limited use to date; reliability unknown; no data from neurodevelopmental groups; no automated version
Tests of Maintenance						
Continuous Performance Tasks (kCPT) (Conners & MHS Staff, 2007)	4–5	7.5 min Computer	Yes	Clinical	ADHD and other clinical norms available; good reliability	–
Sustained Attention to Response (SART) (Robertson, Manly, Andrade, et al. 1997)	–	4.3 min Computer	Yes	Clinical, Research	Good validity	–

Executive Functions

Tower of London (Kaller, Rahm, Spreer, Mader, & Unterrainer, 2008; Senn, Espy, & Kaufmann 2004)	3+	Depends on number of trials given	No	Research	Variety of measures available	Validity uncertain
Self-Ordered Pointing Task (1) (Cragg & Nation, 2007) (2) (Ross, Hanouskova, Giarla, Calhoun, & Tucker, 2007)	5–11	10 min Computer (touch screen)	Yes	Research	Some data from ADHD and autistic groups	Reliability unknown
Comprehensive Trail-Making Task Version B (Reynolds, 2002)	11+	5–12 min Paper and pencil	Yes	Clinical, Research	Good reliability; some data from atypically developing groups Does not require literacy; some data from atypically developing groups; correlated with other versions	Speed measure may be affected by correction of errors
Contingency Naming Task (Mazzocco & Kover, 2007)	6–11	Cards	Yes	Research	–	Reliability unknown; limited data from neurodevelopmental groups (only Turner's syndrome)
Operation Span Mental Arithmetic (1) (Conlin, Gathercole, and Adams 2005)	7–9	Computer	Limited	Research	–	Reliability unknown
(2) (Hitch, Towse, and Hutton 2001)	7–9	Computer	Limited	Research	–	No data from neurodevelopmental groups

Note. This table is not an exhaustive list of all attention tasks, but rather a selected group of tests that have been used extensively with children. Other tasks not described here are mentioned in detail in Chapter 6 (DCCS, Day–Night, Antisaccade, Go–NoGo, Stroop, Eriksen Flanker).

whether norms are available, and the main strengths and weaknesses of each task.

Selective Attention

A number of ingenious tests for selective attention used with very young infants will be described in detail in Chapter 7 (the mobile conjugate reinforcement paradigm, novelty preference, and use of eye tracking methodology). We describe here two types of selective attention tasks: visual search tasks and negative priming tasks.

Visual Search Tasks

We have already described various visual search tasks, both the frequently used single-frame tasks and the continuous search tasks employed in the TEA-Ch and the WATT. Other variations exist, such as *cancellation tasks* (see Lezak, 2004, for descriptions of many variations of these) and the *Trail-Making Task* (Version A) (Reitan, 1958), in which a display of numbered disks is shown and a line has to be drawn from the disk numbered "1" in sequence up to number 20. This latter task incorporates an additional requirement over the TEA-Ch and WATT tasks in that a fixed order of locating targets is required, so in effect, there is only one target present at any time. It incorporates, however, a major weakness in that errors have to be corrected. The experimenter returns the pen to the last correct response, and a correct sequence has to be found. Hence, the performance measure (often time only) is highly dependent on the number of errors made and the speed of correcting them; the discussion of findings from the WATT demonstrates the unsatisfactory nature of such a scoring system.

Negative Priming Tasks

These tasks can be used to indicate the degree to which exclusion of a distracting input (for example, a red item) on one trial carries over and impairs speed of processing of the same item when it is the target on the next trial. Recent studies have successfully demonstrated negative priming effects across a variety of contexts including object and location identification, and specially developed tasks with materials that are suitable for children with word stimuli and pictures (e.g., Frings, Feix, Rothig, Bruser, & Junge, 2007; Pritchard & Neumann, 2004). See Chapter 7 for further details of negative priming tasks and developmental profiles.

NEURODEVELOPMENTAL DISORDERS

Studies using the WATT and TEA-Ch to investigate selective attention performance across different neurodevelopmental disorders have already been described in this chapter, and more detailed findings are also provided in Chapters 8 and 9. Other tasks of selection including cancellation tasks, trail-making tasks, and negative priming tasks have rarely been studied in the context of atypical development.

Divided Attention

Dual Tasks

Tasks that assess divided attention or multitasking, often referred to as "dual tasks," tend to be created anew by each researcher, with the result that we have neither extensive data sets based on a particular combination, nor a clear picture of the effects of different combinations, though we have discussed some of the important features that determine the effects of combining tasks. The NEPSY includes one visual attention task that requires searching for two types of target, and the TEA-Ch offers a combination of Skysearch and Score! (but scores on this combined task were correlated with tasks that test maintenance of attention rather than tasks that form an independent factor testing an independent ability). Miyake et al. (2000) combined maze tracing and generation of words beginning with a specified letter, but none of their factors provided a good prediction for performance in this particular dual task. Bull and Scerif (2001) employed a dual task for children, combining digit recall and a motor task of following a path, and scored the decrement compared with separate presentation of these tasks. Though their other measures of aspects of EF (inhibition, perseverative responses, and working memory span) predicted children's mathematical competence, dual-target performance was unrelated to the latter. Hence, at present no good example of a soundly based dual task is available that predicts other performance. Accordingly, the status of such tasks as measures of EF is uncertain, though they are often used and cited as typical examples of such function, as in the studies above, or of working memory capacity, in the tradition going back to Kahneman (1973).

NEURODEVELOPMENTAL DISORDERS

We know of no published studies that have used dual tasks to examine divided attention in different neurodevelopmental disorders.

Maintenance of Attention

A number of tests of vigilance, including some very well-designed tasks of distractibility, will be described in detail in Chapter 7. Here we focus on tasks that involve ability to respond to rare targets, either occurring in an otherwise monotonous situation, or occurring occasionally in a continuous stream of nontargets. We referred to these in Chapter 2 as low-input and high-input situations.

Low-Input Situations

There are many variations of vigilance tasks used with adults (watching radar screens, clocks, listening for bleeps, etc.), but this type of task, for obvious reasons, is rarely used with young children. The *Vigilance Task* from the WATT battery was designed specifically for children but lasts only some 4 minutes. Since the decline in performance identified in true vigilance tasks occurs in adults after some 20 minutes on task, no decline is expected over a 4-minute period, even in children, and performance is scored over the whole time. This task does measure some aspects of attention that are independent of the visual search tasks in the WATT, but there is no direct evidence to date that this is specifically related to the ability to maintain attention.

High-Input Situations

Methods of recording distractibility in very young children are described in detail in Chapter 7. With older children the most popular task to measure the ability to sustain attention to a continuous input, particularly in children with ADHD, has been some form of the Continuous Performance Task (CPT). This is a very widely used task and there are many versions of it (see Riccio, Waldrop, Reynolds, & Lowe, 2001, for a review of many of these). In most versions, a sequence of stimuli is shown (e.g., letters), and a response is required only when an *X* appears or, in a more difficult version, when an *X* appears following an *A*. In the AX version, as the latter is generally known, responses to *A* are taken to indicate impulsivity, and responses to *X* on its own or failures to respond to a target are both assumed to indicate poor attention, though there are no systematic data to back up these assumptions. However, the most widely used version of the task is that of Conners and MHS Staff (2007), where the task is to respond to all items *except* the *X*. A variety of measures are available. Reliability is claimed to be adequate, though detailed data on this are not provided. A version has now been produced specifically for children aged 4 to 5 years, the Kiddie CPT (K-CPT™ V.5; Conners 2007). This uses pictures rather than letters and lasts only 7 minutes. Note that that the demands of the CPT tasks also require *response inhibition* similar to the Go–NoGo tasks (reviewed below), and therefore, these can also be viewed as tests of inhibitory functioning. It is therefore often difficult to ascertain whether performance reflects sustained attention or inhibition.

In addition, results vary considerably with variations in type of stimulus, duration, rate of presentation, and rule for response, suggesting that a simple interpretation of findings from CPT tasks may be unjustified. Such tasks are commonly presented as tests of sustained attention, but frequently, no attempt is made to measure change in performance over time, and where this has been done, this measure has not proved a consistent indicator of poor attention. More often, poor attention is associated with poor performance overall, implying that it is the overall (possibly inhibitory) demands of the task rather than the effects of time on task that are important (see Corkum & Siegel, 1993, for a critique, and Koelega, 1995, for a reply). As with other relatively complex and lengthy tasks, it seems likely, therefore, that it is the degree of EF involvement that is crucial in determining the ability of the task to reveal attentional weaknesses, rather than simply the duration of the task. Even so, one might expect demands on EF to increase with task duration, leading to a faster decline in performance in groups with impaired EF. The failure to obtain such a difference consistently between atypical and typically developing groups remains unexplained.

The Sustained Attention to Response Task (SART) (Robertson, Manly, Andrade, Baddeley, & Yiend, 1997) was originally used to assess attention impairment in patients with traumatic brain injury (TBI). As in the Conners CPT described above, participants have to inhibit or withhold their response to rare targets, the digit "3" among other digits. Robertson et al. argue that this arrangement is preferable for testing ability to maintain control of an automated response, compared with versions where regular responses are required and may become automated. The authors found that the number of correctly withheld responses to the rare targets was related to self-report of cognitive failures in everyday life, whereas no such relation was apparent for a CPT task that required responses only to targets. False alarms to the target stimuli tended to be preceded by a speeding-up of responses to stimuli, implying a weakening of executive control and increasingly automated responses. The authors argue that such occasional lapses of control may be more important in vigilance performance and a greater source of individual differences than the well-researched decline in performance over time. Robertson et al. also report that performance on the SART correlated significantly with performance on other tasks of sustained attention rather than measures of inhibition. A variety of increasingly sophisticated measures have been developed for this task (Johnson, Robertson, et al., 2007). In addition to the number of false alarms to target stimuli, missed responses to nontargets and short- and long-term variability in response speed over time were shown to provide additional information on the mechanisms involved.

Other tasks of vigilance include the Wilding Vigilan task (described above) and the Early Childhood Vigilance Task (ECVT; Goldman, Shapiro, & Nelson, 2004) and are extensively reviewed from a developmental perspective in Chapter 7.

NEURODEVELOPMENTAL DISORDERS

There is a wealth of data comparing performance on the CPT in its many variants, in children with ADHD (e.g., Epstein et al., 2006; Huang-Pollock, Nigg, & Halperin, 2006; Kieling, Roman, Doyle, Hutz, &

Rohde, 2006; Inoue, Inagaki, Gunji, Furishima, & Kaga, 2008). These studies are described in detail in Chapter 8. In contrast, comparatively fewer studies have used the CPT to assess sustained performance in other neurodevelopmental disorders, but two recent exceptions include a study of children and adolescents with 22q11 DS (Lewandowski, Shashi, Berry, Kwapil, 2007), which reported significant deficits in performance on the CPT that were over and above what would be expected from their general deficit in IQ. The second study was the first attempt, to date, to use the CPT in boys with fragile X syndrome (Sullivan et al., 2007). An important contribution of this study is that it demonstrated a decline in sustained performance over task time compared to typical controls, suggesting difficulties in arousal levels that would be commensurate with the biological mechanisms that underlie fragile X. The concern is that this pattern emerged on only a subsample of the initial sample because, due to task complexity, 39% were unable to complete the task because they either found it too difficult or refused to complete it. It is therefore difficult to know whether the findings represent disorder-specific difficulties in sustained attention or a pattern typical of only the upper end of the ability distribution in this syndrome.

Using the SART, Johnson, Robertson, et al. (2007) reported a dissociation in performance between children with ADHD and children with autism, with performance in the former group characterized by errors of commission (responding to the target) and omission (failure to respond to nontargets), and a waning performance over the course of the task suggestive of sustained attention deficits. In contrast, performance in children with autism was characterized by errors of commission only, suggesting a core deficit in response inhibition. In a preliminary study of children with Down syndrome, a novel auditory and visual version of the SART revealed dissociation in performance with less impairment in the visual compared to the auditory version (Trezise, Gray, & Sheppard, 2008). Furthermore, developmental level predicted outcome on the visual SART but not on the auditory SART.

The Vigilan task has been used successfully to discriminate vigilance performance in children with fragile X syndrome and children with Down syndrome compared to typically developing control children (Munir et al., 2000b). Fragile X children displayed significantly more commission errors than the Down syndrome children, who made more commission errors than a good attention control group but not more than a poor attention control group. The fragile X, Down syndrome, and poor attention groups did not differ in speed of response and were all somewhat slower than the good attention group.

Executive or Control Functions

We now look at tasks designed to test different executive functions. Also, as noted in Chapter 2, in virtually all cases where control systems are involved in organizing nonautomatic behavior, instructions have to be stored in working memory in order to maintain the rules for action. Working memory capacity is conventionally measured using one of the working memory span tasks described under "Updating" below.

Planning

Many examples of failure in organizing sequences of everyday behavior have been observed following brain damage in adults. Schwartz, Reed, Montgomery, Palmer and Meyer (1991) analyzed sequences of common actions such as making coffee or cleaning teeth and noted frequent failure of correct sequencing. Shallice and Burgess (1991) have employed a task of carrying out a sequence of demands with various restrictions; one version was to buy a list of items, entering only appropriate shops within a specified area. However, given the dearth of imaginative tasks to test these functions in children, researchers have commonly fallen back on popular tasks designed for adults.

TOWER OF HANOI AND TOWER OF LONDON

In the Tower of Hanoi (TOH) task, three rings of different sizes are displayed on the left of three pegs; the rings are arranged in order of size with the largest at the bottom and the smallest at the top. The task is to move all three to the right-hand peg, moving one at a time and never placing a larger ring on top of a smaller one. The task can be increased in difficulty by adding more rings. So differences between easy and difficult conditions should, in theory, enable the effect of stronger demands on planning ability to be evaluated; in practice, overall performance is what is assessed most often. The time to make the first move is assumed to measure planning time. Mean time and number of moves required are recorded. This task is assumed to test planning ability, which presumably depends on, among other things, working memory capacity, including ability to visualize future arrangements of the rings.

The Tower of London task is similar in some respects to the TOH, with three colored balls, all the same size, so that there is no restriction on the order in which they can be stacked on the pegs. In the most commonly used version, however, the pegs vary in height and hold (from left to right) three balls, two balls, or one ball, thus restricting the possible moves. Tasks of varying difficulty can be devised, with difficulty measurable in terms of the minimum number of moves required. This enables the effect of specific task demands to be investigated by comparing versions that make different demands.

There is considerable but not conclusive evidence suggesting that these two tasks are not equivalent. In a careful analysis of data from patients with prefrontal lesions, Goel and Grafman (1995) pointed out the many differences in the task restrictions and demands. Bull, Espy, and Senn (2004) also suggest that TOH and TOL may differ in key aspects of their demands. They employed simpler tasks to measure inhibition and shifting of attention (a variation of the Go–NoGo task that is examined later) and short-term memory in 3- to 6-year-olds and found that inhibition was predictive of the number of correctly solved problems on the TOL but not the TOH, and the reverse was true for shifting ability. There was a weak correlation of performance with age. In a particularly striking finding using a very different index of inhibitory ability, Asato, Sweeney, and Luna (2006) found significant relations between TOL performance and ability to inhibit prepotent responses in an antisaccade task (this task is discussed later and requires making eye movements in the opposite direction to that indicated by a cue).

This evidence for relations between inhibition and TOL performance calls into question the assumption that this is primarily a task that tests planning ability. Goel and Grafman (1995) point out that "planning" is an imprecise theoretical term and point to the importance of what they call the "counterintuitive backward move," in which a move is required that makes the current display less similar to the goal rather than more similar to it. Such a move is required, for example, when the ball has to be placed, not on its final peg where it would block a ball that needs to be placed below it, but on a different peg. Miyake et al. (2000) examined the first occurrence of a situation in the TOH task where disks had to be placed in a counterintuitive way, requiring inhibition of the prepotent move, and found that this was, indeed, a major source of error. Kaller et al. (2008) found that such moves ("intermediate steps" in their terminology) caused a big problem for 4-year-olds but not for 5-year-olds. Goel and Grafman argue that such moves are similar to the inhibition of prepotent responses that is required in tasks such as the Wisconsin Card Sorting Task, the Stroop Task (both discussed in detail later), and the antisaccade task, and hence, that TOH and TOL should be classified primarily as tests of inhibition rather than planning.

Neurodevelopmental Disorders   There is an extensive literature that has used the TOL, and its variants, to assess EF in children with ADHD (e.g., Coghill, Rhodes, & Matthews, 2007; Scheres et al., 2004) and in children with autism, with and without comorbid ADHD symptoms (e.g., Geurts, Verté, Oosterlaan, Roeyers, & Sergeant, 2004; Happé et al., 2006; Sinzig et al., 2008). The task appears to be extremely sensitive in discriminating impairments in planning ability in atypical versus typical children but less sensitive in discriminating impairments in ADHD versus autism, with studies displaying mixed findings. Although relatively few studies have examined planning abilities outside of ADHD and autism, one recent study of children with fragile X syndrome found poorer performance on the Tower task compared to typical controls matched on developmental level (Hooper et al., 2008).

Switching

Ability to switch attention or response is obviously involved in organizing many sequences of behavior where one component of the task has to be terminated and another initiated. The most widely used task to test this ability has been the Wisconsin Card Sorting Task (WCST; Heaton, 1981), which has been referred to several times already. This task presents a series of cards varying in shape, color, and number of elements and is also available in computerized format.

In the manual version of the WCST, the participant is shown four cards varying across three features (color, shape, and number of items). There are four values of each feature, and an example of each of these appears in the four cards displayed. Taking further cards from a pack, the participant is required to place them on top of one of the initially displayed cards and to discover from the feedback ("right" or "wrong") the correct sorting rule. So if the rule that the experimenter "has in mind" is to sort by color, a blue card goes on a blue card, irrespective of the values of the other features (their shape or number). However, once a sequence of 10 correct responses occurs, the experimenter changes the rule (to sorting by shape, for example) without giving any indication of the change, and the new rule has to be discovered. A variety of measures are calculated, including the number of solutions achieved and the number of correctly placed cards, and perseverative errors following a change of rule (that is, continuing to sort by the old rule despite negative feedback). It has been claimed that the latter measure is specifically diagnostic of attentional rigidity, supposedly typical of frontal lobe damage (originally proposed by Milner, 1963, 1964).

Is the WCST appropriate for use with children? Developmental studies have consistently reported that the number of detected categories increases with age until about the age of 10 years and is typically associated with a decrease in the number of perseveration errors (see Somsen, 2007, for a recent review of developmental studies). However, despite numerous studies using the WCST in children, a number of questions still remain. The most salient are: To what extent can a child really understand a situation without instructions? Can they construct any strategy other than random choice? Can they carry out the necessary deductions when previously correct responses become incorrect? And, most importantly from our perspective, does the WCST measure one ability or several distinct abilities at different ages? It seems unlikely, however, that the task, as it should be given with no additional or modified instructions, is feasible in young children.

DIMENSIONAL CHANGE CARD SORTING TASK (DCCS)

An alternative to the WCST, and with a proven track record in assessing attentional switching in children as young as 3 years, is the DCCS developed by Zelazo and colleagues (e.g., Muller, Dick, Gela, Overton, & Zelazo, 2006; Zelazo, 2006). For this task, children are required to sort pairs of cards out by one set of rules (e.g., color) and then switch to another set of rules that are incompatible to the first set (e.g., shape). There is a clear developmental progression on the DCCS with advancing age. The findings have been consistently replicated and expanded upon, and are presented in full in Chapter 7.

Neurodevelopmental Disorders The WCST has been used extensively in studies of ADHD. Romine et al. (2004) carried out a comprehensive review of its use in this population and conclude that several measures (e.g., perseverative errors and total errors) discriminate well between ADHD and typical control groups. However Geurts, Verté, Oosterlaan, Roeyers, and Sergeant (2005) compared the performance of children with ADHD combined-type and inattentive-type with typically developing controls and found no group differences on the number of perseverative errors. This suggests that the WCST is less sensitive in discriminating between ADHD subtypes in terms of cognitive flexibility. A recent study comparing scores on the WCST in children with autism and children with ADHD found comparable levels of performance across almost all variables (Tsuchiya, Oki, Yahara, & Fujieda, 2005). The only variable to differentiate the groups was the number of perseverative errors, which were more frequent in the ADHD group.

As indicated earlier, Wilding et al. (2001) and Zhan et al. (2010) employed a computerized child-friendly version of the WCST, the Wilding Monster Sorting Task, in which monsters had to be assigned to teams for the monster sports on the basis of the same features as in the WCST. However, even this version left many children, especially in younger groups, baffled as to the required purpose, and the task was completely unsuitable for use with children with significant cognitive delay, for example, fragile X syndrome or Down syndrome, who simply gave up responding after a few responses.

The DCCS, described above, has shown more promising results in children with autism (Zelazo, Jacques, Burack, & Frye, 2002) and Down syndrome (Zelazo, Burack, Benedetto, & Frye, 1996).

SELF-ORDERED POINTING TASK (SOPT)

Petrides and Milner (1982) have employed the "six boxes" task in which six pictures are shown on each of a series of cards, with the positions

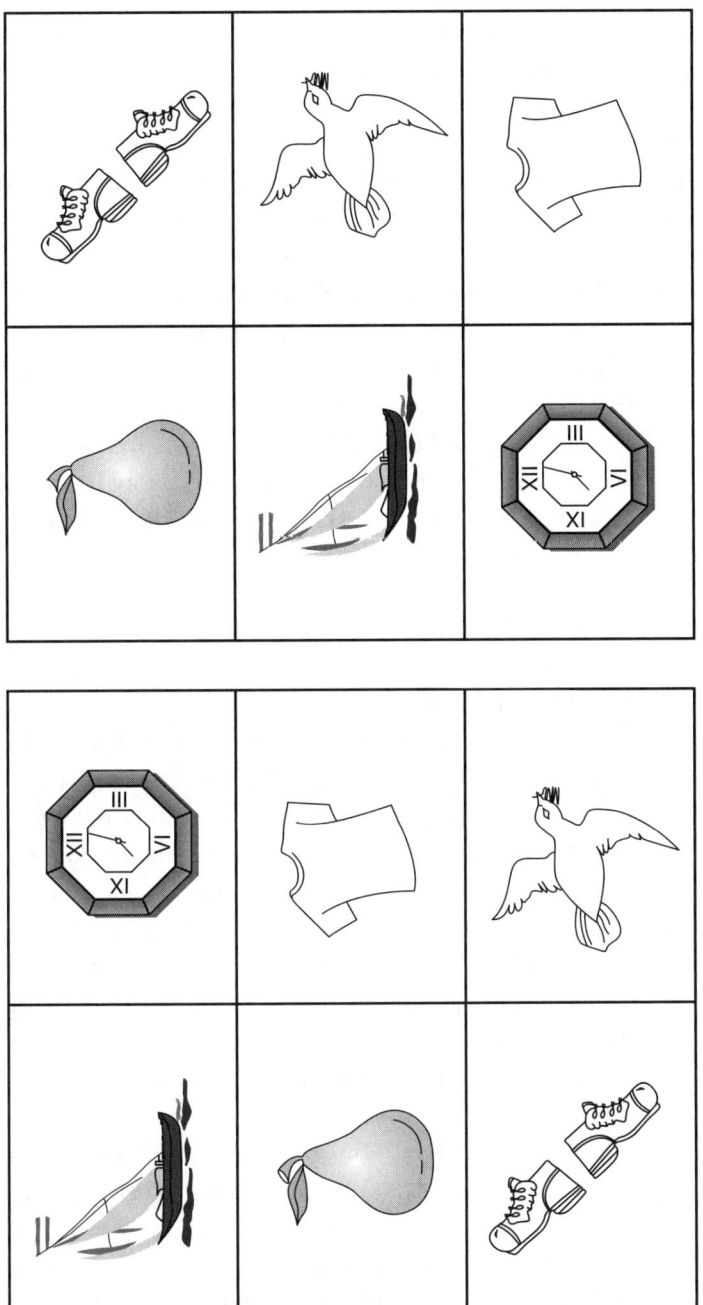

Figure 6-4. The six boxes ordered pointing task.
From Petrides and Milner (1982). Redrawn by permission from *Neuropsychologia 20, 249–262.*

of the pictures on the card varied randomly from card to card. See Figure 6.4. The task is to select a picture on each card, without duplicating the choice. In some versions, repetition on previously chosen locations, rather than pictures, has to be avoided (as in the CANTAB Spatial Working Memory Task). Frontal patients perform poorly, but this could be due simply to failure to recall earlier responses. Petrides, Alivisatos, Evans, and Meyer (1993) used an eight-box version with PET scanning and established that the mid-dorsolateral frontal cortex was involved (see also Diamond, Briand, Fossella, & Gehlbach, 2004). The task possibly includes a large planning element and should not be assumed to be a measure of any specific function. Archibald and Kerns (1999) provide norms for children aged 7 to 12 years and obtained reasonable reliability (r = .76). Although they regarded the SOPT as primarily a test of working memory, it did correlate significantly with a test of ability to switch response rules and with Stroop-type tasks, suggesting that it engages wider functions. In recent adult studies (for an example, see Ross, Hanouskova, Giarla, Calhoun, & Tucker, 2007), reliability on the SOPT was moderate. Ross et al. concluded that performance appears to be more closely aligned with working memory than with EF. See also Bryan, Calvaresi, and Hughes (2002).

There are relatively few development studies that have used the SOPT with the recent exception of Cragg and Nation (2007), who used two versions, one using pictures of familiar objects and the other hard-to-verbalize abstract designs. This study was one of the first to provide normative data against which to compare atypical performance. Performance on this task showed improvement with age from 5 to 10 years but did not reach adult levels by age 11 years. Abstract designs proved more difficult than pictures, and this difference increased with larger displays. Performance was better on the first run than on later ones, presumably due to interference from earlier displays.

Neurodevelopmental Disorders   Given the link with EF, SOPT has surprisingly been underused, with only a few studies to date, namely, children with ADHD (Geurts et al., 2004) and children with autism (Joseph, Steele, Meyer, & Tager-Flusberg, 2005). The former study teased apart ADHD subtypes and compared performances with typical controls and found no group differences. The latter study found a group difference but only on the verbal version of the SOPT. In the nonverbal version, the autism group performed as well as matched, nonautistic control children. In particular, data analysis revealed no evidence of increased perseverative responding in the autism group, suggesting there was no marked impairment of either memory or switching ability.

Shifting Stimulus Focus

TRAIL-MAKING TASK

The Trail-Making Task (Version A) (Reitan, 1958) has been discussed briefly above and comprises version A and B. The task requires switching of attention between the two types of cue, in addition to the demands of following a sequential rule as in Version A, which can be used as a baseline control condition without the switching requirement. However, Gaudino, Geisler, and Squires (1995) pointed out that there are multiple differences between the two tasks, any one of which may be critical in reducing performance on Task B. The most commonly used measures are time to completion, number of errors, or the ratio of completion times for Task B versus Task A. We reiterate our doubts about the validity of the time measure when errors have to be retraced and therefore contribute heavily to time taken.

Reynolds (2002) elaborated the original Trail-Making Task to produce the Comprehensive Trail-Making Task (CTMT), with five conditions; the first and fifth conditions match the original Trail-Making A and B, the second and third conditions have distracting disks added to Task A (either empty or filled), while the fourth condition mixes digits and words in the disks. The Delis-Kaplan Executive Function System (Delis, Kaplan, & Kramer, 2001) also includes a Trail-Making Task with some additional conditions that are explained below.

In a recent review of Reynolds' CTMT, Moses (2004) concluded that it has high reliability over a wide age range from 11 to 74 years but that it is related to other measures of perceptual skills and motor speed. There is, therefore, some uncertainty about whether this test specifically involves attentional switching abilities. The Delis–Kaplan version, which provides norms for ages 8 to 89, attempts to remedy these problems by including tasks that test scanning and motor speed, so that individual differences on these abilities can be controlled for, and this version also scores different types of errors.

Neurodevelopmental Disorders   Surprisingly few studies have examined the Trail-Making Task in neurodevelopmental disorders. A recent study of children with ADHD, divided into subtypes (combined-type and inattentive-type), reports significant impairments on the TMT compared to typically developing controls, but also found that children with ADHD inattentive-type had more off-target errors than children with ADHD combined-type (Chiang & Gau, 2008).

SIMPLER SWITCHING TASKS FOR YOUNG CHILDREN

A variety of simpler tasks have been devised to test different types of switching, many of them specifically for preschool children. Some of these require reversal of stimulus–response pairings established before the experiment during day-to-day experience (e.g., respond "Day" to a picture of night and vice versa). These will be considered in the coming section titled "Opposite Response Tasks" (see Diamond, Kirkham, & Amso, 2002; Gerstadt, Hong, & Diamond, 1994, and Chapter 7 for a detailed explanation of this task and its variants). Other tasks require a frequent response to be withheld or changed when a specified cue is present (see Go–NoGo tasks under "Inhibition" below and Chapter 7). The most frequently used tasks requiring response switching, however, require pairings learned during the experiment to be changed in response to a signal, or sometimes without warning, when the previously success-ful response ceases to be correct (as in the WCST).

The simplest versions of such tasks involve a two-choice selection task and require choice of the previously unrewarded alternative. For children, the *A not B* task (Diamond, 1985) is such a task. A reward is put in one of two containers in full view of the child, followed by an enforced delay before choosing. Switching of reward location occurs after two correct choices in the original version of this task. In related and more difficult versions, the reward may be placed without the child seeing, or it may alternate between the two sides (Delayed Alternation), or it may

---

Box 6.5 Contingency Naming Task

Two examples of stimuli from the Contingency Naming Task (stimuli are presented in three rows of nine items each and instructions differ in different versions):

(a) Baseline requires either the outer shape or outer color to be named.

(b) One-attribute contingency requires color of outer shape to be given when inner and outer shapes match; otherwise, outer shape has to be named.

(c) Two-attribute contingency requires responses in (b) to be reversed when the arrow is present.

Thus, correct responses to the left stimulus below are (a) square/ green, (b) square, and (c) square, while the right stimulus could occur in (c) only and requires the response magenta. Stimulus boxes are shown in color insert of book.

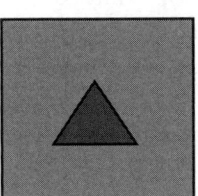 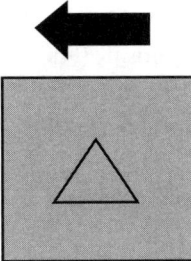

---

be switched after four correct choices (Spatial Reversal) (Senn, Espy, & Kaufman, 2004). However, this last task seems to test some additional functions, as it differs in its pattern of correlations with other tasks and its developmental trajectory.

Obviously there is considerable scope for exploring the effects of different variations of response switching in order to isolate critical features. Ability to switch responses continuously to a stimulus in accordance with cues provided by other aspects of the display is tested by the Taylor Contingency Naming Task (Smidts, Jacobs, & Anderson, 2004). See Box 6.5 for a full explanation.

A recent developmental study by Mazzocco and Kover (2007) is one of the few studies to chart performance on the CNT in a group of typically developing children longitudinally across three time points, 6–7 years, 8–9 years, and 10–11 years. They measured trials to criterion, time to complete, the number of self-corrections and errors and devised an efficiency index from the time and error measures, but concluded that there was sufficient overlap between these scores to suggest that they did not measure distinct functions. There were main effects of age with improvement on all scores. The improvement was faster on the more difficult versions of the task because, at the younger age, children were more affected by increased task difficulty than when they were older.

Neurodevelopmental Disorders   Few studies have used the CNT in neurodevelopmental disorders, but its feasibility in assessing switching in girls with fragile X syndrome was assessed by Kirk, Mazzocco, and Kover (2005). A range of switching tasks has been used to assess attentional switching in different neurodevelopmental disorders, most notably the Day-Night Task in children with ADHD (e.g., Berwid et al., 2005), autism (Russell, Jarrold, & Hood, 1999), Williams syndrome (Atkinson et al., 2003), and most recently, in fragile X syndrome (Hooper et al., 2008). These studies will be discussed in Chapters 8 and 9.

Inhibition

A variety of tasks have been used to study ability to inhibit existing strong responses that depend on past experience or instructions. These will only be described briefly here as studies that have employed them with children are discussed in detail in Chapter 7.

ANTISACCADE TASK

Participants fixate a central point, and a signal is delivered to left or right (e.g., Guitton, Buchtel, & Douglas, 1985; Miyake et al., 2000). Either the natural saccade toward the signal is required or an antisaccade to the opposite side. Time and errors are recorded. Clearly this task is not suitable with young children, but Johnson (1994, 1995) adapted it for use with infants by presenting a relatively uninteresting stimulus on one side just before a colorful and dynamic stimulus on the other side.

Developmental studies using saccadic and antisaccadic paradigms are discussed in detail in Chapter 7 and demonstrate that with ingenuity such tasks can be adapted for use with very young participants.

Neurodevelopmental Disorders   Studies using the saccadic and antisaccadic paradigms in differing neurodevelopmental disorders have produced inconsistent results, but there is evidence for the feasibility of these tasks in studying children with ADHD (e.g., Hanisch, Radach, Holtkamp, Herpertz-Dahlmann, & Konrad, 2006; Karatekin, 2006), autism (van der Geest, Kemner, Camfferman, Verbaten, & van Engeland, 2001), and fragile X syndrome (Scerif et al., 2005). These studies are reviewed in Chapters 8 and 9.

GO—NOGO TASKS

These tasks require response to be made to one stimulus, and either a different response or no response to an (usually infrequent) alternative stimulus. They share similarities to stopping tasks (see below), in which the target stimulus for a response is countermanded immediately after its occurrence so that no response should be made.

Carlson (2005) describes two specific Go—NoGo tasks for preschool children (see detailed descriptions in Chapter 7), in which a bear and a dragon puppet issue orders, but only those coming from one of the two have to be executed. An alternative is the well-known "Simon says . . ." game. A rather different task that can be seen as a variant on the Go—NoGo task requires the child to refrain from touching or opening an attractive gift until the experimenter gives the signal. Ability to inhibit responding while the experimenter's attention is apparently directed elsewhere is measured by time before touching the gift. Carlson (2005) gives a variety of examples of such tasks.

Simpson and Riggs (2006) distinguish "habitual response tasks," such as obeying the bear and not the dragon, and "button press tasks," which require responses to one arbitrary stimulus and no response to another, and note that it is not certain that the latter strongly involve inhibition. The two responses are often equally frequent, making such tasks relatively straightforward choice—response tasks, and results do not correlate highly with other inhibitory tasks. They examined the effects of varying the frequency of Go trials and of time pressure in children aged 3 to 4 years. Effects of variation in frequency were small, but time pressure had rather complex effects. Stimuli were displayed for 1, 2, or 3 seconds, with an interstimulus interval of 1.5 seconds. With a 1-second display time, accuracy was low on Go trials and high on NoGo trials (i.e., misses were frequent, presumably due to slow responding); this pattern reversed with a 2-second display time (false alarms became frequent); accuracy was good on both types of trial with a 3-second display. Accuracy at the two longer intervals was significantly correlated with that on the Day–Night task (a test of ability to inhibit well-learned responses described below). The authors concluded that the 1-sec display time was too fast for this age group to cope with, and that the 2-sec display time was optimal in that it both produced low accuracy on NoGo trials and a correlation with the Day–Night task, which is also assumed to test inhibition. The optimum display time would presumably be lower at higher ages.

From a *developmental perspective*, the studies conducted by Carlson (2005) and Simpson and Riggs (2006) are described in detail in Chapter 7 and so will not be expanded upon here.

Neurodevelopmental Disorders   The best documented studies of attention of impairments in recent years, across a range of neuro-developmental disorders, involve Go–NoGo tasks. Although the main focus has been in children with ADHD (e.g., Derefinko et al., 2008; Wodka et al., 2007), Go–NoGo tasks have been used successfully in other disorders including children with autism, with and without comorbid ADHD (e.g., Christ, Holt, White, & Green, 2007; Sinzig et al., 2008) and fragile X syndrome (Sullivan et al., 2007). These studies and others are reviewed in Chapters 7 and 8.

STOP TASKS

In a typical stopping task, a response is required to a stimulus such as a flash of light, but on a proportion of trials a tone is sounded to cancel the response. The number of responses successfully stopped can be measured for different delays of the tone following the visual stimulus and there are many variants (e.g., Miyake et al., 2000; Nigg, 1999; Stevens, Quittner, Zuckerman, & Moore, 2002). Miyake et al. (2000) used a version of this in which participants were required to categorize words as animal or nonanimal. After 48 trials, all of which required a response, stop trials were introduced in which a tone sounded to indicate that the response should be withheld. The timing of the tone was individually calibrated to each participant's mean response time to make it difficult to withhold the response. The dependent measure was the number of false alarms made on the stop trials. Performance was correlated with several other tasks generally believed to test inhibition (antisaccade, Stroop, Eriksen flanker task, ignoring distracters), all of which loaded on a single factor.

Stop tasks have been used mainly with adult participants and their usefulness with children is uncertain. The few developmental studies to date are reviewed in Chapter 7, but here we will highlight an interesting study in which Scheres et al. (2004) used, along with several other tasks, a novel and potentially promising form of stop task in which children followed a shape on the computer screen with the computer mouse and were given a stop signal at random intervals. Time to stop was recorded. In this study, however, once the effects of IQ were removed, there were no differences between and ADHD and a typical control group in either the novel or a conventional stop task.

Neurodevelopmental Disorders   Stop tasks have been employed most frequently to assess inhibitory control in children with ADHD. See Lijffijt, Kenemans, Verbaten, and van Engeland (2005) for a meta-analytic review of the Stop tasks in ADHD and also studies by Alderson, Rapport, Sarver, and Kofler (2008) and Morein-Zamir, Hommersen, Johnston, and Kingstone (2008) for excellent examples of the importance of teasing apart the cognitive demands of the stop task in children with ADHD. To date, we can find no published studies that have used the stop task in our other five neurodevelopmental disorders.

STROOP TASK

In the classic form of this task, color words are presented written in ink of a competing color. The ink color has to be named, and the effect of the competing word name is assessed by comparing this condition with a control condition that requires naming of simple color patches. Older versions measure the time to sort decks of cards with designs exemplifying one of the conditions, but more recently, computerized presentation permits variation of the conditions within a run of trials and measurement of responses to individual displays. Obviously, the test in this form is often unsuitable for use with children since it depends on the overlearned skill of reading words being dominant over naming of ink colors. However, other equivalent tasks have been devised for children, for example, using arrows in different spatial positions that require spatial responses, or animal bodies with inappropriate heads (see Chapter 7 for a critique).

Berger and Posner (2000) also describe a version for children, in which two houses are shown on the left and right at the bottom of the computer screen, each including a picture. Then a picture is presented on one side at the top of the screen, which either matches the one in the house on the same side or on the opposite side. The picture has to be put in its right home by touching the screen. The task consisted of 4 practice trials and 32 test trials and lasted about 10 minutes. Five-year-olds achieved 100% accuracy and showed a large and significant effect of spatial compatibility.

Neurodevelopmental Disorders   The Stroop and its variants have been studied extensively in ADHD (see Van Mourik, Oosterlaan, and Sergeant, 2005). A full discussion of their conclusions is provided in Chapter 7, but for present purposes we note that they found no consistent evidence for greater interference in ADHD groups on the standard Stroop color-word task using word cards in black to measure reading speed, colored Xs or colored blocks to measure speed of color naming, and color-word cards with color words printed in inks of a conflicting color to measure interference. However, the authors did suggest that computerized versions measuring time on individual responses and other tasks measuring inhibition of interference did demonstrate a weakness in ADHD groups. Hence, the standard card form of the Stroop was not, in their view, a valid measure of inhibition. Goldberg et al. (2005), in a study comparing children with autism and children with ADHD on the Stroop task, failed to find any significant differences between the two groups. The authors suggest that the "verbal" component of the Stroop was not sufficiently sensitive to elicit disorder-specific differences. A recent study using the Auditory Stroop task also failed to elicit a specific interference deficit in children with ADHD compared to typically developing controls (Veugelers, Post-Uiterweer, Sergeant, & Oosterlaan, 2009). Together, these findings warrant caution when attempting to use the Stroop tasks as a clinical tool to identify ADHD phenotypic profiles.

OTHER TASKS REQUIRING THE EXCLUSION OF DISTRACTION

A range of tasks designed to test ability to exclude distraction have occasionally been used. The ANT incorporates such a task, and the Stroop was strongly related to other tasks in the study of Miyake et al. (2000) that required distraction to be ignored, requiring responses to letters, words, or pictures with distracting elements of the same type present (see also Friedman and Miyake, 2004), but there has been no consistent use of these tasks especially with children.

The Eriksen flanker task (Eriksen & Eriksen, 1974) is a more sophisticated version of this type of task. It is a two-choice decision task with responses to a central target letter flanked by two other letters that are either associated with the same response as the target or with a different response. The modified version of the ANT by Rueda et al. (2004) is the best available version of this for children. See our description under the ANT battery above and Chapter 7 for a description of the developmental studies using this task.

Opposite Response Tasks

A number of ways of testing ability to reverse a strong stimulus response association have been devised. They require not merely inhibition of the strongest response to a stimulus in favor of a weaker response, but use of the inhibited response for the opposed stimulus in each case in order to maximize the conflict. Carlson (2005) describes a selection of such tasks.

THE DAY-NIGHT TASK

The Day–Night Task (Gerstadt, Hong, & Diamond, 1994) requires the child to respond to pictures of the sun and moon appropriately with "day" and "night," or inappropriately with the response opposite to the well-learned one. The effect of inhibiting the dominant response is measured by the increase in time required in the second condition, and the number of errors is an alternative measure (hence, the task incorporates a baseline control condition). In this task, as distinct from the Stroop task, the competition that has to be inhibited does not arise from irrelevant features of the stimulus, but from the previously learned association.

Simpson and Riggs (2005a, 2005b) investigated the task in detail in order to determine whether problems in remembering the rule, rather than inhibition difficulty, might be the source of slower responding in the reversed response version, and whether direct naming of designs (black vs. white) still produced the effect. They concluded that inhibition was the main factor and direct naming produced the same result. Berger, Jones, Rothbart, and Posner (2000) describe an opposites task in which pictures of a cat and a dog are shown on the computer screen and a meow or bark is sounded. Children have to touch the appropriate animal or, in the opposites condition, the inappropriate one.

Several developmental studies using the Day–Night Task are described in Chapter 7. Using errors as the dependent measure, Simpson and Riggs (2005a) found improvement from age 3½ to 5 years, at which point performance approached ceiling. However, response times continued to decrease up to 11 years of age, and thereafter there was no difference between normal and opposite naming responses.

Neurodevelopmental Disorders   The Day–Night Task has not been used extensively in the neurodevelopmental literature, but there are an emerging number of studies that highlight its feasibility in children with fragile X syndrome (Hooper et al., 2008). A recent study by Berwid et al. (2005) is especially informative because in their sample of children "at risk" of ADHD, aged between 3 and 6 years, they report age-related changes in performance such that performance improved with age, and more vulnerability to impairment compared to children rated as "low risk" of ADHD.

Updating

OPERATION SPAN TASK

Several updating tasks have been devised to measure working memory capacity, one of the most popular being the Operation Span Task (Engle Tuholski, Laughlin, & Conway, 1999), which requires a number of sentences (or in some cases, simple sums) to be judged as true or false, plus single words. At the end of the sequence of sentences or sums, the words have to be recalled. Task difficulty increases with the number of sentences presented before recall is required. Thus, the set of words in memory has to be updated each time that a new sentence is presented. Conlin, Gathercole, and Adams (2005) explored a number of variations on this task with children. Miyake et al. (2000) found that their Updating factor provided the best predictor of performance, but there was some uncertainty about the analysis, and the result was less convincing than those that they achieved with several other EF tasks, as described in Chapter 2.

There has been some debate over exactly what this task measures. While Towse and colleagues (Towse & Hitch, 1995; Towse, Hitch, & Hutton, 1998; Towse, Hitch, & Hutton, 2000) have argued that the primary determinant of performance is the time that items have to be held in the memory store while carrying out other processing, rather than the capacity of the central executive system. Barrouillet, Bernadin, and Camos (2004) present a convincing case for combining this viewpoint with the operations of the general-purpose executive system in their time-based, resource-sharing model. In this model, the time available for updating and rehearsing material in the memory stores is constrained by the processing load imposed on the executive system.

Running memory/response tasks present a stream of items and require response to targets defined by a relative rule within the sequence. For example, Miyake et al. (2000) presented 15 words drawn from several categories and afterward required recall of the last member of each category. In another task they presented a series of tones of different pitches and required a response to the fourth tone of each pitch. In a third task, they presented a series of letters and then required recall of the last four letters presented.

Developmental studies employing the span tasks have been carried out by Towse et al. (1998) and Hitch et al. (2001), who used three different versions. In Counting Span, children aged 6 to 10 had to count items on cards then recall the totals over an increasing number of cards; performance improved up to 9 years, with little further change thereafter. They used a (presumably) more difficult operation span task to test children aged 8 to 11; this required simple addition and subtraction plus recall of totals or a reading span task that required the final word to be supplied in a sentence, followed by recall of the words. Findings point to continuing improvement over this age range.

Neurodevelopmental Disorders   We have not located uses of the span tasks to study atypical development, probably because they prove too difficult to administer in such cases. However, it is not difficult to envisage simpler versions of these tasks that would prove feasible.

Conclusions

We have now reviewed a wide variety of batteries and tasks that have been designed to test one or more of the three aspects of attention that we identified in Chapter 2 (selective attention, sustained attention, and attentional control). Considering first the task batteries, we have attempted to provide reasonable coverage of these attentional functions, and reliability of the measures is generally (but not universally) adequate. Most of the batteries have been used in one or more studies of ADHD groups, but, apart from the WATT, minimal data are available from studies involving groups with other mainly genetic impairments that may include attentional weaknesses. Consequently, we often do not know whether the task is suitable for such groups and norms are rarely available. It is critical in such cases to be able to disentangle specifically attentional weakness from poor performance due to other factors, such as failure to understand the task demands or slow motor responding. If the task is outside a child's competence, floor effects will be obtained, providing no useful information on the nature of the impairment within the group. Careful consideration of the age-appropriateness of a given task or battery of tasks is critical in developmental studies incorporating atypically developing populations. As clearly demonstrated throughout this chapter, different tasks have varying age cutoffs, and inappropriate usage can produce ceiling/floor effects that fail to tap disorder-specific profiles or signatures. We will be looking in detail in Chapter 7 at a number of tasks that have been developed specifically for use with young (in some cases very young) children that may be more adaptable to the needs we have identified here.

CHAPTER SUMMARY AND KEY FINDINGS

- Careful consideration of the task demands is needed when choosing a suitable paradigm to tap attention profiles in neurodevelopmental disorders. Be aware of the possibility of performance at ceiling or floor levels.
- Many so-called "attention" paradigms involve complex EF processing that makes it difficult to identify whether the source of any impairment is due to a problem in a specific component such as "disengage," or whether it is a more general EF weakness.
- In developmental research, the increased use of experimentally driven paradigms has produced a much-needed understanding of atypical trajectories of attention.
- Future research needs to devise and employ developmentally sensitive paradigms that can be used across a range of neurodevelopmental disorders to facilitate cross-syndrome analysis of attention performance.

Section III: Development of Typical and Atypical Attention

7

Attention Over Development
From Infancy to Adolescence

CHAPTER SNAPSHOT

- The importance of teasing apart attention subcomponents and tracing their pathways from infancy, though to childhood, and then into adolescence.
- What is the best procedure for tracking age-related changes on the same attention subcomponent?
- Improving with age—but some things improve faster than others

The ability to efficiently allocate visual attention resources is a critical component of infant development and emerges surprisingly early. From birth onward there appears to be a gradual developmental trajectory of increasingly sophisticated attention skills that allow the infant to ignore irrelevant stimuli and instead attend to objects and events that are of intrinsic interest. In the first few months of life, the core mechanism through which attention is developed within the context of social interactions is known as *joint visual attention* or more generally as "looking where someone else is looking." Infants' vulnerability to distraction also decreases over time, and their ability to sustain attentional focus and increase inhibitory control slowly matures as they progress from infancy into early childhood. In the past decade alone, we have seen a shift of focus from viewing early infant attention as a tool for understanding the acquisition of learning and sensory development, demonstrated most notably within the domain of visual perception (see the early studies of Fantz, e.g., 1964), to later studies that focused on the role of attention in the development of such skills as visual acuity, saccadic eye tracking of visual stimuli, and the effect of the complexity or novelty of the stimulus. There is increasing acknowledgment that the trajectories and profiles of early visual attention and its subcomponents need to be studied both independently within their own developmental contexts and in terms of their contributions to other cognitive domains (see Colombo, 2001, for a historical review). What brought about this change of perspective? One of the key factors is undoubtedly the emergence of what is now termed *developmental cognitive neuroscience*, the merging of previously disparate disciplines, including cognitive science, neurobiology, neuroscience, and developmental psychology, to study typical and atypical cognitive development.

The quest to understand the developing organism across multiple cognitive domains (perception, attention, memory, learning, language, and executive functions) and multiple systems (from the genetic and brain levels to the cognitive, behavioral, and environmental levels) has culminated in a rich source of multidisciplinary research initiatives that have begun to chart the trajectories of early brain and cognitive development and their impact across the life span (see Karmiloff-Smith, 1998, 2009; Munakata, Casey, & Diamond, 2004). In Chapter 2, we viewed attention in terms of its core subcomponents: selection, maintenance, and control. In the present chapter, we explore the growing number of studies that have sought to identify the trajectory of cognitive performance across these attention subcomponents, beginning in infancy and journeying into late childhood and adolescence. By the very nature of this research, we would expect developmental progression, with infants and young children showing more instability in performance than older children and adolescents. However, a number of emerging studies now suggest that we cannot make the a priori assumption that age-related changes across different attention subcomponents follow identical pathways across developmental time.

Birth to Preschool: The First Four Years

Joint Attention and Attention Orienting

From birth, infants begin to gather information about their environment and the objects around them. Babies are especially drawn to human faces and demonstrate attention and sensitivity to direct eye gaze that increases in sophistication across the first year. Farroni, Csibra, Simion, and Johnson (2002), for example, found that newborn infants (within the first 5 days of life) would look significantly longer at a picture of a face with a direct gaze compared to a face with an averted gaze (eyes averted left or right). Farroni et al. argue that this early preference for mutual gaze arises from a necessity to quickly process visual input dedicated to socially relevant stimuli. Furthermore, preferential attention to direct eye gaze in newborns enhances cortical processing of faces by age 4 months (later replicated by Farroni, Johnson, & Csibra, 2004; Farroni, Massaccesi, Menon, & Johnson, 2007). Other studies have also highlighted the role of direct gaze in the foundation of later social-cognitive development, including social relationships (see Gliga & Csibra, 2007, for a review).

By 3 months of age, an infant is able to avert attention from a face-to-face interaction to follow another person's gaze on a nearby object (Amano, Kezuka, & Yamamoto, 2004; Hood, Willen, & Driver, 1998; Striano, Henning, & Stahl, 2005). The ability to shift attention and look toward an object that is the focus of another person's attention is a central aspect of what is termed "joint attention." *Joint attention* is a fundamental skill that provides the foundations for future social development, most notably communicative skills (including language acquisition), and serves as a critical component of dyadic (infant, person) and triadic (infant, person, object) interactions (see Striano et al., 2005; Vaughan van Hecke et al., 2007). In their most recent study, Tremblay and Rovira (2007) extend these findings to include a triangular interaction (infant, person, person) in which both 3- and 6-month-old infants were able to turn their eyes in the same direction as an adult's gaze during face-to-face interaction to look at another person. Together, these findings, alongside three decades of developmental research, testify that joint visual attention skills represent a critical milestone of early development and social learning. We shall see in later chapters (Chapters 8 and 9) how disruption to this attentional mechanism can severely impact across a range of cognitive skills and even impair neural development.

The development of visual orienting has been explored using tasks similar to the Posner orienting task described in Chapters 2 and 6. The simplest form begins with fixation on a central stimulus, followed by presentation of a target on the right or left of fixation. With very young infants, the direction and latency of eye movements are recorded as indices of switching in attention. Ability to disengage from the central stimulus is studied by varying the attractiveness of the central stimulus, whether or not it remains when the target appears, or the lag between the offset of the central stimulus and the onset of the target. More complex forms of the basic task involve presentation of cues to the target location before target onset as described in Chapter 2.

Johnson, Posner, and Rothbart (1991) tested 2-, 3-, and 4-month-old infants on three aspects of responding in the simple fixation-target version of the task: ability to disengage from the central stimulus and switch fixation to the targets, anticipatory looks to the target location, and learning to predict the target location from the nature of the central stimulus. In all cases these abilities were only apparent in the 4-month-old group. However, the three measures were not significantly correlated, suggesting that they may depend on different mechanisms. Interestingly, ability to disengage from the central stimulus, but not the other measures, was associated with temperament in that greater ability to disengage predicted lower distress levels as assessed by parent questionnaire.

The authors suggested that some anticipatory looking before 4 months may depend on low-level automatic mechanisms, but at 4 months we are seeing the first evidence for self-directed (endogenous) control. Ability to predict the target location was still weak at this age (60% correct), further strengthening the suggestion that the different measures were reflecting different mechanisms.

Hood and Atkinson (1993) studied disengagement by varying the lag between central stimulus offset and target onset (240 or 720 msec or the central stimulus and target overlapped). Participants were aged 1½ to 6 months, and an adult group was also included. Overlap produced the slowest switches to the target in all age groups when eye movements were recorded, implying a similar disengagement process in infants and adults. However, when adults were required to make manual responses to target onset, overlap conditions were not slower, indicating that the disengagement process was specific to eye movements. Zwaigenbaum et al. (2005) have found that slow disengagement measured by eye movements in infants under 12 months of age may be a predictor of autism.

We now explore the mechanisms that infants employ to selectively engage attention across a variety of conditions and environments (selective attention), to maintain attention and avoid distractibility (sustained attention), and to inhibit irrelevant responses to stimuli, regardless of whether they are social or nonsocial stimuli (attentional control). Across all subcomponents, the infant's attentional development is linked in a complex relationship between factors that require *exogenous* control (such as preferential attending to objects or distracters that are salient in nature) and factors that require *endogenous* control (such as the infant's goal-directed behaviors and level of self control). Successful attention requires the interplay between both these factors, and the interaction is complex, changing progressively across the first year. What begins as an early dominance of exogenous control in the first 6 months, when the infant's attention is allocated to objects that are especially salient or familiar and guided by such factors as color, brightness, size, and loudness, progressively develops to a greater reliance on endogenous control as the infant learns to control and direct attention more efficiently.

Selective Visual Attention

The ability to selectively attend to the most important sources of information in the environment and to filter out potentially extraneous stimuli has generated a considerable body of research knowledge. It is now well established that very young infants preferentially look toward a novel visual stimulus but become easily bored with the stimulus once an internal representation has been created. When the novelty is retriggered by the presentation of a new stimulus, then there is a recovery of attentional focus. Although these studies are informative—and we will revisit them later in the chapter when we examine maintenance of attention—they are restricted in what they can tell us about how infants *selectively* attend to stimuli because their primary aim is to examine fixation time to a single isolated stimulus rather than to examine how infants allocate their attentional resources between multiple competing stimuli. Some of the most exciting research to emerge in recent years has been derived from so-called "pop-out" studies that have used novelty preference paradigms (Colombo, Ryther, Frick, & Gifford, 1995), mobile conjugate reinforcement paradigms (e.g., Rovee-Collier, Bhatt, & Chazin, 1996; Rovee-Collier, Hankins, & Bhatt, 1992), and most recently, eye tracking paradigms (Adler & Orprecio, 2006; Amso & Johnson, 2006). *Pop-out* refers to a situation in which the unique perceptual features of a stimulus are detected rapidly in a visual array against a background of similar/ dissimilar distracters, with response time unaffected by the number of distracters. It is the "uniqueness" of the stimulus that is proposed to drive attentional resources to select and maintain focus. This phenomenon is now well documented in infants and adults and is described in detail below. There has also been an intensive exploration of early visual-search processing in infants and toddlers, for example, search efficiency for simple visual features versus conjunctions of features (e.g., Gerhardstein & Rovee-Collier, 2002) and most recently, visual search and the role of target featural salience (Scerif, Cornish, Wilding, Driver, & Karmiloff-Smith, 2004).

Pop-Out Effects

In their pioneering study, Rovee-Collier, Hankins, and Bhatt (1992) assessed sensitivity to visual pop-out in a series of experiments that required 3-month-old infants to discriminate designs composed of two lines (L, T, and +) displayed on a mobile comprising seven wooden pink blocks (3 cm^2). In order for infants to demonstrate that they could recognize a particular target design (e.g., L), they used an operant foot-kick procedure that involved tying one end of a satin ribbon to one of the mobile suspension bars and the other end being tied loosely to the infant's ankle. Infants learned that kicking in the presence of a block bearing a particular design (for example, a display of + signs) would move the mobile and were then tested for generalization or retention of this response. One of the most revealing experiments (Experiment 4) assessed to what extent a single unique target can stand out in an otherwise homogenous background and capture the infant's attention. This experiment used a paradigm known as reactivation in which, after the original learning has been forgotten, the forgotten memory is reactivated by presenting again all or part of the original stimulus. If, however, the reminder contains any novel features, reactivation does not occur. Hence, it was possible to deduce whether novel elements in a reactivating stimulus were being processed or not by testing whether the original learning reappeared afterward. Twelve infants were randomly assigned to two groups, and each received the same procedure during training and testing. One group was trained and tested using the L mobile, and the other was trained with the + mobile. After a delay in which the response was forgotten, the infants were "reminded," using either a mobile containing one familiar stimulus, among six novel ones or one novel stimulus among six familiar ones. Testing for the original response was then conducted across 3-minute sessions. In the test the infants behaved as if in the first case the reminder had consisted entirely of familiar designs (so reactivation occurred) and in the second case as if the reminder had consisted entirely of novel designs (no reactivation occurred). Thus, the single unique design in the reminder display always dominated, irrespective of whether it was familiar or novel, implying that it had "popped out" of the display.

The findings clearly demonstrated that when a symbol is unique in the infant's attention is selectively captured by it, thereby allowing the entire mobile to be recognized as familiar (or unfamiliar). However, to determine whether this effect reflected a specific attention mechanism requiring selective attention to unique local features or whether the effect was more likely due to a simple visual comparison between the trained design presented at test, which did not require selective attention, a further experiment was carried out, this time replacing the unique familiar block in the reminder display with three blocks in a seven-block mobile. Otherwise, the procedure was identical to that described above. Against what one would have anticipated, the infants did not show reactivation in response to the display with three familiar blocks, even though it was more similar to the original display than the previous reactivating display with only a single familiar block. Hence, pop-out did not occur in this situation, and the authors conclude that their findings demonstrate that infant's performance by 3 months old does, indeed, reflect a genuine perceptual pop-out effect for unique items similar to that exhibited in the adult end state.

In contrast to the substantial research using pop-out techniques in infancy, a few studies have compared the infant and adult performance using equivalent procedures (one of the few is described below: Adler and Orprecio, 2006), but there is no research that has taken a truly developmental approach to understanding whether the attention mechanisms underlying pop-out remain stable over time, especially in early childhood when there is a greater development of exogenously driven selection. In infancy, at least in the first 6 months, exogenous factors such as stimulus characteristics and novelty of targets and distracters play a critical role in attention, but as infants approach their first birthday, endogenous factors such as self-regulation will increase in dominance. This is especially important when one looks at the issue of timing when measuring pop-out both in terms of procedure and developmental progression. Techniques such as the novelty preference paradigm and the mobile conjugate reinforcement paradigm examine the rate of pop-out over a course of seconds and minutes (up to 3 minutes in some cases), allowing for a wide range of possible extraneous factors to interfere with the effect (e.g., attentional load, features of the stimuli, memory capacity, and so forth). In contrast, pop-out is typically measured in terms of milliseconds in adults: that "first look." One of the most innovative studies to date to address the issue of timing and comparability of infant and adult performance on pop-out was published in 2006 by Adler and Orprecio. Using a novel eye-tracking paradigm, they were able to assess for the first time the comparability between pop-out as exhibited in infants and pop-out as exhibited in adults. Their paradigm and results are described below.

Measuring saccade latencies to targets in visual arrays (with varying set sizes: 1, 3, 5, or 8 items) in which a target stimulus is present (+ among Ls) or absent (all Ls), Adler and Orprecio were able to assess whether infants as young as 3 months could demonstrate pop-out within a millisecond time frame rather than one using seconds or minutes as calculated in previous infant studies (e.g., Rovee-Collier et al., 1992, 1996), and also to what extent infant performance matched the adult end state. To achieve some degree of comparability, infants and adults completed the same procedures under similar, but obviously not identical, experimental conditions. For example, although adults were told that stimuli would be presented on the screen in a circular pattern, they were not told any information regarding differing set sizes, target present versus target absent arrays, or that they should make eye movements as quickly and accurately as possible.

The experimental trial began with the presentation of a fixation triangle for 1,000 msec followed by a blank interval of 250 msec. This was followed by the visual-search array with a duration of 1,000 msec during which either a target-present array (+ target) or a target absent array (all Ls) was presented. When present, the target appeared at either

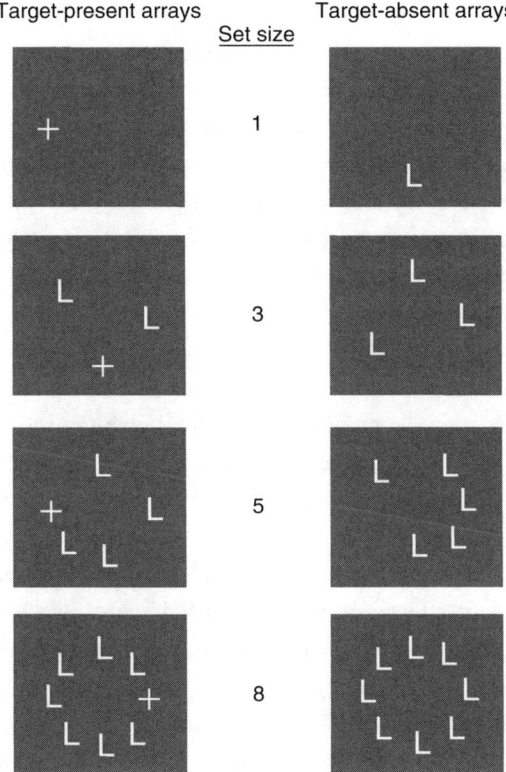

Figure 7-1. Example of stimuli used by Adler and
Orprecio.
From Adler and Orprecio (2006). Redrawn by permission
from *Developmental Science 9*, 194.

the 3, 6, 9, or 12 o'clock position, with 0, 2, 4, or 7 nontargets. Nontarget
trials consisted of 1, 3, 5, or 8 nontargets, and when the single nontarget
appeared, it was at one of the same four positions. See Figure 7.1 for an
example of the visual array configurations.

The latency of the first saccadic eye movement was measured and
counted as a target fixation if it moved at least half way toward the target
from the central fixation position. The movements clearly demonstrated
an expected pop-out effect in infants, although somewhat slower than in
adults, but nonetheless on a millisecond scale. Interestingly, the saccade
latencies showed a nearly identical increase (54.13 msec for adults

and 46.16 msec for children) when set size was increased from one to three items, suggesting some comparability of pathways. Furthermore, infants', as well as adults', saccade latencies to a target stimulus were unaffected by set size from three items upward, an indication of efficient parallel search that was suggested in the data of previous studies but confirmed in this one. Adler and Orprecio conclude that the mechanisms needed to allocate resources for this type of attentional selection may be functionally similar across the life span.

Feature Search Versus Conjunction Search

Although there is evidence of developmental changes in some aspects of feature search in infancy (Bhatt & Waters, 1998), other aspects of feature search appear to display adult-like states (e.g., Adler & Orprecio, 2006; Bertin & Bhatt, 2001; Rovee-Collier et al., 1992). However, at one level the attentional mechanism seems to be fully functional, especially on the observed pop-out effect, and it will always depend on adequate perceptual discrimination and thus will work more efficiently for basic features for which processing is already mature.

Numerous studies that have demonstrated the pop-out effect show a difference in search times for tasks that require *featural search* (presence of a unique single feature) versus *conjunction search* (absence of unique feature), with the latter requiring the more laborious serial process to complete the search. In contrast with the research on pop-out, relatively few studies have examined the developmental profile of both feature *and* conjunction search from the end of the first year to early childhood, thus providing a link between infant start states and adult end states. Using a reaction time (RT) paradigm, Gerhardstein and Rovee-Collier (2002) attempted to provide this link by assessing whether RT on feature and conjunction tasks differentiated infants at 12 months, 24 months, and 36 months. A target stimulus was either a red or green dinosaur cartoon presented on a video monitor equipped with a touch screen. *Feature* search displays were composed of either a red target (a red dinosaur called "Barnette") among green distracters (green "Barnettes") or vice versa and were counterbalanced with age. *Conjunction* search comprised either a red Barnette dinosaur among green Barnettes with the addition of a red target with a different form (a different-looking dinosaur called

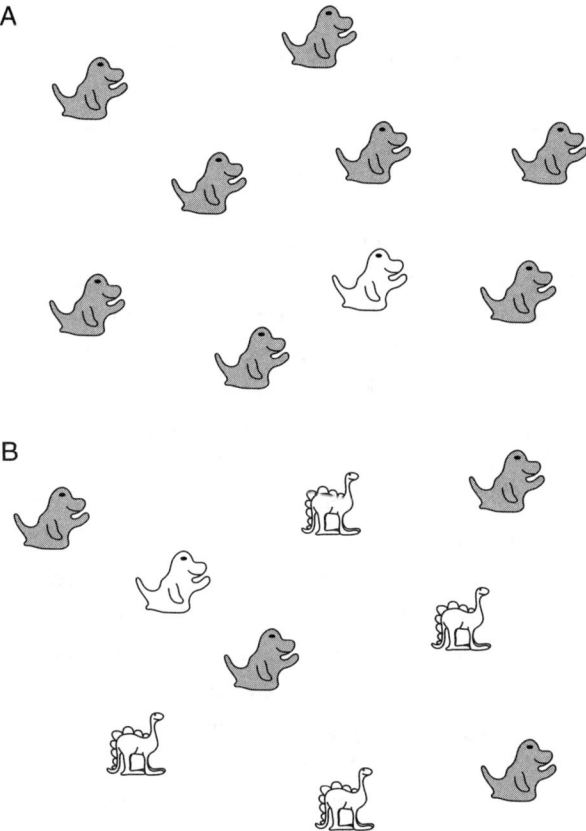

Figure 7-2. Example of stimuli used by Gerhardstein and Rovee-Collier (2002).
Redrawn by permission from the *Journal of Experimental Child Psychology 81*, 200.

"Dino") counterbalanced across age. See Figure 7.2 for a comparison of both stimuli. Across trials, the target could appear equally in one of four segments on the screen.

To avoid the potential confounds of verbal instruction, a nonverbal procedure was adopted to train infants and children to locate and touch a target on screen. This was followed by a training phase and then a testing phase. In training, the infant/child had to accurately touch the target: for example, in the feature search a red Barnette, initially when it appeared in motion and later, once this had been learned, when it was stationary among 24 shapes on the screen. In the testing session, the target had to be located among 2, 4, 8, or 12 distracters. Findings demonstrated a clear difference between RTs for featural search versus conjunction search and in the expected direction. There was an age-related increase in overall response speed commensurate with other studies using speeded task measures (e.g., Cerella & Hale, 1998). Moreover, for featural search targets there was a flat RT over distracter number, indicating that all age groups performed in a similar way, indicative of efficient parallel processing. In contrast, for conjunction search targets, RTs increased significantly with distracter number, taken to be indicative of serial processing. However, in 12-month-old infants, successful target location was dependent on whether the target was actually moving, and they were unable to locate it in the test situation (the authors point out that different methodology may well show successful search at this age, citing observations by Bhatt, Rovee-Collier, and Weiner, 1994, who concluded that at 6 months, infants had the capacity for serial search). Successful performance was achieved in infants at 24 and 36 months, but there was no evidence that visual-search strategy changed developmentally during this time period. Given that 12-month-olds may have had considerable difficulty in completing this task, Experiment 2 comprised slightly older infants in the first age group (18 months, 24 months, 36 months) in order to investigate whether there were any developmental changes in conjunctive search. Using a modified format but with identical stimuli (Barnette and Dino), the authors found an expected age effect on the conjunctive tasks with the older group faster, but they also found no Age × Task interaction, indicating that all three age groups exhibited the same pattern of task-specific differences. They conclude that the infant performance at least by 18 months is identical to that of adults, showing a very similar dichotomy between featural search and conjunctive search.

Visual Search and Target Salience

As we have seen from the studies cited above, a target's featural salience plays a critical role in stimulus selection and this effect is observable in infants as young as 3 months of age. Few studies, however, have focused exclusively on charting the developmental progression of featural search alone and the role of target visual salience. Using a novel, computerized visual-search task (Visearch task) developed initially by Wilding (see Chapter 6 for a full review of this task) for children aged 6 to 11 years old, Scerif and colleagues (Scerif et al., 2004, 2007) recently adapted this paradigm for toddlers to examine whether manipulation of target featural salience would affect search for multiple targets. In brief, toddlers aged 2 to 3 years old searched for target "holes" (black circles of varying sizes) randomly displayed on a touch-screen computer monitor. Distracters were also black holes and were either very small (dissimilar) or medium (similar) in size compared to the targets. Toddlers had to locate eight targets within three conditions: no distracters (baseline), 6 or 24 dissimilar (small) distracters, and 6 or 24 similar (medium) distracters. Thus, both distracter type and number were manipulated. See Figure 7.3. Measures comprised speed (mean search time per hit), search path (mean distance between successive touches), accuracy (total number of errors), and error types (touches on distracters or repetitions of previously found targets).

As expected, Scerif et al. (2004) found 2-year-olds to be slower than 3-year-olds when searching for targets among nontargets. Trials where the distracters were similar in size (medium) to the targets and a large number were present produced longer mean times than trials where distracters were dissimilar in size (small) and fewer in number. Further analysis revealed that the age effect on speed was not simply due to differences in motor speed in 2- versus 3-year-olds but instead reflects an increased age-related ability to deal with search among distracters. This was the first finding to demonstrate that manipulations of perceptual salience for targets can affect toddlers' search speed even for search for a feature within a single dimension, in this case, size. Age also had an impact on search path, with younger toddlers producing longer distances between successive touches than older toddlers. In 2-year-olds, search paths also appeared especially vulnerable to both number and salience of distracters whereas in 3-year-olds, search paths were vulnerable to distracter salience only. This finding suggests that search path may provide an especially sensitive measure of age-related changes in search performance in toddlers that has not hitherto been explored. In terms of search accuracy and error types, 2-year-olds produced more incorrect touches than 3-year-olds and across both error types (touches to nontargets and repeated touches on previously successful targets). Together, these

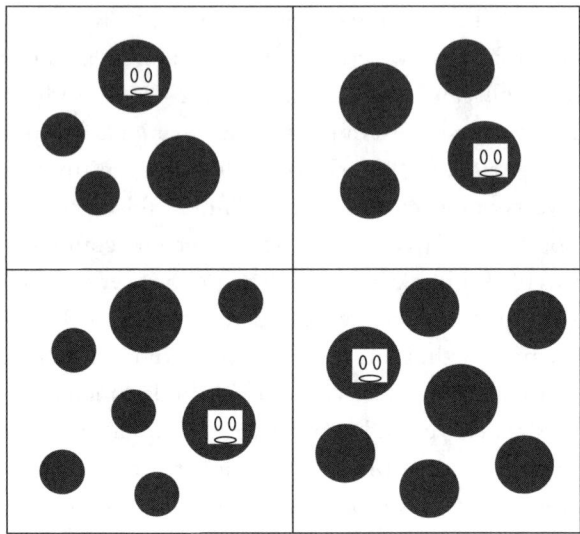

Figure 7-3. Example of stimuli used by Scerif and colleagues (2004).
Redrawn by permission from *Developmental Science* 7, 118.

findings indicate that older toddlers are able to discriminate targets from distracters more efficiently than younger toddlers and can inhibit repetitions on previously found targets more effectively.

Wilding and Burke (2006) used a similar task with 3- and 4-year-olds and required a black circle to be found among black ellipses and smaller black circles. They found no relation between age and time or path, but they did find that verbal mental age was related to the number of errors. Significantly, they also found that teachers' ratings of the children's general attentional ability predicted the number of errors, thus demonstrating that accuracy on the task was related to more global ratings of attention and did not merely reflect ability to discriminate the targets from the foils.

Maintenance of Attention

The ability to maintain focus or orientation when there are *multiple distractions* and when awaiting some significant event *(vigilance)* is a critical component of attention. Clearly, both these abilities involve selective processes, but they require selective focus on a given stimulus, location, or activity to be maintained over a period of time, not simply deployed briefly before switching to a different focus. As we have seen in our survey of studies exploring selective attention, the infant undergoes substantive changes during the first year of life. Attention allocation and maintenance, guided by a complex interaction between attentional state (casual vs. focused) and the physical characteristics (salience) of stimuli, have been studied extensively and predominantly within the context of distraction. Age-related changes across both factors have been well documented and will be discussed below. *Attentional state* can be described as either *focused*, defined as concentrated periods in which there is active attention to an object involving behaviors such as intent facial expressions, enclosing an object of interest and bringing it nearer to the eyes, and either no talking or self-directed soft talking, or *casual*, whereby the infant displays signs of concentration, defined as directing the eyes toward an object but not being actively focused. *Salience* of the stimuli (target vs. distracters, familiar vs. novel) can be altered at the visual level (e.g., color, size, brightness, or shape) and the auditory level (tone, pitch). How infants and children maintain attention to one specific stimulus and resist attention to distracters has been assessed using *distraction paradigms* and to a lesser extent *free-play paradigms*. Both these types of tasks have been used across multiple contexts: watching a videotape (Ruff, Capozzoli, & Weissberg, 1998), playing with toys (e.g., Oakes, Kannass, & Shaddy, 2002; Ruff & Capozzoli, 2003), and in more controlled settings involving structured tasks (Ruff, Capozzoli, & Weissberg, 1998). In contrast, how infants maintain attention over a sustained period of time while awaiting a target stimulus that occurs infrequently is typically assessed by vigilance tasks.

Vigilance

Vigilance refers to the ability to remain focused while awaiting an anticipated event over a concentrated period of time. Vigilance requires a much more sophisticated degree of self-regulated attention (endogenous control) than is typically assessed in the sorts of sustained attention tasks listed above. *Controlled* processing (as opposed to *automatic* processing usually associated with exogenous attention) has been shown to be the locus of vigilance performance, requiring deliberate effort and coordination of attentional resources. Given this complexity it is hardly surprising that vigilance, as a mechanism requiring endogenous control, does not emerge especially early in development, beginning in the second year, but most reliably measured at 4 years, when developmental changes in attentional mechanisms are at their most salient. In contrast to the research on distractibility, there are relatively few studies that have attempted to assess the development of vigilance in early childhood. We highlight two recent studies that have made such an attempt. The first includes toddlers aged 24 to 30 months (a younger sample than traditionally assessed), and the second includes children aged from 3½ years to 5½ years.

In a well-designed study, Goldman, Shapiro, and Nelson (2004) investigated methods of demonstrating vigilance in children aged 2 years. Vigilance was measured as the ability to maintain alertness in the absence of ongoing feedback from target stimuli. Using the Early Childhood Vigilance Task (ECVT) (modified from Ruff, Capozzoli, Dubiner, & Parrinello, 1990), toddlers were required to monitor a blank computer screen in order to wait for cartoon stimuli to appear. With a 5-, 10-, or 15-second interval between cartoons, it was predicted that the more focused ("on task") the toddler is to the screen in the absence of the target stimuli, the more likely they are to visually capture the cartoon when it appears. The task lasted a total of 7 minutes. The percentage of time looking at the screen was recorded both overall and during the blank intervals; the two scores were highly correlated, and the first was used in subsequent analysis. A second free-play task gave the toddler the chance to interact with a number of toys, and the percentage of time spent in focused attention was recorded. Finally, evoked potentials were recorded from a subset of the children while they observed a sequence of stimuli in which a novel stimulus appeared from time to time (the so-called "oddball paradigm").

The ECVT and focused attention scores from the first half of the free-play task were significantly related (scores in the second half were corrupted by increased parental involvement and not analyzed), suggesting that a relatively stable individual difference was being tapped. ECVT was unrelated to parental ratings of the child's overall attentional style but was modestly related to ratings of cognitive development.

Finally, a component of the evoked potentials, the midlatency negative deflection, which is known to be more pronounced to stimuli that
attract attention such as the mother's face, was greater to novel stimuli in
the oddball paradigm, particularly from frontal locations. Recordings of
this component from the right hemisphere frontal site were significantly
correlated with ECVT scores, despite the small sample (n = 14) and were
unrelated to the other measures of attention.

Using quite different tasks to the Goldman et al. study, Akshoomoff
(2002) assessed preschool children between 3½ and 5½ years old on two
vigilance tasks, both requiring the visual monitoring of a succession of
stimuli for a specific target. The first task (picture task) required a motor
response to a specific target stimulus (in this case, a colored drawing of a
duck occurring on 25% trials) and to inhibit a response to a nontarget
stimulus (a colored drawing of a turtle). The second task (selection/
distracter task) was more complex, requiring a response to a specific
target stimulus (in this case, a circle) sometimes accompanied by nontarget stimuli (crosses at either side). The remaining trials were a square with
or without flankers. On each of the two tasks, the stimulus remained on
the screen for 500 msec, and each child was instructed to keep watching
the screen and press the appropriate button as soon as he or she saw the
target stimulus. See Figure 7.4 for an example of the stimuli used across
both tasks. All sessions were videotaped, and measures derived were the
percentage of target detections (hits), false alarms, response times, and
time spent actively engaged with the task.

The findings, however, are less robust than expected due to 11 (46%)
of the original sample of 24 children aged between 3½ and 4½ years failing to meet the criteria for the two vigilance tasks, typically as a result of
too few hits rather than too many false alarms. Looking first at performance on the picture task, those children with the fastest reaction times
tended to make the most hits, and children in the youngest age range
were slower than children in the oldest age range. However, there was no
age effect on the total number of hits or false alarms, indicating that all
groups did eventually locate the target stimulus successfully. On the
selection/distracter task, children in the youngest age group had a significantly reduced rate of hits and were slower overall compared to children in the other age groups. In contrast, false-alarm errors did not differ
between age groups. Overall, the findings are not especially illuminating

A. Picture selection task

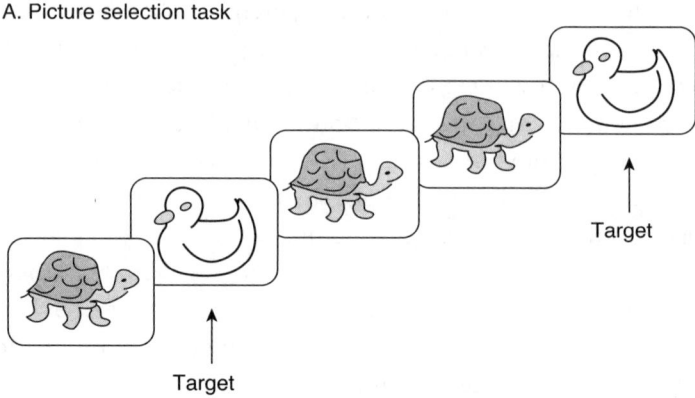

B. Selection/Distractor task

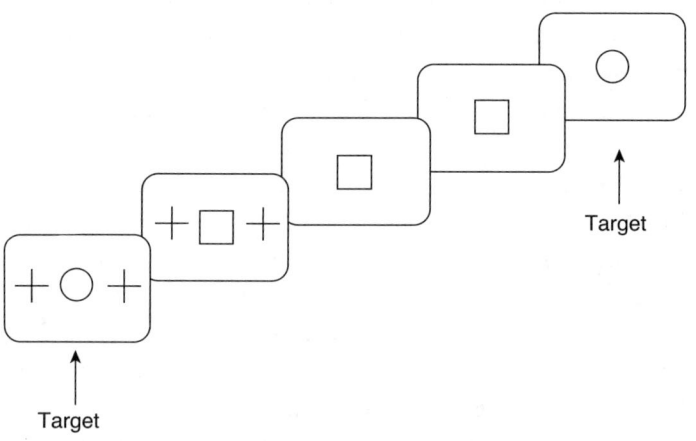

Figure 7-4. Example of stimuli used by Akshoomoff (2002).
Redrawn by permission from *Developmental Neuropsychology 23*, 631.

but differ from previous research by demonstrating that some young
children can efficiently wait and locate a target and respond quickly
when there is a continuous stream of rapidly presented stimuli. In contrast, previous research had required children to actually respond to a
target on a screen before moving onto the next target (e.g., Ruff et al.,
1998; Weissberg, Ruff, & Lawson, 1990). In terms of developmental
changes, the only consistent finding to emerge is in the expected
direction of faster response time with increasing age. The inability of
nearly half the children in the youngest group to complete the tasks

satisfactorily is perhaps the most significant finding, indicating that such ability is only reliably present over the age of 4.

Distractibility

The ability to keep "on task" while avoiding distraction is a fundamental necessity for early learning. We shall see in later chapters how impairment to this skill can impact tremendously on cognitive functioning from infancy onward (see Chapters 8 and 9 for detailed critiques of attention performance in developmental disorders). It is now well documented that both infants and adults are less distractible when their attention is focused rather than casual (see Ruff & Cappozzoli, 2003), that the nature of the activity under exploration impacts on the level of distractibility, and that the characteristics of the distracter(s) (visual vs. auditory) and target objects (familiar vs. novel) can impact on distractibility (see Oakes, Kannass, & Shaddy, 2002; Oakes, Tellinghuisen, & Tjebkes, 2000; Ruff & Cappozzoli, 2003).

In a longitudinal investigation of infants aged 6½ months and 9 months of age, Oakes et al. (2002) assessed whether distraction latencies (time taken to look from a target object, in this case, a single toy, to a distracter, in this case, colored blinking rectangles accompanied by an audio beep) were influenced by age, attentional state (focused vs. casual), and salience of the target toy (familiarity vs. novel). Previous research had consistently shown that the characteristics of a distracter affect distractibility in infants (e.g., Oakes, Tellinghuisen, & Tjebkes, 2000; Ruff & Capozzoli, 2003) and that infants exhibit longer latencies in responding to a distracter when engaged in a state of focused, concentrated attention than when in a state of casual attention (e.g., Ruff, Capozzoli, & Saltarelli, 1996).

In brief, the infant explored each of four target toys for 2.5 minutes, two trials with familiar toys and two trials with novel toys. The distracter trial began when the infant was judged to be fixating on the target toy and commenced every 30 seconds over a 2.5 minute period. As expected and consistent with their previous findings (e.g., Oakes, Tellinghuisen, & Tjebkes, 2000; Tellinghuisen, Oakes, & Tjebkes, 1999), infants were less distractible when in a focused state than when in a casual state. Furthermore, at 6½ months, distraction latencies did not vary as a function of toy familiarity, but at 9 months, latencies were longer when infants were exploring novel toys compared to familiar toys.

Moving along the developmental trajectory, two important studies have examined how changes in attentional engagement continue from infancy into toddlerhood (Ruff & Capozzoli, 2003) and then into the preschool years (Kannass & Columbo, 2007).

Ruff and Capozzoli in their 2003 study assessed three age groups of infants and toddlers (10, 26 and 42 months) across four types of distracters (auditory only [A-O], visual only [V-O]), auditory-visual combined [A-V]), and no distracter), and three levels of attention (focused, settled, and casual). Each infant/toddler within an age group was assigned to one of two task conditions: play with a single toy or play with multiple toys. The distraction tape was composed of intermittent video and/or audio distractions followed by intervals of a blank screen and/or silence. The duration of the distracters was 4 seconds each, and the time between distracters varied randomly. The results were striking. At 10 months, infants were easily distractible across both single toy and multiple toy conditions and across all measures of distractibility compared to the two older groups. At 26 months, although somewhat less distractible than at 10 months, toddlers showed twice as much distractibility in the single-toy condition as in the multiple-toy condition.

Across developmental time, the more focused the attention, the less distractible the infant/toddler (as is consistent with previous research; see, for example, Oakes, Ross-Sheehy, & Kannass, 2004, for a description of their findings of infants aged 6 and 9 months). The duration of *focused attention* increased between 26 and 42 months alongside a significant decrease in *casual attention*. In contrast, *settled attention*, defined in this study as a pause in casual attention in order to look at and manipulate a particular toy, increased across age groups exponentially. The changes in attention highlighted by this study, and suggested in others, point to an emerging system of gradual specialization in which attentional focus is shifted along a pathway that begins with subtle changes in the latter half of the first year and into the second year but then undergoes a major transition from the age of approximately 2 years. Interestingly, the authors had also previously demonstrated age-related changes in sustained attention in infants (30 to 54 months) across different environmental contexts, including watching television, free play with toys, and reaction time experiments (e.g., Kannass, Oakes, & Shaddy, 2006; Ruff, Capozzoli, & Weissberg, 1998; Ruff & Lawson, 1990).

Focusing on the preschool years, Kannass & Columbo (2007) assessed two groups of children (3½ years and 4 years) using two types of distracters (intermittent vs. continuous audiovisual distracters) derived from a children's television show transmitted in Spanish and therefore incomprehensible to the children, across four target tasks (Lego, puzzles, matching, and coloring). Each child was assigned to one of three conditions: intermittent distracters (5-second, randomly ordered segments with intervals of blank tape), continuous distraction (continuous 5-second, randomly ordered distracting segments), or no distracters. Children received each of the four target tasks for three minutes per task and worked independently. The distracter trial began as each child commenced play with a new toy. Examination of both task performance (e.g., the sum of correctly placed Lego pieces) and levels of distractibility (e.g., "on task," "off task," and looking at the distracter) revealed very interesting developmental findings. At 3½ years of age, children were easily distractible, and both distracter conditions impeded performance compared to children in the no-distracter condition. By 4 years of age, however, performance was affected only in those children in the continuous-distracter condition. The authors argue that the impeded performance in the 3½ -year-olds may be due to their relative immaturity in maintaining attention across competing stimuli, but by 4 years they are able to tune out periodic distractions but not constant distraction. This subtle developmental change, hitherto unexplored in previous research, demonstrates the importance of careful investigation of age-related changes across development, not least because the pattern of these findings is in contrast to literature on adults that has shown that intermittent distracters impede performance in adulthood (e.g., Landstrom, Kjellberg, & Byström, 1995) whereas continuous distraction actually facilitates performance (Britton & Delay, 1989).

Attentional Control

The ability to exert effortful control in order to inhibit a dominant or prepotent response, to hold in working memory newly relevant rules that require the suppression or activation of previously learned responses, and to shift attention between tasks, is a key requisite for cognitive and social development. These skills, especially in the mid- to late-childhood years, form the building blocks for later adult outcomes. When there is a dysfunction in self-regulation, as we shall see in the case of ADHD (Chapter 9), then there are repercussions at all stages of the developmental trajectory and at multiple levels, at the brain and cognitive levels, at the academic level, and at the social-emotional level. During the first year, infants undergo a major transition as they move from a reliance on involuntary, stimulus-driven control to a more self-regulated control that sees them select, maintain, and focus attention over longer periods of time.

One of the core prerequisites for efficient selection is inhibition. To be able to successfully select relevant stimuli for processing, one must be able to inhibit irrelevant stimuli. The functionality of this inhibition mechanism has been studied using a range of paradigms and procedures. From the ages of 2 to 5 years, the gains in inhibitory control are remarkable, acting as critical precursors to higher cognitive functioning, and are essential for coping with the varying demands of the academic curricula. One of the most important skills involves the ability to inhibit an inappropriate but prepotent (dominant) response, for example, to enable a preschooler to stay in his or her seat, complete a required task, and to resist the urge to jump up and look outside the window. This ability is often referred to as *response inhibition* and has been studied extensively across development (e.g., Carlson, 2005; Gerstadt, Hong, & Diamond, 1994; Simpson & Riggs, 2005a, 2005b, 2007; see Garon, Bryson, & Smith, 2008, for a substantive review of executive functions in preschoolers).

In infancy, eye movements are used as an indicator of attentional control and can be used to tap inhibition at two levels: inhibition of voluntary gaze to an abrupt visual stimulus (e.g., Nakagawa & Sukigara, 2007; Scerif et al., 2005), and inhibition of a previous stimulus (because of its status as a distracter), which is indicated by inefficient responses when it subsequently becomes the target stimulus, known as the "negative priming effect" (Amso & Johnson, 2005).

Holding in Mind Two Sets of Rules

Inhibition is also required to hold in mind two sets of rules, apply them, and then switch between them in accordance with external or self-generated instructions. Such switching can be from one stimulus or stimulus feature to another, from one stimulus–response mapping to another, between tasks, and so forth. Even the ability to carry out a simple sequence of responses requires that each response is activated then inhibited as activation passes on to the next response in a sequence. An important series of investigations has been carried out using the Dimensional Card Sorting Task (DCCS) (e.g., Diamond, Carlson, & Beck, 2005; Kloo, Perner, Kerschhuber, Dabernig, & Aichhorn, 2008; Zelazo, 2006), which requires switching attention to different stimulus features. In the following section, we describe some of the exciting work to emerge from an intensive decade of innovative research.

The *Dimensional Change Card Sorting (DCCS) task* requires children to sort a set of cards by one dimension (e.g., color) and then to re-sort the cards by a different dimension (e.g., shape). See Figure 7.5 for an example of the task. As is clearly evident in the numerous studies that have used this task, toddlers and children undergo a major transition between the ages of 3 and 5 years in their ability to switch between two competing dimensions on this task. At 3 years of age, most toddlers will continue to sort by the first dimension, struggling to switch to sorting in another dimension even though they know that this is the task requirement (e.g., Kirkham, Cruess, & Diamond, 2003; Kloo et al., 2008; Perner & Lang, 2002; Rennie, Bull, & Diamond, 2004; Zelazo et al., 2003). In contrast, by 5 years of age the majority of children can accurately sort by the postswitch rules and tend to switch immediately when instructed to do so (for detailed reviews, see Zelazo, 2006, and Zelazo et al., 2003).

According to Zelazo and colleagues, this inability to switch dimensions at 3 years of age reflects an immaturity in their cognitive flexibility, specifically, to select from a hierarchical rule order or structure such as "if-if-then" rule needed for selecting the postswitch rule rather than the preswitch rule (referred to as the *cognitive complexity and control theory [CCC]*, Zelazo & Frye, 1998; Zelazo et al., 2003). For example, the rule would be: "If we are sorting by shape, and if this is a rabbit, then it goes here, *but* if we are sorting by color, and if this is a rabbit, then it goes there." Toddlers perseverate because they fail to adopt the use of this higher-order rule.

We would make two points about this theory. One is that it clearly requires attention to the relevant dimension as specified in the rule, so it

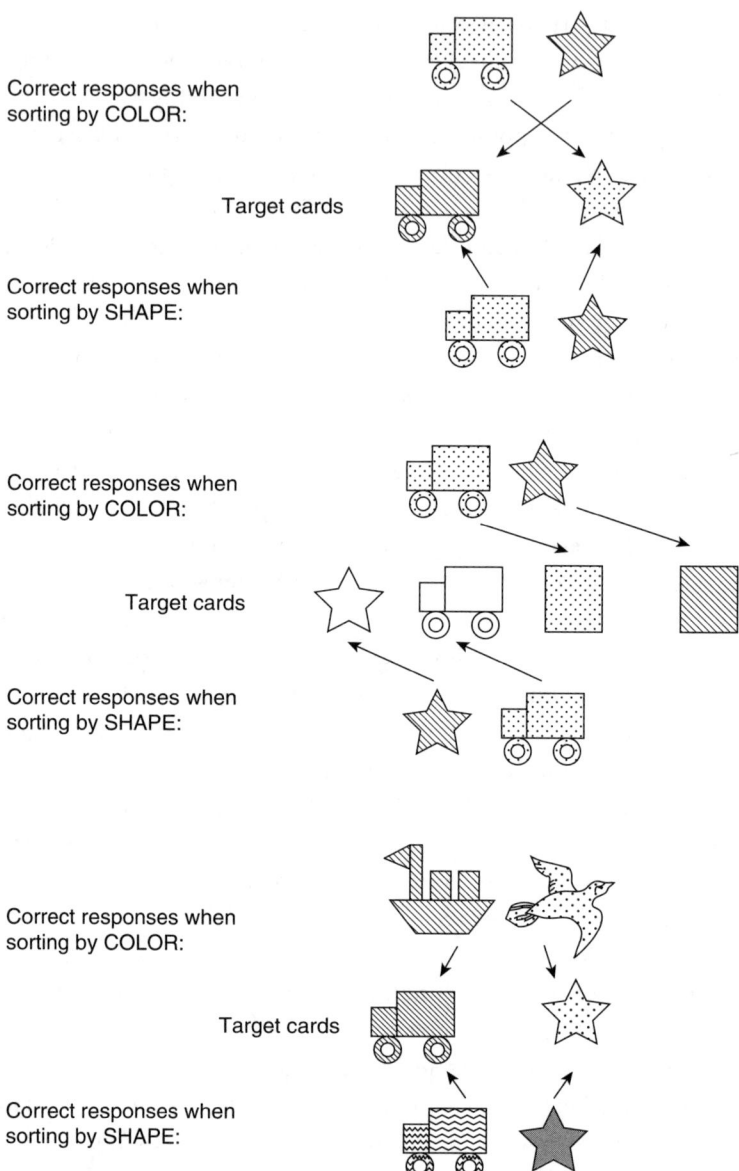

Correct responses when
sorting by COLOR:

Target cards

Correct responses when
sorting by SHAPE:

Correct responses when
sorting by COLOR:

Target cards

Correct responses when
sorting by SHAPE:

Correct responses when
sorting by COLOR:

Target cards

Correct responses when
sorting by SHAPE:

Figure 7-5. Example of stimuli used by Rennie and colleagues (2004).
Redrawn by permission from *Developmental Neuropsychology 26*, 429.

is not in conflict with the attentional inertia theory described below, but rather an elaborated version of that theory. The second is that since the rules refer to stimulus dimensions (shape, color), it would seem to require attention to dimensions rather than simply the specific values of these in the stimuli. This point is not always clear and it is important. If, with a shape-sorting rule, attention is only to rabbit and boat, then changing the shapes after the switch (as is done in some versions of the task) should preclude the persistence of attention to these, whereas if attention is directed to shape in general, such a change will have little or no effect. In view of the way the rule is formulated in the theory, we assume that attention is directed to a dimension rather than just the specific values displayed during the first stage of the task.

Diamond and colleagues offer *attentional inertia* as an alternative explanation (e.g., Diamond et al., 2005; Kirkham, Cruess, & Diamond, 2003). Kirkham et al. (2003) state that, "it is not that [the children] fail to realize that something can be both blue and a truck. Indeed, if queried they can easily state the color and shape of the stimulus. However, having adopted the mindset that blue things go with the blue model card, they have great difficulty switching to think of a blue truck in terms of its shape" (p. 451). In order to move from perseverating in one dimension to sorting in another dimension, the child needs to inhibit this initial mindset and to successfully treat familiar stimuli according to new rules.

In a series of seminal papers, Diamond et al. attest that young children experience difficulties in switching from sorting by one dimension (e.g., color) to sorting by another dimension (e.g., shape) because they are unable to think about an object from multiple perspectives (e.g., as a square that is also the color blue) and thus cannot inhibit one representation from another representation of the same object. They proposed that, by manipulating the inhibitory demands on the DCCS such that the child is not required to think of the same object from two perspectives, for example, as both a shape and as a color, so that children are therefore less likely to become stuck in one dimension (attentional inertia), very young children should be able to successfully switch sorting dimensions. (For an extensive review of their findings, see Diamond et al. 2005, also Kirkham et al. 2003; Rennie, Bull, & Diamond, 2004). Incorporating a developmental approach, the authors assessed performance across three age groups: 2½-year-olds, 3-year-olds, and 3½-year-olds.

There were two conditions (integrated vs. separated dimensions), and each child was tested on both. In the *integrated-dimension* condition, the procedure and task design was identical to the traditional DCCS task already described. In the *separated-dimension* condition, model cards depicted a black car on a green background or a black phone on a yellow background. The sorting cards followed the same design, and each sorting card matched a model card on one dimension (shape or color). In both conditions the correct answer required sorting by one dimension (either color or shape) and then making a switch to sorting by the other dimension.

The findings showed a very clear developmental progression. In the separated condition, 17% of 2½-year-olds succeeded in sorting responses postswitch compared to 10% in the integrated condition. By 3 years, this figure had risen exponentially so that 44% were succeeding on the separated condition compared to 18% on the integrated condition, and at 3½ years, 63% of children succeeded on the separated condition compared to 41% on the integrated position. These novel findings demonstrate that a child as young as 3 years (in some cases, even 2 years) can successfully switch dimensions when the conditions do not require inhibition as was the case in the separated condition and continue to show strong developmental progression from 2 to 3½ years. In contrast, when the conditions are integrated (as in the traditional DCCS), developmental progression is much slower.

In summary, it would appear that at approximately 4 years of age there is a shift in development across a range of critical skills that require holding in working memory two rules, and inhibiting a response to a previously successful response. It is now clear that the original interpretation of the DCCS task as a straightforward index of ability to inhibit an existing rule in favor of a new one was basically correct. However, the circumstances in which this problem occurs have now been clarified and the developmental progression demonstrated. But when the task is altered to remove the need for inhibition, very young children can perform it well.

Response Inhibition

As we have seen already, the ability to suppress unwanted actions and to remain "on task" is a core component of attentional control and undergoes rapid development from infancy through to 4 years. In the DCCS, for example, response inhibition is important for successful task switching (inhibiting the tendency to repeat a response that has been successful previously). We now discuss several further conditions under which prepotent responses have to be suppressed. The first requires suppression or delay of a prepotent or planned response (e.g., Johnson, 1995; Kochanska, Murray, Jacques, Koenig, & Vandergeest, 2006; Scerif et al., 2005). The second requires responding immediately to some commands but inhibiting others (e.g., Jones, Rothbart, & Posner, 2003), and the third requires choosing between two salient but conflicting response options and providing a novel response that is in direct opposition to the prepotent response (e.g., Diamond et al., 2002; Gerstadt, Hong, & Diamond, 1994; Simpson & Riggs, 2005a, 2005b).

SUPPRESSION OR DELAY OF A PREPOTENT OR PLANNED RESPONSE

Saccadic Control   The ability to orient toward visual stimuli that are interesting and novel is present early in development. Oculomotor control, as measured by saccadic eye movements, changes dramatically in the first 12 months. There is a shift from eye movements that are automatic and triggered in response to exogenous factors (e.g., salience, shape, and so forth), as is evident in the first few weeks, to a more endogenous control of saccades beginning at approximately 2 to 4 months of age (e.g., Johnson, Posner, & Rothbart, 1991). This latter ability is measured using the *antisaccadic* paradigm, which requires fixation to a central spot until a peripheral target is presented (to either the left or right of the fixation point) and then an eye movement to the mirror position in the opposite visual field. This differs from the *prosaccadic* paradigm, which requires the infant to look at the target and involves no inhibitory control, since we make prosaccades almost every waking second. Four studies that have made a significant contribution to our understanding of the inhibition of automatic saccades in early infancy are described below.

The first is by Johnson (1995), who assessed 4-month-old infants's ability to inhibit automatic saccades toward peripheral stimuli. In Experiment 1, infants were placed in front of a computer screen that presented a multicolored display. The fixation stimulus was a looming box shape that appeared in the center of the screen. When the infant was

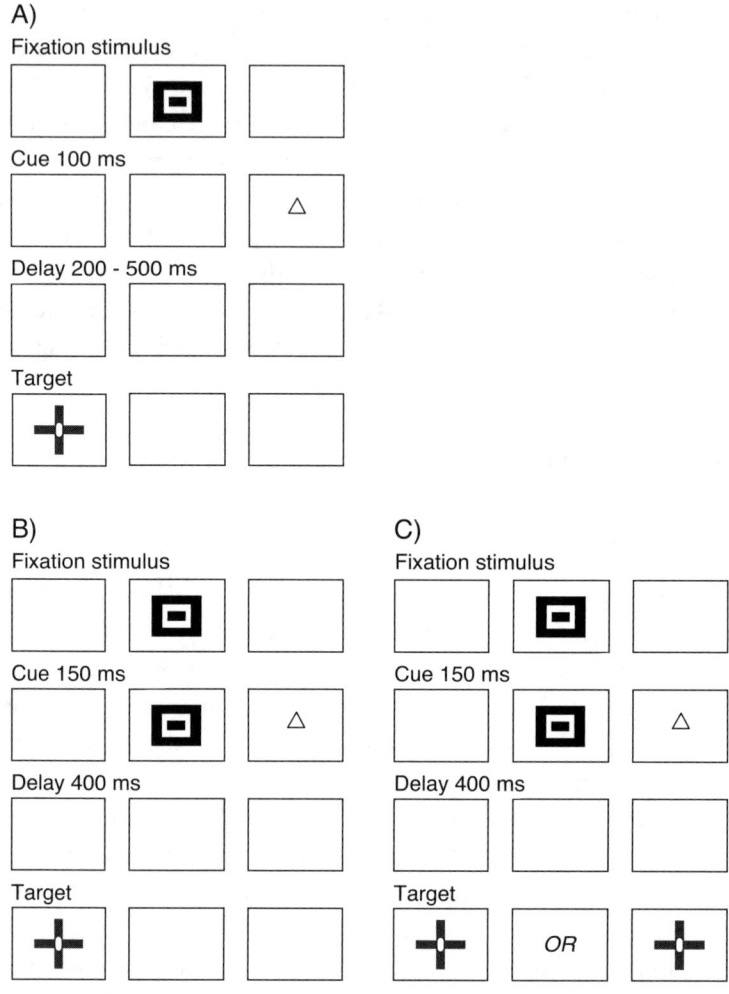

Figure 7-6. Example of stimulus sequence used by Guitton et al (1985) and adapted by Johnson (1995).
Redrawn by permission from *Developmental Psychobiology 28*, 283.

judged to be fixating on the pattern, a peripheral cue stimulus appeared (flashing green diamond) on one of two side screens. Following the onset of the cue, there was a 400 millisecond gap before the presentation of the target stimulus (a multicolored rotating "cogwheel") on the opposite side screen (either right or left). Thus, the peripheral cue always predicted the appearance of the target stimulus. The trial was terminated when the infant looked toward the target shape. See Figure 7.6 for examples of the stimulus sequence.

The aim was to assess whether infants used the cue to predict the location of the target and were able to inhibit prosaccades to the cue and facilitate those to the target. This proved to be the case. Prosaccades to the cue declined over time, and responses directed at the target got faster. However, it was possible that this was due to faster habituation to the simpler cue stimulus than the target. To investigate this possibility, 4-month-old infants were again tested using the identical procedure except that this time the cue appeared in the same location as the target on 50% of trials so was not predictive. No decline in prosaccades to the cue or reduction in response time to the target occurred in this case, so there was no evidence that habituation to the cue was occurring. These findings were the first to demonstrate that infants as young as 4 months old could inhibit prosaccades to an anticipated peripheral stimulus. However, there was no evidence for true antisaccades toward the target prior to its appearance, even though voluntary control of saccades has been shown at this age. In fact, response to the target was faster than to the cue, whereas the reverse is true for adults. This implied that such responses as occurred were anticipatory responses rather than true antisaccades. Johnson argues that the antisaccade task requires a complex oculomotor circuit that inhibits the generation of automatic saccades in the superior colliculus and that this circuit has not yet developed at this age.

Utilizing a much wider age range (between 8 and 38 months), Scerif and colleagues (Scerif et al., 2005) took a developmental approach to charting the age-related changes in saccadic movements. Incorporating a similar sequence of trial events to Johnson's (2005), but a different stimulus sequence (this time the cue was presented *after* the offset of fixation rather than remaining on screen while the cue was presented, as in Johnson's study), Scerif et al. predicted that the production of antisaccades would emerge and gradually improve through toddlerhood, specifically, that older toddlers would decrease looking toward a predictive cue and show improvement in their ability to control (involuntary) saccades. See Figure 7.7.

To test whether infants who showed age-related improvements were not simply producing more predictive saccades (as demonstrated by the Johnson findings), the percentage of "corrective" (anticipatory saccades toward the target preceded by looks toward the cue) and "reactive" saccades (stimulus-driven, exogenous saccades that followed the appearance of a target) was calculated. As predicted, in older infants the mean latency

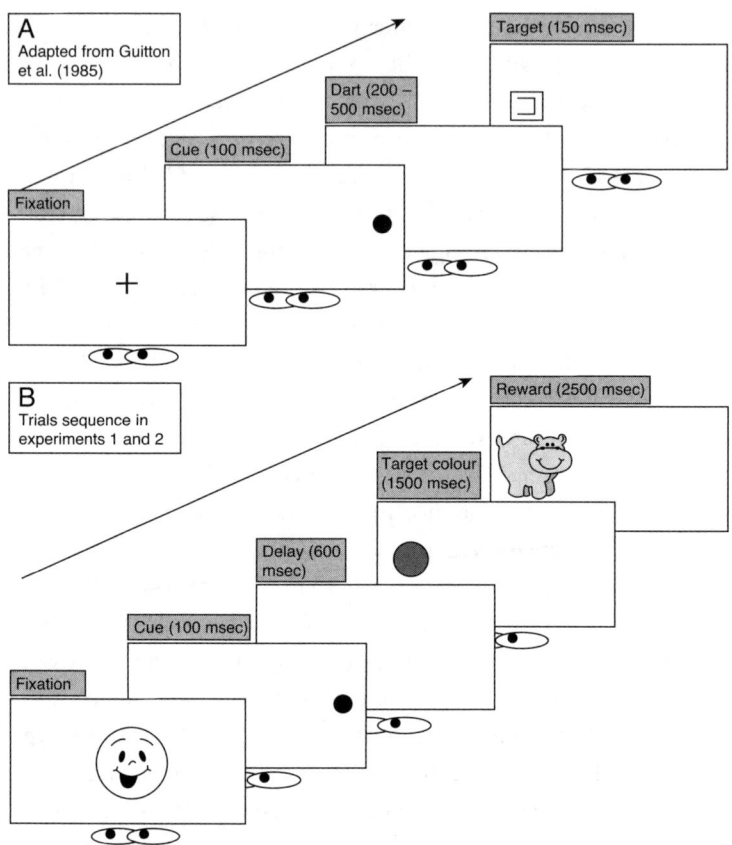

Figure 7-7. Example of stimuli used by Scerif and colleagues (2005).
Redrawn by permission from the *Journal of Cognitive Neuroscience 17*, 593.

of saccades to the cue (prosaccades) was less than the mean latency of saccades to the target stimulus (antisaccades), as found in adults, suggesting that the latter were endogenously controlled and required inhibition of prosaccades. Cue saccades were also faster than both the corrective antisaccades and reactive antisaccades. This is the first study to demonstrate that the inhibition of prosaccades and the generation of antisaccades follow a trajectory from infancy to toddlerhood.

A criticism recently levied at both the Johnson and Scerif et al. studies is that their failure to recognize antisaccades in early infancy may not be due to maturational effects but to the "adult" criteria (developed by

Guitton, Buchtel, & Douglas, 1985) they used to classify infant responses. Nakagawa and Sukigara (2007) assessed whether responses to the opposite side of the cue (antisaccades) could be elicited in infants, aged 3–11 months, if they accepted latencies up to 600 milliseconds post-target onset (i.e., looking toward the target location in the absence of a cue look), as opposed to the adult criterion of 100 milliseconds adopted by Johnson (1995) and Scerif et al. (2005). This study also used, for the first time, head movements as well as eye movements to assess performance during the antisaccade paradigm. They observed longer latencies of response to the target stimulus (antisaccade) than to the cue, arguing that differences in latencies (although marginal in this study) probably reflect the differing mechanisms in coping when an exogenously driven response is required versus an endogenously driven response, which requires more control. Thus, with the less stringent criterion for antisaccades, they were able to show some evidence for these responses at a younger age than in the previous studies; such responses were, however, more consistently present in the older children.

SIMPLE DELAY

In one of the only toddler studies to date to assess delay across a comprehensive range of measures, Kochanska, Murray, & Harlan (2000) assessed toddlers at 22 months and 33 months on three delay tasks described in Box 7.1. Although the reader will appreciate that each task is slightly different, there was a coherent picture with significant age-related changes across all tasks from 22 to 33 months.

The authors conclude that the emergence of more effortful control alongside other factors (e.g., sex differences and parental factors) drives the developmental progression during the second year. However, the independent variables that were included in this study only explained 11–13% of the variance, so much remained to be explained. Undoubtedly, maturation of the frontal lobe system has a major role in this emerging control. See Chapter 4 for a description of the developing brain.

Using a similar delay task to that developed by Kochanska et al., Carlson & Moses (2001, and see also Carlson, 2005), in an extensive investigation of preschool children's inhibitory functions, assessed 3- to 5-year-olds on the "Gift Delay" task (in which the child has to delay

---

Box 7.1   Kochanska, Murray, and Harlan's Delay Tasks

- "Snack Delay" task requires the toddlers/children to sit with their hands on a mat and to wait until they hear a bell in order to retrieve a candy that was visible under a transparent cup.
- "Wrapped Gift" task requires toddlers/children to not peek as a gift was being wrapped. The trial lasts for 60 seconds, and then the experimenter leaves the room to get a "bow," and the toddler is told not to touch the present in her absence.
- "Gift in a Bag" task requires the toddlers/children to not touch a colorful bag placed on a table containing a wrapped gift until the experimenter comes back with a bow.

Scores across all tasks reflected level of inhibitory control and latency.

---

opening an attractive gift) and on the "Pinball" task (Reed, Pien, & Rothbart, 1984). In the Pinball task, children were instructed to hold back the plunger that will eventually release the ball on a tabletop pinball machine in order to land in one of six holes so that a colorful character would pop out. They were instructed to wait in that position until they hear the "Go" command, when they can then release the trigger. Six trials are presented in the following order with delays of 10, 15, 25, 15, 20, and 10 seconds. The average wait time was recorded. The findings show some interesting age profiles. When one simply compares the mean performance of 3- versus 4-year-olds on both tasks, as in the Carlson and Moses study, then there are no significant differences in performance. Both age groups appear to perform at a comparable level.

However, when one looks at performance in terms of percentage of children within each age group who correctly perform the task, as in the Carlson (2005) study, then we see age-related patterns. For example, on the Gift task, there is a definite improvement with age with 42% of young 3-year-olds passing the task (pass criterion: no peek at the gift prior to being told to turn around and open it) compared to 74% at 5 years of age. Likewise, on the Pinball task, 67% of children passed five out of six trials at the young 3-year-old age compared to 95% at the "older" 4 years. Closer examination of these data revealed that the biggest leap in performance occurred within the 4-year-old age group itself, with 81% of the younger 4-year-olds passing the task compared to 95% at the upper end of age 4. It would seem that by the age of 4 to 5 years, children have reached a plateau on tasks that require the inhibition of a simple delay and can perform these tasks with ease.

RESPONDING TO SOME COMMANDS WHILE IGNORING OTHERS

The ability to ignore some commands and respond to others when both are prepotent is typically assessed by the "Simon Says" task and resembles the Go–NoGo paradigm in its requirements that only certain conditions need to be responded to, while others need to be suppressed or ignored. It differs in that the response is to a series of continuous verbal commands. A recent study by Jones, Rothbart, and Posner (2003) assessed three age groups of children (36 to 38 months, 39 to 41 months, and 46 to 48 months) on a standard "Simon Says" task requiring children to inhibit responses from an inappropriate animal and instead respond only to the commands of the appropriate animal. Trials were divided into activation trials and inhibition trials. The former required the child to follow the instructions as directed by the appropriate cue (e.g., elephant says "touch your nose") and the latter to inhibit the instructions from the inappropriate cue (bear says "touch your nose"). Two types of errors were assessed: incorrect response on an activation trial and failure to inhibit responding on an inhibition trial. Reaction times were also recorded separately for correct responses on activation trials and incorrect responses on inhibition trials. It was predicted that children would develop progressively more control over their prepotent responses from ages 3 to 4 years as reflected in their better performance on inhibitory trials and be able to detect their own errors as reflected in increased reaction time for correct activation trials.

   Findings were as predicted with children at all ages performing remarkably well on the activation trials (90–94% success rate). Inhibitory control, as measured by accuracy, on the inhibition trials showed a marked improvement from the youngest group (36 to 38 months, 22% accuracy) to the intermediate group (39 to 41 months, 76%) and a further nonsignificant improvement in the oldest group (46 to 48 months, 91%). In the two older groups, reaction times for inhibitory errors were significantly slower than correct responses on activation trials, as were correct activation responses after an inhibitory error, suggesting that older children became more cautious in their subsequent responding when an error was made. By the age of 41 months, therefore, there was clear evidence of ability to suppress commanded actions. A similar developmental profile of performance on this paradigm has also been reported by Carlson & Moses (2001) and has been extended to 5 years of age using a more complex, anti–imitation version of the task (see Carlson, 2005). One interesting observation by Jones et al. was that physical rather than verbal self-regulation facilitated a child's "holding back" on the inhibition trials and appeared when he or she began to successfully inhibit a response but was rarely seen in the older children. The authors link this age decline in physical strategies (from 3 to 4 years of age) with a similar reported pattern of emergence and then decline for verbal strategies (self-regulatory speech).

In sum, the ability to inhibit a prepotent response across varying cognitive settings undergoes a striking developmental shift from the ages of 2 to 5 years with accelerated growth in the third and fourth years.

RESOLVING CONFLICT

The ability to resolve conflict in terms of responding directly opposite to what is prepotent is a core aspect of inhibition. In an ingenious series of studies by Diamond and colleagues used the "Day–Night" task (described in Chapter 6). In brief, the original task was composed of two conditions, the sun–moon condition (50% of cards depicting a moon and stars on a black background, and 50% depicting a bright yellow sun on a white background), and an abstract (control) condition (50% of cards with a red and blue checkered pattern, and 50% of cards with two red squiggles that formed an X). Children were instructed to say "night" when shown the card with a sun on it and say "day" when shown the card with a moon and stars on it. In the control condition, half the children were told to say "night" to the checkerboard pattern and "day" to the squiggle, and the others were told to say the reverse. Thus, children had to hold two rules in mind and to inhibit the urge to give a verbal response in the opposite direction. Because of the complexity of the task demands (e.g., having to remember two rules), Gerstadt et al. predicted that younger children (less than or equal to 4¼ years) would have considerable difficulty on this task both in terms of correct responses and time taken to formulate a response.

A total of 240 children were tested, of whom 160 children completed the sun–moon task (divided into 8 age groups of 20 children: 3½, 4, 4½, 5, 5½, 6, 6½, and 7 years), and 80 children completed the control task (divided into 4 age groups of 20 children: 3½, 4, 4½, and 5 years). There were 16 trials across each task. The results for the sun–moon task were as predicted with younger children performing less accurately and more slowly than older children. Closer inspection of the data also revealed some interesting patterns. First, younger children were more accurate when they took longer to respond. In contrast, when the latencies were shorter (e.g., they speeded up their responses), they made more errors, and second, when errors occurred they were predominantly of two types and tended to cluster around the 3- to 4½ -year age group. These errors are described in Box 7.2

In contrast, performance on the abstract (control) task did not elicit the same degree of difficulty of error types. Gerstadt et al. suggest that it is immature inhibitory control, rather than having to hold two rules in mind, that explains why younger children perform so poorly on the sun–moon task. In the abstract (control) version of the task, which showed little difference in performance over age, children still had to remember two rules, but they were not required to inhibit a prepotent response.

In a modified version of the sun–moon task, Diamond et al. (2002) attempted to tease out the contributory factors of memory load versus inhibitory load and found that children as young as 4 years were able

---

Box 7.2 Errors Types in the Day–Night Task

- *Matching errors* were the most common and occurred when a child "matched" his or her response to the visual display of the card such as saying "day" to the white sun card and "night" to the black moon card.
- *Alternating errors* occurred when a child repeatedly alternated between "day" to one card and then "night" to another, then "day" to the next, and so forth.

to succeed. They did this in three main ways. First, by training children to use a chunking strategy to remember the two rules, second, by removing the semantic conflict of the tasks such that instead of saying "sun" to a "night" card and vice versa, they were told to say "dog" to a sun card and "pig" to a moon card or vice versa. It was hypothesized that the lack of a semantic relation between the stimuli and the responses would reduce inhibitory load, making the task easier for younger children. Third, they inserted a delay between stimulus and response, thus giving children more time to recall the correct response. Instructing a strategy for remembering the rule had no effect. The delay helped to some extent, but the main factor in facilitating performance was the removal of the semantic conflict. The findings were as predicted, with children as young as 4 years succeeding easily on this revised task. Diamond et al. conclude that the relation between the response to be activated and the response to be suppressed is key in understanding the developmental profile of prepotent inhibition. When a response required is semantically similar to and the direct opposite of the to-be-inhibited response, then children under 5 years will typically fail. However, it is not the case that they do not know how to inhibit saying what a stimulus represents, but they cannot do so when the stimuli and responses to be inhibited are semantically linked as they are in the Day–Night task.

Two recent studies by Simpson and Riggs (2005a, 2000b) examined the task in detail. They demonstrated that the same effect was obtained when direct naming of the stimuli was required (saying "black" and "white" to white and black cards). They then asked whether 3- to 5-year-olds might find the task especially difficult because of what they term a "response set effect" (however, since this term is often used in a different sense, we will avoid it.) They point out that two factors may impair performance in the standard task, where stimuli are from a single category (e.g., color). Responses are drawn from the same category, but reversed, so that each response conflicts both with the stimulus it names and with the other possible (prepotent and correct) response.

These two effects are separable. Using different color names as responses (e.g., blue and yellow) preserves the conflict of the response with the stimulus, but there is no competition from the other response, as it is also incompatible with the stimulus. On the other hand, using a pair of unrelated stimuli (car and book) and requiring the reversed responses "book" and "car" reduces the conflict between the stimulus and the response (at least compared to the color version) but maintains the conflict from the other prepotent and correct response. Simpson and Riggs found that the biggest effect was due to reversing the correct responses, so that there was always a prepotent conflicting response available. They concluded that inhibition was the main factor and direct naming produced the same result. Using errors as the dependent measure, they found improvement from age 3½ to 5 years, at which point performance approached ceiling. However, response times continued to decrease up to 11 years of age, at which point there was no difference between normal and opposite naming responses.

Conclusions

Collectively, a decade or more of research has provided a wealth of information about how attentional control emerges in infancy and its gradual progression into the preschool years. In this section, we have caught a glimpse of how cognitive development itself plays a crucial role in establishing the essential building blocks of inhibitory control. By the time a child enters school, he or she would have progressed significantly in the ability to switch between two simple but incompatible rules, to override prepotent tendencies and substitute a conflicting response in some situations, or to suppress a prepotent response until the appropriate moment and then respond. However, as we have already indicated, attentional control does not exist in isolation (like any component of attention), and there is a well-established body of research that describes the interplay between the development of early inhibitory control and the emergence of other core cognitive skills such as language (e.g., Bialystok & Martin, 2004), theory of mind (e.g., Carlson & Moses, 2001), and working memory (e.g., Davidson, Amso, Anderson, & Diamond, 2006). In the following section, as we move into the school years, we will describe more explicitly these relationships but with a specific emphasis on the relation between inhibitory control and working memory.

Five to Twelve years: The Childhood Years

When children enter school at approximately 5 or 6 years of age (somewhat later in mainland Europe), the majority will begin their academic journey with a solid foundation of basic cognitive skills (attention, memory, language, number, social, and so forth) that will act as precursors or building blocks to the establishment of more defined skills over the next few years. In the domain of attention, we have already seen how an explosion of new skills develops from infancy over the course of 4 years. By the time children begin their fifth year, they have a repertoire of attention skills that reflect a growing reliance on endogenous (self-directed) control and are evident in their ability to efficiently select relevant from nonrelevant information across varying conditions, to search for salient information among nonrelevant information, to override prepotent response tendencies, and to switch between two incompatible rules held in working memory. In the following section, we look explicitly at the developmental progression of attention skills (selective, sustained, and control) from 5 to 10 years of age.

Selective Attention

The ability to selectively attend to specific stimuli while ignoring others is a pivotal component of attention, and we have seen how this ability undergoes a rapid development in the infant and toddler years. The interplay between efficient selection and both maintenance and control of attention continues to develop as the toddler progresses into the childhood years. In this period, we see a finer tuning of the selection and filtering processes guided by a greater development in endogenous control. As is clearly demonstrated in the infancy and preschool years, visual-search paradigms provide a window to observe the efficiency of selection of a specific target from an array of distracters, to elicit so-called "pop-outs." Search difficulty is typically measured in differences in the slopes of search reaction times, plotted against the number of distracters, with steeper slope indicating greater difficulty. Distinguishing targets from distracters on the basis of single features is well documented in the child and adult literature (see Trick & Enns, 1998, for a review of studies pre-2000), and there is some recent evidence of early developmental changes in infants and toddlers (see Scerif et al., 2004, and Wilding & Burke, 2006, for examples of single-feature search in toddlers and young children).

Here we focus on the developmental sequence of conjunctive visual search in which combinations of two or more features distinguish distracters from targets. In a series of recent developmental studies by Wilding and colleagues (e.g.,Cornish, Wilding, & Hollis, 2008; Wilding et al., 2001), visual search involving conjunction of color and shape has been assessed in children aged 6–11 years using a novel computerized task. In a second type of study, Donnelly et al. (2007) assessed the role of top-down and bottom-up mechanisms in the search for targets that were defined first by conjunctions of color, number, and orientation, and second, by being the odd one out.

Visual Search

The Visearch task described in Chapter 6 was devised by Wilding to assess single-target search and dual-target search in mid- to late childhood. Single-target search requires the ability to locate a specific type of single target in a complex display, and the dual-target search requires the ability to locate and switch between specified targets in the same display. The dual-search version has typically been used as engaging attentional control, and the findings are described in the relevant section below (see attentional control). To recap briefly, for the *single-target search*, a display is presented consisting of a river, trees, and eight different kinds of "holes," varying in shape, size, and color, on a green background. Children are instructed that they have to find the king of the monsters who is hiding in the black/pinkish brown holes, but that there are small monsters in most of these holes and they should continue until they find the king. In the *dual-target search*, the children are required to alternate between the black vertical and brown horizontal targets to record hits.

The most informative measures of performance are the mean time of correct responses (hits) and the number of responses (false alarms) to nontarget shapes. A consistent finding over this age range (6 to 11 years) has been that mean time decreased with age over the range tested, while the number of false alarms was unrelated to age but was related to teacher ratings of attentional ability (the latter relation was stronger in more difficult versions of the task but was found even with the simpler versions when attention groups were drawn from the extreme tails of the distribution of ability in the study by Cornish, Wilding, and Hollis 2008). These two measures of performance were largely independent. These results suggested that ability to inhibit responses to nontargets is relatively mature as early as 6 years, but the speed of performing the task improves across this age range. However, a large-scale study recently completed in China (Zhan et al., 2010) employed the task over a wider age range from 6 years to 17 years and refined these conclusions (though a caution about cultural differences is clearly imperative). Results were similar for both the single-target search tasks and the dual-target alternating search task. Mean time decreased from 6 years and reached a floor around 14 years. False alarms were significantly related to age in this study, though less consistently than time was, and appeared to reach a floor around the age of 9 to 10 years (though this was not clear cut). The number of moves made with the mouse, a possible index of efficiency in search strategy, declined up to the age of 8 to 9 years then leveled out.

These results demonstrate, as have a number of other findings that we have discussed, that different aspects of a task may reach adult levels at different ages. In general, it would appear that, in terms of accuracy, performance on this type of task is mature around the age of 9, but speed of performance can go on improving up until adolescence. We consider this issue further below.

Focusing specifically on conjunctive search and the role of top-down versus bottom-up mechanisms, Donnelly et al. (2007) provide a comprehensive evaluation of age-related changes in visual search from midchildhood (6- to 7-year-olds) to late childhood (9- to 10-year-olds) and young adulthood (21 to 35 years). Participants performed two tasks, a "conjunction search" task and an "Odd-One-Out" task. The conjunction search task required the detection of the presence or absence of a known target, a purple oblique rectangle bar, among a set of distracters, comprising red oblique and purple vertical bars. The purple oblique target was present in half of the trials, and participants had to press the appropriate key when it was present and a different key when it was absent. The display remained on the screen until the participant responded. The "Odd-One-Out" task required the participant to detect whether one stimulus was present that differed in any way from all others in an array. Participants were unaware of what features the target or distracters would have before the display appeared, so there was little point in preprogramming a search for specific features. Hence, comparison of the two tasks, it was argued, would indicate whether children could adapt their strategy accordingly.

Across both tasks the number of stimuli within each display was 4, 6, or 8 (referred to as set size). The most informative measure was search reaction time. For the conjunctive task, analysis revealed that participants on both target-present and target-absent trials generated significant search slopes indicative of effortful search across all three age groups. Developmentally, both gradients showed an inverse relation to age, indicating that the rate of search for targets increased with age. This effect was still present when baseline reaction times were taken into account. The authors also report different age patterns of the ratio of absent to present slopes as evidence of developmental progression. Adults, for example, generated parallel absent/present slopes compared to 6- to 7-year-olds, who generated absent/present slopes of 2:1 ratio. The authors suggest that this may be due to a change in search strategy or to an increase in the size of the subset of items that can be examined in parallel.

In the "Odd-One-Out" task, developmental differences also emerged. In the youngest group (6 to 7 years), response time was slowest when a size difference was present and slower for the orientation condition than for the color condition, particularly for larger set sizes. There was no such difference found for the 9- to 10-year-old children and adults. The authors offer a number of explanations for this finding, the most likely being that younger children employed qualitatively different strategies to detect targets, especially if they were unsure of the identity of the target. Color targets in particular may stand out in a display and thus be more effective at guiding search than targets defined by orientation. This explanation is also consistent with earlier research by Lobaugh, Cole, and Rovet (1998), who reported a similar inefficiency in discriminating orientation versus color targets in children aged 7 to 8 years old; this difference decreased to a nonsignificant level in older children and adults.

Maintenance of Attention

The ability to remain alert for relevant stimuli in an otherwise unchanging situation (as in vigilance) or to maintain focus on a given task and ignore irrelevant distracting stimuli is a critical component for successful performance across many cognitive subdomains, for example, inhibiting impulsive, automatic responses, decision making, and learning. We have already seen how this ability is active very early in development and undergoes a gradual progression from infancy into the preschool years as the child is able to utilize more effortful, self-directed (endogenous) control. In brief, for effective maintenance of attention one must be able to sustain attention to sensory signals, minimize distractibility to irrelevant stimuli, and maintain alertness over an extended period of time. At first glance there appears to be quite an extensive literature on sustained attention in the childhood years, but, as we shall also see with attentional control, tasks that tap developmental changes across childhood are often prone to ceiling and floor effects, which, in turn, make comparisons of performance across children and into adulthood extremely difficult to interpret.

Outside the clinical domain, there are relatively few studies that have examined the developmental progression of sustained attention, and fewer still have compared performance to the adulthood profile, mainly for the reasons cited above. In the following section, we describe some approaches to vigilance and sustained attention that have demonstrated age-related changes in performance.

Vigilance

The Wilding Vigilan Task described in Chapter 6 was developed for children aged 6 to 11 years old and employs a similar display to the Visearch task, with intermittent signals that a monster is "at home" in one of the holes. Children have to click with the mouse on the targets and attempt to find the king, who appears when 16 targets have been displayed. The number of responses to targets (hits), mean time for target detection, the number of false alarms to nontargets, and the distance wandered while awaiting targets are recorded. The task lasts about 4 minutes, and preliminary investigations of changes in performance over this time showed no evidence for decline, so these measures have not been pursued. This task discriminates between children with good-rated attention and children with poor-rated attention, even after variance ascribable to the single-target Visearch Task (see Chapter 6 for a review of the task) has been removed, and therefore it measures something additional, which may plausibly, but not unequivocally, be assumed to be ability to maintain attention (Cornish, Wilding, & Hollis, 2008; Wilding et al.,

Figure 3-1. The fragile X facial phenotype.
Permission to reproduce kindly granted by the
photographer, Franklin L. Avery.

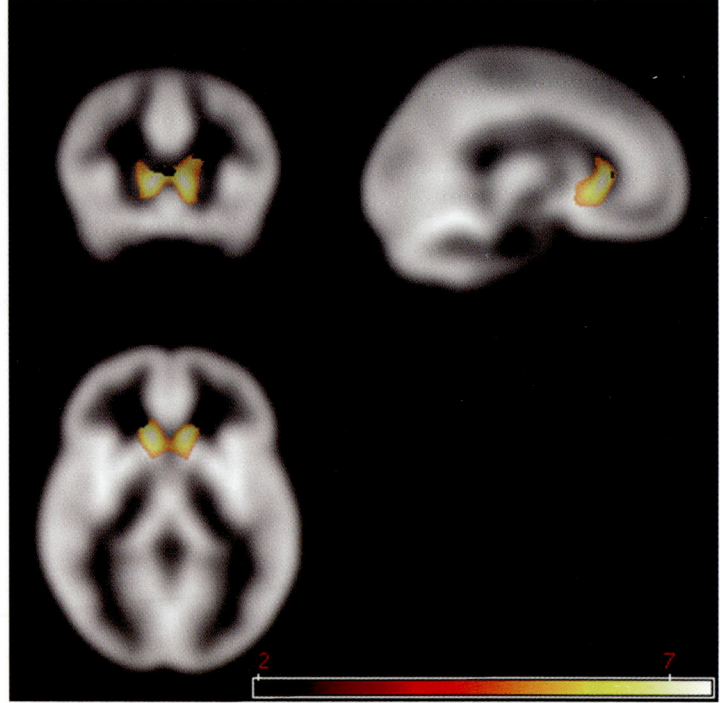

Figure 4-6. MRI images of regions of increased gray matter volumes in individuals with fragile X syndrome compared to typically developing controls.

From Gothelf et al. (2008). Reprinted by permission from *Annals of Neurology 63*, 40–51.

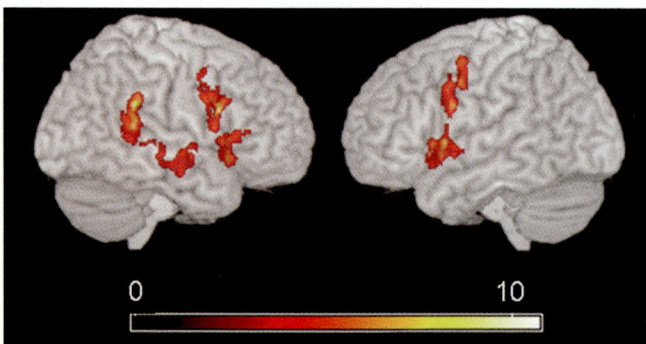

Figure 4-7. fMRI images of the brain activation during a response inhibition task: Go–NoGo paradigm.

From Hoeft et al. (2007). Reprinted by permission from *Human Brain Mapping 28*, 543–554.

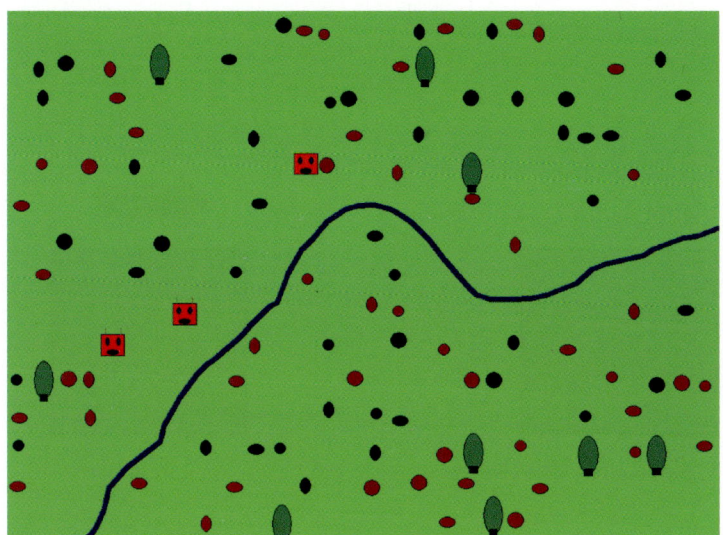

Figure 6-2. The display from the Wilding Visearch task.

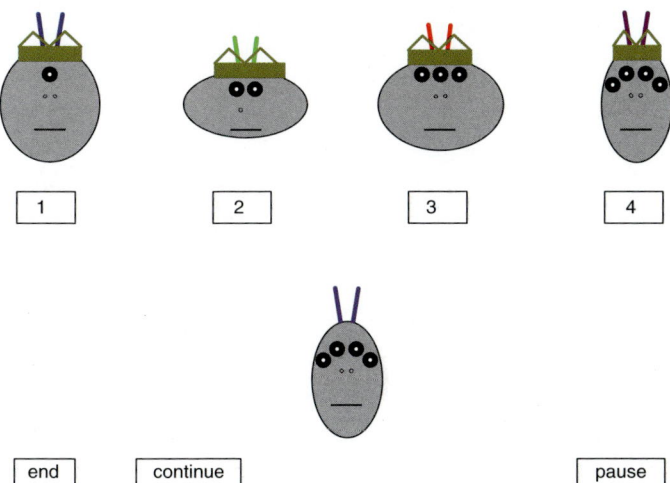

Figure 6-3. The display from the Wilding Monster Sorting Task.

Figure from Box 6.5 Contingency Naming Task.

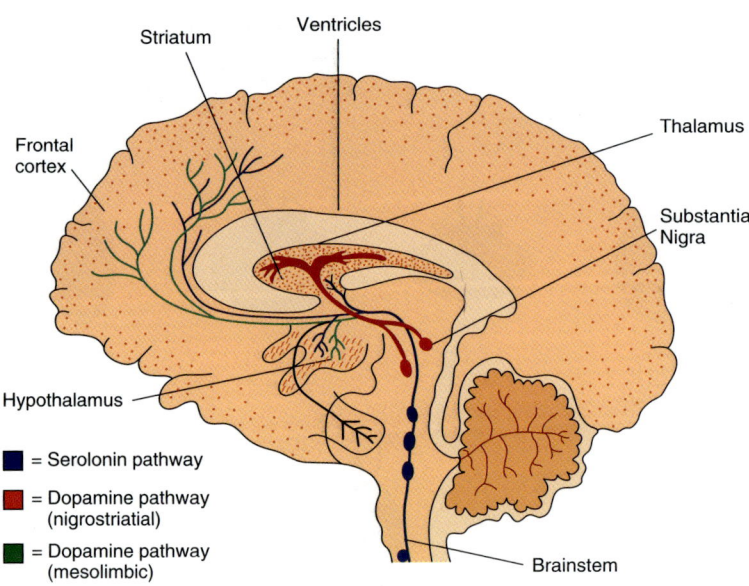

Figure from Box 8.2 Dopaminw and ADHD.

2001). A similar pattern of findings has emerged to those described above for the Visearch data. Mean response time decreases with age but is unrelated to attention rating. False alarms in this task differ somewhat in nature from those occurring in the Visearch task, being responses to nontarget items, normally when no target is present, as opposed to responses to nontargets in Visearch while attempting to find targets that are continuously present. Nevertheless, false alarms in Vigilan were more common in children rated as having poor general attention, as was the distance wandered around the screen with the mouse while awaiting the appearance of a target.

The study of Chinese children referred to above (Zhan et al., 2010) also employed the Vigilan task and produced a clear separation of two types of performance measure: mean time and the number of targets located (hits) on the one hand, and the number of false alarms and distance wandered on the other. Time decreased and hits increased with age up to 14 to 15 years, at which point no further change occurred. False alarms and wandering decreased up to 10 years and then remained constant. It would appear, therefore, that irrelevant behavior while awaiting a target is successfully controlled in the same way as it is by adults by the age of 10, whereas ability to detect targets improves up to adolescence. These results are similar to those found for Visearch in the case of false alarms and speed of hits, respectively, and suggest that two independent developmental trajectories occur in the mastering of these tasks. We again stress that in our view both tasks strongly engage control processes and therefore overlap somewhat with the material to be discussed in the next section.

Maintaining Attention

Three indices of performance are commonly of interest in order to test the ability to maintain attention: *overall sensitivity* (i.e., ability to discriminate targets and nontargets), *decline in sensitivity* over the session, and so-called *vigilance decrement*. Laurie-Rose and colleagues measured the latter by changes in reaction time and caution. The discussion of vigilance in Chapter 2 highlighted the distinction between a reduction in the number of targets detected due to declining sensitivity and due to increased caution in reporting targets and explained how it is possible to separate these two types of change during the duration of a vigilance task. Looking specifically at vigilance decrement over the course of a task, Laurie-Rose (2005) examined sustained attention in children aged 7 to 8 years and adults. She first developed stimuli that yielded very similar performance levels in the two groups. There were 36 events per minute, of which 9 were signals, and the task lasted 16 minutes. In this study, sensitivity did not differ between the two groups and declined over time equally in both groups, but on the measure of caution children were initially less cautious and increased to a level similar to the adults over time; adults did not change significantly. The children were also slower to respond than adults and became even slower over time. Thus, the children showed an increasing reluctance to identify targets as the task proceeded.

In a second experiment, the experimenter varied the rate at which items appeared (10 or 45 events per minute for 16 minutes in all, with targets occurring on a third of the events), since higher event rates commonly induce better performance, often attributed to higher arousal. The results differed somewhat from the first study above. Adults showed superior sensitivity at both rates. Sensitivity was superior at the fast rate, and a decline in sensitivity over time only occurred at this rate. Both groups started at similar levels of caution, which were higher at the fast rate, and children again increased in caution over time. Both groups responded more quickly at the fast rate and children were slower overall. Reaction times also increased over time.

However, it should be noted that only overall statistical interactions are reported, for example, between age groups and time periods. This indicates whether there is a global difference between the groups in the effects of time, but not if this difference is due specifically to differences in linear trends over time; plotting graphs from the table of means offers the only method of determining this and, of course, does not tell us if any apparent differences in linear trends over time were significant. Given this weakness and the differences between the two studies, only cautious conclusions are possible. The young age group appears to have differed mainly in the greater increase in caution and reaction time over the duration of the task. They did not consistently show a greater deterioration in sensitivity over time but did show a greater decline from the first to the second session of testing. The authors interpret their results somewhat speculatively in terms of "optimal arousal theory," in which it is assumed that performance is best at a moderate level of arousal and declines when arousal declines too far (boredom) or becomes too high. However, our discussion of vigilance argued that changes in caution over time could not be explained by changes in arousal, but rather by modification of decision processes following evaluation of ongoing performance. Consequently, it is doubtful whether the obtained differences between the groups can be explained in terms of arousal, and the authors offer no compelling evidence to support their hypothesis. Also, no clear explanation is offered of why increased event rate was not more beneficial to children than to adults. One might have expected that adults were able to modulate arousal more appropriately than children at the slow rate and hence were less in need of additional stimulation.

The tendency for children to show a greater increase in caution might be readily explained in terms of their lower detection rate for targets, leading them to expect fewer targets and therefore to reduce their frequency of positive detections. But this implies a sophisticated control system, despite a great deal of evidence to the contrary that we have been citing.

The *Continuous Performance Task (CPT)* remains one of the most popular measures to assess sustained attention in clinical and nonclinical populations. The initial CPT task by Rosvold, Mirsky, Sarason, Bransome, & Beck (1956) was in two parts. Part 1 required participants to respond by pressing a response key when they saw the letter *X*, and Part 2, somewhat more difficult, required participants to respond only to *X* when it was preceded by the letter *A* (thereafter known as the as the "AX" task). It is interesting to note that in the initial version of the CPT, the occurrence of the target was rare, contrasting with more recent versions that have much greater target frequency, as much as 70% (e.g., Barch et al., 2001). Although used extensively with adults and school-age children, the CPT has only recently been employed in preschool children using such versions as Kiddy-CPT (Connors, 2007). See Mahone (2005) for a comprehensive review of studies that have used CPT in early childhood.

One of the most rigorous longitudinal studies to have assessed age-related change in sustained attention using the CPT is by Rebok et al. (1997). Children aged 8 years at Time 1 were assessed again at age 10 years (Time 2) and again at 13 years (Time 3) on a range of attention measures. The CPT measure included (1) a standard X-only version,

(2) a standard AX version as described above, (3) an auditory distraction version identical to the visual version except for the simultaneous presentation over earphones of the to-be-ignored letters, (4) an auditory version that required the pressing of a button each time that a high-pitched target tone sounded in a random sequence of three tones differing in pitch (e.g., 640, 1,000, and 1,600 Hz), and (5) a visually degraded version of the standard X task in which letter images were blurred. Measures for all five versions included number of correct responses to targets (hits), late correct responses (between 800 and 1,000 milliseconds from stimulus onset), mean reaction time, and error types (misses, "omissions"; and false alarms, "commissions"). However, data were not available on all age groups across all CPT measures. For example, children at age 8 years performed versions 1, 3, and 5; at age 10 years, the entire sample performed versions 1, 4, and 5; and at 13 years, versions 2, 4, and 5. The only common task in all age groups was version 5, the visually degraded version.

Findings show improved performance both in terms of accuracy and reaction times from age 8 to 13 years. Specifically, between the ages of 8 and 10 there was a significant and striking improvement in accuracy with a 10% increase in correct responses over this time period. Between 10 and 13 years, although there was some degree of improvement, it was less striking than in earlier years. The same pattern emerged for error reductions, with the greatest reduction between 8 and 10 years and thereafter a gradual decline. In terms of reaction times, there was a steady decrease from 8 through to 13 years, showing a more consistent improvement with age. The authors conclude that from 8 to 10 years there is a rapid development in sustained attention but then performance plateaus between 10 and 13 years. However, this ignores the continuing decrease in reaction times throughout the whole range. In some respects Rebok's results are similar to those described for Vigilan above with error rates reaching a minimum earlier than reaction times, but the two tasks differed considerably in their demands, so any similarities should be treated with caution.

It is difficult to draw any detailed conclusions about the development of sustained attention between 6 and 12 years, other than that improvement occurs. Better understanding and more adequate theory concerning the impact of a variety of variables on this type of task is needed before their effect on development can be properly assessed. There is some consistency in the tendency of children to increase in caution over the duration of a task, but the interpretation of this finding remains contentious. There also seem to be indications that some aspects of sustained attention reach levels close to those of adults by the age of about 10 years, while others, particularly speed of identifying targets, continue to improve until at least early adolescence.

Attentional Control

Clearly, the first 4 years produce important milestones in the development of attentional control. By age 5, children have a much greater capacity for inhibiting prepotent responses, switching of response mode, and performance of novel actions. At this time, there is also a rapid development in a child's ability to see the world from different perspectives. However, one of the difficulties of assessing attentional control at this age is that so many other individual variables, other than inhibitory control, can have an impact on performance, for example, gender, verbal demands of a given task, level of reading comprehension, learning ability, working memory capacity, and so forth. We are not implying that such variables are absent in the infant and preschool years, but it is likely that their effects are not as great as they are from age 5 upward. It is therefore essential that tasks purporting to tap attentional control do indeed tap this skill and not a range of other cognitive skills that are incorporated into a given task. In order to achieve this goal and be able to provide developmental trajectories that reflect functioning across varying conditions, tasks must be novel, developed specifically for children rather than simply modified adult tasks (i.e., reduced instructions or test items), and be sensitive enough to elicit performance difficulties across a wide age range. In the section below, we focus on two core aspects of attentional control that have shown robust, developmental changes and build upon work from the preschool years: switching across tasks and responses and inhibitory control.

Switching Focus or Response

The ability to flexibly switch back and forth across differing task demands is evident quite early in development. It has a clear developmental progression from toddlerhood onward, and successful performance requires a combination of inhibition (previously relevant stimuli must be suppressed and new rules and responses activated), cognitive flexibility (switching between tasks and rules), and working memory storage. Without a doubt, task switching is one of the most difficult skills to master. In preschoolers, the Dimensional Change Card Sorting (DCCS) task and its variants are the most commonly used paradigm, the findings of which we described in detail above. In the midchildhood years, although research has not been as extensive, a flurry of recent studies has provided a more detailed profile of task switching in childhood and the differing mechanisms that underlie this process.

We have already described (in the section on selective attention) the dual-target visual-search task employed in a number of studies (Cornish, Wilding, & Hollis, 2008; Wilding et al., 2001; Wilding, 2003), which requires continuous switching between two targets. This task has consistently discriminated between children rated as having good attention and those rated as having poor attention. As with the simpler single-target search task using the same type of display, speed of performance improves with age over the age range from 6 to 11 years, whereas false alarm rates are not strongly related to age (see the earlier discussion).

In one of the most comprehensive studies to date, Davidson, Amso, Anderson, and Diamond (2006) assessed children aged 4–13 years and adult controls on a battery of tasks aimed at teasing apart the varying cognitive demands of task switching, for example, switching to an incongruent rule versus switching to a congruent rule, and response site switch (same response site verses opposite response site). Here we focus on one task in particular—the Arrows task—which comprised a single large arrow presented at the left or right of the computer screen and pointed either straight down or toward the opposite side at 45°. See Figure 7.8. There were four types of trials: a *congruent* trial whereby the arrow pointed straight down, and a successful response was to press the button on the same side as the arrow, and an *incongruent* trial whereby the arrow pointed diagonally toward the opposite side, and a successful response was to press on the side opposite the arrow. There were also nonswitch and switch trials, depending on whether the rule in the present trial was the same as in the previous trial. For children over 7 years, stimulus presentation time was set at 750 milliseconds and for children between

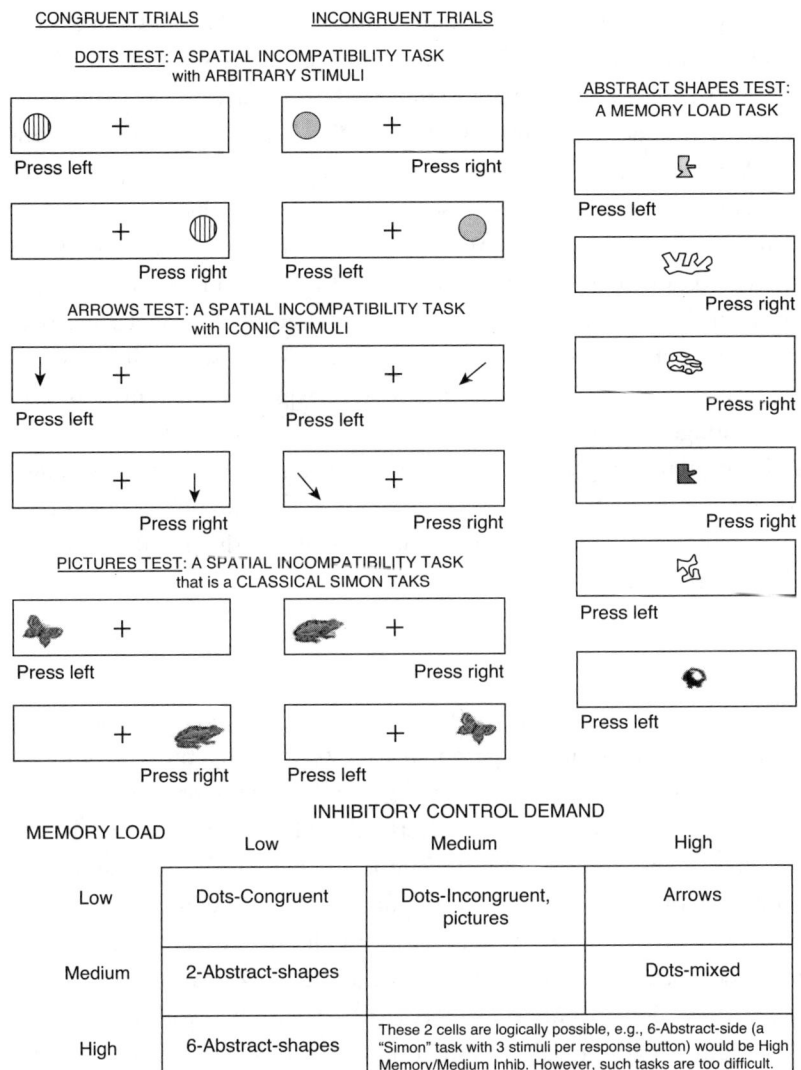

Figure 7-8. Example of stimuli used by Davidson and colleagues (2006).
Redrawn by permission from *Neuropsychologia 44*, 2042.

4 and 6 years, stimulus presentation time was 2,500 milliseconds. The interstimulus interval was 500 milliseconds in all cases.

As predicted, performance improved as a function of age both in terms of accuracy and speed of response. The key finding was that young children could switch successfully provided it was a steady-state switch (e.g., for a block of trials, responses were required on the opposite side to which the arrow pointed), and even then there was a time cost, which adults did not show for such conditions. But the main weakness in children was in cognitive flexibility when required to keep switching back and forth between rules. In other words, it was not so much the ability to inhibit but the ability to control it and keep changing. This showed a much longer developmental progression with children at the age of 13 still not performing at adult levels. Furthermore, adults showed a bigger increase in time and lower increase in errors in mixed blocks than children, obviously compensating for the difficulty by slowing down, which children tended not to do.

Response Inhibition

As we have seen from our review of the infant and toddler literature, the ability to inhibit dominant (prepotent) and often competing responses is a skill that that does not emerge, even in its rudimentary form, until the end of the first year. Its progression is clearly guided by developmental changes in self-awareness and regulation (endogenous control) and requires the inhibition of learned, automatic responses that, although prompted by a current situation, are inappropriate. The ability to deal with and resolve responses that are in conflict with one another is known as interference control and is thought to be mediated in part by working memory. Depending on the task, the resolution of conflict may involve different inhibitory processes. We describe below four types of tasks requiring inhibitory control that are known to produce an interference effect across childhood, with each involving slightly different attentional resources.

SUPPRESSION OR DELAY OF A PREPOTENT
OR PLANNED RESPONSE

The *Stop–Signal* task (along with its sister task, the Go–NoGo task) is one of the most commonly used inhibitory paradigms to assess inhibition of a planned motor response across both typical and clinical populations (most commonly ADHD). Surprisingly few studies, however, have assessed their childhood trajectory (see Carver, Livesey, & Charles, 2001; van den Wildenberg & van der Molen, 2004). At its basic level, the task requires the active suppression or interruption of an activated motor response. In all versions of the Stop task, a speeded response has to be inhibited only upon presentation of a specific Stop signal (either visual or auditory) that can occur at any time subsequent to the Go signal. By manipulating the time interval between the presentation of the Go stimulus and the Stop signal (referred to as the Stop–Signal delay), inhibition can be biased such that short delays make it easier to inhibit the response and longer delays make it harder. This task is interpreted in terms of the mathematical, "horse race" model developed by Logan (Logan & Cowan, 1984; see Band, van der Molen, & Logan, 2003), which speculates that the Go response and the Stop response represent two independent processes that compete for "first place" in the race. If the Go process wins the race, then the response will be executed, but if the Stop process wins the race, then the planned response will not occur. Unlike the reaction time to Go signals, which can be easily measured, the reaction time to Stop signals is not readily measured, and therefore the model derives an estimate of the latency of inhibitory control, the *Stop–Signal Reaction Time* (SSRT). What is especially intriguing about this paradigm is that it mimics an intentional act of control that is similar to many real-life actions we perform everyday, for example, when you start to cross a road and the pedestrian signal changes, or you start to wave to someone then realize it is not who you thought it was or start saying something that you then realize is inappropriate.

A recent study offers some insight into the developmental progression of inhibitory control, as measured by the Stop–Signal task. In an extensive study, over 500 children aged between 4 and 12 years (Tillman, Thorell, Brocki, & Bohlin, 2008) were tested on a newly developed Stop–Signal task. The task was developed as a one-choice paradigm (as opposed to the more standard two-choice) to facilitate testing of very young children. The child is always presented with a car that appears on the screen and is instructed to press the space bar as soon as the car appears. When pressed, the car would drive away. On 25% of trials, a visual stop signal in the form of a road stop sign appears, and the child is informed not to press the space bar as the car would not drive away. The stop delay signal was initially set at 250 milliseconds and then progressively increased by 50 milliseconds after each successful inhibition of a stop signal but decreased by 50 milliseconds with each failure to inhibit. Measures derived included response execution (Go-reaction time), the latency of the inhibition processes (SSRT), and a calculation of inhibition probability. The findings showed developmental effects across the range of measures. For example, SSRTs were longer (indicating poorer inhibition) in younger children, but by 10 years appeared to have stabilized. Likewise, the longer Go-reaction times in younger children appeared to be stabilized at around 11 years. There were also specific periods of development on the SSRT that displayed quite intense activity. Two consecutive periods of improvement were observed, between 5 and 6 years and between 6 and 9 years, suggesting the sensitivity of this measure in detecting subtle developmental changes in young children. Interestingly, a measure of inhibition probability was more variable with age, making it difficult to pinpoint a plateau in later childhood. However, the developmental trajectories for response execution (faster development at the end of the age range) and inhibition of responding (early development) suggest that different systems are involved in these two aspects of performance.

Undoubtedly, this paradigm has potential to provide much-needed information on inhibitory development in younger children, but a potential limitation is the extent to which, as a measure of inhibitory control, it can provide additional information about the developmental progression of older children. The paradigm is limited in this respect primarily because the task design (single-choice reaction time rather than binary-choice reaction time as is typically used in standard versions) may not have been sensitive enough to test for more subtle age-related changes in older children.

Cragg and Nation (2008) used a modified version of the Go–NoGo task to test development of inhibition on NoGo trials. They suggested that inhibition is not necessarily an all-or-none process and the time to inhibit a response on no-go trials may decrease with age. Hence, they required children to hold down the left button on a computer mouse before each trial and release it and press the right button to make a response. Partial inhibition was shown when the left button was released but the right button was not pressed on a NoGo trial. Two groups of children were tested, aged 5 to 7 years or 9 to 11 years. While there was no difference between these groups in the number of failed inhibitions (left button released and right button pressed on NoGo trials), younger children made more partial inhibitions but fewer complete inhibitions (left button held down). Thus, this more sensitive measure showed improvement in inhibition with age.

INHIBITION OF IRRELEVANT STIMULUS INFORMATION

Probably the most famous of all paradigms to measure inhibitory control is the Stroop task (Stroop, 1935). In brief and in its original version, the task comprised two conditions: a "reading color names" condition whereby participants were required to repeat the written meaning of words with differing colored fonts (e.g., say the word "red" when printed in green ink), and "naming colored words" condition whereby participants were asked to verbally identify the color of each printed word, not the word itself (e.g., see the word "red" colored in green ink and say "green"). The task demonstrates how difficult it is, if not impossible, to inhibit automatic, learned responses. It is far easier to read the name of the color words, irrespective of the color of the ink, than it is to name the ink color of a different color word. The former response (reading color words) taps into an automatic cognitive process, but the latter response (saying the ink color of different color words) does not and is therefore susceptible to interference from the stronger conflicting processes, resulting in a delay in response time. The traditional Stroop effect (color Stroop) has been extensively documented in typically developing adults across a wide age range (see MacLeod, 1991, for a review of the early findings; Shilling, Chetwynd, & Rabbitt, 2002; Wuhr, 2007) and is suggested to be a clinical marker of atypical development in conditions such as ADHD (e.g., Lansbergen, Kenemans, & van Engeland, 2007) and schizophrenia (e.g., Henik & Salo, 2004).

In terms of identifying age-related changes across childhood, studies using the standard Stroop task have been limited to older children (8 years and over) in order to ensure that a minimal verbal fluency level has been reached. If a child cannot adequately read the words and does not have the necessary semantic associations, then the interference effect cannot be elicited. An obstacle that many studies come up against, especially when trying to determine developmental progressions, is devising a task that does not immediately produce floor or ceiling effects. Ideally such a task should be testable across a wide age range. A recent study by Leon-Carrion, Garcia-Orza, and Perez-Santamaria (2004) attempted to assess children aged between 6 and 17 years on the standard color-word Stroop task and found, not unexpectedly given that younger children will be less fluent readers than the older children, a nonlinear relationship with an increase in interference from 6 to 8 or 10 years (depending on whether it was response times or errors) and then a decline thereafter. What was intriguing, but not addressed in this study, was when in development the interference effect levels out. There is some indication from the data that age 12 may have been a critical point.

Attempts to develop a Stroop-like task that will tap a prereading age sample of children have met with varying success. There are two exceptions. The first is the Day–Night task (developed by Gerstadt and colleagues in 1994 and described in detail in the previous section), which has been successfully employed in preschoolers and appears to demonstrate a clear developmental progression. A point of caution, however, is that although the Day–Night task is similar in many respects to the Stroop task, the two tasks involve somewhat differing cognitive demands. For example, in the standard Stroop, participants are required to choose between two features: word and color. In the Day–Night task, there is no competing feature, and the competition arises from the prepotent response to the picture. The Day–Night task also has a tightly defined age range and begins to become extremely easy for children 6 years and over, so its applicability is limited.

The second exception is the "Animal-Stroop" task developed by Wright, Waterman, Prescott, and Murdoch-Eaton (2003) to assess the Stroop-like interference effect in a wider age range of children from 3 to 16 years. This novel, computerized task replaces the word items of the standard Stroop with pictorial designs of animals and requires the semantic categorization of facial information. Here, children are shown four animal images (a cow, pig, duck, and sheep). In the conflicting condition, the heads are exchanged between the bodies. The children have to name the animal from the body alone and to inhibit a preferred or prepotent response based on the animal's head, which is assumed to be normally

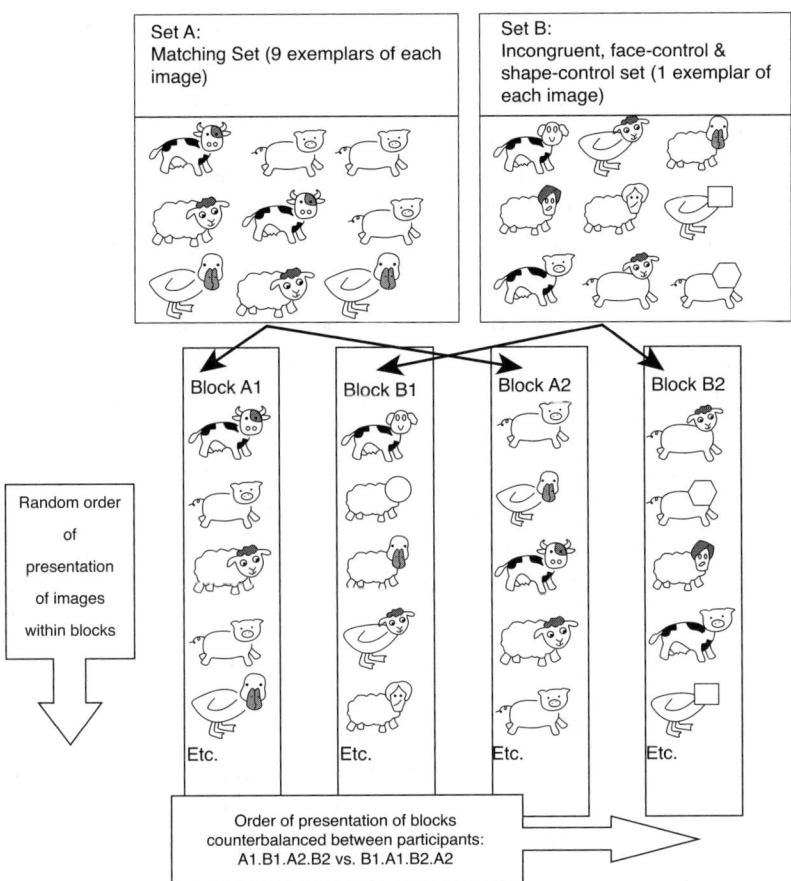

Figure 7-9. Example of stimuli used by Wright and colleagues (2003). Redrawn by permission from the *Journal of Child Psychology & Psychiatry 44*, 566.

the strongest cue for a response. Figure 7.9 shows the similarities between the tasks. Measures included accuracy and error type (e.g., incorrect naming of an animal or an articulation error whereby there is an audible production of an incorrect animal name). It was predicted that response time to the incongruent images (mismatched body/head) would be longer than response time for congruent images.

Findings were as predicted, with an observable cognitive "cost" in inhibition when children had to respond to the animals in the incongruent stimuli compared to other control stimuli. This interference effect also showed a decline with age similar to that reported in the standard Stroop. However, it is noteworthy that this task was still able to elicit a Stroop effect at 16 years, albeit somewhat less marked than in young children. However, it is still unclear when actual adult levels of performance are reached. The pattern of these findings has been recently replicated in a study of 6- to 11-year-olds employing a paper-pencil version of the Animal-Stroop task (Nichelli, Scala, Vago, Riva, & Bulgheroni, 2005).

The "Flanker" Task, originally developed by Eriksen & Eriksen (1974), has been used extensively in the literature on adults to assess the ability to ignore interference from irrelevant stimulus information (distracters) designed to induce response conflict. The Visearch task (Wilding et al., 2001) also assesses inhibition of distracter interference but was developed specially for young children. It requires search for a series of targets in a cluttered scene and is therefore discussed under "Selective Attention," though clearly it requires control processes. The Flanker Task typically comprises two predesignated target stimuli (e.g., letters *J* and *R*, or arrows pointed to the left and right) that require a motor response, either a left or right button press. The target stimuli are also flanked by distracters that are either congruent (similar) (e.g., *JJJJJ* or *RRRRR*) or are incongruent (e.g., *RRJRR*) with the correct response. In adults, the most consistent finding is a tendency to respond to the distracting flanker elements resulting in an interference effect. Relatively few studies have modified the task to examine age-related effects in typically developing children (although it has been used quite extensively to assess interference effects in children with ADHD; see Chapter 8). We highlight below the findings of one of the few studies to trace the developmental progression of performance on a modified flanker task in young children (Rueda et al., 2004).

Using an adapted child version of the Attention Network Test (developed by Posner and colleagues and modeled around the Flanker Task), children aged 6 to 9 years viewed a target array of yellow line drawings comprising either a single fish or a horizontal row of five fish. Children were asked to respond manually based on whether the central fish was pointing to the left or to the right. On the incongruent trials, the flanking fish pointed in the opposite direction, and on the congruent trials the fish were pointing in the same direction. There was also an additional neutral target type. See discussion in Chapter 6 . Each trial began with a fixation cross presented on the screen, and each target was preceded by one of four warning cue conditions: a center cue (an asterisk in the center of the fixation cross), a double cue (an asterisk appearing at the location of the target above and below the fixation cross), a spatial cue (a single asterisk presented in the position of the forthcoming target), or no cue. There was a brief fixation period of 450 milliseconds after the disappearance of the cue, followed by either the simultaneous appearance of the target and flankers, or by the appearance of the target alone. To obtain the conflict scores, a median reaction time for the congruent and incongruent flanker conditions (across four cued conditions) was calculated, and then the incongruent reaction time was subtracted from the congruent reaction time.

Some interesting age-related findings emerged, but the most salient was a remarkable developmental progression from age 6 to 7 years both in terms of better accuracy and quicker reaction time. Thereafter, the profile appeared to remain stable from 7 through to 9 years, and also into adulthood as indicated by their second experiment, which included a comparison of 10-year-old children and adults on both child and adult versions of the ANT. These findings suggest that ages 6 to 7 years may be critical years for the development of attentional control as measured by the Flanker task.

The general conclusion that can be drawn from the studies exploring attentional control in the childhood years is that there exists a rich but varied profile of developmental progression. Different control processes (for example, task switching or inhibition) evoke quite different trajectories, and performance can be dependent on the task itself in addition to the underlying cognitive processes. One source of difficulty in defining more precisely the developmental timing of specific control effects is the lack of sensitivity inherent in many tasks to reliably tap attentional control across a broad age range. This can result in floor/ceiling effects making it very difficult to determine or interpret age trajectories across paradigms. As always, there are exceptions, one being the work of Diamond and her colleagues in tracing trajectories of task switching. We have also highlighted above other stellar work. A further observation is the lack of discussion or evaluation of when, in development, childhood effects might plateau. The emphasis tends to be on explaining findings of early developmental changes, but such studies also yield potentially important information about the end state that is rarely addressed.

Twelve to Eighteen Years: The Adolescent Years

In contrast to the numerous studies that have been conducted in the infant and preschool years, the childhood years and even the adult years (young adults versus older adults), there is a relative absence of studies examining developmental trends during the adolescent years. The reasons for this are somewhat obvious, not least of all being the considerable impact of physical and hormonal changes that occur between the ages of 12 and 18 years. Add to this mixture widening individual differences in gender and cognitive performance alongside major changes in social environment and self-identity, and it is apparent why research is not as extensive in the adolescent period as in other developmental periods. There is also the difficulty in devising tasks that appropriately measure attention and its component parts across a broad age range, and we have red-flagged this previously as an area of concern warranting further investigation. Having said that, the past few years have witnessed an increasing recognition that neural mechanisms are still developing during the late childhood and adolescent period, and that this growth can have an impact on a wide range of cognitive processing including attention, inhibition, and social cognition (for reviews and different perspectives, see Blakemore, 2008; Blakemore & Choudhury, 2006; Paus, 2005). Most notable is the continued development during adolescence of the prefrontal cortex and parietal cortex. It is therefore highly possible that changes in brain development during adolescence are associated with age-related changes in attention performance.

Not unexpectedly, the majority of the studies on attention processing in adolescence have focused on attentional control, and specifically inhibition and task switching (e.g., Cepeda, Kramer, & Gonzalez de Sather, 2001; Huizinga, Dolan, & van der Molan, 2006; Leon-Carrion et al. 2004; Reimers & Maylor, 2005). In terms of selective attention, there is, however, one notable exception by Trick and Enns (1998), who at one level provide one of the most comprehensive assessments of visual search across the life span but who fail to include an adolescent group in their age trajectory, thus jumping 12 years between assessing performance in 10-year-olds and then 22-year-olds. For sustained attention, we highlight findings from Anderson, Anderson, Northam, Jacobs, and Catroppa (2001) on the Codes task and, for vigilance, findings from our recent study of Chinese adolescents on the Vigilan task (Zhan et al., 2010).

Selective Attention

As we have seen throughout this chapter, the term "selective attention" has been used to describe our ability to selectively attend to specific stimuli while ignoring other irrelevant stimuli. Developmental trends in performance in varying components of selective attention are clearly evident across the infant and childhood years and have been described in detail above. One of the most thoroughly researched examples of selection or orienting is visual search, and many developmental studies have emphasized the importance of teasing apart components of attention in the visual-search task, for example, differences between featural versus conjunctive search, to investigate whether there exist within visual search itself differing developmental trajectories dependent upon search type. Returning to a study that we have reported in detail in an earlier section (Trick & Enns, 1998), and one which specifically focuses on age-related changes in featural search (locating a target item on the basis of one feature) versus conjunctive search (locating a target item on the basis of two or more features), the authors examined visual-search performance across three levels of complexity: presenting the target item in a fixed visual location, then presenting the target item in random locations on the screen, and then by varying the number of distracter items presented on the screen to 1, 9, or 17 when the target was present and 2, 10, or 18 when the target was absent. See Figure 7.10. The main aim of this study was to examine whether uncertainty of target location and interference from distracters had differential effects over the age range and whether such effects differed for feature search and conjunction search. As already reported, age groups were composed of 6-, 8-, 10-, 22-, and 72-year-olds, and at age 6 there was a greater increase in time with the addition of only a single distracter than in all the other groups; the three child groups

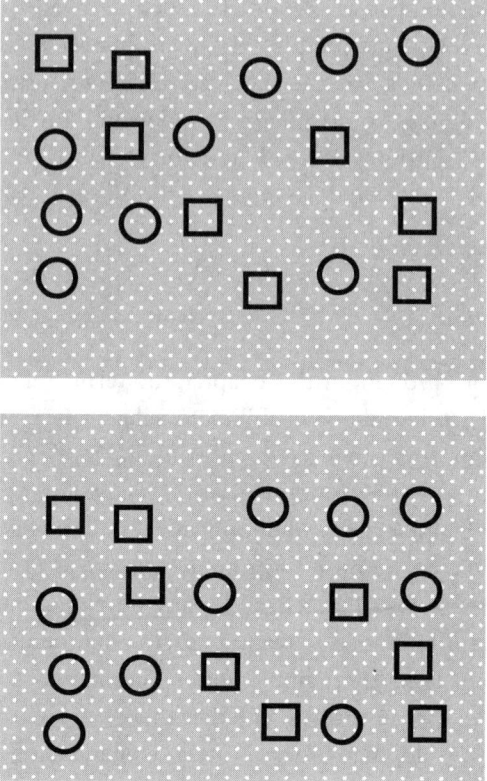

Figure 7-10. Example of stimuli used by Trick and
Enns (1998).
Redrawn by permission from *Cognitive Development*
13, 375.

also showed effects of further increases in the number of distracters.
These effects decreased rapidly up to the age of 10, and there was a further small decrease through the adolescent years up to the age of 22.
These findings suggest that switching between items in complex displays
continues to improve gradually across adolescence into early adulthood.

A similar finding has been reported in a recent large-scale study of
age effects on single and dual-search conditions of the Visearch task
(described in detail in Chapter 6) in a Chinese population of children
and adolescents (ages 6 to 17 years) (Zhan et al., 2010). To recap briefly,
in the single-search version, this computerized task requires search for a

certain type of shape (target). There are three conditions, each administered once. Conditions 1 and 2 are relatively easy, requiring search
for single targets (with different color and shape in the two trials).
Condition 3, however, is somewhat more difficult with 40 additional
distracters very similar in shape and identical in color to the target.
Measures across all conditions included the mean time for correct
responses, the number of false positives, and the number of mouse movements. In the dual-task version, children are required to search alternately for two different types of targets.

Focusing first on the single-search task, surprisingly there was no
clear sign of differences between the three conditions in the age of mature
performance, despite considerable variation in the difficulty of the tasks.
This offers some support to the view that it is the type of demands that
determine when adult levels of performance are reached, rather than the
level of difficulty of a particular variant of the task. In all cases performance in terms of the number of mouse movements was maximally
efficient as early as age 9, and response times were at adult levels by 14.
In the dual search, across all measures, there was improvement with age
such that older adolescents were quicker and more accurate than younger
adolescents, who were quicker and more accurate than young children.
As in the single search, the number of mouse movements reached maximum efficiency quite early in development at age 9 years, while response
times were at adult levels by age 14 years. Statistical analysis did not yield
an age at which false alarms had reached a minimum, but the data also
suggested that age 14 years may be a critical year. Ideally, these age trends
should be compared to those of Western adolescents, but such data are
currently unavailable.

Overall, the available data demonstrate that, while the major improvements in selective attention as indexed by visual search occur before the
age of 10 to 12 years, some aspects continue to improve through adolescence and only achieve optimum performance in later adolescence.

Maintenance of Attention

The ability to maintain attentional focus on a task over an extended
period of time is a core requisite for efficient attention processing. We
have seen how both maintenance of attention and vigilance develop

rapidly during the childhood years, but knowledge of their development across the adolescent years is scarce. We describe only two studies that have attempted to assess sustained attention in the adolescent period, the first using the Vigilan task devised by Wilding and colleagues and the second a paper-pencil test of sustained attention task by Manly and colleagues (known as the "Codes Transmission Task").

We have described the Vigilan task elsewhere in this chapter and in Chapter 6, but in brief, it is a computerized task that requires participants to watch for a yellow border, which appears randomly surrounding one shape on the screen. All 16 targets show up one by one at irregular intervals. The measures include the number of correct targets detected (maximum 16), mean time of clicking on a target, number of false positives, and total wandering distance. In a Chinese population of children and adolescents, we found target detection reached close to ceiling at age 14 years on this task, but in contrast, response times continued declining beyond this point until at least age 15. False alarm responses to nontargets and wandering around the screen with the mouse while awaiting a target declined to a minimum by age 10. These findings suggest that control of irrelevant distracting behavior that may reduce efficiency is achieved by age 10, but target detection itself continues to improve until late adolescence.

In a second type of study, Anderson et al. (2001) employed a well-documented range of executive functioning tasks to examine trends in performance from 11 to 18 years. One of these tasks was the Codes Transmission Task from the Test of Everyday Attention for Children (TEA-Ch; reviewed in Chapter 5). This task requires participants to sustain their attention while listening to a tape recording of 360 numbers (the "code transmission"). During the 12 minutes of the task, they are to listen specifically for two 5s in a row (a total of 40 during the course of the task). When they hear this sequence, participants are instructed to say the number that came immediately before the 5s. Findings did not reveal any consistent age-related profile, although examination of the means showed a small decline in correct responses at age 15 years followed by a slight increase in performance at 17 years. Unfortunately, the pattern of errors is also inconsistent across all ages, suggesting that this task may not be an especially sensitive measure of age trends in sustained attention across adolescence.

Although we have made a valiant attempt to describe some of the age-related changes in sustained attention during the adolescent period, we are hindered in the generalizability of our conclusions because of a dearth of relevant studies. Our findings on the Vigilan task demonstrate the feasibility of eliciting subtle changes in performance, but there is clearly a pressing need for further, more thorough empirical investigations of maintenance of attention and vigilance.

Attentional Control

We have used the term *attentional control* to describe a range of cognitive processes that require the developing child to utilize endogenous (effortful) control in order to inhibit or suppress the urge to make a prepotent response, to hold in working memory newly relevant rules that require either withholding or activating a previously learned responses, and to shift attention between two tasks. Here we describe a series of studies that have employed a range of measures to examine aspects of attentional control in adolescence and early adulthood, namely, response inhibition (e.g., Huizinga et al., 2006; Luna, Garver, Urban, Lazar, & Sweeney, 2004) and task switching (e.g., Cepeda, Kramer, & Gonzalez de Sather, 2001; Reimers & Maylor, 2005).

Switching Focus or Response

Focusing first on the ability to flexibly switch between two different tasks (referred to as task switching), Cepeda et al. (2001) sought to examine age-related changes in accuracy and reaction times from 7 to 82 years. This study was discussed briefly earlier as it provided the basis for the study of younger children by Crone Donohue, Honomichl, Wendelken, & Bunge (2006). Cepeda's study had the additional goals of examining whether "switch costs" (the difficulty in switching from one response set to another) could be attributed to two potential underlying attentional mechanisms-active preparation for a new task and decay of interference from a previously performed task-establishing whether these mechanisms undergo different age-related changes across the life span and assessing the impact of other cognitive processes on task-switching performance, namely working memory and perceptual processing speed.

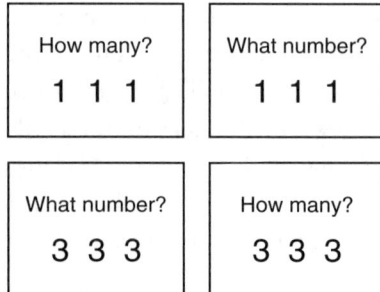

Figure 7-11. Example of stimuli used
by Cepeda and colleagues (2001).
Redrawn by permission from *Developmental
Psychology* 37, 717.

Age groups were composed of 7- to 9-year-olds, 10- to 12-year-olds, 13- to 20- year-olds, and then by decade with the final group comprising 61- to 70-year-olds. The task paradigm used switches every one, two, or three trials, and there were four possible stimuli consisting of either a single digit (1 or 3) or three digits (1 1 1 or 3 3 3). Above each target stimulus, participants saw either the words "What number?" or "How many?" depending on the trial itself. See Figure 7.11 for an example of the stimulus displays. There were two nonswitch blocks and eight switch blocks. The first nonswitch block required participants to identify the number present on the screen (the number 1 or 3), and the second block required identification of how many digits were present (1 or 3 digits). Of the four stimuli, two were response compatible (1 and 3 3 3), requiring the same key press, and two were response incompatible (1 1 1 and 3), requiring differing key presses for each task. In each block, half the stimuli were incompatible and half compatible.

The findings were as follows: overall reaction times were shortest for young adults in the 21 to 30 age group and longest, as would be expected, for the 7- to 9-year age group. The adolescent group continued to improve, and a plateau in performance was not reached until after the end of adolescence. Switch costs decreased from childhood through adolescence into young adulthood and thereafter remained constant until after the age of 60 years. Switch costs were reduced when longer time to prepare was given before stimulus onset (the cue-target interval or CTI);

though this reduction was somewhat less in the middle age range, the effect did not vary markedly with age, indicating that control was effective at the youngest age tested. On the other hand, a longer interval between response and the next cue (the response-cue interval or RCI) did not help children but produced a clear reduction in reaction time from the age of 30 onward (a small benefit began to appear in the adolescent group). The authors suggest that interference decays more slowly in children, which might be explicable in terms of less effective control processes that are needed to "turn off" an otherwise automatic effect.

However, a recent online (Internet) study by Reimers and Maylor (2005) extends and challenges these findings in two important ways. In what appears to be the first attempt to examine task switching using online procedures, Reimers and Maylor assessed a substantive sample size of participants aged between 5 and 109 years on a novel task involving switching face categorization according to either gender or emotion. Task switching was examined in the context of *general* switch costs (the difference between performance on the switching blocks and performance in blocks containing only one task) and *specific* switch costs (the difference between performance on switch trials and performance on nonswitch trials in a switching block). The core aim was to examine age-related switch costs (general versus specific) in task switching.

In terms of general switch costs, the authors report a significant reduction across the adolescent period followed by a gradual increase from 18 to 40 years. In contrast, specific switch costs did not show a similar developmental trend, especially in the adolescent period, and the authors suggest that general rather than specific switch costs are more vulnerable to changes at the brain level during the adolescent period. This latter finding, however, is at odds with Cepeda et al. (2001), who found specific switch costs declined up to young adulthood, whether measured by comparing switch trials with nonswitch trials in nonswitch or switch blocks. However, there are obvious limitations to the study design of Reimers and Maylor, not least the fact that it used a web-based methodology that is still in its infancy, and leaves many factors uncontrolled. The task also differed considerably from that used by Cepeda et al., which appears to be the better-controlled study.

A second type of task switching that explicitly measures the ability to switch cognitive set is the well-documented *Wisconsin Card*

Sorting Task (WCST). We visited this task earlier in its child format, the Dimensional Change Card Sorting Task. In the standard version, four key cards, numbered 1 to 4, are presented at the top of a computer screen. Response cards, presented one at a time, are given at the bottom of the screen. Unbeknownst to them, participants have to match one of three dimensions (color, number, or shape) of a geometrical shape to one the key cards while ignoring the other two dimensions. The rule changes after the participant makes 10 consecutive correct sorts. (See also Chapter 6 for a detailed review of the WCST). This task requires the continuous switching of rules, with the most common error type being perseveration errors that occur when there is a nonswitch to a new rule.

Huizinga et al. (2006) report that performance, as measured in percentage of perseverative errors, on a computerized version of the WCST, continued to develop across adolescence into early adulthood. Using a WSCT–analogue version Crone, Ridderinkof, Worm, Somsen, & van der Molen (2004) had previously reported a more fine-tuned analysis of performance by teasing apart errors that could be attributed to a failure to switch set (*perseverative* errors) versus errors that could be attributed to a failure to maintain set (*distracter* errors). In a sample comprising participants aged 8 to 9 years, 11 to 12 years, 13 to 15 years, and 18 to 25 years, the authors report quite different developmental age changes for the two types of errors. Specifically, perseverative error scores decreased with age across all groups, but the decrease was largest between 8 and 9 years and between 11 and 12 years. This finding is also in accordance with Zelazo and colleagues, who report an accelerated decrease in perseverative errors from early- to midchildhood (see Zelazo, 2006).

In contrast, rule maintenance (as indicated by distracter errors) undergoes a more gradual improvement with distraction errors continuing to decrease until 18 to 25 years. The authors conclude that the two error types follow different developmental trajectories across childhood and adolescence. A recent study by Somsen (2007) extends these findings by emphasizing the significant role of individual differences in performance both within younger and adolescent age groups. Specifically, they argue that high perseverative scores across different ages may be due to a variety of attentional processes that include inefficient reasoning strategies, guesswork, as well as perseverative responding. Finally, the findings from a Chinese population of adolescents (Zhan et al., 2010), using a child-friendly WCST–analogue version, also concur with the above pattern of performance in that both correct responses and perseverative errors change linearly from 6 to 16 years.

Response Inhibition

We have discussed response inhibition and task methodology in detail throughout this chapter and elsewhere so will here only briefly summarize the findings of two studies, one that has employed standard tasks of response inhibition, for example, the Stroop task, and one that has employed eye movement responses such as the antisaccadic task. Both of these studies have assessed the maturation of response inhibition across the adolescent period. In an extremely comprehensive study of executive functions in 7- to 21-year-olds (divided into four age groups: 7-, 11-, 15-, and 21-year-olds), Huizinga et al. (2006) examined performance on three well-documented measures of response inhibition, all of which to some degree require conflict resolution (the Stroop, Eriksen Flanker, and Stop–Signal tasks). In terms of developmental trends, both the ability to inhibit the response to the Go response in the Stop–Signal task and the ability to control response competition in the Eriksen Flanker task were characterized by rapid improvement in performance up until age 15 years and then no significant differences in performance between age 15 and 21 years. The pattern of these findings is consistent with other developmental studies (e.g., Bedard et al., 2002; Ridderinkhof & van der Molen, 1995; van den Wildenberg & van der Molen, 2004). In contrast, performance on an adapted Stroop task (this time incorporating pictures of "smileys" in a color-orientation design in which the two features sometimes indicated different responses) showed that reaction time on interference trials, corrected for basic speed, was still decreasing up to the age of 21. However, these three measures of inhibition, when subjected to Confirmatory Factor Analysis, did not constitute a common factor.

SUPPRESSION OF A DELAY OF A PREPOTENT
OR PLANNED RESPONSE

Using oculomotor measures to assess response inhibition, Luna et al. (2004) gave participants aged 8 to 30 years an antisaccadic task that involved the active suppression of a prepotent tendency to make a saccade

Table 7-1. Attention Tasks and Age Trajectories

Function	Task	Age of First Sign	Source	Age of Maturity	Source
Selection					
Feature	Pop-out	3 months	(Adler & Orprecio, 2006; Rovee-Collier et al., 1992)	10 years	(Trick & Enns, 1998)
Conjunction	Visual search (single target)	18 months	(Gerhardstein & Rovee-Collier, 2002)	10 years (accuracy) 14–15 years (speed)	(Zhan et al., 2010)
Maintenance					
High input	Distracter	6–9 months (?)	Oakes et al., 2002)	No data*	–
	CPT	4 years			
Low input	Vigilance	2 years	(Goldman et al., 2004)	10 years (accuracy) 14–15 years(speed)	(Zhan et al., 2010)
Control					
Switching	Visual search (dual target)	No data*	–	10 years (accuracy) 14–15 years (speed)	(Zhan et al., 2010)
	AB	12 months	(Diamond, 1985)	Post–5 years	(Espy et al., 1999)
	DCCS	4 years	(Zelazo et al., 2003)	5 years	(Zelazo, 2006)

	Task				
	Continual rule reversal	No data*	-	Post–13 years	(Crone et al., 2004; Huizinga et al., 2006)
Suppress or delay prepotent or planned response	WCST	No data*	-	12–16 years	(Zhan et al., 2010)
	Antisaccade	4 months	(Johnson, 1995; Nakagawa & Sukigara, 2007)	14 years (accuracy)	(Luna et al., 2004)
	Attractive toy	Pre–3 years	(Carlson, 2005)	5 years	(Carlson, 2005)
	Stopping response	3 years	(Jones et al., 2003)	7 years (speed) 11 years (speed)	(Johnstone et al., 2007) (Tillman et al., 2008)
	Negative priming	9 months	(Amso & Johnson, 2005)	No data*	
	Simon effect	Pre–2 years	(Gerardi-Caulton, 2000)	3 years	(Rothbart et al., 2003)
	Stroop	3 years Nonverbal	(Wright et al., 2003)	Post–12 years	(Huizinga et al., 2006)
				Post–16 years	(Wright et al., 2003)
	Flanker	No data*	-	7 years 15 years	(Rueda et al., 2004) (Huizinga et al., 2006)
Reverse prepotent responses	Day–Night	3.5 years	(Diamond et al., 2002)	5 years (accuracy) Post–11 years (speed)	(Simpson & Riggs, 2005a,b)

to a suddenly appearing visual stimulus. The task required the fixation on a central stimulus followed by a peripheral stimulus that appeared for 1.5 seconds at one of three locations (8, 16, and 24°) to the left or right of the center fixation. Participants were required to move their eyes immediately to the mirror location in the opposite direction. There was a total of 36 trials. It was predicted that targets with closest target eccentricity would produce the most errors because of the higher demand for response inhibition. Measures included the percentage of trials in which the participant looked toward the peripheral targets (known as response suppression errors), saccade latency, and accuracy of saccades toward the correct location. The primary variable of interest was the proportion of trials with response suppression errors. Findings indicate adult levels of response suppression by age 14 years, but closer inspection reveals that although even the youngest age group could sometimes perform a correct antisaccade (a finding that is consistent with other developmental work on antisaccades in infancy and childhood; see our previous section), the efficiency of correct responses in terms of consistency across trials continued to develop into adolescence. The authors propose that that developmental trends found in response suppression errors reflect an increasing capacity in the adolescent to utilize already existing attentional mechanisms rather than reflecting the acquisition of fundamentally new ones.

Collectively, the findings from the adolescent period suggest age-related trends that are not as intense as in the earlier developmental periods but rather more gradual in nature. However, of all the age sections in this chapter, this perhaps is the least conclusive. The relative scarcity of substantive empirical investigation of age-related effects within the subcomponents of selection and sustained attention preclude any firm overall conclusions about age trajectories from late childhood into early adulthood. In contrast, numerous studies attest to the developmental nature of attention in infancy and childhood and, indeed, from early to late adulthood. Our gap in knowledge remains exclusive to the adolescent years. We have only touched upon some of the possible reasons why this might be the case, including issues related to complexity of task design, onset of puberty, and widening individual differences. However, even with comparatively few studies, it is clear that attention functioning does not necessarily reach adult levels across all subcomponents by adolescence. Instead, improvements in performance, most notably on reaction times, are more gradual and subtle in nature than in the childhood period.

General Conclusions

In a chapter such as this that deals with the merging of a magnitude of studies that have sought to define attention performance across development, it is difficult to provide a neat summary of the precise ages at which maturity is reached. In Table 7.1 we have summarized, from the findings that have been discussed, tentative chronological ages at which performance on different tasks achieves maturity, but as the reader will appreciate, this is no simple matter. For example, if performance on Task A reaches ceiling before performance on Task B, it could arise because the tasks are not matched in difficulty, rather than through their engaging fundamentally different processes. For example, ceiling for Task A may conceal the possibility of further improvement, which would be revealed if the task were made more difficult without fundamentally changing its nature. Such a version might produce later attainment of ceiling in Task A than in Task B. Furthermore, as already indicated, different measures may produce different conclusions. And, as existing research amply demonstrates, development of a new paradigm, or a more sensitive version of an old paradigm, can produce evidence for the initial signs of an ability at an earlier age than indicated by previous studies, or evidence for continued improvement beyond the currently accepted age of mature performance. Moreover, the simpler tasks employed to detect a given function in infants inevitably do not demonstrate that the function is available in more complex contexts: inhibition demonstrated by an attractive toy paradigm does not imply ability to inhibit more complex prepotent responses. Likewise, evidence for ongoing improvement past adolescence in speed of antisaccade responses does not prove that all tasks requiring inhibition of prepotent responses remain at an immature level to the same age. Accordingly, detailed analysis is necessary, preferably using several versions of a task that vary in difficulty and several tasks that are assumed to make different demands on attentional processes. Also, conclusions about a specific function (for example, selection) should not be based on a single task, but only on consistency over a number of tasks that engage that particular function. Only then will it be possible to decide whether any simple conclusions can be drawn concerning the developmental trajectory of performance for different varieties of attention.

CHAPTER SUMMARY AND KEY FINDINGS

- A rich literature exists on age-related changes in the preschool and childhood years that include developmentally sensitive periods characterized by spurts of growth followed by periods of stability. However, different attention subcomponents produce very different developmental trajectories.
- The relative paucity of research in the adolescent period prevents firm conclusions from being drawn about age trajectories from late childhood to early adulthood. This is a critical time period that needs more substantive research.
- Future research needs to include longitudinal designs with developmentally sensitive paradigms that can identify subtle changes in performance. It will also need to adopt more sophisticated methods of identifying and evaluating specific attentional functions rather than relying on just one or two well-worn tasks.

8

Atypical Attention: Attention Deficit/Hyperactivity Disorder (ADHD)

CHAPTER SNAPSHOT

- ADHD as a complex neurodevelopmental disorder—addressing issues of diagnosis, subtypes, comorbidities, and environmental factors
- Cognitive theories of ADHD—Does any one model fit the profile?
- Do impairments in control of attention represent the core signature in ADHD?
- Development of ADHD signatures—tracing the changing trajectories from preschool through to adolescence

---

ADHD: a heterogeneous disorder

Attention deficit/hyperactivity disorder (ADHD) is one of the most common childhood psychiatric disorders and is characterized by a triad of features: chronic and pervasive inattention, impulsive behavior, and hyperactivity. All three behaviors occur to a developmentally inappropriate degree and typically before the age of 7 years. Unlike many other neurodevelopmental disorders (with the possible exception of autism), our understanding of ADHD does not have its starting point in its genetic origin (genotype), but rather in its behavioral and cognitive end state (phenotype). Researchers must first, therefore, identify the ADHD signature or profile in order to explore its possible genetic correlates. ADHD is not a homogenous disorder with discrete characteristics that precisely guide the researcher to its causal mechanisms. Instead, we are presented with an extremely heterogeneous disorder that on the one hand clearly has a substantial genetic component, but on the other hand is defined by multiple pathways that appear, at first glance, to result from differing causal mechanisms that may or may not be genetic in origin. If several genes are involved, as seems likely, the product will be a normal distribution of ability, with ADHD representing one tail of this distribution. Furthermore, different combinations of the causal agents may occur in different cases. Hence, there is no certainty that the wide-ranging behavioral symptoms are always the product of the same cause or that all cases that demonstrate them are identical. Indeed, there is an ongoing debate on the extent to which ADHD is a single syndrome, and two or three subtypes are now commonly accepted (see below).

This chapter cannot possibly capture the sheer volume of research that has attempted to define ADHD at its multiple levels. Our first aim is, therefore, to provide the reader with a "snapshot" of the complexity of the disorder at the brain, genetic, and diagnostic levels. Our second aim is to provide a review of current psychological theories that attempt to isolate mechanism-based pathways that are neurobiologically plausible. We then go on to critique the range of cognitive "markers" that have been highlighted as representing core areas of dysfunction in ADHD and to set these findings within a developmental context beginning in the preschool years.

The Complexities of ADHD

ADHD is a complex disorder with a broad spectrum of involvement. It is also one of the most widely researched of neurodevelopmental disorders, yet pinpointing its causal mechanisms and pathways remains elusive. Although pieces of the puzzle are now in place, the heterogeneity of ADHD demands rigorous and empirically driven research. We highlight below some of the complexities that dominate the ADHD literature. This is not an exhaustive review but is meant to guide the reader to salient issues that have become core research themes and issues of debate in the search for the gene–brain–cognitive correlates of ADHD.

ADHD: Discrete Clinical Disorder or Continuum of Involvement?

There remains considerable controversy as to whether ADHD should be viewed as a discrete clinical entity or as part of a *continuum* of involvement with (in)attention and (hyper)activity as continuous variables representing the extreme end of a normal distribution. The most widely used scales, such as Conners, CBCL, or ACTeRs, all incorporate a more discrete rating profile with an emphasis on severity of inattention and hyperactivity symptoms. These scales produce a skewed distribution because behavior is rated as "not present," "sometimes present" with mild symptoms, or "always present" with more severe symptoms. In a relatively new scale, known as the SWAN, continuous variations in strengths and weaknesses in attention and activity can be measured to produce a normal distribution of the population sampled (Swanson et al., 2005, and reviewed in Chapter 5). This scale has been used in a number of recent genetic and endophenotype studies (see Cornish et al., 2005, 2008; Hay, Bennett, Levy, Sergeant, & Swanson, 2007; Polderman et al., 2007). For example, Polderman et al. collected maternal SWAN ratings alongside CBCL Attention Problem ratings on a large sample of 12-year-old twin pairs. The SWAN scale allows for a 7-point rating score across a broad spectrum with behavior ranging from "far below average" to "far above average." In contrast, the CBCL allows for a 3-point rating score with behavior ranging from "not true" to "very true." The authors found that children who had scored "0" on the CBCL actually displayed variations on the SWAN scale, indicating that inattentive behaviors are rarely completely absent in the general population. The authors concur with the conclusions of Hay et al. (2007), who argue that a core benefit and clinical utility of viewing ADHD as a continuum of involvement is that ADHD phenotypes can be more precisely defined and elucidated.

ADHD: Separable Disorders or One Combined?

Due to the heterogeneity of the behavioral symptoms in ADHD, three different subtypes are distinguished in the DSM-IV classification: ADHD combined subtype, ADHD inattentive subtype, and ADHD hyperactive/impulsive subtype (see Box 5.1. for a description of the symptoms that underlie these subtypes). However, it is still a matter of considerable debate as to whether ADHD can be broken down into subtypes that still share phenotype commonalties. Part of the issue lies in a lack of substantive empirical research directly comparing subtypes across multiple levels of measurement (see Baeyens, Roeyers, & Walle, 2006, for a review).

One recent attempt to address this concern is by Derefinko et al. (2008), who reported quite different cognitive and behavioral styles between children with the inattentive subtype and those with the combined subtype. The former group was characterized by a cognitive sluggish response style that was pervasive in nature and differed completely from the impulsive style of the ADHD combined group. The authors speculate that the intensity of these differences provides further evidence for disorder-specific phenotypes and pathologies that distinguish children with inattentive subtype from children with ADHD combined. We also draw the reader's attention to the work of Nigg and colleagues (Huang-Pollock, Nigg, & Halperin, 2006; Nigg, Blaskey, Huang-Pollock, & Rappley, 2002) for examples of cognitive studies comparing ADHD subtypes. However, what will become clear throughout this chapter is that ADHD subtypes, and specifically their cognitive signatures, are not clear-cut.

ADHD: Rarely a Disorder on Its Own

ADHD rarely occurs in isolation and is comorbid with a range of other psychiatric disorders (Pliszka, 1998). By comorbidity, we refer to the coexistence of two or more discrete disorders that occur in the same individual at the same time. The most prevalent disorders to be associated with ADHD are oppositional defiant disorder (ODD) and conduct disorder (Biederman, Petty, Dolan et al, 2008), learning disabilities, substance abuse disorders (a strong association with cigarette smoking), neurodevelopmental disorders (e.g., fragile X syndrome), anxiety, and depression (e.g., Biederman et al, 1991; Daviss, 2008; Schatz & Rostain, 2006). See Box 8.1 for a description of ADHD comorbid disorders.

The presence of comorbidities in individuals diagnosed with ADHD has been studied most recently from the perspective of response to treatment (Hinshaw, 2007; Ollendick, Jarrett, Grills-Taquechel, Hovey, & Wolff, 2008; van der Oord, Prins, Oosterlaan, & Emmelkamp, 2008), the effect on the course of the disorder (Lara et al., 2009; Spencer, 2006), and the neuropsychological profile (Rommelse et al., 2009; Wahlstedt, Thorell, & Bohlin, 2008). See Spencer (2006) for a stellar review of the issue of ADHD and comorbid disorders in childhood.

However, when one looks at the research as a whole, a main weakness is the paucity of detailed empirical investigations into the impact of

Box 8.1 ADHD and Comorbid Disorders

ADHD rarely exists in isolation. It is often accompanied by other psychiatric or developmental disorders including oppositional defiant disorder (ODD), conduct disorder, anxiety, depression, and learning disabilities. This has prompted some researchers to advocate distinct ADHD comorbid subtypes. We describe below some of the core features of these "comorbid" disorders and their persistence across early childhood into adolescence and adulthood.

- *Oppositional defiant disorder (ODD)* is characterized by argumentative, spiteful behavior with sudden bursts of temper. A recent 10-year longitudinal study of the course of ODD and conduct disorder in ADHD found that those with long-term comorbid ODD were at greater risk for developing major depression (Biederman, Petty, Dolan et al., 2008).
- *Conduct disorder (CD)* is more severe than ODD and is characterized by rule-breaking behavior and physical aggression. Biederman et al. (2008) found that CD, unlike ODD, was selectively associated with increased risks of later substance abuse, smoking, and bipolar disorder.
- *Anxiety disorder* is characterized by excessive worry in the case of generalized anxiety disorder, marked and persistent fear of objects or situations in the case of specific phobias, developmentally inappropriate and excessive anxiety about separation from the caretaker or home in case of separation anxiety disorder. Biederman et al. (1991) found high levels of lifetime risk of anxiety disorders in young adults who had been diagnosed with ADHD and anxiety in childhood.
- *Major depression* is characterized by loss of interest in activities, weight loss or gain, changes in sleep, and feeling of worthlessness. Children and adolescents may display more irritable behavior and have somatic complaints. Fischer, Barkley, Smallish, & Fletcher (2002) found elevated rates of major depression in a longitudinal sample of hyperactive children followed through to young adulthood.

(continued)

Box 8.1 ADHD and Comorbid Disorders *(continued)*

- *Learning disabilities* (LD) and academic difficulties in general have been well documented in children and adolescents with ADHD. Early inattentive symptoms are strongly associated with later difficulties in math, reading, or both (Dally, 2006).
- *Developmental disorders* such as autism and fragile X syndrome show extremely high prevalence rates of ADHD that persist from childhood through to young adulthood (Cornish, Turk, & Hagerman, 2008).

---

co-morbidities on predicting ADHD *phenotypes* and *trajectories*. Unfortunately the lack of detailed empirical investigation makes it often unclear whether differences at the brain and cognitive levels between ADHD and controls are attributable (at least in part) to the comorbid condition rather than ADHD itself. This critical issue needs to be at the forefront of the next generation of research studies if the precise signature of ADHD alone is ever to be resolved.

ADHD: Looking Beyond Genetics to the Environment

Early complications such as hypoxia (inadequate supply of oxygen in the blood) or exposure to chemical or environmental toxins during critical periods of pre- or postnatal development can impact on developmental pathways, inducing permanent changes that can increase susceptibility to various conditions including ADHD. At the genetic level, this process is known as *epigenetics*, defined as the study of heritable changes in gene expression that are not caused by changes in DNA sequence (see Waterland & Michels, 2007, for a discussion of the "developmental origins hypothesis" and Karmiloff-Smith, 2007, for a discussion of atypical epigenesis). The resulting gene dysregulation, caused by environmental exposure, is the mediating link to the phenotypic outcome, in this case to the ADHD phenotype. There is also emerging evidence to suggest that environmental factors may shape the epigenome over the entire life span and may not just be confined to discrete developmental time points. In particular, Mill and Petronis (2008) provide a detailed hypothesis of the role of epigenesis processes in mediating susceptibility to ADHD. Examples of prenatal environmental toxins linked to ADHD include fetal *lead* exposure, which is associated with reduced general cognitive outcomes, poorer performance on tests of executive function and working memory. Fetal *alcohol* exposure is associated with hyperactive, disruptive, and impulsive behaviors in addition to reduced cognitive performance, and fetal *nicotine* exposure is associated with slightly poorer academic achievement. For an excellent critique of these and other environmental factors associated with ADHD, we recommend Banerjee, Middleton, and Faraone (2007).

ADHD and the Role of Dopamine

As is evident from our review of early brain development in Chapter 2, the developing brain undergoes staggering changes in the first year of life and continues to develop in a dynamic fashion throughout infancy and childhood. In ADHD research, there has been tremendous excitement at the possible interactions between candidate ADHD genes and early brain development, in particular, genes that differentially impact the *dopaminergic* pathways such as the *COMT* gene (see Chapter 3 for a detailed description and review of current research on the *COMT* gene and other candidate genes). In brief, the dopamine system has key modulating influences on the functioning of the frontal cortex. Converging studies from multiple sources, including patients with frontal lobe injuries, neuroimaging investigations of typical and atypical populations, and animal studies, attest to the importance of the dopamine system in regulating normal cognitive functions in the prefrontal cortex. During postnatal brain development, substantive changes in neurotransmitter systems occur, including, among others, changes in the dopaminergic system. It is therefore reasonable to assume that early abnormalities in dopamine levels can have a detrimental and cascading impact on the developing brain, making functions that are housed within the prefrontal cortex especially vulnerable, for example, inhibition and working memory. See Box 8.2 for description of the interplay between dopamine, genes, brain development, and cognition; a color version of the illustration in Box 8.2 is located in the color section in the center of this book.

Box 8.2 Dopamine and ADHD

Dopamine is an important neurotransmitter and a member of the catecholamine family. It has been the focus of immense research interest, not least because of its link with disorders such as schizophrenia and Parkinson's disease. Derived from tyrosine, dopamine is produced in many parts of the brain including the arcuate nucleus of the hypothalamus and the substantia nigra. Dopamine plays a critical role in cognition, reward, and emotion and is initiated by nerve impulses that run through dopaminergic neuronal projections originating from the ventral tegmental area (VTA) of the midbrain and terminate in striatal and prefrontal cortical areas. See highlighted brain regions below. As an example of the importance of dopamine in the developing brain, children born with phenylketonuria (PKU), a genetic disorder that decreases the level of dopamine, show severe global brain damage and intellectual delay if left untreated (see Diamond, 2007a).

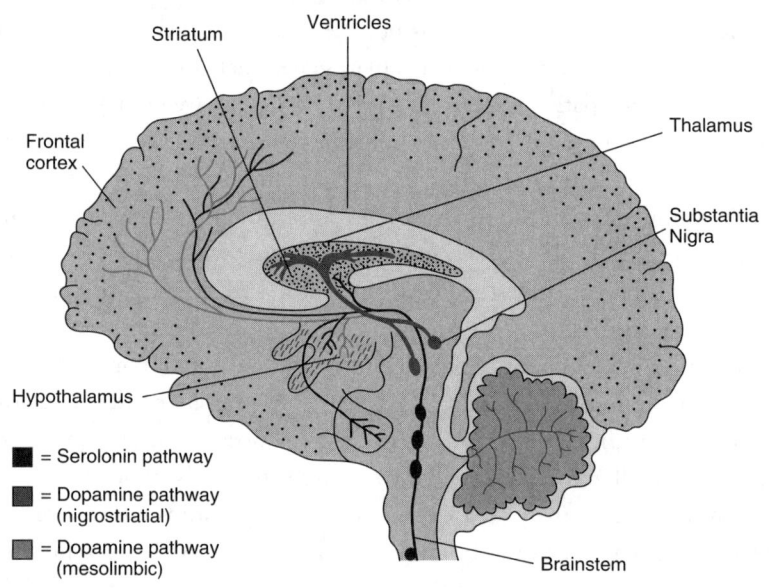

Substantive evidence now points to the role of dopamine in both the cause and treatment of ADHD. Stimulants such as Ritalin (methylphenidate) have been widely used as a core treatment option

Box 8.2 Dopamine and ADHD *(continued)*

and are effective because they significantly increase the levels of dopamine in the brain and allow individuals to maintain attention more effectively. See Levy (2008) for an excellent review of current theories and treatment directions for ADHD. Most recently, researchers have been investigating the involvement of genes functionally associated with the dopamine system in the pathogenesis of ADHD. Genes of interest include the dopamine receptor 4 (*DRD4*), the dopamine transporter (*DAT1*), and the catechol-O-methyl transferase (*COMT*) enzyme. Findings, although promising, are not always consistent, and research is still defining the relationships between candidate genes, ADHD symptoms, and cognitive impairment. Note that the diagram is shown in color insert of the book.

In the case of ADHD, the possibility that dopamine deficiency may play a critical role in the development of the condition, impacting it across multiple levels (brain, genetic, cognition, and behavior), has produced a decade of innovative research. In brief, important differences between brain anatomy and function differentiate children with ADHD from their typically developing peers. A recent anatomical study investigating cortical changes in the developing ADHD brain reported a cortical maturation delay in ADHD that was most prominent in the lateral prefrontal cortex, particularly within the superior and dorsolateral prefrontal regions, as well as a maturation delay in the temporal cortex (e.g., Shaw et al., 2007). Advances in functional neuroimaging tools have further expanded these findings to include abnormalities in key frontostriatal and frontocerebellar networks that are involved in response control and suppression, key areas of weakness in ADHD. These networks are also intrinsically linked to the dopamine pathway, adding one further piece to the puzzle of ADHD.

The critical role of dopamine genes in the neurobiological basis of ADHD is extensively documented. Although a number of genes have been linked to ADHD, the three most widely researched are the dopamine transporter gene (*DAT1*), the dopamine D4 receptor gene (*DRD4*), and the *COMT* gene. We recommend that the reader look carefully at our snapshot of these critical dopamine genes in Chapter 3.

Sagvolden, Johansen, Aase, and Russell (2005) have proposed a theory of ADHD based specifically on dopamine deficiency. In brief, they hypothesize that weakness in the three main dopamine pathways will have a variety of effects on behavior. The mesolimbic pathway to the amygdala and hippocampus involves reinforcement mechanisms, and malfunction will, they claim, produce delay aversion, hyperactivity, and poor sustained attention. Impairment of the mesocortical pathway to the frontal lobes will affect orienting, target selection, and planning, while poor functioning of the nigrostriatal pathway between the substantia nigra and the striatum will affect movement. Thus, this theory invokes a very wide range and variation in possible impairments, the problem being that almost any result can be predicted and testing the theory is problematic.

Cognitive Theories of ADHD

We review here four theories that have been extensively documented in the literature and that provide, albeit not perfectly, theoretically plausible cognitive models for the mechanisms in ADHD: executive dysfunction theory (Pennington & Ozonoff, 1996), inhibition dysfunction theory (e.g., Barkley 1997), state regulation theory (Sergeant, 2005), and the delay aversion theory (e.g., Sonuga-Barke, 2002, 2005; Sonuga-Barke, Taylor, & Heptinstall 1992).

ADHD as an Executive Dysfunction

Welsh and Pennington (1988, p. 201) defined executive function (EF) as "the ability to maintain an appropriate problem-solving set for attainment of a future goal," including inhibition or delay of a response, planning a sequence of responses, and constructing a model of the task. A detailed examination of other possible components of EF was presented in Chapter 2. Pennington and Ozonoff (1996) reviewed the evidence for an EF deficit in ADHD and other neurodevelopmental disorders. Though they identified consistent evidence for weaknesses in EF tasks in ADHD, they admitted that the evidence that these were due specifically to the EF demands of the tasks was not conclusive, noting, for example, that there were few studies that incorporated convincing control conditions for the EF tasks by differing only in the absence of EF demands. The authors drew a general and cautious conclusion that, "it may be that ADHD children have a mix of specific and general deficits: a core EF deficit, perhaps motor inhibition, and some general cognitive inefficiency" (p. 65).

One problem for this type of theory is that it does not really take into account the occurrence of cases of ADHD that show no obvious evidence for deficiencies in EF, even though as a group the ADHD samples show a higher occurrence of EF malfunction (e.g., Biederman et al., 2007; Loo et al., 2007; Nigg et al., 2002), and also findings that show other non-EF functions are impaired in ADHD, for example, basic mechanisms involving timing (e.g., Smith et al., 2002; Toplak & Tannock, 2005), and moment-by-moment variability in task performance (Castellanos & Tannock, 2002). Such findings reinforce the points made above about the heterogeneity of the condition and the possibility that similar behavioral symptoms may arise from different underlying causes. See Castellanos, Sonuga-Barke, Milham, & Tannock (2006) for a coherent and comprehensive review of the existing literature on EF performance and ADHD.

ADHD as an Inhibition Dysfunction

Barkley (1997), pursuing a very similar type of theoretical explanation in a more detailed argument, presents a complex model that includes a huge range of functions and focuses on weak behavioral inhibition as the key to the wide range of impairments observed in ADHD. According to Barkley, inhibition is necessary for inhibiting prepotent responses, stopping ongoing responses, and protecting internal processing from interference from external and internal events. He argues that inhibition thereby allows four types of operation to function but is not itself directly involved in such functioning. Hence, weakness in inhibition results in deficits in working memory (the ability to hold in mind information while performing another task), self-regulation of affect and motivation (the ability to consider emotional responses before responding and to complete tasks that have no immediate rewards), internalization of speech (the ability to use "self-speak" to regulate and direct one's own behavior), and reconstitution of behavior (the ability to initiate and create flexible problem-solving strategies to novel events).

This model, at first glance, holds a lot of promise for explaining the pervasive EF impairments that have been reported extensively by numerous investigators. However, we dispute the precise form of Barkley's model. He sees behavioral inhibition as a distinct process from other functions and one that is positioned at the top of a hierarchy of processes. Its operation permits the uninterrupted functioning of the four processes listed above. The model explicitly reinforces this notion by citing evidence that inhibition is controlled specifically in the orbital-frontal prefrontal cortex, whereas working memory is centered in the dorsolateral prefrontal cortex. However, there is no strong evidence for a major role of the proposed kind in the orbital-frontal region; while the precise functions of the region are still under debate, a major function appears to involve evaluating the significance of reinforcers while planning future actions (Bechara, Damasio, Damasio, & Anderson, 1994; Kringelbach, 2005).

Thus, there is no single center for inhibitory control; rather, such control is intrinsic to all brain processes, either through built-in cross-connections, as in the visual system, or via instructions from anterior control systems to posterior sensory processing systems, for example, as described in Posner's model. Hence, we are unconvinced by the hierarchical nature of Barkley's model, with behavioral inhibition as a form of control center operating independently of a wide range of other processes. As already indicated, we suggest that working memory operates to achieve task solutions by itself implementing balances of excitatory and inhibitory processes to ensure such things as correct response sequences and stimulus–response associations.

Barkley's model also lists a wide range of functions under each of the four main processes identified above (for example, holding events in mind, manipulating or acting on the events, anticipatory preparation for action, and so forth). These detailed lists of functions do not merit detailed elaboration for our present purposes. It is unclear how they were derived independently of symptoms that have been observed in the ADHD literature (if derived from such symptoms the theory would, of course, be circular). We stress that we are not rejecting an explanation of ADHD in terms of inhibitory weakness, but only this particular representation of the hierarchy of processes involved. Indeed, it is a common weakness in several theoretical approaches to ADHD to employ terms such as EF, working memory, and inhibitory function without clear definition of their status (systems or processes, for example) and interrelations. Greater clarity in theorizing is essential if progress is to be made.

ADHD as a State Regulation Dysfunction

The cognitive-energetic model (CEM) proposed by Sergeant (e.g., Sergeant, 2000, 2005; Sergeant, Oosterlaan, & van der Meere, 1999) was described previously in Chapter 2 (Figure 2.6). In summary, the model proposes that the overall efficiency of information processing is determined by both process (computational) and state factors (such as effort, arousal, and activation). Sergeant argues that the hypothesized response inhibition deficit widely reported in ADHD, and suggestive of disinhibition, is dependent upon the state of the subject and the allocation of energy to the tasks at hand as well as weaknesses in control and stimulus–response processing systems.

In the model there are three interactive levels. The top level is the executive control, which feeds into three "energetic" pools. These affect levels of activity and motivation. Effort, which is located in the hippocampus in the midbrain according to the theory, is the energy needed to meet task demands and is influenced by variables such as cognitive load and motivation; it is said to modulate the activity of the other two pools, arousal and activation, and to be directly related to central processing functions such as memory search. A second pool, arousal, is associated with the reticular formation and the amygdala (another midbrain structure) and is involved in encoding the stimulus input; it involves brief phasic fluctuations and is influenced by signal intensity and novelty. The third pool, activation, is related to tonic (long-term) response organization and is associated with the basal ganglia and striatum; it is affected by time of day, time on task, and preparation and is involved in response organization (see Sergeant, 2005, for a detailed review of this model).

Sergeant et al. justifiably point out that any comprehensive account of behavior must take account of both energetic and information processing functions in the nervous system: "the (pre)frontal lobe is the end station of a limbic and midbrain system. State regulation and inhibition are, we suggest, interwoven to such a degree that we wonder whether it is possible to develop tasks that measure purely the one or the other concept" (Sergeant et al., 2002, p. 96). However, as the above quotation demonstrates, they thereby neglect to mention that the prefrontal cortex is also the end station for processing mechanisms originating in the sensory areas of the brain (see Chapter 4).

There are several potential problems with this theory. The energetic pools are apparently distinct from information processing systems, yet evidence from brain studies does not support any such strict demarcation. For example, as Chapter 4 illustrates, there are complex reciprocal links between the prefrontal cortex and reticular systems in the brain stem. These modulate overall arousal and EF activity, which then feeds into the sensory processing systems, adjusting preparation and activation thresholds according to current priorities. The clear demarcations of different energetic and processing systems envisaged in the CEM are an oversimplification of these complex circuits.

Turning now to the application of this model to the explanation of the key deficit in ADHD, it is claimed that the CEM envisages defects at all three levels of the model: cognitive mechanisms such as response output, energetic mechanisms such as activation, and the overall control system. It is difficult to see how such a theory could ever be refuted, and no causal explanation is offered for such a wide-ranging impairment of the whole system. Sergeant accepts that deficiencies in inhibition are firmly established in ADHD but argues that they are not a specific marker for the condition as they also occur in other types of impairment. He therefore infers that there must be an additional impairment in ADHD. It is, of course, equally possible that it is the other neurodevelopmental disorders that include additional impairments, or that future research will reveal subtle differences in the nature of the different inhibitory impairments in ADHD (see, for example, Wilding, Cornish, & Munir, 2002, who demonstrated such differences between children with poor attention and fragile X and Down syndrome groups).

Sergeant presents no consistent, strong evidence that motivational manipulations are critical in modulating differences between ADHD and control groups. In fact, Oosterlaan and Sergeant (1998) found that manipulation of rewards and costs did not ameliorate poor performance in the Stop task in this group. Nigg (2001, p. 571) examined evidence for inhibitory weaknesses in ADHD in detail and concluded that,

"ADHD is unlikely to be due to a motivational inhibitory control deficit; evidence for a deficit in an executive motor inhibition process for the ADHD combined-type is more compelling but is not equally strong for all forms of executive inhibitory control." Shanahan, Pennington, and Wilcutt (2008) conclude that the available studies on the effects of motivation on inhibitory performance in the Stop–Signal task (a task that produces one of the largest effect sizes in group differences) are inconsistent in their results. In their own study, they found no evidence that reward reduced the impaired performance on this task by ADHD children and concluded that the impairment was due to inhibitory weakness alone.

Finally, we offer a clear example of the inability of theories based on arousal or motivation to explain a well-established finding from our own research using the Visearch task (reviewed in Chapters 6 & 7; Cornish et al., 2008; Wilding, 2003; Wilding, 2005; Wilding & Burke, 2006 Wilding & Cornish, 2007; Wilding, Munir, & Cornish, 2001; Wilding, Pankhania, & Williams, 2007) in which children with poor attention consistently made more false alarms to nontargets in the more difficult versions of the task. We have attributed this impairment to inadequate "programs" set up by control systems to differentiate alternative stimulus–response combinations, and hence, we locate it in either the programming or implementation of inhibitory rules. It is difficult to see how these differences in error rates, which were not accompanied by differences in response times and which occurred in a task that children found highly motivating, could be explained by CEM.

ADHD as a Motivational Dysfunction

There are a number of other motivation-based models proposed as an alternative to the cognitive models. These models move away from a focus on inhibitory processes to a focus on suboptimal reward processes in which it is proposed that children with ADHD display "hypersensitivity" to delay, resulting in severe difficulties in waiting for motivationally salient outcome. The most widely researched is the delay aversion theory proposed by Sonuga-Barke (e.g., Bitsakou, Psychogiou, Thompson, & Sonuga-Barke, 2009; Sonuga-Barke, 2005; Sonuga-Barke, Houlberg, & Hall, 1994) which argues that ADHD is not associated with impulsive responding in the sense of inability to inhibit and delay responses, but rather, impulsive (and therefore fast) responding is a choice of response style.

Sonuga-Barke (2005) distinguishes two types of explanation of ADHD, involving so-called "cold" EF (cognitive control functions involving response inhibition) and "hot" EF (motivational and reward functions). His theorizing has developed from envisaging a single main underlying cause of the condition, aversion to delay, to a more complex model in which both types of factors play a role, together with their interactions during development with social factors such as parental reactions and child-raising style. We will focus first on the plausibility of aversion to delay as a major factor in ADHD.

To illustrate this concept, Sonuga-Barke et al. (1994) used two versions of the Matching Familiar Figures Test (MFFT) presented on a computer screen. In the standard version, participants have to provide an exact match for a target within a display of closely similar pictures; the next display appears immediately after a correct choice has been made. In the alternative version there was a fixed interval from response to the next display of 45 seconds, so responding quickly did not shorten the whole task. The prediction was that children with ADHD (more specifically hyperactive children in this study) would respond more quickly and less accurately than typically developing children in the standard version, but these differences would be eliminated or at least reduced in the modified version.

The prediction was only partially fulfilled. Differences were as predicted in the standard condition; hyperactive children made more false choices before selecting the correct picture but nevertheless achieved the latter more quickly than the controls. In the modified condition both groups performed more slowly and the groups did not differ significantly, but error rates declined only slightly and the group differences remained. Despite this finding, the result is claimed as a key piece of evidence in favor of the delay aversion hypothesis, but in fact, it demonstrates that higher error rates in the hyperactive children were not due primarily to fast impulsive responding, so aversion to delay cannot provide a complete explanation of the underlying cause of ADHD. Poor inhibitory control is a plausible explanation for the higher number of false choices before selecting the correct picture.

Solanto et al. (2001) provide further evidence. They employed both a Stop–Signal task and a Choice-Delay task in which choices were offered between a small immediate reward and a larger delayed reward. Success in stopping responses in the Stop–Signal task, widely accepted as an index of inhibitory efficiency, was not correlated with choice of delayed rewards. This suggests that ability to inhibit responses on the occurrence of a Stop signal and ability to inhibit premature responses in the Choice-Delay task reflect different mechanisms, and it supports Sonuga-Barke's claim that impulsive responding in the latter task is not due primarily to poor inhibition. Successful stopping responses and choice of the delayed rewards were lower in an ADHD group than in a comparison group. However, this tendency correlated only with hyperactivity/impulsivity ratings and not with attention ratings. While measures from each task discriminated between the ADHD and non-ADHD groups to a modest degree, in combination they offered excellent discrimination, demonstrating that the two main features of ADHD are related to different underlying mechanisms that can be measured in different tasks and that the ADHD group are heterogeneous.

Sonuga-Barke (1994) hypothesized that inattention, hyperactivity, and impulsivity may all result from delay aversion, but these subsequent results suggest independence of attention impairment and the hyperactivity/impulsivity factor. As already indicated, he now advocates a more complex set of causal relations, including disturbances of reward mechanisms, delay aversion, and EF, together with feedback effects from failure and impoverished learning experiences. We accept that the evidence does demonstrate that delay aversion is an important factor in the condition and agree that it is not the complete explanation. In general, the evidence suggests that weakness in inhibition and/or control is the key factor in impaired attention, while delay aversion and impulsive responding are more closely related to hyperactivity. See, for example, the account in Chapter 7 of our own work with the Visearch task in young children, in which attention ratings predict false alarms but not speed of performance, and altering speed through instructions left the difference in accuracy between a poor and a good attention group unchanged.

Cognitive Theories at the Crossroads

A number of conclusions and several questions emerge from this brief survey of the major theories about ADHD. First, all the theories examined agree that no single mechanism of those proposed to date can explain the condition, and recent developments all involve development of more complex models. Of these, the recent proposals of Sonuga-Barke are the most explicit, acknowledging a role for both an inhibition and a motivational factor and then attempting to relate these to different elements of the ADHD profile. However, it seems clear that the motivational element is specifically related to delay aversion and is not a general motivational abnormality, since manipulations through reward and instruction have little effect on differences between ADHD and control groups. Second, despite agreement that more than one causal factor must be operative, there appears to be little or no debate on why the two or more causal agents should co-occur so frequently. We might assume that the neural circuitry involved is closely related for some reason such as overlapping functions, biochemical commonalities, or physical proximity, but more evidence is required to answer this question. Third, as is already apparent and will continue to emerge, theorizing is generally imprecise, tending to use constructs such as EF, working memory, and inhibition, without clear acknowledgment of distinctions and relations between them. Working memory and EF are postulated systems that carry out specified functions, whereas the term inhibition covers a range of processes, some of which are highly specific and anchored in physiological research, whereas others are more in the nature of theoretical constructs derived from behavioral data. For our present purposes, we note that the weight of the evidence firmly supports weakness in the operation of EF, and particularly in the inhibitory functions necessary for its operation. We therefore envisage EF as operations in the working memory system, which manipulates inhibitory and excitatory connections in order to meet task demands. However, as yet it is unclear whether this should be ascribed to relatively low-level weaknesses in inhibitory connections or in higher level instructions that set up patterns of excitation and inhibition needed to implement a specific task. Further research will be needed to clarify these alternatives.

The Developing ADHD Signature

The ADHD literature begins to flourish in the midchildhood years, reaching a crescendo by early adolescence. The numerous studies that constitute this period of development represent a wealth of research endeavors that have sought to define more precisely the unique ADHD profile at the genetic, brain, and behavioral/cognitive levels. The enormity of this task and some of the reasons for its complexity have been discussed above but, in brief, research is attempting to define a neurodevelopmental disorder from the "top down", in other words, from its observed phenotype to its genetic correlates. In this section, we expand upon our initial argument that ADHD is a weakness of control systems and make three specific claims: (1), that weakness in attentional control represents the core impairment in ADHD; (2), that development itself plays a crucial role in the emergent control systems such that key aspects of ADHD behaviors, especially at the cognitive level, only become apparent when these control systems are sufficiently developed to be adequately monitored by appropriate tasks; and (3), that attention constructs such as selection or maintenance will show no effects when they are routine or simple. It is only when task performance requires additional "control" systems that difficulties will emerge.

ADHD: The Preschool Years

Not surprisingly, the literature is not especially extensive for this early period of development for a number of key reasons. First, clinical diagnosis of ADHD is not usually made under the age of 7 years. That is not to say that a diagnosis cannot be given in the preschool years; it is just more complex and the prevalence rates are much lower. As a result, the majority of studies have relied on the presence or absence of ADHD "symptoms" as measured by attention rating scales such as the Conners Rating Scales (CRS) or the Child Behavior Checklist (CBCL) as rated by parents and/or teachers. See our review of the various attention rating scales and discussion of the issues related to measuring ADHD in preschool children in Chapter 5. We also recommend the review by Greenhill, Posner, Vaughan, & Kratochvil (2008). Second, there is a wide variability in inattentive behaviors in the preschool years that can be classified under the "typical range," and this makes identifying children who may be at high risk for later development of ADHD a challenging task. Third, there are major inherent difficulties in developing attention control tasks that are sufficiently sensitive to discriminate a unique ADHD signature, such as a core deficit in response inhibition, in such young children. This latter issue is further compounded if development itself plays a major role in elucidating the ADHD *signature*. With increasing age, and as critical brain networks develop, it is highly possible that a more refined profile emerges that mirrors that seen in later childhood.

In this section we review the recent studies that have sought to elucidate *age-related* changes in performance on attention measures. We include preschoolers with a clinical diagnosis of ADHD, and preschoolers who have been identified as "at risk" of ADHD. As we noted above, as well as in previous chapters (Chapters 3 and 5), the debate is still on going with regard to the clinical utility of an early-onset diagnosis, especially in relation to the stability of diagnosis over age.

Our focus in this chapter is not to critique every individual study but rather to highlight the salient findings that emerge with regard to putative attention correlates, as measured by tasks that tap selection, maintenance, and attentional control, and their link with ADHD symptoms. We use a slightly broader age range than used in Chapter 7, encompassing children between 3 and 6 years, following the practice of most current research studies.

Selective Attention

The ability to selectively attend or orient to some aspects of the environment and to avoid distraction is a critical attention skill that can be observed from very early in infancy. In terms of neural substrates, in Posner's attention networking model, skills of visual orienting are associated with posterior brain areas, including the superior parietal lobe and temporal parietal junction. There exists a rich and well-documented literature charting age-related changes in the typical development of visual orienting in infants and toddlers using a range of innovative tasks, for example, visual search and saccadic control. It is somewhat surprising then to find no equivalent published studies that have assessed selective attention or visual orienting in preschool children with ADHD or at risk for ADHD. One plausible explanation for the paucity of studies may be that findings are generally nonsignificant for this young age group and thus go unpublished. These null findings would be expected if, as we hypothesized, ADHD reflects a core difficulty in the control of attention rather than simple selection or maintenance.

Maintenance of Attention

The ability to stay focused ("on task") when there are multiple distractions and when awaiting some significant event *(vigilance)* are both critical components of attention. In Posner's model, skills that tap the "alerting" system are associated with thalamic as well as frontal and parietal regions of the cortex. In the literature on typically developing children, we see a gradual improvement with age in a toddler's ability to focus over longer periods of time and to resist distraction. For a detailed profile of these studies, see Chapter 7 and the excellent critique by Mahone (2005). Given the wealth of these data, it is somewhat disappointing to find so few published studies that have assessed age-related changes in sustained attention in preschoolers with ADHD or at risk for ADHD. Of those that have, the majority have used a version of the Continuous Performance Task (reviewed in detail in Chapter 6). In brief, the CPT requires participants to respond to predesignated targets among stimuli that are presented at a rapid fixed rate over a period of minutes while simultaneously withholding a response to nontarget stimuli. Targets are normally presented infrequently so participants must wait for the stimulus to appear. Although often used as a measure of sustained attention or vigilance, the task also requires inhibitory control, and it is therefore often quite difficult to dissociate the two components.

Perhaps a reason for the scarcity of studies is the difficulty in the ability of very young children to sustain their attention over duration of 5 to 8 minutes as required by many traditional CPTs. Boredom with the actual task demands may indeed be the CPT's greatest weakness as a tool for research on children of preschool age. Why alternatives to the CPT are not employed as measures of sustained attention is still a puzzle.

As an example of one of the few studies to focus on the preschool years, Berwid et al. (2005) assessed 3- to 6-year-olds rated as "at risk" for

ADHD using the DSM- IV Symptom Checklist (similar to the ADHD-RS IV reviewed in Chapter 5). Using cartoon pictures as stimuli, children were instructed to press a button in response to a target stimulus and to refrain from pressing to a nontarget stimulus. Forty-eight trials were given at each of four target/nontarget ratios, but no indication is given of task duration or length of breaks between blocks. The ratio of targets to nontargets varied from 5:1, 2:1, 1:2, to 1:5, with omission errors (nonresponses to targets) expected to increase, and commission errors (response to nontargets) to decrease as the frequency of targets declined. These predictions were confirmed.

In order to examine any age-related changes, the authors further divided their sample into three age groups: a "young" group (3.44 to 4.30 years), a "middle-aged" group (4.31 to 4.88 years), and an "older" group (4.89 to 5.99 years), but sample size numbers were exceedingly low with Ns of 4, 5, 5 in the three high-risk groups. Findings revealed that the youngest group performed worse than the two older groups on both omission and commission errors and reaction time; the two older groups did not differ significantly. In the case of omission errors and reaction time, this appears to have been due to floor effects, suggesting perhaps that the task was too easy to reflect further improvement by the age of 5 years.

From one perspective the findings are encouraging in that they show an ability to discriminate between preschool children at high and low risk of ADHD, but we question whether it was testing sustained attention as claimed because there was no examination of change on any measure over the course of task duration, and we suspect it may have had poor discriminative validity for children older than 4½ years.

Attentional Control

Executive control is the most studied of all attention subdomains in preschoolers, children, and adults with ADHD. A quick glance at Chapter 7 will alert the reader to the enormity of research undertaken in the typically developing literature. Clearly, attentional control matters for preschoolers and undergoes rapid development from the third year of life. When it is underdeveloped, as is the case in ADHD, the potential consequences especially for efficient and effective goal directed behaviors are severe.

As the child progresses through the preschool years, there is an inter-play between the development of early attentional control and the emer-gence of other core executive function skills, most notably working memory (e.g., Davidson, Amso, Anderson, & Diamond, 2006). Working memory, as we discussed in Chapter 2, is a system that permits temporary storage of information while simultaneously processing the same or other information. The most influential model of working memory is by Baddeley (1986; See Chapter 2, Box 2.5), in which short-term storage systems are separated into two "slave" subsystems: a *verbal* (phonological) subsystem and a *visuospatial* subsystem. The central executive executive capacity is a third component that acts as a cognitive control mechanism that manipulates the contents in active storage, and most recently Baddeley (2001) introduced a fourth component, the episodic buffer, which provides temporary storage of information held in a multimodal code, and which is capable of binding different types of sensory informa-tion into a single integrated structure or episode. There is now a well-documented literature that has teased apart verbal and spatial working memory deficits associated with ADHD. In their 2005 meta-analysis of 26 empirical papers, Martinussen, Hayden, Hogg-Johnson, and Tannock (2005) concluded that working memory is impaired in ADHD, but with greater impairments relative to typically developing children in visu-ospatial working memory than in verbal working memory. However, few studies have assessed this relationship in preschool children.

Response inhibition is by far the most widely researched component of attentional control in the preschool years. It involves a child's ability to inhibit a powerful tendency or urge to make a response that is inappropriate. Two of the most widely used tasks are the Go–NoGo and the Day–Night tasks.

INHIBITING A PREPOTENT RESPONSE

Links between inhibition measured by the Go–NoGo task and impair-ment in the frontal lobes in ADHD are supported by Booth et al. (2005), who found reduced brain activity in the frontostriatal network in children with ADHD during performance of a Go–NoGo task. A recent longitudinal study by Brocki, Nyberg, Thorell, & Bohlin (2007) assessed a large community sample of preschool children at Time 1 (mean age 5.5 years, age range 4–7 years) and then again 2 years later at Time 2 (mean age 7.6 years, age range 7–9 years). Children were included on the basis of being at risk for ADHD and/or oppositional defiant disorder (ODD) (described briefly in Box 8.1). Oddly, the authors do not specify how many children in their sample were classified as ADHD, only how many were classified with a dual diagnosis of ADHD and ODD. Parents and teachers completed the DSM-IV RS (DuPaul et al., 1998) at both Time 1 and Time 2. The Go–NoGo task (developed by Berlin and Bohlin, 2002) comprised a blue square, a blue triangle, a red square, and a red triangle presented one at a time on the computer screen. In the first part of the task when the children saw a frequent stimulus (a blue figure), they were instructed to press the "Go" key but to make no response when they saw an infrequent stimulus (a red figure). In the second part of the task, the same stimuli were used, but this time when the children saw a square they had to press the "Go" key and make no response when they saw a triangle, irrespective of color. There were 60 stimuli. The number of commission errors to nontargets was used as a measure of inhibitory ability. This study also assessed spatial working memory and verbal working memory.

Findings indicated that prepotent response inhibition made a unique contribution to the prediction of an increase in ADHD symptoms (not ODD symptoms) over time, indicating an early association of inhibition with ADHD. A similar study using the identical task, also on a community sample of preschoolers, found impairments in performance to be related to inattention symptoms, providing some of the first evidence of a possible dissociation between ADHD symptom domains and response inhibition as early as the preschool years (Thorell, 2007). Conversely, there were no significant correlations between the two types of working memory and ADHD symptoms at either Time 1 or Time 2. One possibility, as argued by the authors, is that working memory skills are not sufficiently developed in the preschool years to expose the weakness that is evident in later childhood. However, other evidence (e.g., Kane & Engle, 2000, 200; Lui & Tannock, 2007) suggests that it is the central executive of working memory rather than the storage systems tested by Thorell that is implicated in ADHD, so this conclusion may be unjustified. Another possibility is that ratings at 3 to 4 years of age may have depended primarily on hyperactivity rather than attention symptoms, which is why there were no significant findings. In studies of older children with ADHD, an association between working memory difficulties and ADHD appears to be confined to inattention symptoms, not hyperactivity/impulsivity symptoms.

INHIBITING A RESPONSE AND INITIATING A DIFFERENT WEAKER LEARNED RESPONSE

The Day–Night task is one of the most widely used tasks of response inhibition in preschool children. Chapter 7 provides a detailed description and review of this task in the context of typical development. The task was originally developed by Gerstadt, Hong, & Diamond (1994) and required children to say "night" when shown a card with a sun on it and say "day" when shown the card with a moon on it. Children had to hold two rules in mind and to inhibit the urge to give a conventional verbal response. This task has been extremely sensitive in detecting age-related changes in typically developing children from age 3 years. Similar findings have emerged in children with ADHD symptoms.

In a recent longitudinal study, Wahlsted, Thorell, and Bohlin (2008) assessed a community sample of preschool children over two time points using a subsample of participants recruited at Time 1 (age range 4–6 years) for testing at Time 2 (age range 6–8 years). In a somewhat complex design, ADHD symptoms from DSM–IV in Time 1 were rated only by teachers and at Time 2 by both parents and teachers. Response inhibition was assessed using a modified version of the Day–Night task alongside a Go–NoGo paradigm (Time 2 only), measures of verbal and spatial working memory, and a measure of verbal fluency. The main finding was that early EF impairment (as measured by errors in the Day–Night task at Time 1) served as a predictor for later inattentive difficulties, measured by a combined score derived from both tasks at Time 2, and early ADHD ratings predicted later ratings for a range of disorders. Hence, each type of measure was predictive of similar measures 2 years later, but there were no predictions across types of measure. It is therefore unclear what conclusions to draw from this pattern of results. See also Berlin and Bohlin (2002), Brocki et al. (2007), and Thorell (2007), who have used a similar Day–Night modification and methodology.

Conclusions

Taken together, the findings from the relatively sparse literature (selection, maintenance, and control) suggest that any impairments are confined to certain measures that tap control processes. Most of the evidence that we have cited indicates impairments in inhibition, but where several measures have been taken, ADHD impairments have emerged in omission errors, reaction time, and variability, as well as in commission errors, the main index of impaired inhibition. Hence, we argue for a general weakness in control processes, rather than one specifically in inhibition, and this weakness seems to be apparent at an early age, contrary to our hypothesis that it would be difficult or impossible to observe at an early age when control functions are rarely operating.

ADHD: The Childhood Years

The majority of ADHD research has focused on the middle childhood years, when a clinical diagnosis can be made and treatment efficacy can be evaluated. Not unexpectedly, this has generated a substantive volume of research, the findings of which are too extensive to be discussed here and instead are explored in a later chapter that focuses specifically on treatment outcomes, notably the impact of stimulant medication on behavioral and cognitive outcomes (Chapter 10). We will review here those studies that have used participants who are medication free at the time of testing. This will allow us to capture what is termed "baseline" performance, that is, performance prior to the influence of any medication.

   Children entering school bring with them an increasing repertoire of cognitive and behavioral skills that serve as the foundation or "building blocks" for future academic and social functioning. We have looked at this in detail in Chapter 6. The changes we see over this period reflect a child's increasing mastery of effortful control and directed attention. In a condition such as ADHD, when control systems appear to be impaired quite early in development (see our previous section), there is going to be some impact on the development of other interlinking cognitive functions, in particular, working memory, which allows a child to retain and manipulate information contained in task instructions in order to generate appropriate actions and to inhibit unnecessary actions. We provide here a snapshot of findings that have focused on delineating ADHD performance across attention subdomains.

Selective Attention

In the typical literature, the development of selective attention is frequently assessed using visual search paradigms designed to provide a window to observe the efficiency of selection of a specific target from an array of distracters. A variety of visual search paradigms exist with sometimes quite different task demands and complexity. As we have seen in Chapter 7, search efficiency, for example, can be dependent on the display, which can be manipulated in terms of number of items (e.g., 4 vs. 8 vs. 12), size (e.g., small vs. medium. vs. large items), color (e.g., blue target vs. green nontarget), or shape (vertical vs. horizontal lines). Important distinctions are also made between "feature" search, in which a single feature distinguishes distracters from targets, and "conjunctive" search, in which combinations of two or more features distinguish distracters from targets. In the latter type of search, reaction times are much slower because targets are much harder to discriminate from nontargets. In a comprehensive review of existing visual search paradigms and their use in children with attention problems, Wilding (2005) distinguished two types of visual search tasks described in Box 8.3.

SINGLE-FRAME VISUAL SEARCH

Wilding concluded from the single-frame studies of visual search that there is "a large but inconsistent body of evidence supporting impairment in processing accuracy and/or speed in attention-impaired

---

Box 8.3 Wilding's Visual Search Distinctions

- *Single-frame visual search*, in which a series of displays is shown, each of which does or does not include a target
- *Continuous visual search*, in which several targets have to be found in a single complex display

children. However, tests on many specific aspects of such tasks, such as an increase in the number of distracters, produced no significant difference between attention-impaired and control groups in the effects of such manipulations." (p.498). However, he also noted a number of suggestive findings. Karatekin and Asarnow (1998) found that attention-impaired children took longer to change focus, and Mason, Humphreys, and Kent (2004) found that a preview display sharing features with the target in the main display caused more problems for their ADHD group. Longer stimulus intervals also adversely affected attention-impaired groups (van der Meere, Vreeling, & Sergeant, 1992; Zahn, Kruesi, & Rapoport, 1991), and adjusting speed in accordance with different task demands also appeared to be difficult (Mason et al., 2004; Taylor, Sunohara, Khan, & Malone, 1997). All these findings are suggestive of impaired ability when required to adjust to specific task demands such as switching attention, preparing to inhibit prepotent responses, or irrelevant inputs. In other words, EF would appear to be inefficient. It is noticeable that only one of the studies surveyed by Wilding examined relations between age and performance.

CONTINUOUS VISUAL SEARCH

The second type of visual search identified by Wilding (2005) involved search of a complex display for several specified targets. Manly and colleagues provide two examples of this type of task (Manly et al., 2001). In one task (Mapsearch), the participant has 1 minute to search for restaurant (knife and fork) symbols on a map. The number of items found is recorded, which is essentially a measure of speed; no measure of false alarms to nontargets is taken, if such errors occur. In the second task (Skysearch), pairs of matching spaceships have to be found as quickly as possible in an array of multiple pairs, the majority of which are nontargets as the two spaceships in the pair are different. The participant decides when to stop searching and time per target is calculated. Again, the basic measure is speed, though the time taken may also be affected by individual differences in willingness to stop searching. False alarms to nontargets are recorded, but little use has been made of this measure.

Using the Map search task, Wilding, Munir and Cornish (2001) and Heaton et al. (2001) found no differences between children with impaired attention and typically developing control children. Cornish et al. (2005) did find that a group with high ADHD symptoms drawn from a community sample performed worse on this task. With Skysearch, Wilding et al. and Heaton et al. and also Manly et al. (2001) again found no differences between groups.

These consistent results appear to suggest that selective attention as tested in visual search tasks of this type is unaffected in groups with impaired attention. However, a large body of data obtained by Wilding, Cornish, and colleagues using the Wilding Visearch task (see Chapter 6) and discussed by Wilding (2005) demonstrates a more complex pattern of results. These studies employed teacher ratings on the ACTeRs or SWAN scales (see Chapter 5) to define good and poor attention groups. In one study (Cornish, Wilding, & Hollis, 2008), the bottom and top 10% of boys from a large population sample were selected, giving a poor attention group that differed little from a clinically diagnosed ADHD group. In most studies groups with less marked rating differences were studied, and a similar pattern of differences in performance emerged, as would be expected if attentional ability were continuously distributed throughout the population.

These studies were discussed briefly in Chapter 7. They employed mainly children in the age range of 8 to 10 years (though one study tested preschool children and one included some teenagers) and have demonstrated a strong association between age and time to locate targets and between rated attention and the number of false alarms to nontargets (Cornish et al., 2008; Wilding, 2003; Wilding & Burke, 2006; Wilding & Cornish, 2007; Wilding, Munir, & Cornish, 2001; Wilding, Pankhania, & Williams, 2007). In general, no relations have been found between attention and speed or between age and accuracy, though with preschool children Wilding and Burke (2006) did find that children rated as having poor attention by their teachers were slower, and in a very large Chinese sample ranging in age from 6 to 17 years, Zhan et al. (2010) did obtain a reduction in false alarm rate with increasing age. However, this relation was much weaker than that between age and speed and presumably has not been found in the majority of the studies because it is relatively weak and the age range studied has been much narrower. The relation between rated attention and false alarm rate has appeared most consistently in more difficult versions of the search task, with the inclusion of nontargets that are very similar to the targets or the requirement to search alternately for two targets, which represent situations where precise control operations are most clearly involved.

Wilding (2005) examined possible explanations for the absence of any (marked) relation between impaired attention and speed on continuous search tasks, particularly in view of the common assumption that impaired attention is closely related to impulsivity, which would predict fast inaccurate responding. Several findings from the single-frame studies, however, suggest that children with impaired attention operate more *slowly* than control groups, a finding that conflicts both with this prediction and the results from the continuous visual search studies. Wilding noted again the indications that children with impaired attention appear to have problems in preparing responses, especially if stimulus–response relations are unusual or a delay or a shift in focus is required, and pointed out that such adjustments have to be refreshed for every new trial in single-frame studies, whereas only one major "set-up" process need occur at the beginning a continuous visual search for several targets. Increased response times in the former type of task may be due to slowing of such preparatory processes, but such processes would only make a minor contribution to overall time in the continuous search tasks. Such an explanation of the slower responding of attention-impaired children is consistent with indications that search rate itself is not impaired, shown by the similar increase in time for attention-impaired and control groups as the number of distracters increases.

Wilding et al. (2007) also tested explicitly the possibility that attention-impaired children may make frequent false alarms due to responding impulsively before stimulus processing (which may itself be less efficient) is complete. In a visual search task, participants were instructed to concentrate on either responding quickly or accurately. The authors argued that if impulsive responding were the source of errors in the poor attention group, imposing speed instructions would impair the good attention group more than the poor attention group, and imposing accuracy would help the poor attention group more than the good attention group. In the event, both groups adjusted response speed according to the instructions, and no differences in speed occurred between groups in either condition, just as with neutral conditions. Also, as in the other studies with this task, the poor attention group made more false alarms than the good attention group, but both groups adjusted the number of these in accordance with the instructions by the same amount, leaving the difference between the two groups unchanged (see Figure 8.1).

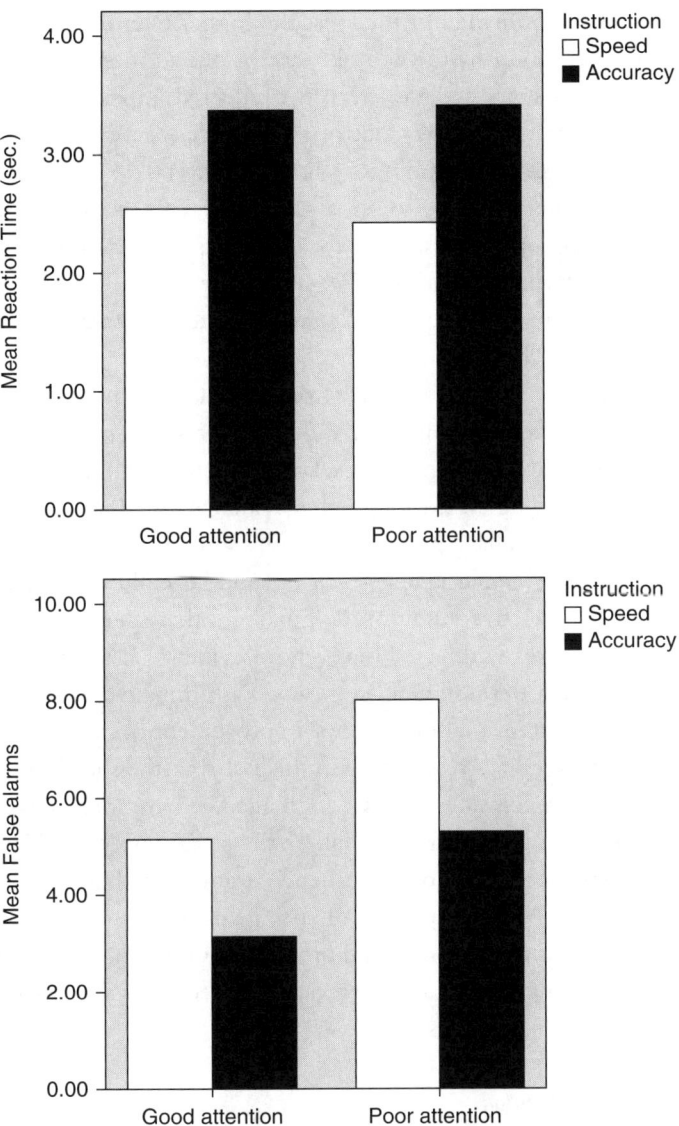

Figure 8-1. Mean times to find targets and mean number of errors in the Wilding Visearch task for good and poor attention groups under speed and accuracy instructions

This result confirmed further that the higher rate of false alarms in poor-attention groups with this task was not due to impulsive responding. The authors suggested, rather, that it reflected impaired preparation or pretask programming by control systems that set up the required pattern of excitatory and inhibitory links for the task. Such inadequate control leaves the possibility of cross-talk between processing channels that yields inappropriate responses to nontargets. The nature of the task does not reveal whether the other possible type of error, missed targets, is also more frequent, as such misses are covert, with targets passed by during search on the way to the next target detection.

While there was some evidence from these tasks for slower responding in poor-attention groups, the difference was not marked. Rather, speed of responding appears to be related directly to chronological age (and possibly intelligence, though the evidence on this is uncertain). Hence, these studies have provided the only substantial evidence on the nature of developmental changes during middle childhood in performance on this selective-attention task, namely that speed of executing the search improves with age. However, no evidence has emerged for a difference in such development in groups with impaired attention. The evidence is consistent that only minor improvement occurs in processes responsible for accurate responding in this task, but the speed of executing the search process does increase with age. We have interpreted these results as indicating inaccuracy in either the setting up or activation of stimulus–response associations and hence argue that they support our interpretation of ADHD in terms of an impairment in control systems. The task was highly motivating, and timing was under the child's control, so the state regulation and delay aversion theories offer no obvious explanation for the findings.

Maintenance of Attention

The ability to focus and remain alert on task and at the same time to ignore distracting stimuli or to remain alert for relevant and often infrequent stimuli is a requisite skill for effective, lifelong learning. Ability to maintain attention also serves as a foundation for the development of more complex, control skills including decision making and inhibiting impulsive, automatic responses. In Chapter 2, we summarized the history of research on maintenance of attention, highlighting the difference between situations where targets occur infrequently in an otherwise unchanging background and those where events occur at a high rate, only some of which are targets. In the first case, readiness to respond to target events declines over time, whereas in the second case it is ability to discriminate targets from nontargets that deteriorates. The majority of research on sustained attention in ADHD has employed the latter type of task.

In ADHD, a lack of focus or concentration is a core behavioral symptom that is well documented. At the cognitive level, findings are less consistent, and there is a heavy reliance on one particular type of sustained task: the Continuous Performance Task (CPT). There are, in fact, numerous studies that have used the CPT, and its many variants, as a measure of sustained attention, and we have described this task at length in other chapters (see Chapters 6 and 7) and earlier in the present chapter, so we will not do so again here. There is ongoing debate, however, as to its effectiveness as a discriminating measure of sustained attention in ADHD, especially on the basis of error types alone (e.g., number of omission errors as an index of lapses in attention).

Instead, we argue here that one of the most sensitive measures of attention impairment captured in tasks such as CPT and in other similar paradigms such as the Sustained Attention to Response Task (SART) is *response variability* from moment to moment over task performance. Some exciting new studies have attempted to tease apart response variability into component parts: a progressive slowing of reaction time over the course of a task and shorter-term reaction time variability over time. The former is argued to reflect impairments in vigilance (maintaining alertness for target stimuli) and the latter to reflect impairments in sustained attention (maintaining focus over task duration). We now highlight some studies that have examined the association between response variability and ADHD.

FOCUSING ON RESPONSE VARIABILITY

The Sustained Attention to Response Task (SART) has been used quite extensively to measure sustained attention in studies of brain-injured patients (e.g., Robertson, Manly, Andrade, Baddeley, & Yiend, 1997; Whyte, Grieb-Neff, Gantz, & Polansky, 2006). The task has been described in detail in Chapter 6 but, in brief, requires active attention over long and, in the original version, unpredictable intervals between targets over numerous trials lasting approximately 5 minutes. Another version uses a repeating ascending sequence of digits from 1 to 9 and requires responses to all but the digit 3. The most extensive investigation of performance on these tasks in children with ADHD has been conducted by Robertson and colleagues. In their 2007 study, using the ascending sequence version (Johnson, Robertson, et al.), children with ADHD (age range 8 to 15 years) were compared to typically developing control children (age range 9 to 15 years) on the SART. The measures taken were omission errors (nonresponses on the Go trials), commission errors (responses made to the NoGo digit 3), and the mean and standard deviations of the reaction time. The commission errors were later used to further subdivide the ADHD group into "impaired" and "unimpaired" groups.

We highlight three findings that the reader may find especially interesting. First, the impaired ADHD group were initially faster in their responses than the other two groups, but became progressively slower in mean reaction time over the duration of the task, and in the second half of the task were slower than the other groups. Overall, this resulted in no mean difference between groups in reaction time. Conversely, the mean reaction time in the unimpaired and typically developing controls remained constant over task duration. The authors hypothesized that this progressive slowing of reaction time in the impaired ADHD group may be indicative of a decline in arousal levels over task duration that reflects a deficit in the bottom–up subcortical arousal system. Second, the impaired ADHD group showed greater variability in response time from trial to trial and, even though all three groups became more variable as the task progressed, the impaired ADHD demonstrated the greatest increase in standard deviation of response time over task duration. Third, the impaired ADHD group made more omission and commission errors throughout; all groups showed some increase in errors over time, but there were no group differences in this increase. The authors suggest that the greater variability in response time from trial to trial in the impaired ADHD group, together with the higher error rates, indicates a deficit in sustained attention from the beginning of the task that is not affected by time on task (although it is unclear why this should be classified as a deficit specifically in sustained attention as there were no differential effects of time in different groups). This, they argue, implies a deficit in the top–down control systems associated with the frontoparietal attention networks.

Two points are worth noting about the Johnson, Robertson, et al. study. Their method of subdividing the ADHD group, originally selected on the basis of ratings, revealed that only the subgroup that also showed impairment on the basis of error rates demonstrated other abnormal characteristics of performance. This clearly supports the suspicion that ADHD groups selected by conventional rating methods are not homogeneous, and it implies that objective performance measures may eventually enable us to distinguish subsets of children with similar impairments who can be studied more intensively to uncover the precise nature of these impairments. The nature of any impairment in the subgroup that did not exceed the error rate of the control group was, of course, not revealed in this study. Second, we note that two impairments were suggested in the ADHD impaired group: a problem in maintaining arousal and a deficit in control systems that maintain focus on the task. However, no decisive evidence is presented that these were in fact distinct (the relevant measures were, we suspect, correlated) or any suggestion as to why these two impairments should frequently coexist. We consider it more likely that these are two aspects of a common impairment.

MAINTAINING ATTENTIONAL FOCUS

The ability of CPT tasks to discriminate between children with ADHD and children without ADHD is at best a somewhat weak association. Yet despite mixed findings, it remains one of the most widely used attention measures. In an exceptionally detailed study, Huang-Pollock et al. (2006) assessed children (aged 7 to 12 years) with ADHD combined-type, ADHD inattentive-only type, and typically developing comparison children on the A-X version of the CPT. In brief, this task requires a response to be made only after a specific sequence of letters is shown on a computer screen, in this case, the letter *A* followed by an *X*. No response is required for any other sequence.

Findings demonstrated an overwhelming impairment in performance in children with ADHD compared to control children, in terms of all types of errors, a steeper increase in errors over task duration, response time variability, and perceptual sensitivity (d'). Differences between the

ADHD subtypes only emerged with respect to response bias, with those children with the ADHD combined subtype showing a greater tendency to respond when the target was present compared to children with the ADHD inattentive subtype who were equivalent to controls. At one level, these findings indicate a pervasive impairment in sustained attention in the childhood period that almost certainly characterizes the ADHD combined subtype and to a lesser extent, the inattentive subtype.

Replications of these findings are essential, and although in its early stages, this type of research holds strong promise for understanding the detailed pattern of performance in ADHD and possibilities for comparison of ADHD with other neurodevelopmental disorders. We note, however, that only one measure out of seven suggested any difference between the ADHD combined and inattentive types, with the former group having a lower (more impulsive) beta value. Though this is what would be predicted, it is not a compelling demonstration of any difference between the two subgroups. We also note that, as every measure except hit response time yielded a difference between the ADHD groups and the control group, either the different measures did not, in fact, reflect distinct processes as claimed, or a single underlying weakness impacts on all aspects of performance at this task. Once again, as with the findings with younger children, we are inclined to the view that weakness in the control processes needed to set up a program for carrying out the task, excluding distraction and maintaining focus and arousal level, offers a more plausible and parsimonious theoretical approach than postulating multiple weaknesses.

Attentional Control

Efficient and goal-directed behavior is a process that depends on multiple levels: selective attention, vigilance, working memory, conflict monitoring, and response inhibition. It is this latter skill that we focus on here, in part because it has generated the most intensive research activity of all EF functions in the childhood period. The Stop–Signal task is one of the most widely used measures of response inhibition in ADHD. Its popularity is matched only by its sister task, the Go–NoGo task. Both tasks have been described extensively in previous chapters but are described in brief here to illustrate their differences and commonalities.

In the Go–NoGo task, a response to an infrequent nontarget has to be inhibited, while in the Stop–Signal task a speeded response to a "Go" stimulus has to be inhibited only upon presentation of a specific Stop signal (either visual or auditory). In contrast to the Go–NoGo task, the delay between the presentation of the Go stimulus and the Stop signal (referred to as the "stop–signal delay") can be manipulated. Although similar in some respects (for example, both tasks have a similar end point, which is the active suppression of an activated motor response, and both require working memory to be efficiently active in holding in mind the single rule of when to inhibit a response to the stimuli), performance on these tasks requires somewhat different inhibitory processes, restraint in the Go–NoGo task and cancellation in the Stop–Signal task (Crosbie, Pérusse, Barr, & Schachar, 2008).

SHOWING RESTRAINT

There is a wealth of research studies that have assessed response inhibition using a Go–NoGo task in children with ADHD. Findings in the childhood period consistently show that ADHD is associated with more common commission errors, indicating poorer ability to inhibit responses relative to non-ADHD control children and most recently, intrasubject variability (e.g., Klein, Wendling, Huettner, Ruder, & Peper, 2006). However, given the relatively low inhibitory load required on the simple Go–NoGo paradigm, recent studies have begun to look at the differential impact of working memory load on task performance. Specifically, does increasing working memory load impair response inhibition performance in ADHD? A recent study by Wodka et al. (2007) addressed this question using three "levels" of Go–NoGo paradigms in children with ADHD and typically developing control children aged between 7 and 16 years. The first task was a simple Go–NoGo, which involved responding manually to a green spaceship and withholding for a red spaceship. The familiar components of the tasks (i.e., red = stop and green = go) minimized the demands on working memory. The second task also used a similar design of red or green spaceships, but this time the child was told to press quickly for green spaceships and also for red spaceships, but only if they were preceded by an even number of green spaceships. They were to inhibit pressing a red spaceship if it was preceded by an odd number of green spaceships. The third task incorporated a motivational component, which, although similar in design to the simple Go–NoGo task, indicated after every response whether the child had lost or won points based on his or her performance.

Findings were threefold: first, the ADHD group made a greater number of omission errors than controls across all three tasks irrespective of working memory demand; second, not all tasks demonstrated differences in response time variability, with the working memory Go–NoGo paradigm showing no group differences; and third, significant age effects were found for all three tasks and across both groups, with performance improving with age on percentage of commission errors, reaction time, and response variability (with the exception of the motivation Go–NoGo paradigm). However, the only task to report a significant age by group interaction was the simple Go–NoGo paradigm, with younger ADHD children more vulnerable to inhibitory errors than control children, but no such difference after the age of 10 years. The authors conclude that response inhibition as measured in the simple Go–NoGo task represents a primary deficit in ADHD that is unaffected by working memory load or motivational factors.

CANCELLATION

As the reader will observe as they continue this chapter, the Stop–Signal task is a widely used measure of response inhibition in children with ADHD. Its robustness and sensitivity in teasing apart group differences is well documented. In a recent meta-analytic review of 24 studies, Alderson, Rapport, Sarver, and Kofler (2008) sought to examine possible moderator effects related to sample composition (e.g., age, diagnostic grouping) and task variables (e.g., target frequency, task trials). They found that three core variables differentiated children with ADHD from typically developing control children: first, a significantly slower mean reaction time (MRT) (moderate effect size .45[1]), indicating less efficient cognitive control in task processing; second, greater reaction time

---

[1] Effect sizes indicate the magnitude of effect of a given statistical difference. Generally, effect sizes are classified as small if they are below .30, medium if between .30 and .67, or large if above .67 (taken from the classifications employed by Alderson et al.).

variability (SDRT) to the primary "Go" stimuli (large effect size .73), indicating a greater number of lapses in attention; and a slower stop–signal reaction time (SSRT) (moderate effect size .63), indicating poorer inhibitory control. However, it is unclear whether the slower SSRT simply reflected the slower processing evident in the MRT or indicated an additional impairment of the inhibition process. Because no group differences were found in the time between Go and Stop signals at which the 50% criterion for inhibited responses was achieved, the authors argue that the main differences between groups was in the speed and variability of stimulus processing. Interestingly, in a study that did not include an ADHD group, Johnstone et al. (2007) concluded that response inhibition did not improve with age while the efficiency of response execution did, supporting the suggestion that these two processes are distinct. This conclusion is reminiscent of the findings from the Wilding Visearch task, reported earlier, in which speed but not accuracy improved with age.

Tillman and colleagues looked specifically at the development of motor inhibitory control using a novel version of the Stop task in a community sample of children from 4 to 12 years of age (Tillman, Thorell, Brocki, & Bohlin, 2008). This version of the Stop task has been described in a previous chapter (Chapter 7), so, briefly, the child was presented with a car on the screen and instructed to press the space bar as soon as the car appeared. When the bar was pressed, the car would drive away. On 25% of trials, a visual Stop signal in the form of a road stop sign appeared, and the child was informed not to press the space bar as the car would not drive away. The task took approximately 10 minutes to complete. Four measures were included under two broad headings: measures of inhibition capacity (Stop signal reaction time and a probability of inhibition) and two measures of response execution (Go reaction time and response variability). Performance was also compared to measures of vigilance (as measured by omission errors and response variability on the CPT) and interference control (as measured by correct responses on a Stroop-like task).

There were no interactions between age and symptom domains, hence differences due to ADHD did not change with age. In terms of an association between ADHD symptoms and Stop performance, the four measures were differentially linked with inattentive and hyperactive/impulsivity behaviors. Furthermore, though the task aimed to achieve a success rate on the Stop trials of 50%, in the event, rates well above this occurred, with sufficient variability to permit correlations between this variable and the rating scales to be calculated. There are, therefore, some doubts as to whether all participants were adopting the intended strategy and whether the distribution of success rates is valid for this purpose. See also Johnstone et al. (2007) for an example of the typical trajectory of inhibition versus response execution processes on the Stop–Signal task.

In summary, these studies suggest that a number of measures from the Go–NoGo and Stop tasks discriminate between ADHD and control groups, but the balance of the evidence suggests that it was speed and variability of responding rather than the inhibitory process itself that was impaired in the former group. This is not consistent with several other findings that we have described from other tasks and casts some doubt on purely inhibitory theories of ADHD. It suggests, rather, that problems in control required for maintaining focus and arousal level may be the main factors operating in the Stop task, which is consistent with our general argument that overall control, rather than specific inhibition processes, is the key impairment in ADHD.

ALTERNATING VISUAL SEARCH

The alternating version of the Wilding Visearch task requires the child to search alternately for two different targets and therefore requires complex control to ensure continual switching of attentional focus. As already indicated, children rated as having poor attention skills consistently make more false alarms to nontargets on this task than those with good attentional skills (Cornish, Wilding, & Hollis, 2008; Wilding, 2003; Wilding & Cornish, 2007; Wilding et al., 2001, 2007). The absence of marked group differences on the simpler versions of the task that require search for a single target suggests that the critical factor was the greater demand on control systems in the alternating search. Groups did not differ in the speed of search, so the higher error rates were not attributable to more impulsive responding in the poor attention groups, a conclusion confirmed by Wilding et al. (2007) in their study of the effect of speed versus accuracy instructions.

Conclusions

This brief glance at the ADHD childhood literature provides evidence that clearly demonstrates that developmental changes occur in children's performance over this period, with expected improvements in the ability to select and sustain attention and to inhibit prepotent responses with age. However, there is no compelling evidence to indicate that these age-related changes differentially affect children with ADHD.

We argue that differences in performance, irrespective of age, associated with ADHD are explained most economically by impairment in overall control systems, rather than postulating several different impairments, the co-occurrence of which is not explained. Impairment in control can affect a variety of processes such as setting up a "program" to match the demands of a task, especially in the case of more difficult tasks (i.e., simple forms of planning), and maintenance of a consistent task focus for the duration of the task. No consistent evidence for fast, inaccurate responding or abnormal motivation in ADHD has emerged.

ADHD: The Adolescent Years

Comparatively less cognitive research has focused on ADHD during the adolescent years, which we define here as ages 12 to 18 years. The scarcity of current research focusing on the teenage years is surprising given that neuroimaging findings indicate persistent structural brain abnormalities into the adult years (see Castellanos et al., 2002, for developmental trajectories in brain volume abnormalities in children and adolescents with ADHD aged from 4 to 18 years; see also Makris et al., 2008). Of the current research, the majority has focused almost exclusively on trajectories of executive function (e.g., inhibition, working memory, set shifting), with relatively few studies examining age-related patterns of selective, sustained, and attentional control in the format used in the childhood literature. Our review here will therefore be brief.

Selective Attention

We can find no comparable visual search studies in adolescence that would complement those described above in the childhood period.

Yet, given that impairments are clearly evident in the preschool and childhood years, it is important to trace developmental trajectories into the adolescent years to establish the stability of visual search performance with increasing age, and to establish when in development performance converges and diverges from typical development.

Maintenance of Attention

Of those studies that have assessed sustained attention in adolescents with ADHD, the CPT has been the task of choice. From preschool onward, we have seen how difficulties begin to emerge on this task, as reflected by a greater number of omission errors (lapses in attention), slower mean reaction times, and atypical levels of response variability, perceptual sensitivity (d'), and response bias (β), all becoming more salient with increasing age. In the adolescent years, there is some indication that difficulties in sustained attention remain stable throughout this period (e.g., Loo et al., 2007, but also see Seidman et al., 2005, for a comprehensive review of the impact of age and gender on a broad range of neurocognitive tasks, including CPT, in ADHD in the preteenage and teenage years), but generally research is limited, so conclusions can only remain somewhat tentative.

One of the few studies to provide data on a longitudinal sample of adolescents with ADHD investigated age-related changes in performance on the CPT (among other measures) in adolescents originally tested in childhood between the ages of 7 and 11 years and then followed up approximately 9 years later (Halperin, Trampush, Miller, Marks, & Newcorn, 2008). A main aim of this research was to examine the relation between any age-related changes in ADHD symptoms and neurocognitive performance. To achieve this goal, the ADHD group was divided into *persisters*, comprising those adolescents who presented with at least six ADHD symptoms in multiple settings, and *remitters*, comprising those adolescents who presented with three or fewer ADHD symptoms. Sustained attention was assessed on a standard CPT. Findings revealed two profiles. One profile concurred with other studies that have reported differences on sustained attention measures in adolescence, in this case, between adolescents with childhood ADHD and non-ADHD controls. However a second, more informative, profile revealed a differential pattern of performance between the two ADHD subgroups. Together, both ADHD subgroups were impaired relative to controls in their ability to detect targets and their response time variability, and exhibited more ankle movements. However, ADHD persisters were the only group to differ from controls in working memory, hits and false alarms, and response bias (β) in the CPT task, indicating a greater tendency to respond that the target was present (lower value of beta). The reduced value of beta is an interesting finding that links with a similar finding we reviewed above in the ADHD childhood section (Huang-Pollock et al., 2006) in which lower levels of beta characterized the ADHD combined subtype, but not the inattentive subtype.

This interesting study highlights the need for further research to address the changes that appear to occur from midchildhood through to adolescence in terms of ADHD subtype and cognitive strategies. The differences between persisters and remitters observed by Halperin et al. also need further investigation, with direct comparisons of sustained performance between persisters and remitters that were not available in that study. The authors suggest that the persisters demonstrated impairments in EF that were not present in the remitters, but the pattern of differences is not as yet sufficiently clear to draw such a conclusion.

Attentional Control

Focusing on response inhibition, which is the most widely researched component of cognitive attention in adolescence, we describe findings from two of the most popular tasks in the childhood ADHD literature, the Stop–Signal task and the No–GoNo task. Although lacking the intensity of research found in the childhood period, one of the most consistent findings is that impairment across these measures of inhibition is relatively stable over time into late adolescence (e.g., Greimel, Herpertz-Dahlmann, Gunther, Vitt, & Konrad, 2008; Loo, et al., 2007; Martel, Nikolas, & Nigg, 2007; Schulz et al., 2004).

One recent study by Bitsakou, Psychogiou, Thompson, and Sonuga-Barke (2008) is one of the few to assess age-related changes in ADHD children aged between 6 and 17 years on both the Stop–Signal task and the Go–NoGo task. Both tasks have been described in detail above so will not be discussed here. Findings demonstrated poorer performance by the ADHD group relative to control children and adolescents on almost all measures on the Stop–Signal task (e.g., slower *stop–signal reaction time*, or SSRT) and the Go–NoGo task (e.g., lower probability of a correctly inhibited response to the Go response). Younger children, collectively, were slower and more variable in performance than older children, but the lack of status by age interactions indicated that differences between the ADHD and control groups apparent in childhood were maintained into adolescence. However, the authors note that the cross-sectional nature of the study may not have been sufficiently sensitive to identify age-related differences in trajectories.

However, given the paucity of longitudinal studies for this period, we can go no further in our discussion of developmental age changes in relation to the Stop–Signal task and the Go–NoGo task and perhaps conclude that any changes in ability are likely to be minimal during the time course of adolescence given the already documented stability of response inhibition deficits from childhood to adolescence.

Of course, one also needs to look at the stability of ADHD itself as a diagnosis across childhood into adolescence and adulthood. This is a complicated issue because changes in expression of ADHD symptomatology occur with increasing age, for example, hyperactivity symptoms often reduce with age, and ADHD adolescents and adults are less fidgety compared to ADHD children. See Faraone, Biederman, and Mick (2006) for a meta-analytic review of the persistence of ADHD symptoms into adulthood and Monuteaux, Mick, Faraone and Biederman (2010) for a longitudinal study of sex differences in ADHD symptoms and comorbidity.

However, some insights regarding longitudinal outcomes might be gained from the findings of a series of studies by Seidman and colleagues, who assessed the stability of executive functions from the childhood years through to early adulthood and also studied these separately by gender. Biederman et al. (2007), for example, assessed males across a broad range of executive functions including sustained attention, as measured by an auditory CPT. Other EF measures included response interference (from the Stroop task) and set shifting (Wisconsin Card Sorting Task perseverations). Individuals were categorized as having an executive function deficit (EFD) if they performed at least two tasks at 1.5 standard deviations below the level of controls. At the first testing time point, ADHD males ranged from 9 to 22 years and had aged approximately 7 years at follow-up. Findings indicated a stability of EFDs across development into adulthood such that, of those individuals who exhibited EFDs in late childhood/early adolescence, a majority continued to exhibit weak executive functions in early adulthood. See also Seidman et al. (2005) for a similar design, but this time teasing apart the possible impacts of gender on EFDs in the preteen and teenage years. In brief, the authors found that EFDs in ADHD were present irrespective of age or gender. Stability of the EFDs in girls with ADHD initially tested between the ages of 6 and 18 years and then tested again at 5-year follow-up has recently been reported by Biederman, Petty, Doyle et al. (2008). Over two-thirds of their sample (79%) continued to show executive functioning deficits from baseline to follow-up.

Taken together, studies from the ADHD adolescent period with a focus on cognitive inattention are surprisingly sparse, although there is an emerging literature on the development of more general executive deficits. This contrasts with the numerous studies that comprise the childhood years. The tentative conclusion to be drawn from the existing data is that attention impairments exhibited in childhood ADHD appear to remain stable into adolescence for a majority of individuals but certainly not for all. ADHD symptoms are also variable across development, and it is plausible to assume that in those individuals whose symptoms wane with age from childhood to adolescence (as indicated by rating scales), there will be a similar pattern of improved attention functioning, especially in relation to performance on tasks that involve cognitive control.

Conclusions

Findings from over a decade of exciting and innovative research have produced a much clearer, if not complete, picture of the ADHD signature at the neurobiological, brain, behavior, and cognitive levels. There is, undoubtedly, still much to learn, but existing research has made tremendous advances toward an understanding of this condition at its multiple levels. We know, for example, that cognitive deficits lie at the core of

ADHD, but it is certainly not the case that children with ADHD cannot inhibit or use goal-directed behavior because they do but instead appear to be less efficient than their non-ADHD peers.

Although the reader can access many more studies than those reviewed in the current chapter, a number of important and salient findings have emerged. First, that impairments in cognitive attention are persistent from preschool onward with the most consistent impairment reported for skills that require the use of control processes such as the ability to plan ahead, inhibit prepotent responses, and maintain task focus. Second, it should be stressed that, while poor inhibition is a prevailing feature of ADHD, it is not a defining signature of this condition as it occurs in several other disorders; see Sergeant, Geurts, and Oosterlaan (2002) for a detailed review of relevant comparisons and the descriptions in Chapter 9. Third, where multiple measures of performance have been taken, it is clear that it is not only specific measures of inhibitory control that are affected in ADHD; other measures also show effects, for example, variability in reaction time for measures of sustained attention. This suggests that the condition is due to a general impairment of control systems rather than a specific impairment of inhibition only; and fourth, there is little evidence that poor performance is associated with impulsive (fast, inaccurate) responding, or with impaired motivational processes.

At the beginning of our discussion of experimental evidence on ADHD, we made three claims: that weakness in attentional control represents the core impairment in ADHD; that development itself plays a crucial role in the emergent control systems such that key aspects of ADHD behaviors, especially at the cognitive level, only become apparent when these control systems are sufficiently developed to be adequately monitored by appropriate tasks; and that attention constructs such as selection or maintenance will show no effects when they are routine or simple. It is only when task performance requires additional "control" systems that difficulties will emerge.

As already indicated, the evidence strongly supports the first of these claims. However, in relation to the second claim, with suitable experimental techniques it has been proven possible to demonstrate weaknesses in attentional control (even in preschool children; while control weaknesses are more readily exposed in school age groups, it is certainly not true that they are absent at a younger age). Third, we have cited a range of evidence suggesting that differences between ADHD and control groups are demonstrated most clearly when tasks are more demanding, requiring some form of control.

There remain, however, important gaps in current research, of which the most notable is a surprising lack of longitudinal research that focuses on the development of inattentive behaviors and cognitive inattention from preschool into adolescence. To date, although a few studies have attempted to capture developmental changes, they are not sufficient to allow firm conclusions to be drawn about ADHD and its developmental trajectory. This issue is further compounded by the lack of developmental research delineating the trajectories or pathways of ADHD preschoolers, children, and adolescents with and without comorbid disorders. Despite the consistent finding that ADHD has strong comorbidity with a range of psychiatric disorders (e.g., conduct disorder, oppositional defiant disorder, anxiety, depression, etc.), few studies have compared selective, sustained, and inhibitory functions in ADHD samples and other diagnostic groups, or evaluated the impact of ADHD plus additional comorbidities on cognitive attention functioning across age.

One essential study would be to compare trajectories of attention performance (selective, sustained, and control) in ADHD populations with and without comorbid disorders, to establish whether developmental pathways differ according to ADHD diagnosis and comorbidity (e.g., ADHD + Conduct Disorder vs. ADHD + Anxiety vs. ADHD alone). It is also possible that different ADHD comorbid disorders also have different genetic correlates that need to be explored within a developmental perspective.

Furthermore, no clear explanation is apparent for the frequent co-occurrence of inattention and hyperactivity. While we have noted some evidence that these two symptom domains may be related to different cognitive correlates that change with age, research is needed that further elucidates the common factor(s) responsible for this predominant symptom cluster (e.g., common biochemistry).

The three key features attracting a diagnosis of ADHD, as stated at the beginning of this chapter, are inattention, impulsivity, and hyperactivity. In general terms, it is easy to postulate associations between these behaviors, and the weakness in executive function that we have suggested is the main underlying weakness responsible for the disorder.

In terms of our yoghurt example, when the control systems are weak, irrelevant aspects of the display or extraneous noises will attract attention, maintenance of the main goal will be disrupted, sudden impulsive selections of other items will occur, and attention and movement will tend to switch continually rather than pursue a single goal with persistence. (Note that impulsivity in this sense implies switching without necessary planning or good reason, not necessarily overrapid responding; the latter has not been apparent in the evidence cited, while the former has.) We leave you with the exercise of elaborating on this example.

In conclusion, the heterogeneity of ADHD warrants thorough and empirical investigations that are theoretically driven and situated within a developmental framework that begins in preschool and proceeds through to adulthood. Identifying early precursors or markers to later cognitive impairments will have tremendous benefits for targeted interventions that recognize the unique and dynamic signatures associated with ADHD and its phenotypic outcomes.

CHAPTER SUMMARY AND KEY FINDINGS

- ADHD is a complex and varied disorder, and uncertainty still exists over the number and nature of subtypes of the disorder that may exist.
- Comorbidities and subtypes remain under-researched and need to be carefully considered in research designs.
- It is increasingly agreed that ADHD should be seen as the extreme of a continuum of variation in abilities rather than a discrete disorder due to the failure of one specific mechanism.
- Weakness in EF processes is the most plausible explanation for the disorder, particularly in planning and inhibitory mechanisms, which are required for excluding distraction and maintaining focus.
- The relation between inattention and hyperactivity and the reasons for their frequent co-occurrence remain unclear.

9

Specific Neurodevelopmental Disorders and Attention

> ### CHAPTER SNAPSHOT
>
> - Current complexities and issues facing research on neurodevelopmental disorders: gene expression and cognition, shared phenotypes, disorder heterogeneity, and methodological pitfalls
> - How commonalities in behavioral problems, notably persistent inattentive and distractible behaviors, can often mask important disorder-specific differences in cognitive inattention and executive dysfunction.
> - Charting atypical trajectories across different neurodevelopmental disorders: Does age make a difference?
> - Cross-syndrome comparisons and the search for disorder-specific attention signatures

Major advances accrued throughout the past decade have resulted in a significant improvement in our understanding of the causal mechanisms of many forms of developmental delay. Collaboratively, the seemingly disparate fields of molecular genetics, developmental cognitive neuroscience, neurobiology, and brain imaging have provided scientists and clinicians with a unique "window" through which to observe the gene–brain–behavior correlates across many neurodevelopmental disorders, including those disorders for which genetic etiology is not in doubt, but their causal mechanisms are not yet determined (e.g., autism) and those for which genetic etiology is well established (e.g., fragile X syndrome, Down syndrome). On the one hand, this window has enabled researchers to explore hitherto unknown effects of gene dysfunction on cognitive impairment and their impact across development. The most significant outcome of this research has been to enable developmental neuroscientists and practitioners (educators and clinicians) to go beyond the impact of generalized intellectual impairment on cognitive development. This involves, in essence, not making the a priori assumption that intellectual impairment equals global cognitive delay but instead identifying *signature profiles* that distinguish children with differing neurodevelopmental disorders from each other and from typically developing children. This is a potentially powerful tool for early identification and treatment of disorder-specific proficiencies and deficiencies. To place these developments briefly within a historical context, previous research in the 1960s and 70s had tended to ignore the role of etiology in explaining cognitive and behavioral impairments in disorders of mental retardation and instead had proposed that a common cognitive impairment accounted for mental retardation across different disorders (see Hodapp, Burack, & Zigler, 1998, for a seminal review).

Although the past 20 years have produced a wealth of innovative work and numerous studies attesting to gene–behavior correlates, one of the major lessons learned from this period of intensive research is an appreciation of just how elusive and complex this relationship actually is and how many additional factors play an interacting role from genotype to phenotypic end state. Our first aim is, therefore, to provide the reader with a discussion of the core complexities currently under debate and highlight common misconceptions. Our second aim is to situate the reader within the attention research as we search for disorder-specific signatures at behavioral and cognitive levels. We then go on to address one of the most complex of issues, namely, the extent to which symptom overlap implies common developmental trajectories or etiologies, or whether so-called "commonalities" in overt phenotypic behavioral outcomes, in this case, inattentive behaviors, actually reflect different underlying cognitive and brain processes that diverge from the normal trajectory over developmental time and across disorders.

Comparison of the attentional signatures of different disorders is far from easy. There are relatively few studies carrying our direct comparisons between syndromes. Different disorders have been studied using different tasks, and even when the same task has been used in different studies, details of the methodology often differ. And impaired performance on a specific attentional function may differ in degree or pattern or underlying cause between different disorders, so merely to note impairment is not adequate. Consequently, we have not provided a simple checklist of impaired functions as this would be misleading, but we have tried to interpret the overall findings in each case. For each neurodevelopmental disorder examined, we highlight any clear differences from the prevailing pattern of ADHD impairment and report direct comparisons with other disorders. After describing each disorder, we conclude with sketches of the overall indications of disorder-specific differences.

Current Complexities and Concerns

We highlight below four core issues that demonstrate the complexity of the task facing researchers as they attempt to elucidate the gene–brain–behavior pathways and their interactions on the developing cortex: gene expression and cognition, shared phenotypes, disorder heterogeneity, and methodological issues. We focus here on our five targeted neurodevelopmental disorders (described in detail in Chapter 3): autism, fragile X syndrome, Williams syndrome, Down syndrome, and 22q11 DS. We also recommend Hodapp (2004) for his thoughtful critique of some of the complexities and issues facing this field of research.

Gene Expression and Cognition

All too frequently, researchers make the assumption that a simple linear relationship can be drawn between a specific gene and a specific set of behaviors. This is a misconception. Take, for example, the case of fragile X syndrome, in which a single gene is turned off, the *FMR1* gene. As a consequence, its protein, FMRP, is totally absent in fully affected individuals but also partially reduced in those individuals with the *carrier* status. This results in a "continuum" of involvement with differing levels of gene expression selectively impacting across early brain and cognitive networks. This stands in stark contrast to the prevailing view that fragile X represents a "discrete" disorder with clear genotype–phenotype correlations. See Cornish, Turk, & Hagerman (2008) for a review of the current findings that challenge these assumptions. In other cases, multiple genes are known to be involved, for example, in Williams syndrome, in which between 25 and 30 genes are deleted on one copy of chromosome 7. Identification of the genes within this *critical region* has the potential to provide important information on which genes contribute to different subsets of the Williams syndrome phenotype. To date, however, it has proven extremely difficult to isolate specific gene–behavior correlations, with the only exception being the Elastin gene shown to be responsible for the vascular abnormalities associated with Williams syndrome (Tassabehji et al., 1999). The lack of correlations has recently been proposed as being the result of a cumulative effect of gene dosage, rather than a single dosage effect, across a *family* of genes involved in the Williams syndrome critical region. For example, Tassabehji et al. (2005) highlight the TFII-I gene family (described in Chapter 3) as being critical regulators for craniofacial and neurological development and propose that, although it is unlikely that any one gene is solely responsible the associated deficits, the *GTF2IRD1* gene may contribute to the craniofacial abnormalities and the *CYLN2* gene to the neurological abnormalities associated with Williams syndrome.

At the cognitive level, the *LIMK* 1 gene and its encoded protein, kinase, has been hypothesized to play a significant role in the development of normal visuospatial cognition and therefore has been seen as an attractive candidate gene for explaining the spatial impairments reported in Williams syndrome (e.g., Frangiskakis et al., 1996, and see Meyer-Lindenberg, Mervis, & Berman, 2006, for a comprehensive review of this proposed relationship). However, not all studies have reported a correlation. Gray, Karmiloff-Smith, Funnell, & Tassabehji (2006), for example, failed to find any association with deletion of the *LIMK* 1 gene in individuals with only *partial* deletions that allow the *LIMK* 1 gene to be relatively isolated with regard to function, and spatial impairment. The authors argue that the lack of consensus between studies may in part be due to the use of different methodologies to assess spatial cognition itself, in particular, the use of standardized versus experimental paradigms. In the context of interpreting gene–behavior relations, for Williams syndrome in particular, standardized paradigms often fail to capture qualitatively different neurocognitive processes underpinning seemingly unimpaired behaviors and therefore confound interpretations of gene–behavior relationships. This is currently a hot topic of debate and is discussed below under methodological complexities.

In Down syndrome, it is generally assumed that the complex cognitive phenotype results from an "overdosage" of genes on chromosome 21 caused by the trisomy. Innovative work on patients with partial trisomy 21 and mouse models has led to a decade of exciting discoveries that have identified a specific critical region known as the Down Syndrome Chromosomal Region-1 (DCR-1) on the distal part of the long arm of chromosome 21 (see Olson et al., 2007; Rachidi & Lopes, 2008, and Chapter 3 for a review of candidate genes). Within this critical region, a number of possible candidate genes, for example, *DYRK1A*, have been identified as playing a possible causative role in the spatial and memory deficits associated with Down syndrome, but as yet these and other findings, although holding tremendous promise, provide a clearer but not complete picture of the impact of triplicated genes on early brain development and the resulting phenotypic expression. The rarity of partial–trisomy 21 patients, who account for less than 1% of the Down syndrome population, means that research has to rely predominantly on mouse models. Although extremely informative, the next step must be to develop appropriately sensitive, experimental paradigms that can be transferred to the mouse from human studies of cognitive development in Down syndrome. In turn, paradigms that are sensitive to human developmental change in Down syndrome, for which there is a wealth of information, need to influence the design of analogous mouse paradigms. Until then, we cannot definitively link the specific cognitive deficiencies associated with Down syndrome, as assessed in human studies, with specific brain–gene expression found in mouse models.

In 22q11 DS, intense research interest has focused on the catechol-O-methyltransferase (*COMT*) gene as a possible candidate gene linked to ADHD. *COMT* is located at 22q11 and deleted in children with 22q11 DS. However, despite numerous studies there is as yet no clear consensus or association between *COMT* and ADHD behaviors. Any association is weak at best. See Chapter 3 for a critique of the *COMT* gene as it relates to 22q11 DS, and Levy (2007) for a critique of the *COMT* gene as it relates to ADHD.

In autism, research has yet to identify any one gene that acts as a sole contributor to one or more of the cognitive impairments associated with the disorder. Although the *FMR1* gene (fragile X syndrome) is one of the few genes that can result in autism (between 2% and 6% of children with autism will have the FMR1 mutation; see Chapter 3), there is no research that has mapped any of the autism cognitive phenotypes to FMR1. The association is with the behavioral commonalities. When you closely examine the cognitive "similarities," findings indicate quite different profiles of error types and accuracy rates that serve to differentiate the two syndromes (see Cornish, Turk, & Levitas, 2007, for a detailed analysis of behavioral and cognitive profiles in autism and fragile X). Other potential genes of interest include the Reelin gene located on chromosome 7 at q21–q36 (e.g., Serajee, Zhong, & Huq, 2006) and the *SHANK3* gene located on chromosome 22 at q13 (e.g., Gauthier et al., 2009; Moessner et al., 2007), but currently these genes, among others, remain only possible autism risk genes, and their contributions are likely to be subtle. It should also be remembered that autism, alongside other neurodevelopmental disorders, has additional complex gene–environment interactions that will impact across developmental time. The nature of these interactions is only just beginning to be explored.

Associations Across Genetically Disparate Syndromes

Commonalities in cognitive and behavioral outcomes across different neurodevelopmental disorders, despite each having arisen from disparate genetic causes, can, at first glance, add considerable confusion to the whole concept of disorder-specific signatures that we advocate throughout this book. Take a second glance, however, and these so-called "commonalities" in phenotypic end states, most notable at the behavioral level (for example, ADHD symptoms), do not imply common cognitive mechanisms or etiology. The reader will see a clear example of this when we discuss attention processing across our targeted disorders, all of which by late childhood share a similar degree of inattentive behaviors, such as distractibility, lack of concentration, and impulsivity, but present at the cognitive level with very different profiles of attention impairments and trajectories. See Cornish, Scerif, and Karmiloff-Smith (2007) for an example of attention trajectories in toddlers and children with fragile X syndrome, Williams syndrome, and Down syndrome.

One intriguing similarity is the "autistic-like" symptoms that often seem to be present across many neurodevelopmental disorders and appear to be associated with two pathways: one has a genetic link and a possible shared neural basis, such as that found between fragile X and autism (Bailey, Raspa, Olmsted, & Holliday, 2008; Loesch et al., 2007; but see Cornish, Turk, & Levitas, 2007, for a profile of cognitive differences); a second pathway appears to be linked to more severe forms of cognitive impairment and occurs sporadically in disorders that are not strongly associated with autism, such as Williams syndrome (Gillberg & Rasmussen, 1994) and Down syndrome (Kent, Ivans, Paul, & Sharp, 1999; Lowenthal, Paula, Schwartzman, Brunoni, & Mercadante, 2007). In the case of 22q11 DS, the autism link is more controversial with recent research studies reporting an elevated rate of comorbidity of between 40% and 50% (e.g., Antshel Aneja et al., 2007; Vorstman et al., 2006), but the extent to which this figure represents a true comorbidity is uncertain.

Dissociations Within Disorders

Clearly, not every individual with a given disorder will present with identical characteristics to the same degree of severity. All too often, researchers can make the false assumption that their findings based on a relatively small, skewed sample are applicable to the wider community of all affected individuals. However, moderator variables such as gender, age, IQ, or developmental level can play critical roles by interacting together or acting individually to influence phenotypic trajectories and outcomes, and therefore, these variables need to be empirically controlled and evaluated. Looking at the example of fragile X syndrome, one of the most widely researched of genetic disorders, its X-linkage status makes it essential that research controls for the differential impact of gender on performance. At first glance, intellectual disability and behavioral difficulties characterize many children with fragile X, but careful examination of the gender profiles reveals overlapping but nonetheless distinct profiles. In almost all boys, IQ is in the mild-to-moderate range of impairment, with profiles emerging as young as 3 years of age (Skinner et al., 2005). In contrast, females present with more phenotypic variation, with some girls only showing subclinical learning disabilities (Bennetto & Pennington, 2002) while approximately 25% display more significant cognitive impairment (most with borderline to mild mental retardation). In terms of the cognitive phenotype, at first glance there appear to be similar profiles differentiated only by severity of intellectual impairment, with males displaying greater levels of deficit than females. However, subtle gender differences emerge when specific domains of cognitive function are assessed, such as visual-spatial cognition (e.g., Cornish, Munir, & Cross, 1998, 1999; Cornish, Sudhalter, & Turk, 2004) and number cognition (e.g., Mazzocco, 2001; Murphy & Mazzocco, 2008a, 2000b; Roberts et al., 2005). Also, females are less likely to be diagnosed with autism than males but will nonetheless display quite significant social difficulties (see Cornish, Levitas, Sudhalter, 2007). Genetic variation in the form of X-inactivation (when one of the two X chromosomes remains inactive and the other active) is seen as the major contributor to the heterogeneity of intellectual disability and the broad range of cognitive deficits seen in females with fragile X. For obvious reasons, this is not an issue of concern in fragile X males whose impairment, without the protection of X-inactivation, shows greater severity.

Given the pattern of these findings alongside recent data demonstrating a fragile X premutation ("carrier" status) signature that is also gender specific (see Cornish, Turk, & Hagerman, 2008, for a review), it is perhaps more appropriate to view the fragile X syndrome as a *continuum of involvement* rather than as a discrete, all-or-none, genetic disorder in which the gene responsible for the syndrome (in this case, the *FMRI* gene; see Chapter 3 for further details) is either switched on or off. The profile appears to be far more complex than the simple dichotomy often suggested in the literature. We have already touched upon this concept of a *continuum* in the context of ADHD symptoms (see Chapter 8). This all-or-none view needs to instead be replaced by the remaining empirical challenge: understanding the multiple factors influencing where, on such a continuum, individuals find themselves.

In autism, IQ, and perhaps to a lesser extent, gender, are also important determinants of phenotypic variations (e.g., Banach et al., 2008) that can significantly impact on early cognitive development (e.g., Dietz, Swinkels, Buitelaar, van Daalen, & van Engeland, 2007). Obviously, children with higher intellectual functioning will perform much better on cognitive tasks than those with greater intellectual impairment. Thus, the majority of empirical studies that include autism as a targeted disorder will differentiate between those children classified as *low-functioning* (IQ < 70) and those classified as *high-functioning* (IQ > 70).

Converging evidence also indicates chronological age as an important contributor to disorder heterogeneity. In recent cross-syndrome studies of fragile X, Williams syndrome, and Down syndrome, Scerif, Cornish, and colleagues have found attention profiles to be identical at some time points in development but then to diverge away from each other at other time points (e.g., Cornish, Scerif, & Karmiloff-Smith, 2007; Scerif, Cornish, Wilding, Driver, & Karmiloff-Smith, 2004; Scerif, Cornish, Wilding, Driver, & Karmiloff-Smith, 2007). Together, these findings suggest that development itself may be a source of syndrome variability.

Methodological Inconsistencies

One of the prevailing complexities in research focused on neurodevelopmental disorders is the inevitable difference in methodology across studies. Differences in task measures and design can often be sufficient to produce quite different findings even when research aims are equivalent. We focus here on four areas of controversy that have been resolved to varying degrees but still remain prominent issues of concern. See Box 9.1 for a summary of the core issues.

Longitudinal Versus Cross-Sectional Research Designs

The past few years have seen an awakening of interest in exploring developmental trajectories of cognitive functioning in different neurodevelopmental disorders. To date, a significant number of studies now attest to the fact that development is neither static nor simply "delayed" relative to typically developing peers. Instead, current research provides us with a *glimpse* of disorder-specific profiles that are dynamic and interactive, converging with typical pathways at some stages in development then diverging at other stages. In other words, they are "atypical." However, we use the term "glimpse" because, although there has been some innovative research that has provided much-needed "snapshots" of performance within and across cognitive domains, isolating different developmental periods (referred to as *cross-sectional designs* or *trajectories*) (see, for example, studies by Ansari, Donlan, & Karmiloff-Smith, 2007; Cornish, Scerif, & Karmiloff-Smith, 2007; Paterson, Girelli, Butterworth, & Karmiloff-Smith, 2006), relatively few studies have used a truly

Box 9.1 Core Methodological Issues

- Longitudinal versus cross-sectional designs
- Cross-syndrome versus single-disorder perspectives
- Standardized versus experimental paradigms
- Chronological versus developmental age comparison groups

longitudinal approach that includes following the same children tested across different time points. The lack of such studies can be attributed to the nature of the research itself. For example, it is time consuming and laborious, often taking a number of years from onset to completion, the financial and time costs inherent in such designs may make them less attractive to grant agencies (although, in our experience, this is becoming less of an obstacle than it was in previous years), and there is a high dropout rate from participating families from one time period to the next. However, without a longitudinal perspective it is impossible to draw any firm conclusions on the nature or pattern of developmental changes that occur across time periods in performance, in our case, cognitive measures of attention, and how such profiles differ in their trajectory from other neurodevelopmental disorders and also from the typically developing trajectory.

In a nutshell, cross-sectional research, although important and informative, merely allows us, as researchers, to glimpse at changes over time studying different individuals in a range of age groups. This, in turn, raises the question of how well the mean of these age groups reflects individual patterns of change, and this can only be addressed by tracking the same individuals over time. Longitudinal research allows us to observe atypical development as a "live," dynamic process that reflects an individual's progression over time on a task or skill, rather than simply reflecting an average score derived from a group of individuals tested at one particular age band. See Thomas et al. (2009) for a detailed critique of the utility of the trajectory approach in understanding neurodevelopmental disorders and, in particular, their discussion on the issues of variability and null (nonsignificant) findings that are inherent in this research.

A recent series of studies by Baranek and colleagues (Baranek et al., 2008, Roberts et al., 2009) provides good examples of the importance of longitudinal assessments in observing subtle developmental changes that begin in infancy and become more problematic with increasing age. In the Baranek et al. (2008) study, infants with fragile X syndrome were observed at various time points between 9 and 54 months on a range of sensory processing responses, for example, hyperresponsiveness. As infants increased with age, even by a few months, performance became more variable and delayed compared to the typical age trajectory.

Cross-Syndrome Versus Single-Disorder Perspectives

To date, the majority of research has tended to focus on single-disorder designs to isolate cognitive proficiencies and deficiencies that differentiate a specific disorder (e.g., Down syndrome) from typically developing children matched either on chronological age or developmental level. On the one hand, this approach can provide some important information about the degree of *developmental delay* associated with task performance. For example, at a rudimentary level it addresses, albeit not completely, the question of whether delay is generalized to all tasks (e.g., all cognitive measures of attention) compared to typical controls, or whether there are "degrees" of delay that are dependent on task demand. For example, Lanfranchi, Cornoldi, Drigo, and Vianello (2008) assessed visual and verbal working memory performance in children with Down syndrome and compared performance to a sample of typically developing children matched on mental age to the Down syndrome group. Both groups had a mental age of approximately 5 years, but the children with Down syndrome were chronologically much older than the control children (11 years and 5 years, respectively). In line with previous research, the authors found that children with Down syndrome had difficulties in working memory, a finding that is to be expected given the degree of intellectual impairment associated with this condition. However, upon closer examination, it was not the case that all working memory performance was impaired to a similar degree. Two patterns emerged: first, children with Down syndrome were much better at performing visual rather than verbal working memory tasks, and second, performance was comparable to typical children when the task required only a low level of control, such as recalling green squares on a board. However, when a more complex level of control was required, such as remembering a specified pathway and at the same time tapping when a frog stimulus appeared in a specific location, the groups diverged, with the Down syndrome group significantly inferior to the controls on both types of working memory tasks.

What can we conclude from these findings? The first is the importance of matching on mental or developmental age rather than chronological age. To have matched solely on chronological age would have produced highly significant group differences across all measures, and we would have concluded, quite wrongly, that children with Down syndrome were globally impaired on all working memory tasks irrespective of domain (verbal or spatial) or task demand (simple or complex levels of control). However, by comparing performance to developmentally matched controls, the authors were able to show a more detailed profile of group similarities and differences that appear to be task specific. Other studies across different cognitive domains and disorders also attest to the importance of comparing atypical performance to typical children matched on developmental age (e.g., Cornish, Munir, & Cross, 1999; Laing & Jarrold, 2007; Preissler, 2008). The hitherto unresolved issue relates to the matching procedure itself, namely, do you match on verbal or nonverbal mental age? It will primarily depend on the strengths and weaknesses of the disorder under investigation. We will address this issue below.

The second conclusion to be drawn from the Lanfranchi et al. study is that we cannot infer a disorder-specific profile of working memory performance that differentiates Down syndrome from other disorders. It is highly possible that any differences could be due to general developmental delays, rather than disorder-specific impairments. The only way to differentiate between these two is to incorporate a cross–syndrome perspective that compares performance across two or more disorders on the same tasks. Using the example of working memory, numerous studies have demonstrated that when compared to typically developing peers, performance in children across a range of different disorders is significantly impaired, suggestive of a generalized cognitive deficit. However, when disorders are contrasted with each other, for example, fragile X syndrome versus Williams syndrome (Scerif et al., 2004, 2007), Down syndrome versus fragile X syndrome (Munir, Cornish, & Wilding, 2000a,b), and Down syndrome versus Williams syndrome (Jarrold, Baddeley, & Hewes, 1999; Vicari & Carlesimo, 2006), important qualitative differences emerge that suggest disorder-specific working memory signatures that cannot readily be attributed to generalized cognitive delay. Such profiles are crucial to the development of intervention programs that allow clinicians and educators to target the unique signatures that represent differing neurodevelopmental disorders. See Chapter 10 for more details of these interventions and programs.

Standardized Versus Experimental Paradigms

Several different methods are available for comparing children with a specific neurodevelopmental disorder to typically developing children, but all present with certain problems. The most general form of comparison will employ one or more of the rating scales that were discussed in Chapter 5 for assessing attention, plus a variety of other scales to assess constructs such as intelligence or executive functioning. This method presents a global picture of differences and similarities between the two groups, such as distractibility or hyperactivity, providing a useful starting point for further investigation of the precise impairments that characterize the neurodevelopmental disorder under investigation.

However, such ratings offer a very limited perspective on which cognitive, motivational, or other processes may underlie the behavioral manifestations. Distractibility could, for example, be a consequence of rapid habituation to a stimulus or of poor ability to exclude other distracters, while hyperactivity could result from high arousal, or poor motor inhibition, or various other causes. To understand the nature of the disorder, we need far more detailed information than can be provided by comparing general ratings of behavior with normative data.

Investigations can also employ standardized tasks such as those described in Chapter 6, and in the present context, particularly those tasks included in batteries specifically designed to test different aspects of attention. Such investigations will tell us whether an individual or a group differs from the typically developing group on which norms for the task are based or from the control group if one is included in the study. Investigations of this kind, particularly when used to test a range of functions, can provide valuable information on the cognitive strengths and weaknesses that typify a given disorder.

There are, however, several inadequacies in research that is based entirely on such standardized tasks, particularly when studying neurodevelopmental disorders. First and foremost, as we have continually stressed, standardized tasks are exactly that: tasks designed to be appropriate in their demands and level of difficulty for a typically developing population. Participants are assumed to be able to understand the instructions, to be motivated to carry out the tasks, to focus for the full duration of the task, and to match various other assumptions. There is no certainty that such assumptions hold when the task is given to participants with a neurodevelopmental disorder; poor performance can occur for a variety of reasons in such cases, not only the inadequacy of the component process or processes that are engaged by the task, and the task may simply be too difficult for the target population. Conversely, a task that is appropriate for a group with cognitive delay may be too easy for a typically developing group. There is an urgent need for tasks specifically designed for or shown to be appropriate for developmentally disordered groups such that floor effects can be avoided in those with cognitive delay and also sufficiently sensitive to avoid ceiling effects in the control groups.

A second problem with standardized tests is that desirable control conditions are frequently lacking. We have several times made this point in relation to tasks purporting to test EF (see Chapters 2 and 6), where no control conditions are included that are matched in all respects apart from the EF demands (a difficult and sometimes impossible requirement, but one that needs to be addressed more consistently). A related problem is that standardized tasks are extremely difficult to modify in a way that facilitates testing novel hypotheses about neurodevelopmental disorders.

In conclusion, the future expansion of research on neurodevelopmental disorders will necessarily depend on the design of appropriate hypothesis-driven tasks that enable manipulation of critical parameters in order to reveal the specific nature of cognitive weaknesses in different conditions. Relatively few examples of such research exist at present. Some examples of informative studies are Lui and Tannock's (2007) investigation of working memory efficiency as a unique predictor of rated attention, Mason, Humphreys, and Kent's (2004) examination of ability to exclude preview stimuli in ADHD and control groups, and Wilding, Pankhania, and Williams's (2007) study testing predictions of the effects of speed and accuracy instructions on good and poor attention groups.

Chronological Versus Developmental Age Comparison Groups

In order to tease out the precise nature of any related changes that may be disorder specific, performance is often compared to a typically developing group of children matched on chronological age and/or matched on developmental age. The former approach, we argue, tells us very little about how performance, or the development of specific processes, such as those required for efficient attention, differs between groups other than the obvious finding that children with mental retardation will perform more slowly and less accurately than their age-equivalent typical peers. This is hardly an unexpected finding and may only indicate that general intellectual development is proceeding more slowly than is typical for that age group. This conclusion would be supported if performance does not differ from that of the developmental age–matched control group.

There are, however, a number of qualifications to this simple conclusion. First, it is obviously necessary to show that improvement does occur with increasing mental age and that performance has not reached an age plateau. Second, we need to know whether the highest level achieved matches the normal mature performance or falls short of it; in the latter case, developmental delay is clearly not a sufficient explanation. Third, we need to know whether a group with a specific developmental disorder (e.g., fragile X syndrome) that involves weaknesses in attention shows the same pattern as a group without the disorder that also demonstrates attentional problems (i.e., a group that falls at the lower end of the normal distribution of the ability). To this end it is desirable to include a control group matched on rated attention to the group with a specific neurodevelopmental disorder. Subtle differences may emerge that provide clues to the specific nature of the impairment in the latter group. The series of attention studies of fragile X and Down syndrome carried out by Munir et al. (2000b), Cornish, Munir, & Wilding (2001), and Wilding, Cornish, & Munir (2002) illustrates this point.

Third, some caution is desirable in relation to the matching process. For example, a sample of children with autism may have a developmental age of say 5 years but will have a higher chronological age than a typically developing group with the same developmental age (whose chronological age will on average match their developmental age, in this case, 5 years). Consequently, if the typically developing group is matched to the developmental age (DA) of the autism group, the latter group will have more experience and will tend to score more highly than the typically developing group on task components that depend on knowledge and measure crystallized intelligence such as vocabulary. Conversely, they may score at a lower level on components measuring fluid intelligence, such as problem solving, producing a similar overall developmental age score but one comprising differing components. Ideally, matching should be achieved on both components but this ideal is difficult to achieve. When it is not achieved, inferior performance by the developmental group on a task that demands problem-solving abilities may merely reflect inadequate matching, and it will not be justified to conclude that a specific disorder has been revealed that is independent of overall developmental level.

This is an important point because current research activity on neurodevelopmental disorders is increasingly directed toward isolating specific weaknesses in cognitive processes, rather than assuming a global and undifferentiated general weakness in all such cases. Critical information about the nature of a disorder is most clearly revealed by differences in the *pattern* of performance, such as the type of errors made or the variability of reaction time, rather than general measures of performance, such as the number of errors or the speed of performance. Such differences between atypically and typically developing groups, and still more, differences between groups with different developmental disorders, can give key information about the nature of a disorder, indicating whether there is a difference in cognitive processes and not simply delayed development. As we have already noted, this also means that, in addition to chronological and developmental age–matched typically developing control groups, comparison between different neurodevelopmental disorders is critical in order to obtain a true understanding of the nature and specificity of disorder-specific signatures.

Attention and Neurodevelopmental Disorders

Collectively, the wealth of research now available at the molecular, brain, and cognitive levels converges to demonstrate the complexity of geno-type–phenotype correlations and the necessity of research that is empirically and theoretically driven. In this section we highlight some exciting and innovative studies that have sought to identify and trace the cognitive signatures across our five neurodevelopmental disorders. The reader will see how commonalities in behavioral problems, notably persistent inattentive and distractible behaviors, can often mask important disorder-specific differences in cognitive inattention and executive dysfunction. Both need to be carefully investigated. Given that we are describing multiple disorders in this chapter, we will discuss each one separately, focusing on their unique behavioral and cognitive phenotypes, and where possible discuss any age-related changes.

One of the core problems facing researchers as they attempt to identify cognitive profiles in children and adolescents with significant developmental delay is devising tasks that are sufficiently sensitive to tap different aspects of cognition without creating an immediate floor effect, meaning that the task is too difficult. We have discussed already how standardized tasks, although informative at one level, often fail to capture the cognitive lacunae in atypical populations precisely because they are measuring the normal distribution of typical performance. In this chapter, we report findings from standardized tasks, so the reader can see both the difficulties and strengths of these paradigms, along with findings from recently developed experimental paradigms designed especially to tease apart attention functioning in atypical populations. Undoubtedly, and depending on the population tested, both types of paradigms can produce rich data that can shed new light on phenotypic outcomes and trajectories.

As in all our previous chapters, cognitive attention will comprise sections on *selective attention*, *sustained attention*, and *attentional control*. To avoid repetition, we provide here snapshot definitions of each of our attention subdomains. Selective attention is defined as the ability to selectively attend or orient to target information and to avoid distraction. Sustained attention is defined as the ability to maintain the focus of attention when there are *multiple distractions* or when awaiting some significant event (*vigilance*), and attentional control is defined as the ability to exert effortful control in order to inhibit a dominant response, to hold in working memory newly relevant rules that require the suppression or activation of previously learned responses, and to shift attention between tasks.

It will be impossible to consistently divide studies into distinct age periods (i.e., toddlerhood, childhood, adolescence) in this chapter as we were able to do in Chapters 7 and 8. We have made these distinctions where possible, but as the reader will see, the majority of research has focused almost exclusively on late childhood, with only an emerging literature on toddlers and adolescents. The findings from cross-syndrome investigations and from studies examining age trajectories will conclude this chapter.

Given the breadth of genetic and brain imaging studies already covered in Chapters 3 and 4, we only provide here brief descriptions as they relate to each of our targeted disorders. Table 9.1 provides a "snapshot" of the attention profiles and trajectories, and Table 9.2 provides a summary of the genetic, brain, and behavioral profiles.

Autism

Autism is a complex and severe neurodevelopmental disorder that is characterized not at the genetic level by a deletion to a certain gene or by a trisomy but by its behavioral-cognitive phenotype. Recent epidemiological studies indicate that the rate of autism is much higher than previously thought, with approximately 30 to 60 cases per 10,000 (see, for example, Rutter, 2005). The disorder represents the prototypical pervasive developmental disorder (PDD) and is traditionally characterized by a "triad of impairments" that include a severe disruption of social cognitive functions and impaired social interaction and communicative skills alongside unusually restricted and repetitive stereotyped patterns of behaviors and interests. However, see Mandy and Skuse (2008), who recently challenged the assumption that social-communication impairments and restricted behaviors and interest always co-occur in autism.

Table 9-1. Differences in Attention Between Disorders and
Typically Developing Controls

Attention Subdomain	Autism	Fragile X	Down	Williams	22q11
Selection:					
Orienting (exogenously controlled)	0	–	0	–	0
Visual search	–	–	–	–	No Data
Maintenance	0	–	0	0	–
Control	0	–	–	0	–

– = inferior to typically developing control group
+ = superior to typically developing control group
0 = no difference

Autism is undoubtedly a lifelong condition with an onset in the first 3 years of life. However, symptomatology does not remain static with age and instead undergoes dynamic changes from toddlerhood and across childhood (see Charman et al., 2005, for a longitudinal investigation of the stability of autistic symptomatology from 2 to 7 years).

A dimensional approach to autism includes classification based upon cognitive measures or verbal ability, with lower-functioning individuals tending to have an IQ in the lower range (<70) and demonstrating more severe symptomatology. Individuals with autism who have IQs in the normal-to-superior range are classified as higher functioning and tend to have fewer or less severe autistic symptoms.

Behavioral Phenotype

Attentional problems are a core behavioral concern and appear early in autism. A recent study of children aged from 1.5. to 5.8 years found attention problems to be among the most widely reported behavior difficulties, reaching clinical significance on the Child Behavior Checklist (CBCL; reviewed in Chapter 5) (Hartley, Sikora, & McCoy, 2008). The degree of attention impairment also appears to relate to the level of nonverbal ability displayed by autistic children (see Estes, Dawson, Sterling, & Munson, 2007). ADHD has been reported as a comorbid

Table 9-2. Summary of Genetic–Brain–Cognitive Correlates by Neurodevelopmental Disorder

Neurodevelopmental Disorder	Genetic Basis	Brain Pathology	ADHD Behaviors
Autism	Polygenetic with susceptibility genes linked to chromosomes 7, 15, 16, 22	Increases in occipital activation and frontoparietal activation associated with enhanced search efficiency Dysfunctional cerebello-frontal spatial attention system associated with impairments in spatial attention	30–55% meet ADHD diagnostic criterion
Fragile X syndrome	Single gene switched off, *FMR1* gene	Impaired functioning in frontostriatal regions and deactivation in the ventromedial prefrontal cortex found in females associated with inhibitory deficits Reduced activation in the right ventrolateral prefrontal cortex but increased activation in the left ventromedial prefrontal cortex in males	50–80% of boys meet ADHD diagnostic criterion; many girls meet ADHD inattentive subtype
Down syndrome	Trisomy 21	Decreased frontal lobe volumes associated with cognitive functions associated with frontal lobes (executive function, inhibitory control . . .)	8% meet ADHD diagnostic criterion; teaching ratings indicate 60%

Williams syndrome	Microdeletion on chromosome 7 involving 25–30 genes	Reduced gray matter concentration in left parieto-occipital region, associated with impairments in visuospatial construction and visual selection	25–50% meet ADHD diagnostic criterion
		Problems in engaging frontostriatal systems (dorsolateral prefrontal cortex, dorsal anterior cingulate cortex, and striatum) associated with basic inhibitory control (e.g., Go–NoGo paradigm)	
22q11 deletion syndrome	Microdeletion on chromosome 22 involving 30–40 genes	Frontal lobe enlargement/preservation associated with cognitive ability as measured by a borderline IQ	Up to 55% meet ADHD diagnostic criterion
		Reduction of parietal lobe associated with visuospatial deficits and lower performance in abstract reasoning tasks (arithmetic) and learning difficulties	

disorder in over a third of autistic children. For example, a recent study by Leyfer et al. (2006) assessed children and adolescents with autism aged between 5 and 17 years (mean CA 9.2 years) on the Autism Comorbidity Interview – Present and Lifetime Version (ACI-PL). They found that 55% percent of their sample had significantly impairing ADHD symptoms with 31% meeting the DSM-IV criteria for ADHD. A further additional 24% fell just short of meeting DSM-IV criteria. When the authors teased apart ADHD subtypes, the majority of children were classified as ADHD inattentive type only. Surprisingly few children were classified as ADHD combined type (23%).

Cognitive Phenotype

As we have noted already, autism is a disorder defined by its behavioral and cognitive phenotype. Research has consistently demonstrated that level of cognitive functioning (high- vs. low-functioning autism) impacts on later cognitive outcomes. In the domain of language, for example, semantic expression, characterized by an ability to understand and use nonliteral language, is one of the most affected areas of language impairment in low-functioning children with autism but is only moderately impaired in high-functioning children with autism, although not spared (see Boucher, 2003, for a review). Relative strengths in verbal working memory (e.g., Williams, Goldstein, & Minshew, 2006) and visual-spatial processing (e.g., Edgin & Pennington, 2005) contrast with a pervasive weakness in expressive language (see Groen, Zwiers, van der Gaag, & Buitelaar, 2008, for a review), and relative weaknesses in spatial working memory (e.g., Koczat, Rogers, Pennington, & Ross, 2002; Steele, Minshew, Luna, & Sweeney, 2007), and executive functions, including inhibiting an ongoing response (e.g., Christ, Holt, White, & Green, 2007), planning (see the seminal study by Hughes, Russell, & Robbins, 1994), and cognitive flexibility (Verté, Geurts, Roeyers, Oosterlaan, & Sergeant, 2005).

However, due to the wide phenotypic heterogeneity associated with autism, the reader needs not only to pay close attention to the sample composition across cognitive studies in terms of whether they include high- or low-functioning individuals as participants, but also to observe the quite wide variations in performance *within* a given domain such as EF. See Russo et al. (2007) for discussion of some of the methodological issues associated with understanding the broader EF deficits associated with autism.

Attention Studies

Selective Attention

ORIENTING

There is a large body of research on visual orientation in autism, stimulated by the observations of early abnormalities in gaze behavior in this disorder. The main questions investigated have been whether automatic (*exogenously controlled*) orienting and voluntary (*endogenously controlled*) orienting are both abnormal or whether they differ and whether any observed deficiencies are specific to orientation to cues derived from the faces of others or whether they are more general. Unfortunately, it is difficult to draw any consistent conclusions from the available evidence. Apparently small differences in methodology seem to produce different outcomes and samples are generally small and heterogeneous.

Studies have often employed adaptive versions of Posner's orienting task, which has been described and discussed several times already. To recap briefly, to test automatic orienting, following fixation on a central point, responses are cued to a location on the left or the right of fixation by (for example) a brief flash of light, and speed of responding to the subsequent stimulus is measured. The stimulus may be presented at the cued location (valid trials) or the opposite location (invalid trials), usually with equal probability so that the cue is nonpredictive. Response times on the two types of trial can be compared; faster times on valid trials indicate automatic orientation in response to the cue (alternatively neutral trials with no cue or cues on both sides may be employed as a baseline). To test voluntary control of orientation, a number of variations may be used. Either a central cue that requires interpretation is employed (such as an arrow), or the cue may predict the correct location with only a low probability. In the latter case, ability to override the automatic tendency to orient to the cued side will be reflected in the relative speed of response on the valid (but rare) trials and the invalid (but frequent) trials; differences are more likely at longer cue–target intervals than for automatic orienting at short intervals.

Using variations on the Posner task, Townsend et al. (1999) found substantially intact exogenously controlled orienting in autistic adults, provided that more time was given between cue and stimulus, and Iarocci and Burack (2004) found no impairments in younger samples (mean CA 11.6 years, mean MA 7.2 years). However, Renner, Klinger, and Klinger (2006) tested a group aged 7 to 17 years (mean IQ 112) and found a reduced difference in the autism group between valid and invalid trials. No obvious reason for the difference in results is apparent.

Landry and Bryson (2004) used a different method with a central screen that participants (mean CA 5.6 years, mean MA 3.4 years) were told to watch. Time to switch attention to occasional displays appearing on other screens to left or right was measured. When the central display continued throughout, but not when it switched off at the same time as the peripheral stimulus appeared, the autistic group showed fewer and slower switches, suggesting disengagement from an existing attentional focus was less flexible (so-called "sticky fixation," also observed in Williams syndrome and discussed below). Elsabbagh et al. (2009) drew a similar conclusion from a study of infants (8–12 months) with an autistic sibling. When a central stimulus overlapped the onset of a peripheral stimulus, this group was slow to switch attention. In addition, when the central stimulus disappeared 200 msec before onset of the peripheral target, this group was slow compared to a control group, suggesting they had difficulty in using the cue to prepare for switching. The weight of the evidence is therefore for some impairment of exogenously controlled orienting in children with autism, probably involving the disengagement function.

Tests of endogenously controlled orienting in autism are also inconclusive. These results should strictly be considered under control functions but are reported here as often studies report on both exogenously and endogenously controlled orienting. Senju, Tojo, Dairoku, and Hasegawa (2004) compared the effects of a central arrow cue and a face cue with eyes directed to the right or left on a simple detection of a stimulus on the left or the right by children with autism (mean CA 10.11 years) and a typically developing control group (mean CA 11.1 years). Central cues are assumed to engage endogenous control, but cueing effects were obtained at shorter cue–stimulus intervals than would be expected if this were the case, so interpretation of these results is uncertain. Though the autism group was slower overall, the effects of cueing were similar in both groups, and there was no indication of impaired responding to facial cues in the autistic group. A similar finding for facial cues is reported by Kylliäinen and Hietanen (2004).

In a second experiment, when cues only predicted the stimulus location on 20% of the trials, the autistic group continued to respond more quickly on correctly predicted trials when the cue–target interval was 300 msec or less, indicating an automatic effect of the cue that was not being overridden by endogenous control processes. The control group, on the other hand, responded more rapidly on invalid trials when the cue–target interval was 300 msec or more, indicating that they were making use of the overall probabilities (i.e., they responded to the side where the target was more likely to appear).

However, Renner et al. (2006), in the experiment described above, also used central arrows as cues and found similar patterns of performance in their autistic and control groups.

Collectively, these results suggest that some children with autism can make use of central predictive cues (though it is uncertain that the effects found by Senju et al. reflected endogenous processes) but have low ability to make use of predictive contingencies that run counter to more automatic cueing influences. Thus, some control functions appear to be impaired but not others. Clearly further research is needed to clarify these possibilities.

VISUAL SEARCH

Interesting results have been obtained from studies of visual search in autism, and we highlight some innovative studies that have demonstrated that autism is associated with *superior* performance on this task. O'Riordan, Plaisted, Driver, and Baron-Cohen (2001) assessed conjunctive and featural search in children with autism (age range of 6 to 9 years, mean CA 8.5 years) and typically developing control children matched on CA and on a nonverbal measure (Raven Colored Progressive Matrices). Children were placed in front of a computer screen and were presented with a display comprising 5, 15, or 25 stimuli, in this case, letters. The stimuli each had two dimensions: color (red or green) and form (S, T, or X). In Experiment 1, the featural condition required the discrimination of targets from nontargets based on shape only, irrespective of color, whereas in the conjunctive condition each nontarget shared one feature with the target (e.g., a green *T* among red *S* and green *X* distracters). Findings revealed a number of interesting patterns. First, children with autism were faster than typical controls on the conjunctive search task and also when the displays comprised larger set sizes. Second, there were no group differences in terms of accuracy, with both groups demonstrating relatively low error rates, 3.0% versus 2.7% for controls and the autism group, respectively. In Experiment 2, the authors explored in more detail the mechanisms involved in feature search using a more complex task design than in Experiment 1, which both groups found comparatively easy. To avoid ceiling effects, the authors used a paradigm that has previously been reported as difficult for adults and involves a reversal of target between Condition 1 (tilted vs. vertical) and Condition 2 (vertical vs. tilted). A similar pattern of findings emerged with superior performance by the autism group compared to controls.

A noteworthy finding is that the autism group was especially quick when searching for a target that was absent from an array, suggesting that the perceptual salience of a target is important, and thus, its absence triggers more efficient search. The authors make two intriguing suggestions to account for this superior performance. First, that search efficiency is guided by an enhanced ability in children with autism to discriminate between display items irrespective of set size, and second, that children with autism may have a superior *inhibition of return* mechanism such that they can keep track, more efficiently than controls, of previously inspected search locations, thus speeding up search efficiency. However, no evidence for the latter hypothesis was found in a study by Rinehart, Bradshaw, Moss, Brereton, and Tonge (2008). We should also note a recent finding from an fMRI study of reduced prefrontal involvement in autism in identifying a shape embedded in a complex figure (Lee et al., 2007), which implies that superior performance in visual search may depend on different mechanisms from those operating in typical groups. See also the excellent investigations by Jarrold, Gilchrist, and Bender (2005), and O'Riordan (2000, 2004).

NEGATIVE PRIMING

Negative priming allows an indirect view of how our attentional mechanisms process irrelevant but competing stimuli. Two trials in quick succession are presented; the first is referred to as a "prime" and the second as a "probe." In each trial, the distracter must be ignored and the target located as quickly as possible. Negative priming occurs when the ignored distracter in a previous trial becomes the target stimulus in a current trial. There is a slowing down in response to the probe target, suggesting that a distracter stimulus may have been ignored but is not forgotten. The two most frequent types of negative-priming tasks are identity-based tasks and location (spatial)-based tasks. In one of the most detailed studies to date, Brian, Tipper, Weaver, and Bryson (2003) assessed children and adults with autism with an age range of 7 to 33 years and a typically developing control group aged between 10 and 35 years. In brief, a display containing four boxes was presented centrally on the screen, which was immediately followed by a small square in between the four boxes that indicated the color of the target stimulus, which was always an X. In the prime trial, two Xs would appear on the screen but only one in the target color. Participants had to locate the correct target X as quickly as possible while ignoring the distracter X. In the probe condition, which followed immediately from the prime response, the target now shared certain features (location or color) with the previously ignored distracter in the prime display.

Findings showed that performance on all aspects of the task was within normal limits with an observed location-based and object-based negative priming effect. Although a location-based negative priming effect was not predicted, given previous findings of difficulties in children with autism in orienting attention in space, these findings do complement those of other studies of negative priming in autism (e.g., Ozonoff & Strayer, 1997; O'Riordan, 2000).

No simple conclusions are apparent from the evidence we have considered on selective attention in autism. Disengagement from an existing stimulus in order to switch to a new one and ability to override automatic orienting processes seem to be impaired. Visual search may engage different mechanisms from those operating in typically developing groups. Experimental work targeting specific questions is urgently needed.

Maintenance of Attention

To date, there have been surprisingly few empirical studies that have assessed the ability of children with autism to maintain attention over task duration. Of these, the two most commonly used paradigms are the Continuous Performance Test (CPT) and the Sustained Attention to Response Task (SART).

There are now numerous adapted versions of the *Continuous Performance Test*, and many of these have been discussed extensively throughout this book. However, central to all versions is that the participant has to monitor a stream of stimuli while waiting for an infrequent stimulus to appear. These tasks require a considerable degree of concentration and are probably a main reason why they are so rarely used in populations that have attention problems. But they are necessary for precisely this reason. One early study by Garretsen, Fein, and Waterhouse (1990) examined sustained attention using a version of the CPT developed by Rosvold, Mirsky, Sarason, Bransome, & Beck (1956) but modified to include everyday, familiar pictures of objects and animals as the stimuli rather than letters. Performance in children with low-functioning autism (age range 4 to 19 years, mean CA 12.4 years, mean nonverbal MA 6 years) was compared to typically developing children matched on nonverbal MA. The target stimulus in this case was a photo of a chair, and participants were required to press when they saw the target in a continuous presentation of nontarget stimuli. The impact of task complexity was assessed by rate of presentation across two conditions, fast (stimulus every 0.7 sec) and slow (stimulus every 3 sec).

The results revealed that the autism group followed a similar trend in performance to that expected in a vigilance paradigm, notably that accuracy deteriorated over time and especially on more difficult tasks as measured by the slow versus fast conditions. There were no group differences in accuracy rates or in the number of false alarms, suggesting that, although delayed, sustained attention may be a relative strength in autism.

The *SART* task differs from the traditional CPT by requiring participants to withhold a response to an infrequent target and to respond to all other stimuli. This difference is argued to place greater demands on sustained attention by having to interrupt an ongoing repetitive action (see Robertson, Manly, Andrade, Baddeley, & Yiend, 1997). Johnson, Robertson et al. (2007) assessed performance on the SART in children with high-functioning autism (mean CA 12.2 years), children with ADHD (mean CA 10.5 years), and typically developing children (mean

CA 11.1 years). We will describe the SART in brief here as the task has been reviewed in detail in Chapters 6 and 8. Participants sit in front of a computer screen and are shown a series of digits (1–9) across two conditions: fixed and random. In the former, digits are presented in a fixed sequence and in the latter in a pseudorandom order. Each version lasts approximately 5.5 min.

Findings demonstrate a clear autism advantage on this task compared to the ADHD performance. As described in the CPT task above, the autism group seemed to show a similar profile to that of typical controls in terms of errors (e.g., more omission errors made in the second half of the task), slower reaction time on the random SART compared to the fixed SART, and maintenance of SART reaction time on hits across both the fixed and random conditions. These results contrast with the poorer performance of the ADHD group. The authors conclude that sustained attention is not a deficit equally shared by the two disorders. We agree. These findings illustrate the importance of taking a closer look at patterns of cognitive inattention even when behaviorally two disorders appear to share common ADHD symptomatology.

Attentional Control

Dysfunction in inhibitory control is a well-documented feature of autism, but the picture is far from crystal clear. There is ongoing debate as to whether inhibition is globally or selectively impaired and whether any such pattern or signature of impairment can differentiate autism from other neurodevelopmental disorders that also demonstrate a core inhibitory impairment, notably ADHD. Dissociating inhibitory control into component parts has been successfully accomplished in the typical developmental literature from preschool onward (see Chapter 7 and Garon, Bryson, & Smith, 2008). Not surprisingly, the most investigated area of inhibitory weakness is *response inhibition*, and although there are numerous studies that attest to this being a core weakness in autism, a number of recent studies have begun to disentangle quite subtle profiles of deficiencies and proficiencies that suggest that response inhibition may be differentially impacted depending on the actual inhibitory demand of the task. We highlight findings from these exciting studies below.

WITHHOLDING A PREPOTENT OR DOMINANT RESPONSE

The ability to delay or stop altogether a prepotent response is a well-documented deficit in many neurodevelopmental disorders and most especially in ADHD. Using the Stop Task, Ozonoff and Strayer (1997) found no impairment in high-functioning autism. However, the Go–NoGo paradigm is undoubtedly the task of choice in the majority of studies of autism and requires quick responding to a continuous stream of target stimuli and withholding of responses to nontarget stimuli. Different rates of presentation can have a considerable impact on arousal levels, and one of the key questions is to what extent difficulties reflect an inability to modulate arousal and therefore maintain attention on task and to what extent they reflect deficient inhibitory abilities. Unfortunately, attempts to address this question have been inconclusive.

A recent study by Christ et al. (2007) used a simple Go–NoGo paradigm to assess response inhibition performance in children with autism (age range of 6 to 12 years, mean CA 8.2 years), their biological siblings (mean CA 10.2 years), and a typically developing control group of children (mean CA 11.3 years). The task involved four basic stimuli, all shapes (triangle, square, diamond, circle) in which three shapes on any one trial were designated targets and one a nontarget. Children were asked to press the button for all target stimuli but to withhold a response for the nontarget stimuli. If a response was given less than 100 ms after stimuli presentation, the child heard a brief tone and saw a visual message stating "early response." If a response was not made within 1,500 ms, then a tone and a message of "too slow" was given. Finally, if a response on "NoGo" trial was made, then the message "no response needed" was given. Measures included reaction times and error rates. Specifically, errors due to poor inhibition are reflected in increased error rates on NoGo trials, and errors due to poor sustained attention are reflected in greater response variability in reaction time.

Findings revealed that children with autism performed at a comparable level to control children in terms of Go trial reaction times and on NoGo error rates even when IQ and age were accounted for. The only difference was a slight increase in variability in response time on correct Go trials in children with autism compared to controls. One possibility is that difficulties in sustained attention, typically associated with variability in reaction time, may underlie performance rather than inhibition per se. The contributory role of ADHD has also been explored in a recent study by Sinzig, Morsch, Bruning, Schmidt, and Lehmkuhl (2008), who found inhibitory performance on a Go–NoGo task to be more severely impaired in children with autism and comorbid ADHD symptoms compared to children with autism without ADHD. These findings underscore the need to dissociate the impact of ADHD symptoms on performance in children with autism, given its high comorbidity. In addition, the pattern of findings offers the intriguing possibility that there may be two separate pathways of cognitive functioning in autism, one that is less impaired and perhaps comparable to children of an equivalent developmental level, and one that shows a deviant profile with greater impairment on functions known to be especially vulnerable in children with ADHD.

INHIBITION OF IRRELEVANT COMPETING INFORMATION

The ability to ignore conflicting and irrelevant stimuli and focus on target information is typically measured by such tasks as the Stroop Color-Word Task and the negative priming task (both reviewed in Chapter 6 and mentioned in previous chapters). In the study by Christ et al. (2007) described above and using the same samples, the authors employed two versions of the Stroop task. The first was Golden's (1978) traditional card version and the second a computerized version, comprising congruent, incongruent (inhibitory), and neutral trials. Errors on these trials were of three types: an early response (<100 msec after presentation), a too slow response (>3,000 msec after presentation), and an incorrect response. Findings were unequivocal. There were no differences between the autism group and typically developing control groups, indicating comparable performance on this type of inhibitory task.

In a comparison study of children with high-functioning autism and children with ADHD, Goldberg et al., (2005) also found no group differences in children with high-functioning autism, children with ADHD, and typically developing control children. This result remained constant even when verbal IQ was controlled for. Likewise, a similar finding was also reported in an earlier study by Ozonoff and Jensen (1999), which compared performance on the Stroop in three neurodevelopmental disorders (autism, ADHD, and Tourette's syndrome) matched on chronological age to a typically developing control group. The only group difference to emerge was between the ADHD group and control group, with the ADHD group performing significantly worse on this task than controls. The autism group did not differ in performance from the typical controls, indicating that the ability to inhibit irrelevant and competing stimuli is relatively intact in autism.

REVERSAL OR SWITCHING OF RESPONSE FROM
ONE STIMULUS TO ANOTHER STIMULUS

The ability to hold information in mind and switch from one learned response to an opposite response is a complex skill. One of the most innovative paradigms to tap this inhibitory function has been developed by Diamond and colleagues and is known as the *Day–Night Task* (Gerstadt, Hong, & Diamond, 1994; reviewed extensively in Chapters 7 & 8). Russell, Jarrold, and Hood (1999) in a seminal study examined performance on this task in children with low-functioning autism (mean CA 13.76 years, mean verbal MA 7.33 years), children with mild learning disability (mean CA 11.90 years, mean verbal MA 7.29 years), and a typically developing control group matched on verbal MA (mean CA 7.33 years). In brief, the Day–Night Task requires the opposite response to a target stimulus such that when shown a card depicting a bright yellow sun on a white background, the correct response is "night," and when shown the card with a moon and stars on it, the correct response is "day." Thus, children have to hold two rules in mind and to inhibit the urge to give the standard type of verbal response. Measures on the Day–Night Task included correct responses, error rates, and reaction time. Findings demonstrated no group differences across any of the measures, indicating that performance on this type of response inhibition in children with autism is at least at a comparable developmental level to that of control children.

There are also a number of interesting studies that have looked at switching attention from one rule to another, typically assessed by such paradigms as the Wisconsin Card Sorting Task (WCST). However, this task is quite complex and was originally designed for adults, so its use as a reliable tool to assess inhibitory performance in neurodevelopmental disorders must be questioned (see Chapter 6). In brief, shifts between three sorting dimensions (shape, color, number) are required, with the participant having to maintain a sorting set until the next one is revealed. Measures include perseverative errors and categories achieved. A recent study by Kaland, Smith, and Mortensen (2008) reported no significant differences in performance in a sample of high-functioning adolescents with autism (mean CA 16. 4 years, mean IQ 109.0) compared to typically developing, control adolescents matched on age and IQ. Sample size was quite low (n = 13) in each group and the mean well within the normal range, suggesting that normal levels of performance had been achieved in this group. However, when this paradigm is used in younger populations with lower IQs, a clear difference is observed.

Conclusions on Autism

Together, the current findings, at first glance, indicate that performance on many aspects of attention is relatively good, even superior to typically developing children in the case of visual search and superior to ADHD in maintaining attention. Unimpaired control is particularly apparent when ADHD symptoms are not present. Combined with the observed impairment in some orienting processes, these findings indicate that there are marked differences between the attention signatures of autism and ADHD.

However, autism is a disorder of heterogeneity, and this adds a significant level of complexity in the interpretation of findings. Level of functioning (high vs. low) plays a critical role in the cognitive end state, and unfortunately, the majority of studies have tended to focus on performance in individuals with high-functioning autism, who tend to have IQs within the normal range. This is particularly evident in studies investigating attention control. It is therefore likely that a lack of statistical difference on some measures, especially those related to executive functions (e.g., inhibition) are impacted by higher IQs rather than representing an autism-specific profile. Future studies need to include paradigms that can tap sustained attention and control in low-functioning children with autism. The visual-search task used in the Jarrold et al. (2005) study provides an excellent example of how an experimental paradigm can be adapted for use with low-functioning children with autism. Without this data it is difficult to ascertain if there is an attention signature of proficiencies and deficiencies in autism irrespective of degree of intellectual functioning. One clue to the range of putative deficits has come from the findings of recent neuroimaging studies that have identified atypical activation in the inhibition circuitry in autism (e.g., anterior insula, cingulate cortex) (Kana, Keller, Minshew, & Just, 2007). However, the breadth of abnormalities precludes one region of specific impairment (McAlonan et al., 2005), and there is significant impact on the frontostriatal and parietal networks.

Developmental Issues

There is a virtual absence of studies of attention functioning in very young children with autism, with the majority of research focusing on late childhood and adolescence. At one level, the age of diagnosis can preclude children less than 3 years of age, but nonetheless, a diagnosis of autism is often made between 3 and 5 years of age, though no studies, to the authors' knowledge, have assessed attention functioning in preschool children less than 5 years old. This is an area of urgent research enquiry given the importance of development in defining disorder-specific profiles. We cannot assume that the late childhood/adolescent profile mirrors that of the toddler/preschool profile. However, there are some interesting recent findings that indicate age-related changes in attention performance from midchildhood through to adulthood. For example, Luna, Doll, Hegedus, Minshew, & Sweeney (2007) tracked eye movement responses to an antisaccade task in high-functioning individuals with autism with a broad age range of 8 to 33 years. In terms of response inhibition, developmental improvements were greatest in younger children with adult levels reached by age 15 years. In typically developing controls, adult levels were reached marginally earlier at age 14 years. Happé, Booth, Charlton, and Hughes (2006), in a comparison study of young and older children with autism (without ADHD) (mean CAs 9.2 years and 13. 2 years, respectively) matched in age to an ADHD control group and a typically developing group, found that older children with autism outperformed younger children on the majority of Go–NoGo measures (e.g., percentage of commission errors). This same trajectory was not observed in children with ADHD who showed minimal age improvements, suggestive of more pervasive impairments.

In contrast, Solomon, Ozonoff, Cummings, and Carter (2008) in a sample of children with autism aged between 8 and 17 years found no evidence of age-related improvements in performance on an executive control task. Unlike typically developing children who showed some improvement on task performance with age (errors reduced in number), no such effect was observed in the autism group, and in fact, there was a slight worsening in performance with age. Differences in levels of functioning in autism groups across studies, the prevailing use of cross-sectional rather than longitudinal designs, and differing cognitive task demands all contribute to the current state of minimal knowledge on the developmental trajectories of attention in autism.

Fragile X Syndrome

Fragile X syndrome is a well-recognized cause of hereditary developmental delay in males and, to lesser extent, females. It is one of the most widely studied of genetic disorders worldwide affecting an estimated 1 in 2,500 males and females (Hagerman, 2008). The genetic cause of this disorder was described in detail in Chapter 3. Briefly here, fragile X is caused by the silencing of a single gene on the X chromosome, the fragile X mental retardation gene (FMR1). The gene encodes a CGG repeat sequence, which in the general population ranges from 6 to 60 repeats. In some cases, the number may increase up to 199. These individuals are referred to as fragile X "carriers." When the number of repeats exceeds 200, then the FMR1 gene is turned off, and the encoded protein, FMRP, is not produced, resulting in the full mutation of the fragile X syndrome and its characteristic phenotype. One of the most intriguing advances in the field in recent years is the finding that carriers of fragile X, originally assumed to display no observable symptoms and thus thought to be unaffected by fragile X at the clinical and cognitive levels, actually have their own unique phenotypes and trajectories (e.g., Cornish, Li et al., 2008; Grigsby et al., 2008; Loesch et al., 2003; and see also Cornish, Turk, & Hagerman, 2008, for a snapshot of these recent discoveries).

In the fragile X full mutation (>200 repeats), X-linkage means that males are especially vulnerable to the full effects of the condition at the brain, behavioral, and cognitive levels. Almost all boys will present with moderate intellectual disability compared to girls, who display a more variable phenotype, the reasons for which are described above and in Chapter 3. In males, FMRP level accounts for about 75% of the variance in IQ, but in females the proportion is much smaller (Lightbody, Hall, & Reiss, 2006).

Behavioral Phenotype

Attentional problems are the most frequently cited behavioral characteristics in fragile X, affecting both boys and girls but to quite different degrees of involvement depending on whether there is an associated developmental delay. In the largest nationwide parent survey to date of children with fragile X, Bailey and colleagues report findings on 976 males and 259 females with the full mutation. Of these, parents identified inattentive behaviors as a significant problem in 84% of males and 67% of females (Bailey, Raspa, Olmsted, & Holiday, 2008). However, further analysis revealed that of those females who were reported as having a developmental delay, 82% had also been diagnosed with attention problems, a figure comparable with the incidence reported in males. This elevated incidence of attention problems associated with fragile X is striking, and further detailed assessments are needed to establish the range of inattentive behaviors across the life span and whether they are gender specific or even disorder specific. Currently, elevated rates of attention difficulties have tended only to be reported in fragile X boys using both standardized rating scales (Child Behavior Checklist [CBCL], reviewed in Chapter 5) (Cornish et al., 2001; Hatton et al., 2002; Sullivan et al., 2006) and clinical interview (Parental Account of Childhood Symptoms Interview; Taylor, Schachar, & Hepstinall, 1993, as used by Turk, 1998). Other behavioral features associated with fragile X include eye gaze aversion (indicating social anxiety), hand flapping, and perseverative speech (e.g., Cornish, Levitas, & Sudhalter, 2007; Sullivan et al., 2007), and in many cases, autistic-like symptoms (for a review, see Cornish, Turk, & Levitas, 2007; also Carter, Capone, Gray, Cox, & Kaufman, 2007; Hooper et al., 2008).

Cognitive Phenotype

At the cognitive level, the observed pattern is complex and is characterized by uneven abilities within and across cognitive domains. By late childhood, relative strengths in vocabulary (Roberts et al., 2007), long-term memory for meaningful and learned information (Munir, Cornish, & Wilding, 2000a), and visual-perceptual skills (Cornish et al., 1999) are accompanied by relative weaknesses in the storage and manipulation of complex information in working memory (Lanfranchi et al., 2008; Munir et al., 2000a), linguistic processing (Abbeduto, Brady, & Kover, 2007; Belser & Sudhalter, 2001), visuospatial cognition (Cornish et al., 1998, 1999), and inhibition (Munir, Cornish, & Wilding, 2000b; Wilding, Cornish, & Munir, 2002).

We highlight below the recent studies that have attempted to specify more precisely the key weaknesses in attention and EF. Disappointedly, there is an imbalance in the ratio of male to female studies, with a greater proportion of research focused on the male phenotype, especially within the domain of attention. However, we recommend the reader to visit the work of Mazzocco and her colleagues, who have produced some of the world's finest empirical work to date on fragile X females and their performance across and within two core domains: spatial cognition and numeracy (e.g., Mazzocco, Singh Bhatia, & Lesniak-Karpiak, 2006; Murphy & Mazzocco, 2008a, 2008b).

Attention Studies

Selective Attention

We shall focus here on two important components of selection that have been documented in fragile X. The first component is termed "orienting" and describes a recent study that has investigated the ability of fragile X infants to orient attention to suddenly appearing stimuli in the environment, and the second describes a series of studies that have investigated the ability of fragile X toddlers and children to visually search for specific targets among an array of distracters, using so-called "visual-search" paradigms.

ORIENTING

Flanagan et al. (2007) assessed reflexive and voluntary orienting in fragile X adolescents and young adults with an age range of 11 to 24 years (mean CA 17.15 years, mean nonverbal MA 5.86 years) subsequently divided into low- and high–MA groups. A Posner-type cueing task was employed in which one of four locations of a forthcoming stimulus (boxes in the four corners of the screen) was cued correctly or, on invalid trials, a different location was cued. For reflexive (exogenously cued) orienting, the cue was a darkening of one of the four stimulus locations. For voluntary (endogenously cued) orienting, an arrow was present at the fixation point indicating one of the four possible locations. While the higher MA group showed a clear advantage in reaction time to the target after valid exogenous cueing, the lower MA group did not, though in typically developing children such an advantage is present well before an MA of 4 years. For endogenous cueing, neither group showed a significant advantage with valid cues. These results suggest delayed development of reflexive orienting and continuing impaired control of endogenous orienting. It should perhaps be noted that the endogenous cueing condition was slightly unusual in that the cue was not presented following initial fixation in the same way as the exogenous cue but was present in place of the fixation point, so it may not have attracted attention in the same way as the sudden darkening of the forthcoming stimulus location. Unfortunately, this study also included an unspecified number of fragile X females, which, as we have discussed above, would have resulted in a sample bias given their much higher intellectual functioning due to their X inactivation.

A recent study by Cornish, Scerif, and Karmiloff-Smith (2007) assessed voluntary orienting in fragile X infants and toddlers (mean CA 35 months, mean MA 19 months) using a Posner paradigm. They were compared with a typically developing group approximately matched on MA (21.6 months); due to the differences, MA was used as a covariate in all analyses. To test orienting, a cue was presented on the right or left of fixation, followed by a target after a delay ranging from 150 to 1200 msec on either the same side as the cue (valid trials) or on the opposite side (invalid trials). While typically developing infants showed slower orienting to invalidly than to validly cued targets, fragile X infants showed no difference. Though the authors point out that limited statistical power due to the small sample size prevents any firm inferences, this finding suggests that the fragile X infants were not significantly aided by exogenous cueing.

VISUAL SEARCH

As we have seen in our previous chapters, the development of efficient visual search is one of the hallmarks of selective attention. Here, we highlight findings from a series of innovative studies with our colleagues (Scerif and Karmiloff-Smith). In these studies we have attempted to assess search proficiency in toddlers and children with fragile X syndrome from age 2 years to adolescence (MA range from 2 to 6 years), using versions of the Visearch Task described in detail in Chapters 6 and 7.

Toddlers   Scerif et al. (2004, 2007) used a modified version of the computerized Wilding Visearch task (described in detail in Chapters 6 and 7) in fragile X boys aged between 34 and 50 months (mean CA 43.5 months, mean MA 29.1 months) and a typically developing control group of MA-matched boys (mean CA 29.1 months). In brief, toddlers were placed in front of a computer screen and told that funny monsters were hiding behind the big target holes on the screen but were not hiding in the small holes (nontargets). Toddlers had to touch only the target holes, and when successful a colored, square-shaped face appeared and stayed on the screen. When a toddler touched a nontarget shape, nothing appeared. In addition to collecting data on accuracy rates and reaction times, this task also examined, for the first time, the effect of similarity between targets and distracters and also the number and homogeneity of the distracters (see an example of the visual display in Chapter 7, Figure 7.3).

The findings were remarkable in two respects. First and most strikingly was that the data corroborated observations previously made in studies of older fragile X boys (see below), namely, a high probability of repeating responses, either immediately or by subsequent returns, to already-located targets, even though these were clearly marked as already found. The authors interpreted this as indicating a problem in inhibiting previously successful responses in order to proceed to locating the next target. The tendency was more marked when targets and distracters were more similar, so the next target was harder to find. Scerif et al. (2007) further extended these findings to show that, even though no problem was apparent in discriminating targets from distracters when presented in pairs, in the complex search display greater similarity and salience of distracters induced more erroneous responses. Also, poor motor control was apparent in the number of inaccurate touches made on the background close to targets. Together, these data indicated a clear and precise deficit in inhibitory control in fragile X boys from toddlerhood onward. The second is that the findings demonstrated the feasibility of testing visual search performance in toddlers with significant developmental delay and attention difficulties using novel, experimental paradigms.

In late childhood, a series of studies using similar methodology to the toddler study described above but incorporating a more complex visual display (described in Chapters 6 and 7) has been conducted by Wilding, Cornish, and their colleagues (Cornish et al., 2001; Munir et al., 2000b; Wilding et al., 2002). In this version, children were required to click with a mouse on the target shapes. Search performance in fragile X boys, aged between 8 and 15 years (mean CA 10.88 year, mean verbal MA 6.77 years), was compared to two typically developing control groups matched with the fragile X group on MA. The control group was subdivided into "good" attenders or "poor" attenders as rated by their teachers on the ACTeRS scale (reviewed in Chapter 5). These two attention control groups were included to determine whether the fragile X group would perform in a similar way to the poor-attention group or whether the pattern of performance might differ between these groups, indicating different underlying weaknesses.

The findings in this age group were also striking. Older boys with fragile X when searching for a single target were slower, moved around the screen more, and made more errors than both control groups, thus demonstrating more comprehensive impairment than the younger children in the Scerif studies. However, detailed analysis of error types (false alarms to targets vs. false alarms to nontargets) showed that all errors in the fragile X group were repetition on already-located targets, indicative of perseverative difficulties. In contrast, there were no false alarms to nontargets, indicating an absence of perceptual confusion, adding further support to the relative strength of visual-perceptual skills in fragile X (Cornish et al., 1998, 1999). In the version of the Visearch Task that required alternation between two targets, fragile X children also showed extreme difficulty in switching between targets, thus providing further indications of a problem in inhibiting a previously successful response.

In terms of selective attention, the pattern of the findings in fragile X toddlers and children indicates relative proficiency in their ability to select targets from nontargets; that is, they did not make especially high numbers of perceptual confusions between targets and nontargets. Instead, errors occurred when control processes were required to inhibit a previously successful target response and move on to the next one.

Maintenance of Attention

Sustained attention has been measured in fragile X children using two main types of tasks: an experimental paradigm similar to the Visearch Task described above and an adapted Continuous Performance Task (CPT).

VIGILANCE

Munir et al. (2000b) also employed the Wilding Vigilan Task, described in detail in Chapters 6 and 7. In brief, this task uses the same design format as the Visearch Task, but the child has to withhold clicking on a shape until an infrequent target is shown, and then the child has to click on it within 7 seconds. The task takes approximately 4 minutes to complete, and measures include the number of targets detected within the criterion time, mean target detection time, the number of false alarms to nontargets, and the distance travelled around the screen while awaiting a target. The same children who performed the Visearch task described above—fragile X and typically developing controls—also performed the Vigilan task reported here. Some interesting findings emerged. First, the fragile X group was no slower in detecting targets than the poor–attention control group, though inferior to the good–attention control group, but there were no differences in the number of targets detected. The most striking finding was in the number of false alarms to nontargets made while awaiting the appearance of targets, again an indication of poor inhibitory control. Unfortunately, decline in performance over time was not explored in this relatively brief task, and to date performance has not been explored in a younger fragile X age group.

MAINTAINING ATTENTION

An adapted visual and auditory CPT was employed recently by Sullivan et al. (2007) in which fragile X boys aged between 8 and 13 years (mean CA 10.1 years) were matched to nonverbal MA control boys. Unfortunately, the authors provide no details regarding mental age or the age range of the MA–matched typically developing control children. This lack of information is disappointing because in this study over a third of fragile X children were unable to complete the CPT tasks, and thus, matching becomes a critical issue for the remaining children. Also, given that 1 in 3 of the fragile X group could not carry out these tasks, probably being related to MA level, the findings represent a somewhat skewed sample.

Both visual and auditory CPTs were adapted for use with children with cognitive delay and were much shorter in length (177 seconds and 192 seconds, respectively) than the more traditional CPT that can range from 10 to 15 minutes duration. Targets also took the form of meaning-ful stimuli, such as a rabbit in the visual CPT and the word "dog" in the auditory CPT. Those fragile X children who were able to perform the task successfully detected fewer targets overall, but decline in the detec-tion rate over time was no different from the control group. Correct rejections of nontargets, however, did show a greater decline in the frag-ile X group, indicating a declining inhibitory efficiency as the task pro-ceeded. Despite the reduced sample size making generalization to the fragile X phenotype difficult, these findings demonstrate the necessity of assessing performance over task duration rather than just relying on accuracy and mean reaction time. See Chapter 8 for our review of recent ADHD studies that have used a similar approach.

Attentional Control

As already indicated, the most prominent feature of fragile X perfor-mance, at least by late childhood (see Scerif et al., 2004, 2007; Wilding et al., 2002), is a weakness in inhibitory control processes, as shown by frequent repetitions and failures to switch targets in visual search. Woodcock, Oliver and Humphreys (2009) have also recently demon-strated problems in switching attention in fragile X, using a simon spatial interference task. Also Scerif et al. (2005) examined ability to inhibit inappropriate orienting responses using the antisaccade paradigm in infants and toddlers aged 8 to 38 months. A cue was presented to the left or right of fixation, followed by an interesting stimulus on the opposite side. Typically developing infants and toddlers showed a decrease in saccades toward the cue from the first to the second half of the experiment (71.7 to 52.8% trials) while the fragile X group did not (74.4 to 73.9%), indicating difficulty in inhibiting the exogenously cued response. The failure of infants and toddlers with fragile X to inhibit looks toward cues could not simply be explained by a failure to learn that cues predicted the appearance of interesting stimuli on the opposite side. Both groups made similar numbers of anticipatory saccades directly to the interesting target stimulus, but children with fragile X continued to orient (perhaps automatically) to the cue.

Recent studies have also assessed the broader EF profile of children with fragile X. Most recently, Hooper et al. (2008) assessed performance across a wide range of EF tasks, tapping inhibition (Day–Night Task and Contingency Naming Task [CNT], both reviewed in Chapter 5), working memory (Memory for Words subtest and the Auditory Working Memory subtest from the Woodcock-Johnson Tests of Cognitive Abilities [WJ-III]), cognitive flexibility (Subtest 3 from the CNT), and planning (the Tower subtest from the NEPSY, reviewed in Chapter 6, and the Planning subtest from the WJ-III). Measures of processing speed were also included. Performance of fragile X boys (mean CA 10.1 years, mean nonverbal MA 5.3. years) was compared to a MA-matched typically developing group of boys. As in their previous study of CPT performance, reported above (Sullivan et al., 2007), children with fragile X struggled to complete all tasks, the most difficult being the cognitive flexibility task (25.9% completion) and the easiest being a task of working memory (Memory for Words) (94.4% completion). Overall, the authors found pervasive impairments in EF functions compared to their MA-matched typical control group but surprisingly not in speed of performance in which both groups were comparable. This finding is reminiscent of the conclusions drawn by us from our studies of visual search where we conclude that speed and accuracy depend on distinct mechanisms, with speed related principally to age and accuracy to the efficiency of EF. There was further partial support for this position in the Sullivan et al. study where MA was a significant predictor of performance in only two EF subdomains: cognitive flexibility and working memory. From these data, however, it is difficult to isolate the fragile X *signature* from that of general delay. The comparison would have been richer with the inclusion of another developmental disorder to help tease apart the impact of developmental level and cognitive delay on EF performance.

Munir et al. (2000a), in an earlier study, assessed performance across different working memory subdomains (verbal memory, visual-spatial memory, and executive capacity memory) in children with fragile X syndrome (mean CA 10.59 years, mean verbal MA 6.77 years) compared to typically developing children and found that performance was influenced primarily by the cognitive load of the task itself, irrespective of working memory subdomain, a similar finding to that recently reported by Lanfranchi et al. (2008).

Conclusions on Fragile X Syndrome

Together, the current findings suggest that toddlers and children with fragile X when compared to typically developing children have significant impairments across all attention subdomains. However, closer inspection reveals a specific and pervasive deficit on tasks that require inhibition of previously correct responses. This may account for the wide-ranging pattern of impairment. Interestingly, recent brain imaging studies in fragile X individuals using a well-recognized measure of inhibition, the Go–NoGo task, have found unusual activation patterns in prefrontal cortex. For example, Menon, Leroux, White, and Reiss (2004) found that adolescent fragile X females demonstrated impaired functioning in inhibitory circuits in frontostriatal regions of the cortex, and also deactivation in the ventromedial prefrontal cortex, an area involved in self-monitoring. Hoeft et al. (2007), using a similar paradigm, found adolescent fragile X males to have reduced activation in the right ventrolateral prefrontal cortex, while there was some evidence of increased and possibly compensatory activity in the left ventromedial prefrontal cortex.

It is as yet unclear whether the observed attentional deficit in fragile X is simply an extreme version of ADHD or whether it has unique characteristics. We incline to the latter view. Munir et al.'s (2000a, 2000b) fragile X group performed at a much lower level than their poor attention–matched control group, especially in making repetitive errors in visual search, but a direct comparison with a diagnosed ADHD group is still needed.

Developmental Issues

We have only been able to briefly describe data on fragile X separately for toddlers and children because there is minimal information available from matched tasks at different ages, and the major differences between CA and MA add additional uncertainty over how to partition the studies. The studies by Flanagan et al. (2007) and Scerif et al. (2005) demonstrated impairments in basic control processes at a very early age, and the work of the present authors and colleagues (Munir et al., 2000b; Scerif et al., 2004, 2007; Wilding et al., 2002) shows a common pattern of repetitive errors from age 2 upward.

It has been claimed that intellectual ability in fragile X actually deteriorates from childhood onward (Fisch et al., 1992, 1999; Fisch, Simensen, & Schroer, 2002). Cornish, Turk, Wilding et al. (2004) have argued that this decline might be due to increasing problems in maintaining and developing successful cognitive strategies that keep pace with their typically developing peers, rather than an actual regression in intellectual level or failure in neural development. Cornish and Wilding (2006) reported a study of both males and females with fragile X ranging in age from 6 years to 34 years, divided into two separate groups distinguished by CA (under 16 years and over 16 years). Intellectual level ranged from 3 years to 13 years in affected males and from 3 years to 20 years in affected females. Neither males nor females displayed a reduction in performance with increasing age in a range of verbal and spatial tasks (though no specific tests of attention were included). Indeed, performance on all tasks improved with verbal MA to the same degree in both fragile X and typically developing groups. Lightbody et al. (2006) found that females with fragile X aged 5 to 23 years improved more slowly on the Contingency Naming Task, a task of EF described in Chapter 6, than typically developing controls, but at the same rate on the Spatial Relations Test (Woodcock-Johnson Tests of Cognitive Ability) that requires construction of a shape from component pieces and verbal fluency, producing words beginning with a specified letter. Further work is needed to clarify these issues and determine whether different cognitive skills show different patterns.

Cornish, Scerif, and Karmiloff-Smith (2007) have compared developmental trajectories across three disorders (fragile X, Williams syndrome, and Down syndrome), using the data from studies described above. Comparisons between syndromes will be more appropriately considered in discussing the evidence on the specific signature of fragile X in relation to other neurodevelopmental disorders, which we will consider after examining the evidence on each syndrome in turn. However, we note here their findings on fragile X. On the antisaccade task employed by Scerif et al. (2005), the number of antisaccades in the second half of the task, compared with the first half, provided a measure of learned ability to inhibit inappropriate saccades toward the cue on the opposite side of the screen. This measure was related to CA and MA in the typically developing group but not in the fragile X group, indicating no development in this ability over the age range tested (CA 14 to 55 months, MA 12 to 30 months). On the Mapsearch from the TEA-Ch battery, the number of targets detected improved with both CA and MA, as did the number of hits (but not the time to detect targets) on the Wilding Vigilan Task (Munir, Cornish, & Wilding, 2000b). No improvement occurred on the Walk/Don't Walk task from the TEA-Ch, which is a type of Go–NoGo task requiring inhibition of some responses; this is consistent with the above finding on the absence of any development in the ability to inhibit saccades. These results suggest improvements in visual discrimination and speed of performance with age, but not in the more critical control abilities. However, it should be noted that sample sizes were small, resulting in some lack of sensitivity.

Down Syndrome

Down syndrome is the most common genetic cause of developmental delay (96% of cases) and results from an extra copy of chromosome 21, known as *trisomy 21*. In the Down syndrome critical region (DSCR), there are approximately 33 genes, and it is proposed that at least one or more of these genes is responsible for the phenotypic features we associate with Down syndrome (see Delabar et al., 1993). The incidence rate is high with an estimated 1 in 700 (see Sherman, Allen, Bean, & Freeman, 2007, for a review). Almost all children with Down syndrome will have some degree of developmental delay.

Behavioral Phenotype

Behavioral problems in Down syndrome have been well documented, and although they are normally less pronounced than in other developmental disorders, there is an increased risk for certain behavior problems compared to typically developing children. In a recent review, Dykens (2007) cites a variety of evidence that children with Down syndrome still demonstrated higher rates than typically developing children of stubbornness, oppositional behavior, inattention, speech problems, concentration difficulty, attention seeking, and impulsivity. She further notes the following facts. In 6–8% of cases, ADHD is diagnosed (although see Cornish et al., 2001, for a much higher prevalence of symptoms: approximately over 60% as rated by teacher surveys), and up to 7% of cases may have autistic spectrum disorder. At adolescence, hyperactivity declines and withdrawn behavior increases. Depression and early onset Alzheimer's disease are common over the age of 40. Age-related changes in behavioral profiles are also common (see Visootsak & Sherman, 2007, for a review).

Clark and Wilson (2003) used the Reiss Psychopathology Scale to evaluate the range of behavioral problems in children with Down syndrome between the ages of 4 to 21 years (mean CA 9.55 years). The highest aggregate scores were found in the categories that measured attention deficit (e.g., distracted, disobedient), anger/self-control (temper tantrums, impulsive, impatient), and psychosis (communication problems). Though the assignment of items to categories is questionable in several cases (for example, why is disobedience a feature of Attention Deficit and communication of psychosis?), the overall indications of poor control of attentional focus and maintenance are clear. Cornish et al. (2001) also examined behavioral problems in boys with Down syndrome (with a mean age of 11.7 years) using the Child Behavior Checklist (CBCL, reviewed in Chapter 4) and found a mean CBCL T-score of 59, a score slightly less than that reported for typically developing controls (T-score 63) who had been rated previously by their teachers as being "poor" attenders, but much higher than the score reported for control children rated "good" attenders (T-score 50).

Cognitive Phenotype

At the cognitive level, isolating specific profiles of cognitive strengths and weaknesses in Down syndrome can be quite complex, and there is a prevailing assumption that general delay across multiple domains characterizes this disorder. However, close examination of the literature reveals that by late childhood and adolescence, relative strengths in receptive vocabulary (Abbeduto et al., 2003), visual short-term memory (Visu-Petra, Benga, Tincaş, & Miclea, 2007), and inhibition (Munir et al., 2000b) are accompanied by weaknesses in expressive language (see Ypsilanti & Grouios, 2008, for a review), verbal short-term memory (Brock & Jarrold, 2005; Jarrold, Thorn, & Stephens, 2009), and visual perceptual skills (Vicari, Bellucci, & Carlesimo, 2006).

Attention Studies

Selective Attention

ORIENTING

Flanagan et al. (2007), in the study described above, also tested a Down syndrome group with an age of 11 to 24 years (mean CA 17.15, mean nonverbal MA 5.86 years), subsequently subdivided into two subgroups comprising high- and low-MA groups. Both Down syndrome groups showed unimpaired exogenously controlled orienting but poor endogenously controlled orienting, suggesting that the mechanisms responsible for changing focus were intact but the relevant control systems were weak. Brown et al. (2003) also examined automatic orienting of saccades in toddlers with Down syndrome with a mean age of 29 months and compared them with a CA control group (mean CA 30 months) and an MA control group (mean CA 15 months), matched using the Bayley

Scales of Infant Development II. They found no impairment in the Down syndrome group, supporting the conclusion above.

VISUAL SEARCH

Munir et al. (2000b) and Wilding et al. (2002), in their study of continuous visual search (described above in the fragile X section) also included a group of boys with Down syndrome (mean CA 11 years, mean verbal MA 6 years) (see also Wilding & Cornish, 2004, for a focused review of the findings). Compared to typically developing boys (including both those with good and those with poor attention), Down syndrome boys found fewer targets, searched more slowly, travelled further around the screen, and made more false alarms when searching for a single target than typically developing boys. Compared to fragile X boys, they found more targets and made fewer false alarms, but did not differ in speed and distance travelled. Examination of error types showed that repetitions on already-located targets were more frequent than in the control groups but less frequent than in the fragile X group, forming 73% of all errors, while confusions with nontargets were relatively rare (15% of all errors for Down syndrome, 49% for controls, and 0% for fragile X). The absolute number of such confusions was very similar in the Down syndrome group and control groups. These results demonstrate a considerable problem in inhibiting repetitive responses in the Down syndrome group, though not as severe as those reported in the fragile X group, and no impairment of visual discrimination in this task for the Down syndrome group.

Maintenance of Attention

Using the Vigilan Task in the study described above on fragile X children, Munir et al. (2000b) found that children with Down syndrome did not differ from the typically developing control group with poor attention but were slower to detect targets and made more false alarms than the control group with good attention. Thus, there was no indication of a specific pattern of impairment other than a generally poor level of attention. The Down syndrome group did not differ in mean time taken to detect targets from the fragile X group, but the latter made many more false alarms (29.88 compared to 10.32 for Down syndrome).

Attentional Control

In the alternating-target version of the Visearch Task (Munir et al., 2000b), while perseveration errors on the previous target type were more common in Down syndrome than in the control groups (on average, 1.01 perseverations following a successful response, compared to 0.1 in the control groups), the rate was much lower than in the fragile X group (5.43 perseverations per hit). Thus, children with Down syndrome again presented a less drastically impaired profile of poor inhibitory control compared to fragile X.

A similar profile was found on the Walk/Don't Walk task from the TEA-Ch attention battery (reviewed in Chapter 6), which measures ability to inhibit a rare NoGo response. Munir et al. (2000b) found a difference only between Down syndrome and the good-attention control group. However, children with Down syndrome in comparison to children with fragile X made substantially fewer errors overall (3.5 vs. 8.6).

Conclusions on Down Syndrome

Together, these findings suggest a global impairment across attention measures, though not always when compared with typically developing control children matched on attention level as well as MA. There appears to be no one specific area of vulnerability that would indicate a *signature* profile or difference from the typical ADHD profile. There are currently no published functional brain imaging studies of children or adolescents with Down syndrome, but one of the most detailed anatomical studies to date by Reiss and colleagues reported a range of atypical brain structures in individuals with Down syndrome aged between 5 and 23 years (mean CA 11.3 years), compared with a typically developing control group matched for gender and CA (Pinter, Eliez, Schmitt, Capone, & Reiss, 2001). Specifically, they found frontal lobe volumes to be significantly smaller in individuals with Down syndrome, which the authors suggest may be sufficiently underdeveloped to impact on cognitive functions typically associated with frontal lobes and impaired in Down syndrome. See Schaer and Eliez (2007) for a recent review of brain imaging studies across multiple disorders, including Down syndrome.

Developmental Issues

At a developmental level, Cornish, Scerif, and Karmiloff-Smith, (2007) found that on a measure of selective attention, the Mapsearch Task from the TEA-Ch battery (reviewed in Chapter 6), children with Down syndrome improved in speed of search as CA increased but this measure was unrelated to MA. On the Wilding Vigilan Task, detections became faster with increasing MA and CA. On the Walk/Don't Walk task, the number of successful inhibitions improved with CA. However, due to small sample size, this study may have failed to detect some other significant relations.

Williams Syndrome

Williams syndrome is a relatively rare disorder that results from a micro-deletion on one copy of chromosome 7 involving between 25 and 30 genes. Earlier prevalence rates indicated 1 in 20,000, but the most recent estimate is 1 in 7,500 (Stromme, Bjornstad, & Ramstad, 2002). Almost all children with Williams syndrome will have some degree of developmental delay, but this can vary, with some children showing mild impairment and others being more severely affected.

Behavioral Phenotype

Children with Williams syndrome have a well-documented behavioral profile that includes obsessive behavior, high degrees of anxiety, extreme friendliness and sociability, with an excessive display of empathy, together with hyperactivity, distractibility, inattention, and lack of persistence. Greer, Brown, Pai, Choudry, and Klein (1997) report that 73% of their sample showed clinically significant attention problems, with a further 14% on the borderline. Some intriguing findings recently reported by Leyfer, Woodruff-Borden, Klein-Tasman, Fricke, and Mervis (2006) indicate a possible age trajectory in ADHD diagnoses in children and adolescents with Williams syndrome. Of their sample, 55% of young Williams syndrome children (4 to 6 years) had received a diagnosis of ADHD combined-type compared to 25.7% in their midchildhood years

(7 to 10 years) and 10% in the late-childhood/early-adolescent years (11 to 16 years). In contrast, a diagnosis of inattentive-type only increased with CA, suggesting a selective maturational effect of reduced hyperactivity with increasing age. These are important findings, but unfortunately, given the cross-sectional nature of these reported changes, it is extremely difficult to know whether they reflect real developmental changes, or whether they simply reflect a sample bias. They are, in fact, somewhat similar in pattern to recent reports of children and adolescents with ADHD (see our discussion of these findings in Chapter 8). Support for the first alternative is provided by the findings from an earlier longitudinal study by Einfeld, Tonge, and Rees (2001), who report that hyperactivity but not inattentive behavior declined with age in Williams syndrome from childhood to adolescence. For further evaluation of the Williams syndrome behavioral phenotype, see the recent reviews by Martens, Wilson, and Reutens (2008), and Paterson and Schultz (2007).

Cognitive Phenotype

At the cognitive level, there has been tremendous debate in recent years as to the nature of the strengths and weaknesses that define Williams syndrome. It was previously assumed that a number of cognitive abilities are actually "intact" (i.e., within the normal range) in Williams syndrome, for example, face-processing skills (see for example, Wang, Doherty, Rourke, & Bellugi, 1995) while others are profoundly impaired, for example, visual-spatial skills. However, more recent investigations have challenged this assumption, and there is now convincing evidence to indicate that so-called "proficient" performance in face processing (alongside language) in children with Williams syndrome is achieved by cognitive processes very different from those employed by typically developing children and indeed follows a deviant rather than typical trajectory. Karmiloff-Smith et al. (2004) make the following argument:

> Scores in one domain (face processing) may outstrip scores in the other domain (visuospatial cognition), both domains may be affected by similar deficient processes but one reveals this impairment more subtly than the other. Seeming dissociations in the outcome, then, do not necessarily entail dissociations all the way along the developmental pathway. (p. 1272)

These findings underscore the absolute importance of research that teases apart the actual processes by which children with neurodevelopmental disorders perform a given task. In other words, do not assume that "equivalent" performance indicates equivalent proficiency in cognitive processing.

With regard to other cognitive domains, children with Williams syndrome display poorer visual-spatial working memory than visual-object working memory (see Vicari, Bellucci, & Carlesimo, 2006) and better perceptual than visual-construction skills (e.g., Rondan, Santos, Mancini, Livet, & Deruelle, 2008; Rondan, Mancini, Livet, & Deruelle, 2003). This latter dissociation has led to the hypothesis of a dorsal visual stream–deficit theory of Williams syndrome (see Atkinson et al., 2003, 2006).

Attention Studies

Selective Attention

ORIENTING

Brown et al. (2003) in the study described above in the section on Down syndrome, included a group of Williams syndrome children aged 2 to 3 years (mean CA 29 months), but no precise MA could be obtained from the Bayley scales. Two stimuli were shown in rapid succession in order to test efficiency in recalibrating the control system following the first saccade in order to move accurately to the second stimulus. Williams syndrome toddlers made fewer looks to the first target than controls or Down syndrome toddlers and frequently failed to make a second saccade toward the second target or made an inaccurate response. This suggested that they were impaired compared to both the typically developing and the Down syndrome toddlers in basic control systems required for orienting to targets.

Atkinson et al. (2003) examined the timing and direction of saccades in response to a visual cue in children with Williams syndrome aged between 8 months and 70 months and found that if a central fixation display remained on the screen rather than vanishing when the cue appeared on the right or the left of fixation, then children showed increased latency as if they had difficulty in disengaging attention (so-called "sticky fixation"). Older children with Williams syndrome and typically developing children above 6 months in age were not affected by this manipulation, suggesting delayed maturation of this mechanism in younger Williams syndrome children. Difficulty in disengaging from an attended location has also recently been reported in toddlers with Williams syndrome (Cornish, Scerif, & Karmiloff-Smith, 2007).

VISUAL SEARCH

Scerif et al. (2004) and Cornish, Scerif and Karmiloff-Smith (2007) used a modified version of the Wilding Visearch Task to test visual search in toddlers with Williams syndrome (mean CA 45.8 months, mean MA 27.9 months). See a description of this task in the fragile X section above. As expected, toddlers made more errors than typically developing control children, but closer inspection revealed a distinct error type. In contrast to toddlers with fragile X whose preponderant errors were repetitions on already-located targets, toddlers with Williams syndrome made predominantly more touches on the nontargets, confusing them with targets. This is despite the fact that all toddlers with Williams syndrome could discriminate targets and nontargets when presented in pairs. This pattern points to a distinct error profile in Williams syndrome that involves specific difficulties in the selection of targets due to impaired perceptual discrimination rather than inhibitory or working memory difficulties.

Most recently, search efficiency has been investigated in a sample of children and adults with Williams syndrome (aged between 8 and 41 years) using an eye-tracking procedure. Montfoort, Frens, Hooge, Lager-van Haselen, and van der Geest (2007) recorded eye movements during search for a target circle in a display that was presented for 5 seconds. The authors found that compared to typically developing controls, individuals with Williams syndrome were less efficient and consistent in their search with only two thirds (67%) able to locate the target in 5 seconds compared to 99% of controls. Duration of fixation of stimulus elements was significantly longer in the Williams syndrome group, who also needed more fixations to reach the target than controls. However, the main limitation of this study is that it did not control for developmental level. Some attempt was made to include a small subgroup (n = 5) with lower IQ to try to balance out the discrepancy in IQ levels between the groups but this procedure is hardly adequate. The inclusion of a very broad age range also makes it difficult to identify any age-related trends in performance. In other words, did the youngest children make the most errors, show the slowest overall performance, and so forth? The only conclusion one can really draw from this study is that individuals with Williams syndrome demonstrate inefficient search capabilities compared to CA-matched typically developing controls. This is hardly a surprising finding given the level of developmental delay associated with Williams syndrome.

Maintenance of Attention

To date and to our knowledge, only one study has explicitly assessed sustained attention in Williams syndrome. Brown et al. (2003; Experiment 2) recorded the duration of attention by toddlers with Williams syndrome (mean CA 29 months), Down syndrome (CA 29 months), and CA- and MA-matched typical controls to three novel toys presented separately in 45-second segments. Measures included duration of attention and number of periods of attention with each toy. See Chapter 7 for a review of the typically developing literature using this methodology.

In contrast to expectations, given the high rate of ADHD reported in this condition, no differences in performance were found between typically developing control groups and the Williams syndrome groups. Down syndrome toddlers, as already reported, showed shorter periods of attention. The authors suggest that attention deficits in toddlers with Williams syndrome may not emerge until later in development, when increasing demands on other cognitive skills interact to create a burden on attention capacity.

However, some specific abnormalities have been reported previously in maintenance of attention in Williams syndrome. For example, Mervis et al. (2003) studied attention to a stranger's face in Williams syndrome children aged 8 to 43 months and found intense and persistent fixation on the face and disregard of other objects in the environment. This is an early indication of the unusual attention to faces that is typical of older children with the syndrome. Note also the "sticky fixation" phenomenon observed by Atkinson et al. (2003) and described above, which represents an early, more general form of inability to switch attention appropriately.

Attentional Control

Atkinson et al. (2003) used three tests of attentional control with a broad age range of toddlers and children with Williams syndrome, aged between 2 and 15 years. Note that not all children performed all three tasks, with age ranges differing across individual tasks. On a measure of the ability to inhibit a prepotent spatial orienting response, as measured by an antisaccade task (age range 4.7 to 14.7 years, mean age 9.4 years), the child has to point as quickly as possible to the side opposite the displayed cue, in this case, bright and dark stripes. Performance was compared to a typically developing control group matched on CA. The Williams syndrome group performed the task more slowly than controls and made more errors. Of particular interest, the authors note that a number of children continued to make false starts pointing toward a target and then having to self-correct. On a measure of the ability to inhibit a familiar verbal response, as measured by the Day–Night task (reviewed in Chapter 6) (age range 4.8 to 15.3 years, mean CA 10.0 years), the Williams syndrome group performed quite well, with only minor impairments compared to published norms for this task. The third task was a complex EF task, the Detour Box Task, which requires the ability to inhibit a prepotent spatial response. Children can see an attractive colored ball, through an aperture, but if they try to grasp it, they interrupt a photobeam and the ball falls out of sight. In the lever condition, they learn an alternative detour strategy of operating a paddle that pushes the ball down a chute and makes it accessible. In the more difficult switch condition, they can deactivate the photobeam by operating a switch, but there is no physical connection apparent other than a red light going off to signal that the ball is now accessible. Typically developing children master the lever method by about 2.5 years and the switch method by about 3.5 years.

Williams syndrome children (age range 2.7 to 13.5 years; mean CA 7.8 years) had no difficulty with the lever method, but the switch method proved extremely difficult for a number of children who performed at floor level; half the group did not perform at the equivalent of typically developing 3.5-year-olds. Whereas the latter group showed a sudden improvement in performance at this age, a similar improvement in Williams syndrome did not occur until the age of 7 years. The authors suggest that response inhibition in Williams syndrome is not a uniform weakness but, instead, is domain-dependent with a specific weakness for responses that involved spatially organized and directed actions. However, we question this explanation. Scerif et al. (2004), in their study of visual search, found that Williams syndrome toddlers, unlike fragile X toddlers, showed no greater tendency to repeat responses on already-located targets than typically developing toddlers, indicating no difficulty in inhibiting previously successful responses.

Conclusions on Williams Syndrome

Together, the pattern of attentional deficit in Williams syndrome emerges as very different from the other disorders that we have examined. The syndrome is characterized by reduced gray matter concentration in the left parieto-occipital region, consistent with observed impairments such as those in visuospatial construction (Reiss et al., 2004). Visual selection is difficult primarily due to problems in discriminating targets from non-targets in a crowded environment, and this may also be related to these observed brain abnormalities. Response inhibition is also impaired (although not to the same degree of severity as that found in fragile X syndrome), and findings from a recent functional brain imaging study suggest that these difficulties may arise from a failure to engage frontos-triatal systems (the dorsolateral prefrontal cortex, the dorsal anterior cingulate cortex, and the striatum) when performing tasks that require basic inhibitory control (e.g., Go–NoGo paradigm) even when performance is at a comparable level with that of typically developing controls (Mobbs, Eckert, Mills, et al., 2007).

In contrast to many other neurodevelopmental disorders, no pervasive weakness in EF has as yet been identified.

Developmental Issues

To date, there is minimal information in the above studies about the developmental trajectory of different components of attention in Williams syndrome. Atkinson et al. (2003) showed that ability to disengage from a stimulus and direct attention to another, which is efficient in typically developing infants at about 6 months, was still inadequate in Williams syndrome up to the age of 6 years. Their findings also showed a delay in the ability to master complex spatial problems.

22q11 Deletion Syndrome

22q11 DS (also referred to as DiGeorge syndrome or velo-cardio-facial syndrome) results from a microdeletion of one copy on chromosome 22. Prevalence estimates indicate approximately 1 in 4,000 (e.g., Botto et al., 2003; Goodship, Cross, & LiLing, 1998; but see Shprintzen, 2008, for a more recent estimate of 1 in 1,600). Developmental delay is a core characteristic feature of 22q11 DS but varies in severity from mild/borderline to moderate impairment.

Behavioral Phenotype

22q11 DS has received tremendous interest in recent years, not least because of its association with a range of complex psychiatric disorders such as schizophrenia, anxiety disorders, obsessive-compulsive disorder, and ADHD. The most widely researched link is that between 22q11 DS and the development of schizophrenia, which is indicated as being over 35 times more frequent than in the general population (e.g., see Gothelf, Schaer, & Eliez, 2008; Murphy, 2005, for reviews). 22q11 DS is also strongly associated with ADHD, inattentive and combined types, being found in up to 55% of children and adolescents (e.g., Niklasson, Rasmussen, Oskarsdóttir, & Gillberg, 2001, 2005, 2009). Of note, a recent study comparing 22q11 DS children with comorbid ADHD reported a different profile of ADHD symptoms (less likely to be hyperactive or impulsive) when compared to children with ADHD only (Antshel, Faraone, Fremont, et al., 2007). This is an important finding, and we agree with the authors in their conclusion that different ADHD symptom profiles (e.g., ADHD alone vs. ADHD + Genetic Disorder) may reflect different "at risk" trajectories for the development of later psychopathology that are disorder specific. In this case the pattern of ADHD symptoms observed by Antshel et al. in 22q11 deletion syndrome may serve as an early indicator of the later development of schizophrenia, at least in a subset of affected individuals. One could also extrapolate this reasoning to other genetic disorders in which there is also high comorbidity with ADHD, such as fragile X.

The link between the *COMT* gene, which is deleted in 22q11 DS, and ADHD symptoms and cognition is somewhat controversial, but nonetheless an important topic of research and debate. Given our already extensive discussion on this issue, we will not address it again here, but instead refer the reader to Chapters 3 and 8. See also the findings of Glaser et al. (2006) and Michaelovsky et al. (2008) as recent examples of the complexity of this association in 22q11 DS.

Cognitive Phenotype

At the cognitive level, there is a well-documented profile of strengths and weaknesses but also a wide variability in performance. Individual differences such as gender, developmental age, and parental origin of deletion have all been proposed to account for some of the variation. However, by late childhood, the most consistent profile is one of relative strengths in reading and writing skills (see Antshel, Fremont, & Kates, 2008) and short-term verbal memory (e.g., Lajiness-O'Neill et al., 2006) and with specific weaknesses in short-term visual-spatial memory (e.g., Sobin et al., 2005a), numerical cognition (e.g., Simon et al., 2008), and sustained attention (Lewandowski, Shashi, Berry, & Kwapil, 2007). This "signature'" has led some researchers to postulate a specific correspondence with the pattern of proficiencies and deficiencies described in nonverbal learning disability (e.g., Lajiness-O'Neil et al., 2006).

Attention Studies

Selective Attention

Sobin et al. (2004), assessed *visual orienting* performance on the child version of the ANT (see Chapters 6, 7, and 8) in a sample of affected girls aged between 5 and 11 years (mean CA 7.6 years) and control siblings (mean CA 8.3 years), and Bish, Ferrante, McDonald-McGinn, Zackai, & Simon (2005) assessed a sample of affected boys and girls aged 7 to 14 years (mean CA 9.2 years) and typically developing control females matched on CA. Neither study found any group differences in exogenous orienting. Sobin et al. used the median reaction time, and accuracy, and missed trial means as their measures, but Bish et al. used a compound measure of RT / (1 - % Error), which is difficult to interpret if time and accuracy reflect different processes in this task as they do in continuous visual search. At one level, these results suggest comparability in orienting efficiency between 22q11 DS syndrome and typically developing controls. However, one reason for concern is the wide variability in scores within the 22q11 DS group reflected by their unusually high standard deviation. For example, when Bish et al. computed an orienting index (calculated by subtracting the reaction time of the neutral cue condition from that for the valid orienting cue condition), they report a mean difference of 140.64 for the 22q11 DS group, but with a standard deviation of 186.81. This is compared to a mean difference of 121.49 in the control group, but with a more reasonable standard deviation of 45.93. The high variability in the 22q11DS group may well be attributed to differences in intellectual functioning impacting on performance or may imply that there are different subgroups concealed within the disorder.

Maintenance of Attention

There are surprisingly few studies that have assessed the ability of children with 22q11 DS to maintain focus on a given task. To date, only two studies have published findings that suggest vulnerability associated with this disorder, and both offer only a glimpse rather than a comprehensive overview of the nature of any impairment. The first study is by Lewandowski et al. (2007), who examined performance on two versions of the CPT: a traditional A–X version, described in Chapters 6 and 8, and an identical pairs (IP) version comprising four conditions (numbers fast, numbers slow, shapes fast, and shapes slow). It is not specified whether the CPT tasks were modified versions of the originals. A measure of perceptual sensitivity (d') was the only measure recorded for both versions. Performance in children with 22q11 DS (mean CA 9.3 years) was compared to a typically developing control group matched on CA and gender. The age range of both groups combined was 7 to 16 years. Findings demonstrated an overwhelming impairment in performance in children with 22q11 DS compared to control children. This difference remained even when differences in IQ were accounted for with a mean Full Scale IQ of 70.7 in the 22q11 DS group compared to 108.9 for the control group. Unfortunately, aside from the measure of perceptual sensitivity (d'), no other data are provided that might have illuminated this profile further, such as the number and pattern of omission errors, response bias (measured by beta), or response time variability. We can therefore only conclude that children with 22q11 DS in this study had impaired ability to detect CPT targets (d'), which might be due to inadequacies in perceptual function or lapses of attention, among other possibilities. Until more extensive data are provided, the findings from this study do not offer convincing evidence of a disorder-specific deficit in sustained attention.

In contrast to the those studies that have reported some deficiency (albeit very tentative) in the ability of children with a 22q11 DS to maintain attention focus, using the ANT paradigm, Bish et al. (2005) found no group differences in performance on the alerting component of this task. The alerting response is assumed to be an automatic arousal process that is activated with the presentation of a cue signaling the onset of a target.

Attentional Control

Attentional control is by far the most intensively studied area of cognitive attention in 22q11 DS. Deficits in inhibitory control are well documented as are general EF impairments, especially on standardized measures that involve attentional switching (e.g., Wisconsin Card Sorting Task; Trail-Making Task), working memory (e.g., Auditory Set Response subtest from NEPSY), and planning (e.g., Tower subtest from NEPSY) (see Lewandowski et al., 2007; Niklasson, Rasmussen, Oskarsdóttir, & Gillberg, 2002; Sobin et al., 2005a, 2005b; Woodin et al., 2001, for examples of these studies). We focus here on describing recent findings that have demonstrated a core deficit in control processes, notably those that involve active disengagement of current attentional resources in order to reallocate and those that are required to withhold inappropriate responses.

RESPONSE INHIBITION

The ability to withhold or inhibit an inappropriate response has been assessed in a variety of contexts in children and adolescents with 22q11DS.

Aside from data derived from the ANT paradigm, the two other studies we describe below provide excellent examples of why it is important to fully describe the sample, matching criterion, and task measure. The lack of detail in these studies unfortunately allows us but a glimpse, if that, of any possible disorder-specific signature, and indeed raises more questions than it answers.

INHIBITION OF IRRELEVANT STIMULUS INFORMATION

As cited above, Sobin et al. (2005b) and Bish et al. (2005) both assessed cognitive attention in children and adolescents with 22q11 DS using modified versions of the ANT paradigm. We focus here on their findings on the executive component of this task. The executive response of inhibiting interference is activated when incongruent stimuli are processed (e.g., the central fish is pointing in a different direction to the flanker fish). Both studies found clear evidence of considerable difficulty in inhibiting the processing of irrelevant information as evidenced by poorer performance (slower reaction times and increased error rates) by the 22q11 DS groups on trials that contained incongruent flankers compared to those with congruent or no flankers. To further explore the nature of this deficit, Simon and colleagues assessed the extent to which children with 22q11 DS benefited from the inhibition generated from a previous incongruent trial when performing a new incongruent trial. This effect, known as the *Gratton effect*, predicts that the resource allocation needed to respond to distracting stimuli will be maintained on subsequent trials that require the same processes, thus producing more efficient performance on the second trial. Simon, Bish, Bearden, et al. (2005) found that, unlike control children who demonstrated a clear Gratton effect, performance by children with 22q11 DS actually worsened and became slower. The authors suggest that an inability to divert and control attentional resources to deal with conflicting or distracting stimuli is a specific impairment and may explain the children's poorer performance across other cognitive subdomains that require, at least in part, efficient executive control processes (e.g., arithmetic). See also Simon (2008) for a new account of the neurocognitive phenotype associated with 22q11 DS.

One cautionary note, however—and we have alluded to this in our review of the ANT in Chapter 6—is that the task duration is long (approximately 20 minutes) and requires continued fixation over this period. In attention-impaired groups such as those with 22q11 deletion syndrome and the other core disorders described throughout this book, the cognitive demands required to maintain attention and respond efficiently to task demands may show a differential decline that is not necessarily due to a disorder-specific inhibitory dysfunction, but rather the fatiguing effects of task duration. One possible solution is to include a measure of response variability (see the Sustained Attention to Response Task as an example) that will allow researchers to track performance at different time points within and across task conditions. As we have noted above, there is as yet no good evidence available on the ability specifically to *sustain* attention in this group.

The Stroop task is one of the most widely used measures of inhibition and requires the ability to inhibit familiar but distracting information. However, because this task requires some proficiency in basic reading, it is rarely used in populations that present with significant intellectual impairment. Using Golden's (1978) version of the Stroop (see our description above), Lajiness-O'Neil et al. (2006) found no differences in performance between their 22q11 DS group (mean CA 12.6 years) and a sibling control group (mean CA 11.4 years). At first glance, this finding seems to imply that some inhibitory functions may remain relatively intact in children with 22q11 DS. However, we raise some concerns about this study. First, the two groups do not appear to be equivalent in reading ability, with performance on the reading subtest of the Wide Range Achievement Test (WRAT) significantly poorer in the 22q11 DS compared to the control group. Thus, it is possible that there is a significant within-group variability in reading ability in the 22q11 DS children that is not addressed. Second, only the analysis for the interference effect is presented. No other details regarding any other variable are given, so we do not know whether group differences emerged across one or more of the three conditions. Finally, there was a 30-point difference in IQ between the two groups, and given that the Stroop is a language-based task, at the very least, verbal IQ could have been controlled for in order to see if there was an IQ-by-group interaction.

IMPACT OF THE COMT GENE ON INHIBITORY PERFORMANCE

As already indicated in Chapter 3, the *COMT* gene is particularly of interest with respect to the attention profile of 22q11 DS given its location in the deleted region of chromosome 22. The COMT genotype comprises a polymorphism related to high (Val) and low (Met) activity of the COMT enzyme and is associated with dopamine degradation. Given this intriguing association, it is not surprising that studies have attempted to elucidate the link between the Met/Val genotype and EF performance in 22q11 DS. Unfortunately, the association remains elusive. One the one hand, Bearden et al. (2004) have reported a Met superiority on performance on an EF task (the Trails B task) in a sample of children and adolescents with 22q11 DS (mean CA 11.1 years). However, Glaser et al. (2006) found no evidence of either a Met/Val advantage or disadvantage on performance on the Stroop task in individuals with 22q11 DS. These inconsistencies, alongside those reported in other studies, emphasize the need to look closely at variables that might reduce the power to isolate significant findings associated with 22q11 DS such as sample size, IQ, and age distribution, and even the experimental paradigm itself.

Not surprisingly, intense research activity has surrounded the possible association between COMT genotype and its relation to schizophrenia. For further debate on this issue, we recommend an excellent recent review by Prasad, Howley, and Murphy (2008) that examines the current evidence for an association between schizophrenia and COMT and also discusses other susceptibility genes located on the 22q11.2 region (e.g., proline dehydrogenase [PRODH] and Gnb1L).

Conclusions on 22q11 Deletion Syndrome

Together, the findings on attention functioning in 22q11 DS indicate only an emerging profile of strengths and weaknesses, with executive control processes especially vulnerable. Precise differences between this syndrome and ADHD are as yet unclear. As indicated in Chapter 4, white and gray matter are reduced in frontal areas and the cerebellum in this syndrome (Campbell et al., 2006). There is some suggestion from recent imaging studies that specific EF difficulties may be accounted for through impairment in their ability to orient to visual cues resulting from parietal dysfunction (Bish et al., 2005). See also Gothelf, Hoeft, Hinard, et al. (2007) for a recent imaging study of response inhibition and possible parietal compensation in performance on a Go–NoGo paradigm in adolescents with 22q11 DS.

Unfortunately, at the cognitive level there is a heavy reliance on standardized paradigms from such attention batteries as NEPSY, which, although informative at one level, do not tap subtle attention components that can be easily explored in atypical populations. Many of the published studies also incorporate numerous measures of cognitive functioning (10 or more tasks in some cases), and we wonder whether the heavy task load weakens the performance outcomes. The lack of appropriate comparison groups, the broad age ranges within a single study, and the wide, often unacknowledged, variability of IQ and cognitive ability within the disorder itself may weaken any interpretation of findings and their application to the wider population of individuals with 22q11 DS.

Developmental Issues

As far as we are aware, there is minimal information in the above studies about the developmental trajectory of different components of attention in 22q11 DS. No longitudinal studies or age-related studies have been published, but the importance of tracing the developmental profile appears to be quite critical if, as some researchers have alluded to, the neurocognitive phenotype seen in schizophrenic patients may be present in nonpsychotic children "at risk" of developing schizophrenia (a so-called cognitive *endophenotype* of schizophrenia; see Lewandowski et al., 2007). Thus, the earlier the phenotype can be established, the more predictive it may be as an early marker of later psychopathology in individuals with 22q11 DS.

Cross-Syndrome Perspectives

One of the most important findings to emerge from this extensive review of current studies of attention and EF across our five neurodevelopmental disorders is that attention deficits so prevalent at the behavioral level (e.g., ADHD-type symptoms) do not infer identical profiles or signatures at the cognitive level. In this section we highlight some of these important distinctions but also identify existing gaps in knowledge that warrant further investigation.

Autism and ADHD

Understandably, given the overlap and comorbidity rate of ADHD symptoms in children with autism, there is a growing literature seeking to identify core commonalities and differences between these two disorders. The main focus of this research has been situated broadly in executive functions and more specifically in inhibitory control. One of the key findings to emerge from a decade of exciting research is that children with autism with and without comorbid ADHD show quite different attention profiles. Furthermore, when these two groups are compared to children with ADHD alone, we see three signatures emerge. For example, Sinzig et al. (2008) compared performance on a Go–NoGo paradigm in children with high-functioning autism with and without ADHD symptoms (ASD+ and ASD-, respectively) to children with ADHD alone. In terms of error types, accuracy rates, and reaction times, the ADHD group displayed the most significant weakness, followed by the ASD+ group and then the ASD- group. This profile appears to be specific to inhibition because when performance is compared on a measure of cognitive flexibility (as measured by the Intra-Dimensional/Extra-Dimensional Shift Task of the CANTAB, reviewed in Chapter 6), a different profile emerges with the ASD+ group most impaired and the ASD- group least impaired. These findings are extremely interesting and need further exploration, but they provide the intriguing suggestion that different cognitive pathways can differentiate autism and ADHD even though at the behavioral level there is considerable overlap. See also the findings of Happé et al. (2006) and Johnson, Robertson, Kelly, et al. (2007). Furthermore, the Sinzig et al. study is the first to provide some evidence that ADHD symptoms in autism may act as a precursor to the development of poorer inhibitory performance compared to autism without comorbid ADHD. This finding warrants urgent research investigations, not least because of the clinical and educational implications.

Autism Versus Fragile X

There are currently very few single-gene disorders for which there is a certainty of the involvement of autism, but fragile X is one of those disorders. An estimated 40% or more of boys with fragile X will receive a clinical diagnosis of autism (see Bailey et al., 2008) and between 60 and 80% will fulfill the diagnostic criteria for ADHD (Bailey et al., 2008; Sullivan et al., 2006). Given the strong ADHD link and the possible pathways associated with autism and fragile X, it is surprising and somewhat disappointing that there are currently no attention studies outside the behavioral domain. Given that the findings from studies of general cognitive and linguistic functioning in fragile X with and without autism suggest that children with a dual diagnosis (Autism + Fragile X) show greater impairment that children with fragile X alone (see Philofsky, Hepburn, Hayes, Hagerman, & Rogers, 2004), we predict the following pattern. Children with fragile X with autism will demonstrate a pervasive impairment across all attention domains (selective, maintenance, and control), followed by fragile X alone, and then low-functioning autism. Teasing apart disorder-specific inhibitory weaknesses will pose the greatest challenge, but the experimental paradigms currently in use are likely to produce, in our view, dissociable differences that will ultimately guide interventions.

Fragile X Versus Down Syndrome

The most distinctive feature of cognitive performance in fragile X is the major difficulty in inhibiting prepotent responses, which is apparent from an early age. Thus, in continuous visual search, fragile X participants found it very difficult to move on, even from a successful response, to search for another target (Scerif et al., 2004, 2007; Wilding et al., 2002). This apparently simple operation of inhibiting the response just made in order to proceed to the next one is crucial in any chain of behavior and is disastrously impaired in this disorder. Wilding et al. (2001) compared fragile X performance directly with Down syndrome and found a similar pattern in the types of errors, with frequent repetitions on already-located targets, and rare responses on nontargets. However, the degree of impairment was less extreme in the Down syndrome group, and the difference was particularly marked when alternating between two target types was required. The fragile X group found this task almost impossible, whereas the Down syndrome group did manage to achieve such alternations from time to time, often after one or more failures to switch to the new target type. No single distinctive impairment of attention has been identified in Down syndrome, rather, a general pervasive weakness. Similar conclusions can also be drawn from comparative studies of fragile X and Down syndrome in the domain of working memory (Munir et al., 2000a).

Fragile X Versus Williams Syndrome

Scerif et al. (2004, 2007) compared these two syndromes in toddlers on a simple continuous visual–search task and demonstrated marked differences in the weaknesses displayed. While the majority of errors in fragile X were repetitions on already-located stimuli, as discussed above, toddlers with Williams syndrome rarely made such errors but frequently made false alarms to nontargets, particularly when these were similar to targets. However, the evidence does not indicate a straightforward weakness in visual discrimination, but rather points to inadequacies in the programming of search strategies due to such factors as impaired visual orienting and disengagement from a current focus. These appear to be part of a general weakness in visuospatial processing, which has led to the hypothesis that the main feature of this syndrome is a deficit in dorsal visual stream processing. Of note, vulnerabilities in dorsal stream functioning have also been reported in other neurodevelopmental disorders, including fragile X syndrome (Kogan, Bertone, et al., 2004; Kogan, Boutet, et al., 2004). What is currently unclear is to what extent these impairments reflect a common etiology that is attributable to the general effects of cognitive delay, or whether they represent the impact of specific gene–brain interactions that are unique to a given disorder. In the case of fragile X, Kogan and his colleagues found that reduced levels of the encoded protein, *FMRP*, had a greater impact on magnocellular and dorsal stream functioning than on their parvocellular and ventral counterparts. This finding was the first to identify a specific neurobiological cause for this dysfunction in fragile X. Future studies need to isolate commonalities and differences across syndromes known to have impairments that may reflect dorsal stream functioning. Early identification can help to guide disorder-specific interventions that can reduce the impact of such difficulties at the cognitive level, for example visual-motor coordination.

Williams Syndrome Versus Down Syndrome

There is now a well-documented literature that has identified a range of similarities and differences in cognitive functions in individuals with Down syndrome compared to individuals with Williams syndrome. In the domain of working memory, see the stellar work of Jarrold and colleagues (e.g., Jarrold, Baddeley, & Phillips, 2007), and in the domain of number processing, see the innovative work of Karmiloff-Smith and colleagues (Paterson et al., 2006). In the domain of attention, a growing number of studies now attest to important syndrome differences in sustained attention and inhibitory abilities in toddlers with Down syndrome and Williams syndrome (Brown et al., 2003; Cornish, Scerif, & Karmiloff-Smith, 2007). Specifically, Brown et al. found that toddlers with Williams syndrome showed comparable levels of sustained attention with that of typically developing control children (matched on MA), suggesting an early strength in this attention skill. In contrast, toddlers with Down syndrome appeared to have a deficit in sustained attention, often displaying fewer periods of focused attention and reduced task duration compared to all other groups. Employing a modified version of the Wilding visual-search paradigm, Cornish et al. (2007) found striking error profiles in toddlers with Williams syndrome and fragile X syndrome that contrasted with a more global impairment in toddlers with Down syndrome such that performance across many measures was often comparable with typical controls matched on MA. In contrast, the Williams syndrome group displayed frequent distracter errors such that they touched more distracters than any other group, suggesting that their ability to deal with perceptual similarity between targets and distracters when under high attentional demands is lower than expected given their developmental level. As described above, the fragile X group displayed an unusually high number of perseverative errors, reflecting a core difficulty in moving on from a previously successful target to a new target. These cross-syndrome differences demonstrate the importance of tapping finer-tuned profiles of attention weaknesses and possible strengths. Performance in Down syndrome is often commensurate with their developmental level with no one specific area of difficulty, whereas Williams syndrome is associated with a specific weakness in perceptual discrimination that may be attributed to their underlying visual spatial difficulties.

Tracing the Developmental Pathways of Fragile X Versus Willliams Syndrome Versus Down Syndrome

Our own research on attention over the past decade, alongside that of Karmiloff-Smith and her team, has taken us along a journey of syndrome comparisons from childhood through to adulthood and then into infancy. This research has produced an array of exciting and fruitful observations of atypical developmental trajectories. Collectively, beginning with *selective attention*, these results suggest that, while older children with Down syndrome displayed poorer performance than control children and children with fragile X, their earlier performance falls within what is expected given their developmental level. In contrast, toddlers with Williams syndrome suffered striking difficulties in selecting targets among distracters, suggesting difficulties in selective attention that need to be further mapped in later childhood. On the other hand, performance by toddlers with fragile X (unlike that of any of the other groups) displays striking perseverative errors, a marker of the *inhibition* difficulties that characterize the syndrome throughout the life span. Toddlers with Down syndrome did not display this type of error, as might have been predicted from their later childhood performance. Finally, infants and toddlers with Down syndrome revealed shorter and fewer periods of sustained attention, a finding that would not have been predicted given the fact that, in mid- to late childhood, they are capable of sustained attention within what is expected given their overall developmental level. Therefore, early performance by children with these distinct disorders illustrates both differences and continuities with the attentional profiles that characterize them later in childhood and adulthood. These findings underscore the critical importance of charting developmental trajectories across different disorders instead of relying on the childhood end state to inform research and practice.

Other syndromes versus 22q11 Deletion Syndrome

To our knowledge there are no current studies that have compared performance in children with 22q11 deletion syndrome with other syndrome groups. As yet, no distinctive profile of attentional impairment is apparent for this disorder from the limited evidence available, and it is not yet possible to define exactly how it might differ from the other neurodevelopmental disorders we have described.

Conclusions

We are in an exciting decade of research innovation on neurodevelopmental disorders. The trajectory of knowledge from simple behavioral observations to a more finely tuned understanding of the neurocognitive signatures represents a significant advance in our understanding of atypical development. Undoubtedly, the information gathered from these numerous studies has the potential to make a significant and long-lasting impact on both clinical and educational domains. Here are some highlighted core findings: first, that focusing on generalized cognitive delay can mask important disorder-specific signatures that, if recognized early, can promote and facilitate intervention programs that target unique weaknesses as well as strengths (see examples from our own work (e.g., Cornish, Scerif & Karmiloff-Smith, 2007; Wilding et al., 2002) and the work of Karmiloff-Smith (e.g., Karmiloff-Smith et al., 2004; Paterson et al., 2006); second, that development itself plays a key role in shaping later cognitive and academic outcomes across all disorders; and third, that commonalities in behavioral phenotypes (e.g., ADHD symptoms and autistic-like symptoms) may actually reflect different cognitive processes that diverge from the normal trajectory and across disorders. This is why it is so important to conduct cross-syndrome research that is situated within a developmental framework. At present the majority of empirical research in this field is cross-sectional, which, on the one hand, has allowed us to explore important age-related changes across different time points but, on the other hand, is not sufficient to provide a complete picture of subtle, disorder-specific trajectories of attention difficulties and proficiencies.

Four areas of concern remain unresolved in our review of the current literature. The first relates to the persistent use of standardized measures (for example, the traditional IQ measures such as WISC or Stanford Binet) to explore attention and EF in disorders that include intellectual disability as part of their phenotype. Reliance on standardized measures of cognitive function appears to be most prevalent in the research on 22q11 deletion syndrome and makes it almost impossible for the reader to disentangle what is the disorder-specific *signature* and what is the general effect of delay. When studies have adopted a more experimental approach to teasing apart attention profiles, this has often resulted in quite insightful and informative findings. The second area of concern relates to the low sample sizes that dominate many studies across all our five disorders. This is especially surprising in those disorders that are relatively common, such as autism and 22q11 deletion syndrome, and the observant reader will note that these studies are also the ones that have used more standardized measures of attention. It is highly likely that these small samples represent a skewed population of children and adolescents with higher functioning than is typically the case. In such cases we cannot extrapolate the findings to the wider community of affected individuals, only to the subgroup tested. The third issue relates to matching criteria. It is clearly more appropriate to compare performance in children with cognitive delay to typically developing children of an equivalent mental age or developmental level. When studies do not match in this way, and instead match on chronological age, very little knowledge is gained other than the obvious finding that children with developmental disorders are developmentally delayed compared to their unaffected peers. Burack and his colleagues have produced some excellent critiques of this important issue, and we recommend strongly that the reader take on board their recommendations (e.g., Flanagan et al., 2007; Russo et al., 2007).

The fourth issue relates to disorder heterogeneity and the need to recognize the impact of moderator variables such as gender (as in the case of fragile X) and level of functioning (as in the case of autism) on cognitive outcomes. As we have demonstrated throughout this chapter and in our previous chapters, it is rarely the case that all individuals with a given disorder will present with the identical phenotype. This is particularly true in the case of disorders for which no specific genetic cause has yet been identified and which are behaviorally defined; as we have seen, there are doubts about the homogeneity of both ADHD and autism as distinctive syndromes. But this possibility must not be discounted even in the case of genetically defined disorders. Genetic variations will impact very early in development, impacting at the brain and cognitive levels, and so need to be empirically controlled for and meticulously investigated.

In conclusion, we can now state with some certainly that identifying attention profiles across different neurodevelopmental disorders, although fraught with challenges, provides the researcher with important glimpses into the complex set of processes that comprise the general concept of attention and the extent to which these processes are independent of each other. Careful analysis can also uncover the extent to which compensation is possible for basic impairments and the ability of the brain to achieve apparently identical performance by different strategies. Greater insight into successful compensation strategies can also be used in devising effective educational strategies and resources.

CHAPTER SUMMARY AND KEY FINDINGS

- Commonalities in behavioral symptoms across different neuro-developmental disorders do not necessarily infer identical cognitive mechanisms or etiology.
- Attention profiles need to be empirically investigated using developmentally sensitive cognitive paradigms that can tease apart attention across its varying subcomponents.
- Research is currently hindered by methodological stumbling blocks that include small sample sizes, inadequate matching criteria that fail to control adequately for the effects of general delay on performance, and convenience sampling that includes the more "capable" individuals as participants.
- Age-related changes are apparent, and the available data provide a glimpse of how genes interact with the developing brain and the environment to produce disorder-specific attention signatures.
- Future research needs to continue to chart developmental trajectories of attention at the behavioral and cognitive levels, recognizing the critical role of the environment.

Section IV: Treatment Approaches and Avenues for Future Research

10

Treating Attentional Impairments

CHAPTER SNAPSHOT

- Are there effective treatments for attention impairments?
- Stimulant medication and issues of complexity: comorbid disorders, subtypes, dosage, and adverse effects
- Use of stimulant medication in ADHD and its impact on reducing inattentive behaviors and cognitive inattention across development
- Psychosocial approaches in ADHD: parent-based, cognitive-based, and computer-based approaches
- The Multimodal Treatment Study of Children with ADHD (MTA)
- Effectiveness of stimulant and psychosocial approaches in reducing inattention in other neurodevelopmental disorders

---

Throughout this book we have made a bold attempt to guide the reader through one of the most exciting decades of research and to do so from multiple perspectives—from the genetic level, the brain level, and the cognitive level. Numerous studies attest to the critical role of attention in guiding behavior from infancy onward, and most recently, considerable efforts have begun to isolate atypical trajectories of attention in children and adolescents with known neurodevelopmental disorders. The key questions we ask in this chapter, based on the wealth of new knowledge are: First, what are the implications of these findings for facilitating treatment of attentional disorders, and second, how can these research discoveries translate to the public domain to help clinicians and educators target early interventions that improve the long-term academic and social outcomes of young children with significant attentive difficulties across their developmental trajectories? We recommend to the reader the recent insightful discussion of possible avenues of early identification and treatment of ADHD that take into account the need to develop and target treatment approaches that recognize early ADHD "signatures" and their developmental pathways (Sonuga-Barke and Halperin, 2010).

We begin by focusing on the efficacy of current treatment approaches and address the extent to which they impact on both the behavioral and cognitive attention impairments associated with ADHD. Uniquely, we look at treatment approaches and their goals across multiple targets: the child, the parents, and the teachers. We also review the emerging literature that has begun to investigate the efficacy of treatment approaches for ADHD symptoms in other neurodevelopmental disorders such as autism and fragile X. Throughout our chapter we highlight the need for research-based interventions and resources that reflect and are guided by scientific discoveries in developmental cognitive neuroscience.

ADHD Symptoms

There has been a tremendous surge in research assessing the effectiveness of different treatment approaches in alleviating ADHD symptoms in terms of behavioral and cognitive inattention difficulties alongside other known impairments closely linked to ADHD, such as executive function difficulties. We know, for example, that early inattentive symptoms (e.g., distractibility, disorganization, impulsivity) can have a deleterious impact on later academic functioning such as literacy and math (e.g., Dally, 2006; Dobbs, Doctoroff, Fisher, & Arnold, 2006). Inattentive symptoms, but not necessarily impulsive-hyperactive symptoms, are also strongly linked with working memory difficulties throughout development. See the seminal work of Tannock and her colleagues for examples of this association (e.g., Lui & Tannock, 2007; Martinussen, Hayden, Hogg-Johnson, & Tannock, 2005; Martinussen & Tannock, 2006; Tannock & Martinussen, 2007). Given the long-term academic and social consequences of untreated ADHD, it is important that intervention options target known behavioral and cognitive weaknesses as early as possible in development. A critical goal of any intervention program is to reduce the impact of immediate symptoms in order to promote optimal functional growth, but it is equally important that, in a pervasive condition such as ADHD, immediate effects turn into sustainable effects that can be built upon across the developmental trajectory. Not surprisingly, the most intensive research activity has focused on the efficacy of pharmacological interventions, namely stimulant medication, on symptom reduction. However, other treatment approaches, although less extensively documented, are showing considerable promise. Most notable of these are the psychosocial interventions that are specifically directed toward training the child, parent, and teacher in the management of symptoms in different environments such as the home, the classroom, and the community. Throughout, we highlight findings from two landmark projects that have directly focused on treatment efficacy in ADHD: The Multimodal Treatment Study of Children with Attention Deficit Hyperactivity Disorder (MTA), and the Preschool ADHD Treatment Study (PATS). A main strength of these projects is that both are housed within an interdisciplinary, multisite framework and incorporate what is considered to be the "gold standard" approach to treatment efficacy: the Randomized Controlled Trial (RCT), in which participants are randomly assigned to one of a number of conditions, be they the targeted treatment/s or control. In a further extension of this design, the double-blind crossover RCT, participants can be exposed to all treatments and can also act as their own control.

These studies alongside all other approaches have their weakness as well as their strengths. The questions that guide this chapter and indeed this book concern the developmental outcomes of attention disorders. Here we ask whether such interventions (pharmacological, psychosocial, or a combination of both) produce sustainable improvements in the domain of attention at the behavioral and cognitive levels, and over development itself. Accordingly, we will examine treatment efficacy for the core symptoms of ADHD in the preschool, school age, and adolescent years.

Is Stimulant Medication Beneficial in Treating Attention Impairments?

There has been extensive publicity regarding the use of psychostimulants in treating ADHD. There is good reason for this, notably that stimulant medications do have a markedly beneficial effect on alleviating inattentive and hyperactive behaviors. The stimulant *methylphenidate* (MPH) is the prescribed stimulant of choice worldwide. MPH, a catecholamin-ergic stimulant, increases dopamine levels in the brain by blocking their reuptake. We saw a glimpse in earlier chapters of how lower levels of dopamine in the prefrontal cortex can impact on behaviors strongly associated with ADHD, including hyperactivity and attentional control. But such relationships are never simple, and research is only just begin-ning to map these complex interactions. We recommend Diamond (2007) for an insightful critique of how variations in dopamine genes, for example, *COMT* and *DAT*, selectively impact on prefrontal cognitive functioning.

Issues and Complexities

The beneficial impact of MPH on reducing symptoms of ADHD is well documented (see for examples, Biederman & Spencer, 2008, and Levy, 2008), but there are a number of issues that need further research explo-ration. We highlight briefly four of these issues in Box 10.1.

We turn now to address our main questions: Does MPH impact on attention functioning at the behavioral and cognitive levels, and does it do so across development?

The Preschool Years

Inattentive Behaviors

A cursory glance at the current literature will inform the reader that diagnosing ADHD in the preschool years is extremely difficult, not least because of the transitory nature of ADHD behaviors during these early years. Although there are no precise estimates, recent studies indicate a prevalence of between 2 and 6% (e.g., Keenan & Wakschlag, 2000). For more details of the prevalence, behavior, and cognitive profiles, comorbidities, and trajectories of preschool ADHD, see Chapter 8. Treatment with MPH of severe and pervasive symptoms of ADHD in preschoolers is controversial. Over the past decade alone, and even though stimulants have not received federal approval for children 6 years and under, there has been a substantive and alarming rise in the number of stimulant medications prescribed to young children. Yet, empirical

Box 10.1 MPH and Complexities of Treatment

ADHD Comorbid Disorders

The critical importance of recognizing the impact of ADHD comorbid disorders in treatment response to MPH has been demonstrated in a number of recent research studies. One very good example is by Bedard and Tannock (2008), who investigated working memory performance in ADHD children with and without comorbid anxiety disorder. When working memory was teased apart to reflect different modalities (auditory-verbal vs. visual-spatial) and processing demands (storage vs. manipulation of information), findings revealed MPH to have a selective impact on working memory. Specifically, improvement occurred on skills required for auditory-verbal manipulation as well as storage (e.g., backward digit span) but not for those skills requiring auditory-verbal storage of information (e.g. forward digit span). The reader may want to visit Chapter 7 & 8 for a more detailed discussion of the relation between working memory and attention.

ADHD Subtypes

Early work by Barkley and colleagues had reported that children with inattentive-only subtype needed lower dosages than children with ADHD combined-type in order to archive maximal benefit (Barkley, DuPaul, & McMurray, 1991). A more recent study by Stein and colleagues confirms this finding (Stein et al., 2005 and see also the discussion by Diamond, 2007) but the literature is comparatively sparse and in need of further exploration.

Dosage Level

The differential impact of MPH dosages (low, medium, and high, with actual dose level depending in part on body weight) may also play an important role in reducing severity of ADHD symptoms. However, this issue remains partially unresolved with some studies finding a linear dosage-related response, such that higher dosages are more effective in reducing ADHD behavioral symptoms while others

(continued)

Box 10.1 MPH and Complexities of Treatment *(continued)*

have found all three levels to be equally beneficial. See King et al. (2006) for a systematic review of over 60 studies. In terms of cognitive benefits, there is some indication that dosage levels may play a selective role in performance improvement and be dependent on the actual cognitive demands of a given task (e.g., McInnes, Bedard, Hogg-Johnson, & Tannock, 2007).

Adverse Effects

Research clearly demonstrates that the *short-term* use of MPH does significantly impact on ADHD symptoms, and studies generally indicate a good tolerance for its side effects, especially with careful monitoring of dosage amounts. There has been considerable debate, however, as to whether *long-term* treatment with MPH increases or decreases the risk of later substance abuse in adolescents with ADHD, and the issue remains controversial, although a recent longitudinal study suggest no long-term influence of MPH on nicotine dependence (Huss, Poustka, Lehmkuhl, & Lehmkuhl, 2008).

Less well documented are the long-term effects of prolonged MPH treatment on the developing brain, especially when treatment begins in early childhood or even in the preschool years. As noted in Chapter 4, the developing brain undergoes continual and critical changes, and therefore, the dopamine system that is explicitly targeted by MPH will likely be vulnerable to some long-lasting, developmental effects.

---

data that evaluate the efficacy and safety of pharmacological approaches in preschoolers is limited. One pivotal study is the Preschool ADHD Treatment Study (PATS), which is a multicentered, blind, randomized efficacy trial designed to evaluate short-term efficacy and safety of stimulants in reducing ADHD symptoms in preschoolers. The study findings have now been widely published and indicate that immediate-release MPH in preschoolers does reduce clinical symptoms of ADHD (e.g., Greenhill et al., 2006) and produces some functional improvements, although to a lesser extent than those reported in school-aged children

(Abikoff et al., 2007). Although there are relatively few studies that have assessed the long-term treatment efficacy of MPH in preschool ADHD, one recent study in children aged between 3 and 5 years found that symptom improvement was maintained from the initial treatment phase to follow-up at 10 months (Vitiello et al., 2007).

Cognitive Inattention

Evidence on the impact of MPH on cognitive inattention in preschoolers is scarce. One of the few studies to date to examine treatment efficacy on attention performance in ADHD preschoolers was an early study by Byrne, Bawden, DeWolfe, & Beattie (1998). In this study, performance of preschoolers with a diagnosis of ADHD aged 4 to 5 years and treated with MPH was compared with a typically developing control group matched on age and gender. Measures included the Continuous Performance Task—visual and auditory—which taps sustained attention. Children were tested at two points: Time 1, in which both groups were medication free, and then at Time 2, following 6 months of MPH treatment in the ADHD group. Findings were impressive, but some caution is necessary given the small sample size of 8 participants per group. Performance at Time 1 showed significant differences in group performance, with the ADHD group exhibiting more errors (omission and commission errors) on the visual CPT compared to controls, but performance did not differ significantly on the number of commission errors on the auditory CPT. After 5 months of MPH treatment, the ADHD group appeared to show significant changes in performance, with no group differences reported across any of the attention measures. These findings still await replication.

The Childhood Years

Given that ADHD is most consistently diagnosed over the age of 7 years, and that MPH is FDA approved for children aged 7 years and upward, we see the majority of studies focus within the childhood years. We describe first the findings from the National Institutes of Health (NIH)– funded project, the Multimodal Treatment Study of Children with ADHD (MTA). The MTA stands out as a pivotal study in both its design and outcomes. It is a multisite, multitreatment, randomized clinical trial that compared children aged 7 to 9 years across four recognized treatments for ADHD (medication management, behavior management, combined medication and behavior management, and a routine community treatment) over a 14-month treatment phase.

Inattentive Behaviors

Focusing here on the specific impact of medication on reducing ADHD behavior symptoms (we will address other aspects of this study in later sections), findings indicated that children in the medication management group maintained sizable improvements in symptom reduction at 14 months from treatment onset (e.g., MTA Cooperative Group, 1999) and at 24 months (MTA Cooperative Group, 2004), but with reduced effect; by 36 months from treatment onset, the initial benefits of intensive medication treatment were negligible (Jensen et al., 2007). The authors suggest that a reduction in the effects of medication may be due, in part, to age-related changes in ADHD symptoms over the 36-month period, and also to a lack of maintaining an·effective intervention after the initial treatment phase. For a more extensive update and interpretation of these findings, see Swanson et al. (2008). However, an alternative interpretation of these findings is that MPH, although very effective in the short-term, is not especially effective in sustaining a reduction in behavioral symptoms over an extended period of time.

Cognitive Inattention

We turn now to address the cognitive benefits of MPH treatment and ask the following questions: Is it the case that short-term benefits accrue but then diminish with age? And are all aspects of attention improved with medication, or is there a differential effect on different functions? We examine current findings across our three attention subcomponents: selection, maintenance, and attention control.

SELECTIVE ATTENTION

The ability to selectively attend to target information and to ignore distraction is a fundamental cognitive skill, and one that is significantly impaired in children with ADHD. Unfortunately, it is the least researched of all attention components in terms of the effects of stimulant medication on performance. The few studies to date suggest that MPH does have a positive although selective effect on performance. Looking specifically at visual search paradigms, studies suggest that MPH does improve performance in the short-term by facilitating a more efficient search that is characterized by faster reaction time compared to performance off medication, but this may be dosage dependent, with higher dosage levels producing greater efficiency (Berman, Douglas, & Barr, 1999). Two other studies have included search-type tasks that assess selective attention in larger batteries of tasks testing the effects of MPH in this age group (Hanisch, Konrad, Günther, & Herpertz-Dahlmann, 2004; Tucha et al., 2006) and found evidence of improvement, but no indication that this was specific to these tasks.

MAINTENANCE OF ATTENTION

The ability to maintain attention over a period of time and to avoid distraction is a core attention skill that develops exponentially across the childhood years. In ADHD, this ability is significantly compromised. Here we describe the findings from studies that have used two widely documented measures of sustained attention: the Continuous Performance Task (CPT) and the Sustained Attention to Response Task (SART). Both tasks were described in detail in Chapters 6 and 8.

The Continuous Performance Task (CPT) is undoubtedly the paradigm of choice for many clinical studies that assess treatment efficacy and cognitive inattention (see Riccio, Waldrop, Reynolds, & Lowe, 2001, for a meta-analysis of the effects of stimulant medication on CPT performance). In brief, participants are required to wait for a target stimulus to appear on a screen and to respond only to those stimuli, ignoring any nontarget stimuli. The task is deliberately long, lasting approximately 20 minutes, and so places a considerable burden on attentional capacity.

Findings are invariably mixed. Not all studies have shown improvements with MPH (see Riccio et al., 2001, for examples), but other studies show some improvement, albeit selectively. A recent study by Schachar et al. (2008) found a reduction in omission errors (lapses in attention) but not commission errors (response inhibition) in their double-blind, crossover, placebo-controlled study of 6- to 15-year-olds.

The SART (described in previous chapters) is a relatively new task and lasts approximately 5 minutes. In brief, participants have to respond continuously to the digits 1 to 9 presented in repeated ascending order but to withhold a response to the digit 3. Note that responding is required on most trials, and withholding of a response is required only infrequently, so that the dynamic processes will almost certainly differ. In the first study to date to assess performance on the SART in children with ADHD treated with MPH (mean age 9.0 years) compared to typically developing children (mean age 8.7 years), findings indicated that after 6 weeks of treatment, the ADHD group improved their performance but only on selective aspects of the task, namely errors of commission and fast moment-to-moment variability (reflecting moment-to-moment variability in reaction time). In contrast, omission errors did not improve with MPH (Johnson, Barry, Bellgrove, et al., 2008). There were no differences between groups in slow changes over time, such as decreased performance with longer time on task. The authors argue for a differential impact of MPH on sustained attention that reflects those brain processes most at risk from abnormal levels of dopamine. Specifically, they suggest that MPH may be more effective at improving top-down attention control components but may be less effective at improving arousal mechanisms.

ATTENTION CONTROL

The abilities to exert effortful control in order to inhibit or withdraw a dominant or prepotent response, to hold in working memory newly relevant rules that require the suppression or activation of previously learned responses, and to shift attention between tasks are essential components of learning and are fundamental to academic success. In ADHD they all suffer a core, pervasive impairment. In examining the efficacy of MPH in improving performance, studies have tended to focus on improvements in response inhibition, a skill that requires the active suppression of a verbal or motor response. Obviously, other aspects of attentional control have been evaluated, and in particular, the role of working memory in producing attentional impairments in ADHD has been well documented (see the work of Tannock and her collaborators, e.g., Bedard, Jain, Hogg-Johnson, & Tannock, 2007). Although our focus here will be limited to response inhibition, we argued in Chapter 2 that control systems involve complex patterns of excitation and inhibition that are held in and operate through working memory.

Inhibition of Prepotent Responses

The ability to inhibit a prepotent response is typically assessed using paradigms that require the suppression of an activated motor response as well as efficient working memory. The two most commonly used paradigms are "Go–NoGo" and "Stop–Signal," both of which were extensively reviewed in Chapters 6, 7, and 8, so our reader will be very familiar with these tasks at this stage in the book! Although similar in some aspects of their task demands (i.e., suppression of a response), the two tasks involve quite different inhibitory processes: restraining a response in the case of Go–NoGo, and cancellation of an already initiated response in the case of Stop–Signal (Crosbie, Pérusse, Barr, & Schachar, 2008).

On the Go–NoGo task, studies are inconsistent, and taken together, they provide evidence of a minimal impact of MPH on performance. For example, two studies by Rhodes and colleagues examined performance in ADHD boys aged 7 to 15 years using a prospective, randomized, placebo-controlled, crossover design (Coghill, Rhodes, & Matthews, 2007; Rhodes, Coghill, & Mathews, 2006). Both studies failed to find any convincing evidence for an effect of MPH on Go–NoGo performance. In contrast, Hood, Baird, Rankin, and Isaacs (2005) found performance on the Walk/Don't Walk subtest of the TEA-Ch battery to be improved immediately by MPH within 60–90 minutes of treatment in boys aged 7 to 11 years. However, given the small sample size of this study compared to the number of tasks administered, this finding failed to reach significance using a corrected alpha level of $p < 0.005$.

On the *Stop–Signal task*, which has received more extensive investigation in terms of treatment efficacy, findings suggest that performance is generally improved by MPH especially on its core outcome measure, the SSRT (Stop–Signal Reaction Time), which provides an estimate of inhibitory control (see Scheres et al., 2003; Tannock, Schachar, & Logan, 1995). Interestingly, all three of these studies examined MPH dosage responses, and no evidence of a linear relationship was found between dosage and performance, although Tannock et al. (1995) reported an inverted U-shape effect, with the medium dosage being optimal. Other studies also confirm that speed of stopping as measured by the Stop–Signal task or its variants is improved by MPH (e.g., Devito et al., 2008; Lijffijt et al., 2006).

The Adolescent Years

In comparison to the childhood literature, there are relatively few published treatment studies on adolescence. Instead, many studies have incorporated a wide age range that includes children from aged 7 years to adolescents aged 15 years as part of one sample. Few, if any, of these studies assess the impact of age on performance, which is somewhat puzzling given that ADHD symptoms, both at the behavioral and cognitive levels, can change with increasing age. See Chapter 8 for a full review. At the other extreme, there is now an emerging adult literature that attests to the efficacy of stimulant treatment postadolescence (e.g., Wilens, Spencer, & Biederman, 2002). The scarcity of adolescent-specific studies may reflect a concern over the possible relation between stimulant treatment and later substance abuse in the teenage years. However, there is now a growing body of studies that indicate that, though there is always the potential for substance misuse, evidence points to the fact that long-acting stimulants may in fact reduce the likelihood of substance abuse. The ADHD adolescents most at risk from abusing substances are those with a comorbidity of ADHD and substance disorder (see reviews by Faraone and Wilens, 2007; Wilens, Faraone, Biederman, & Gunawardene, 2003).

Findings from the few available studies and reviews indicate that MPH is safe and well tolerated in adolescents producing short-term clinical improvements in ADHD symptoms (e.g., McGough et al., 2006; Smith et al., 2000; Wilens et al., 2006). However, evidence on the long-term effects of MPH treatment in terms of dosage and reduction in behavioral symptoms is still needed.

Age-Related Changes of MPH Efficacy on Attention Impairments

Few studies have examined the efficacy of stimulant treatments over long periods of time. The MTA is one of the exceptions, and we have highlighted their findings above. In brief, the data indicate minimal long-term benefit with reduced benefits at 10 months past cessation of the treatment and a virtually negligible impact 14 months later. However, it is difficult to know whether this effect is due to the diminished impact of MPH over a time period or due to a lack of consistent treatment management after the initial intensive treatment phase of the study. In terms of longitudinal analysis of effects of MPH treatment in reducing cognitive behaviors, there are currently no existing studies. However, clearly such data would provide a rich source of information on the trajectories and sustainability of cognitive improvements from childhood through to early adulthood. Combined with longitudinal behavioral data, we could begin to address hitherto unanswered questions regarding the correlation between MPH efficacy and developmental timing in the treatment of both behavioral and cognitive symptoms. For example, do behavioral reductions in ADHD symptoms impact differentially on specific aspects of cognition, and if so, does this change with development? Also, what aspects of early inattention (behavioral and cognitive) are predictors of later academic outcomes?

Conclusions on Stimulant Medications

This brief overview of current research suggests that MPH efficacy in treating ADHD behavioral symptoms, at least in the immediate or short-term, is consistent across numerous studies that encompass preschool, childhood, and the adolescent years. There is, however, less consistency of results in its impact on the long-term sustainability of symptom reduction, and the few studies to date suggest a waning effect by approximately 24 months, but this needs further systematic investigation. In terms of reducing cognitive attention impairments, MPH would appear, at first glance, to be effective is restoring performance on aspects of inattention that are known to be especially weak in ADHD, including sustained attention and response inhibition. At second glance, however, these findings are often inconsistent and lack replication. Furthermore, any benefits appear to be dependent on the cognitive demands of the task, and where there is improvement, performance is rarely equivalent to the levels achieved by the typically developing comparison groups.

We turn now to examine the role of other treatment designs in ADHD, namely psychosocial treatments for children and adolescents with ADHD.

Are Psychosocial Interventions Beneficial in Treating Attention Impairments?

Alongside pharmacological interventions, a range of additional treatment approaches is also available, and these play a critical role in management of ADHD symptoms across multiple environments that include the classroom, home, and community settings. One of the core advantages of behavioral treatments is that they address a broader range of difficulties than those directly related to the clinical symptoms of hyperactivity, impulsivity, and distractibility. For example, the deleterious impact of ADHD on academic functioning, on social and family relationships, and on adherence to societal rules is well documented. It is therefore important that treatment, if it is to be of maximal effect, include the child, their teachers, and parents, and that techniques can be transferable across different settings and at different ages because symptoms do not remain static with age but instead change across developmental time.

We focus here on findings from two well-documented psychosocial approaches to treatment. The first is behavioral-based interventions with a focus on family-based interventions and school-based interventions. In the former, modifying parent behavior to better adapt to the challenges faced by their child with ADHD has proven to be very effective in reducing ADHD behaviors (see Chronis, Chacko, Fabiano, Wymbs, & Pelham, 2004, for a comprehensive review of the multiple parent and child factors that promote efficacy). In the latter, promoting classroom intervention approaches that directly involve a child's teacher and peers have been shown to have positive effects on ADHD behaviors (see Pelham and Fabiano, 2008, for a recent review). The second is cognitive-based interventions that focus on improving academic and cognitive functioning known to be especially weak in ADHD, for example, working memory (Klingberg et al., 2005) and organizational skills (Langberg, Epstein, & Graham, 2008).

As in our evaluation of MPH treatment efficacy, we follow the trajectory from preschool, childhood, and adolescence.

The Preschool Years

As expected, there are relatively few studies that have assessed psycho-social treatment efficacy in preschoolers with ADHD symptoms, compared to the number of published studies in childhood and adolescence. However, with a growing recognition that ADHD can be manifest in the preschool years, an emerging literature has begun to demonstrate the importance of both early identification and treatment of preschool ADHD in order to modify early behavioral symptoms.

To date, however, the majority of these few studies have focused on establishing the safety and efficacy of MPH treatment (see the findings from the PATS study described above). In contrast, the efficacy of psychosocial treatments has not received comparable attention, yet it would seem an essential step forward in reducing the adverse outcomes of later ADHD symptoms as demonstrated so effectively by a new program known as "Tools of the Mind," specially aimed at targeting executive functioning (EF).

Tools of the Mind

Diamond, Barnett, Thomas, and Munro (2007) have tested a detailed 2-year program, Tools of the Mind, for training preschool children in various executive functions. The program consists of a number of activities built into a normal school curriculum directed at teaching specific aspects of executive function, such as inhibiting talking while listening to another child reading a story (children hold a model ear or a mouth to indicate their correct role as listener or reader), cleaning up the classroom without being distracted into irrelevant activities (a song is regularly played, during the duration of which the task has to be completed), and planning a scenario then acting it out (such as taking a sick child to the hospital). Ratings of behavior were taken at an average age of 5.1 years and were supplemented with objective indices of inhibitory ability derived from a Simon-type task and a flanker task, both testing ability to inhibit effects from irrelevant aspects of the input. The effects of the Tools of the Mind program were compared with those obtained from another program—the Balanced Literary program run over the same period but one that did not stress EF. The superiority of the Tools program was evident in all measures taken. The potential of this program and others that focus on core attention and inhibition impairments is clearly evident and warrants further investigation.

The Childhood Years

The most extensive research on the efficacy of psychosocial treatments has been in the 7-to-9-year age group. We focus briefly on four approaches, all of which provide some benefit. The reader will note that more extensive research is needed across all four approaches.

Parent-Based Training Approaches

Targeting interventions that can modify parental expectations and improve strategies for coping with maladaptive behavior (e.g., aggression, defiance) has been demonstrated to be an effective tool that can promote positive outcomes in parent–child relationships and reduce parental stress (see reviews by Chronis, Jones, & Raggi, 2006; Pelham & Fabiano, 2008). In brief, such programs generally comprise weekly training sessions that aim to provide parents with skills to recognize and address problem behaviors effectively in the home environment (e.g., monitoring problem behaviors, setting rules, reinforcing positive behaviors). We refer the reader to Antshel and Barkley (2008) for a description of this type of procedure. However, the efficacy of this approach in reducing the core ADHD symptoms of inattention and hyperactivity is less well established (see Barkley, 2004), and findings suggest that parent training programs may be most effective when housed within a more comprehensive behavior treatment package, as in the MTA study (see below), and when combined with other treatment approaches, notably pharmacotherapy. See our section below.

Cognitive-Based Training Approaches

As mentioned previously, untreated ADHD behaviors can have a detrimental impact on long-term academic outcomes, notably literacy and math development. The interested reader may want to visit Langberg et al. (2008) for a discussion of interventions designed to improve organizational skills, a core deficit in ADHD. However there is a virtual absence of studies that have focused on investigating the effects of cognitive-based programs specifically targeting attention in terms of selection, maintenance, and control other than those studies indirectly linking cognitive inattention with working memory or organizational skills.

Computerized Training Approaches

In recent years, an emerging body of research has begun to explore how intensive computerized training methods can improve the amount of information that children with ADHD can attend to and are able to retain and process. Posner and Rothbart (2007), for example, report that typically developing 4-year-olds who participated in an intensive 5-day attention training program showed improved performance, and children were able to complete the varying levels of task complexity. Recognizing that their sample size was small, the authors make the provocative statement that training may have altered brain activity in the anterior cingulate, as measured by EEG, to a level comparable to adult performance.

Similar gains in attentional functioning, as well as additional academic improvements, were also reported in a recent study by Shalev, Tsal, and Mevorach (2007) using a similar methodology with children aged 6 to 13 years who had been diagnosed with ADHD. They used computerized training directed at four aspects of attention: a CPT for sustained attention, conjunctive search for selective attention, an orienting task with flanking distracters, and switching between identification of large and small letters in Navon letters for control functions (Navon letters are large letters made up of smaller letters, such as a large capital *H* composed of small capital *H*s, known as a compatible condition, or a large capital *H* made up of small capital *T*s, known as an incompatible condition). Eight weeks of practice was given, and the matched control group received a similar period of exposure to playing computer games. At the end of this period, tests of academic performance on reading comprehension, copying a written (Hebrew) text, math tasks, together with parent ratings, showed benefits from the training for all but the math tasks.

Klingberg et al. (2005; see also www.cogmed.com) tested the efficacy of a 25-day computerized training program consisting of visuospatial tasks (referred to as "Robomemo"). Children aged 7 to 12 years diagnosed with ADHD were divided into treatment and control groups;

the treatment group was trained with increasing levels of task difficulty while the control group remained on the lowest level throughout the training period. At the end of the training period (5 to 6 weeks after initial assessment), and again 3 months later, the children were tested on abilities unrelated to the training task (forward and backward spatial span, digit span, Stroop performance, and Raven's matrices testing nonverbal intelligence). The treatment group was superior on all measures at both testing points, and parents, but not teachers, rated them as having reduced inattention and hyperactivity/impulsivity. Holmes, Gathercole, and Place (2008; see also Gathercole, 2008) also assessed the efficacy of Robomemo and found substantial benefits for children with ADHD.

Classroom-Based Training Approaches.

Behavioral classroom management techniques that target specific ADHD behaviors using various strategies (e.g., daily report card, time out) have consistently demonstrated efficacy in reducing the impact of problem behaviors in the classroom (e.g., Fabiano & Pelham, 2003; see also Pelham & Fabiano, 2008, for a review). Continuous feedback and daily reinforcements of goals are the key objectives of this approach. Findings indicate that this approach does improve on-task behaviors in the classroom (e.g., Fabiano & Pelham, 2003). However, sample sizes in many studies are often low, and in some cases, limited to a single case study.

A main obstacle preventing long-term success of classroom management programs is that teachers often have no access to, or knowledge of, appropriate tools that can help to decrease the effects of ADHD behaviors in the classroom. One of the most innovative projects to address this gap between a new generation of research knowledge about ADHD and the uptake and utilization of this knowledge by educators is the "TeachADHD" resource program developed by Tannock and her colleagues at the Hospital for Sick Children in Toronto, Canada. The program is designed to provide classroom teachers and other related professionals with knowledge about ADHD, its manifestations in the classroom, and evidence-based teaching strategies that can help reduce the impact of inattention and promote optimal behavioral and academic functioning (http://research.aboutkidshealth.ca/teachadhd).

MTA Study

The MTA study compared three treatment strategies in children aged 7 to 9.9 years with ADHD: medication management, behavioral management, and combined treatment, plus a community care regime as the control condition. We have described above the findings as they related to treatment efficacy of MPH and the study's long-term follow-up at 10 and 24 months after the end of treatment. The authors noted significant immediate improvement in symptom reduction, but by the final assessment 3 years after the beginning of treatment, there was little evidence of sustained improvement (see Swanson et al., 2008, for a comprehensive discussion of the study findings). We examine whether a similar profile emerged for children enrolled in the behavior treatment condition of the MTA.

The behavior modification training comprised intensive parent training (partially based on the methods proposed by Barkley, 1987), a child-focused treatment (developed by Pelham and Hoza, 1996) based at a therapeutic summer camp, and a school-based intervention organized during the school year that consisted of biweekly teacher consultations that focused on classroom management strategies and part-time assisted aid by paraprofessionals working directly with the child in the classroom (partially based on the methods devised by Swanson, 1992).

Initial findings at the end of the treatment phase showed that, although benefits in the form of symptom reduction did accrue from intense behavioral modification, they were not as substantive as those observed in the medication condition, as measured by parent and teacher ratings of inattentive behaviors (MTA Cooperative Group, 1999). However, one of the most striking findings was that improvement in this treatment condition remained consistently above baseline at both follow-up assessments (10 months and 24 months after the end of treatment) (Jensen et al., 2007; MTA Cooperative Group, 2004).

The Adolescent Years

Given the changing nature of ADHD symptoms from childhood to adolescence, appropriate psychosocial treatments that target adolescent behavior and cognitive functioning are essential. However, few empirically based studies have assessed the efficacy of treatment programs, especially those geared toward functional improvements, in this age group, and so our review here will be relatively brief.

In terms of behavioral improvements, we begin with an early seminal study by Barkley, Guevremont, Anastopoulos, and Fletcher (1992) in which a sample of ADHD adolescents aged between 12 and 18 years were randomly assigned to one of three treatment conditions: behavior management training, problem solving and communication training, or family therapy. Treatment comprised multiple sessions, and families were assessed on three occasions: pre- and posttreatment, and at 3-month follow-up. Findings were unambiguous. All three treatment conditions demonstrated efficacy in behavior improvement from baseline, with no significant difference between treatment conditions. However, posttreatment, few of these gains actually translated into clinically significant changes, with 70-87% of families showing no reliable recovery. A later study by Barkley and colleagues found similar results, this time in ADHD adolescents with comorbid ODD (Barkley, Edwards, Laneri, Fletcher, & Metevia, 2001; see also Barkley, 2004). Together, these findings suggest that parent training programs may have minimal clinical benefit in reducing ADHD symptoms, and as we suggested in a previous section, these programs tend to benefit problem behaviors associated with a diagnosis of ADHD rather than the actual symptoms of inattention and hyperactivity themselves.

Other behavioral intervention approaches that are not reviewed here but that the reader may wish to explore further include neurofeedback training and cognitive-behavioral training. See Toplak, Conners, Shuster, Knezevic, and Parks (2008) for an excellent review of the current limitations and potential of these treatment strategies to produce beneficial and long-lasting behavioral and cognitive outcomes.

Longitudinal Studies of Psychosocial Treatment Efficacy on Behavioral and Cognitive Impairments

To our knowledge, there are virtually no longitudinal studies that have explicitly assessed the time course of psychosocial interventions outside of a short follow-up that typically occurs within a few weeks posttreatment. There are two exceptions. The first is the MTA study, which incorporated a 10- and 22-week follow-up and found that, although improvements were not significantly maintained by 22 weeks posttreatment to a degree observed immediately following treatment, improvement in ADHD behaviors from baseline was still evident. However, we note that this treatment group also comprised, at follow-up, children who had switched to MPH after treatment cessation (Jensen et al., 2007; MTA Cooperative Group, 2004), and this might have influenced the degree of observed improvement. The second study, which focused on children with ADHD inattentive-type, provides a more reliable indication of psychosocial treatment efficacy with initial improvements in behavior and organizational skills still maintained at 2- to 5-month follow-up in children with ADHD inattentive-type (Pfiffner et al., 2007). The lack of developmental research is quite surprising, given the potential importance of early intervention and increasing sensitivity of early diagnosis.

Conclusions on ADHD

This is a somewhat difficult literature to draw any firm conclusions from. On the one hand, there is an emerging body of research that attests to the efficacy of psychosocial treatments for ADHD, and findings demonstrate the importance of involving multiple partners in treatment approaches: the child, parent, and teacher. On the other hand, methodological weaknesses, such as heterogeneity of study designs and small sample sizes, reduce the statistical power of any research findings and therefore limit any conclusions to be made about treatment efficacy. What we can conclude is that psychosocial treatments do seem to have an immediate impact on reducing core behavioral problems associated with ADHD, but not necessarily symptoms of inattention and hyperactivity. What is less conclusive is the long-term impact of these treatments.

Other factors that impede efficacy when used as a stand-alone technique include moderator variables such as the severity of ADHD symptoms, parenting style, and the presence of comorbid disorders. A second conclusion is that psychosocial treatments appear to not be as effective at reducing ADHD symptoms as pharmacological treatments or combined medication and behavioral treatments. Future investigations need to identify more carefully the key factors in the efficacy of these treatments.

Efficacy of Combined Treatment Approaches

The efficacy of psychosocial treatment linked with stimulant treatment has been evaluated in a number of studies, most notably in the MTA study, described in detail above. In brief, the findings indicated that an intensive combined approach to reducing ADHD symptoms (as measured by parent and teacher ratings of inattention and parent-rated hyperactivity/impulsivity symptoms) was as effective as stimulant treatment alone in its immediate effects, but superior in terms of measurable outcomes to both stand- alone psychosocial treatment and routine community care comparison (MTA Cooperative Group, 1999). At 14-month follow-up, the combined treatment group was still benefiting from the initial treatment intervention, and these improvements were superior to those observed in the psychosocial treatment group. At 21-month follow-up, although improvements over baseline were still apparent, there were no longer any treatment differences on outcome measures. Together, the MTA findings suggest that the short-term efficacy of a combined treatment approach, while clearly advantageous in reducing ADHD symptoms, does not appear to translate into substantive long-term benefits.

The effectiveness of such treatments for improving academic performance is much less impressive. A recent meta-analysis of studies of school-age children with ADHD (6 to 12 years) investigated the efficacy of stimulant treatment, psychosocial treatments, and combined treatments on outcome measures that included ADHD symptoms and academic functioning. Of the current studies, 26 met inclusion criteria. Findings demonstrated that both stimulant treatment and a combined treatment approach were equally effective at reducing ADHD symptoms, but that neither treatment approach was effective in improving academic functioning (Van der Oord, Prins, Oosterlaan, & Emmelkamp, 2008).

Efficacy of Treatments for Attention Impairments in Other Neurodevelopmental Disorders

Psychopharmacological Treatments

The reader will note immediately the lack of comparable treatment efficacy studies in neurodevelopmental disorders outside of ADHD. This discrepancy, although striking, is not surprising and reflects justified concerns about the use of stimulant medication for treating ADHD symptoms in children with specific genetic disorders or who have significant developmental delay. At the core root of these concerns are the issues surrounding safety and tolerability of medication and, in particular, the noted elevated risk of developing adverse effects to MPH above and beyond that seen in children with ADHD alone (e.g., Research Units on Pediatric Psychopharmacology Autism Network, 2005). There is also a concern that the intellectual impairment coupled with disorder-specific psychiatric or motor anomalies may produce more variable and inefficient responses to MPH than for children with ADHD symptoms in the general school population; hence, studies are relatively scarce. Of note, however, are the findings from Pearson and her colleagues, who assessed the effects of MPH on the behavioral and cognitive performance of children with familial intellectual impairment and ADHD using a placebo-controlled, double-blind, crossover treatment trial. Findings indicated that MPH can help improve inattention and hyperactivity symptoms (Pearson et al., 2003) alongside improved cognitive outcomes on measures that tap selection, maintenance, and control of attention (Pearson et al., 2004). In the section below, we highlight the current findings from studies that have assessed the short-term efficacy of stimulant medications in our targeted disorders: autism, fragile X syndrome, Down syndrome, Williams syndrome, and 22q11 DS.

Autism

There is now an emerging literature that demonstrates that stimulant medication may improve ADHD symptoms in a subsample of children with autism, namely those with high-functioning autism and comorbid ADHD. In their retrospective and prospective study of children and adolescents that were diagnosed as having ADHD alone versus autism and ADHD, Santosh, Baird, Pityaratstian, Tavare, and Gringras (2006) found that ADHD symptoms did significantly improve with stimulant medication in both groups but with a trend toward a worsening of obsessionality in the combined autism and ADHD group, possibly as a side-effect of medication. A second retrospective study also reported a favorable response rate to psychostimulants in reducing target ADHD symptoms in a population-based study of children with autism (Nickels et al., 2008). Although both studies suggest a positive effect of stimulant medication in reducing symptoms of inattentiveness and hyperactivity, they are both limited by their study design, and as yet (at least to our knowledge) there is no definitive "gold standard" clinical trial that has compared children and adolescents with autism and comorbid ADHD on and off medication, and none that has looked at the efficacy of stimulants to reduce cognitive inattention in autism.

If we broaden the autism spectrum to include pervasive developmental disorders, then research indicates that MPH in particular may be a useful therapy in reducing symptoms of hyperactivity and impulsivity (Di Martino, Melis, Cianchetti, & Zuddas, 2004). See in particular the findings from one of the few clinical efficacy trials of MPH in pervasive developmental disorders (Research Units in Pediatric Psychopharmacology Autism Network, 2005).

Fragile X Syndrome

The current paucity of efficacy studies of MPH or other stimulant medication in fragile X syndrome is surprising given the cited frequency of stimulant use in treating ADHD symptoms in this disorder (see reviews by Berry-Kravis & Potanos, 2004; Tsiouris & Brown, 2004; and the empirical investigation by Sullivan et al., 2006). An early study by Hagerman and her colleagues remains one of the only studies to date to assess the effectiveness of MPH in children with fragile X using a placebo, double-blind crossover trial (Hagerman, Murphy, & Wittenberger, 1988). Findings indicated that MPH was tolerated by over two-thirds of participants, with beneficial effects noted on ratings of inattentive behaviors and social skills. There were, however, no significant improvements in ability to delay a response or on performance on a Go–NoGo task (the duration of this task is not reported, and the only measure presented is the number of correct responses to the target).

Given the possible side-effect of increased mood lability and irritability with stimulant medication in ADHD (Hagerman et al., 1988), alternative pharmacological treatments for reducing ADHD symptoms in fragile X have recently been published. For example, Torrioli et al. (2008) used a double-blind parallel study design to assess the effect of L-acetylcarnitine (LAC; a nonstimulant agent) versus placebo in children with fragile X and comorbid ADHD aged between 6 and 13 years (mean age 9.18 years). All children were assessed at baseline and after 1, 6 and 12 months of treatment. These initial findings demonstrated that LAC did significantly reduce hyperactive behavior and increase attention in boys with fragile X across all time points. However, it is noteworthy that those fragile X boys on placebo also showed reduced hyperactivity during the course of the study but not to a level comparable to the LAC-treated boys. The authors suggest that the enriched environment (e.g., extra stimulation and support strategies) provided by the study treatment (whether on medication or placebo) probably contributed to the decrease in behavioral symptoms across both groups. The findings are extremely interesting but need to be further replicated in future studies, perhaps using a more stringent clinical format similar to the "gold standard" used in the MTA ADHD studies. Unfortunately, because Torrioli et al. did not assess performance at the cognitive level, it is unknown whether LAC would help in reducing deficits in sustained attention and inhibitory functioning, core weaknesses in fragile X. However, the reader may be interested in the findings from a recent innovative study by Berry-Kravis, Sumis, Kim, Lara, and Wuu (2008), who have identified a potential battery of cognitive measures (e.g., commission scores on the CPT and performance on the Nepsy Tower task) that might serve as useful, short-term outcome measures in future clinical trials in fragile X. See also Valdovinos (2007) for a general review of the pharmacological interventions currently used to address the range of behavioral problems in fragile X.

Down Syndrome

At the time of writing, we were unable to locate any published studies that have assessed pharmacological medication in reducing ADHD symptoms in children or adolescents with Down syndrome.

Williams Syndrome

There is currently limited information regarding the efficacy or tolerability of stimulant medication or any pharmacological treatments in reducing ADHD symptoms in Williams syndrome. Two early studies represent the core research findings to date, both published in 1997. In the first study, Bawden, MacDonald, and Shea (1997) conducted a series of four double-blind placebo-controlled case studies of children and adolescents with Williams syndrome (ages 9.1, 10.2, 10.10, and 13.3 years). Findings revealed that 2 out of the 4 participants responded well to a single dose of MPH, with ratings indicating a reduction in inattentive and hyperactive behaviors during the weeks on medication compared to the weeks on placebo. One participant, in particular, demonstrated a sharp reduction in the number of errors of omission and commission on performance on a standard CPT when in the MPH treatment phase. In the second study, Power, Blum, Jones, and Kaplan (1997) report findings from two double-blind, placebo-controlled case studies of children with Williams syndrome aged 7 and 8 years. Both children were administered two dosages of MPH (5 mg and 10 mg) and appeared to respond well to MPH, with ratings showing a reduction in inattentive and hyperactive behaviors during the treatment phase compared to placebo. Performance on a simple vigilance task was also at its highest level during the MPH phases. In both cases, the higher dosage (10 mg) seemed to be more effective at reducing symptoms of ADHD.

22q11 Deletion Syndrome

There are currently no available studies that have assessed the efficacy of psychopharmacology for reducing ADHD symptoms in children or adolescents with 22 q11 DS, although studies indicate that a percentage of individuals do use stimulant medication (e.g., Aneja et al., 2007). This lack of efficacy data is surprising given the emerging phenotype of 22q11 DS and comorbid ADHD (see Antshel, Faraone et al., 2007). Perhaps one reason is that studies tend to report a more inattentive subtype in 22q11 DS that may not respond especially well to stimulants such as MPH. Nonetheless, given the increased number of studies focusing on ADHD in this disorder, it is timely to conduct clinical efficacy studies to ascertain the tolerability and safety of pharmacological interventions in children and adolescents with 22q11 DS and ADHD.

Together, the findings across our five neurodevelopmental disorders suggest an overall lack of clinical research targeted at ascertaining the efficacy of psychopharmacological approaches to reducing ADHD symptoms. We have already highlighted some of the possible reasons for this, not least of which is a concern over elevated adverse side-effects associated with stimulant medication in children with intellectual impairment compared to children with ADHD alone. One also needs to take into account the disorder-specific psychiatric and behavioral profile and the potential interaction of stimulants that may exacerbate other known psychiatric effects such as sterotypies in autism and mood disorder in fragile X. All these are undoubtedly justifiable reasons for exercising caution in providing pharmacological interventions for ADHD symptoms. However, our review of the current literature reveals that many children and adolescents, including those with autism, fragile X, and 22q11 DS, already use stimulant medication. The point is that there is currently an insufficient number of efficacy studies that have assessed the tolerability, safety, and effectiveness of these medications across different disorders. Given the relatively large numbers of children using stimulants, research to address these current gaps in knowledge should be a priority.

Psychosocial Treatments

The past decade has provided a wealth of new information on the cognitive and behavioral profiles that characterize differing neurodevelopmental disorders. We have outlined in this book the current published signatures as they relate to the domain of attention. As the reader will be aware, the literature is often substantial, especially in well-documented disorders such as autism and fragile X. It is therefore disappointing to see virtually no studies that have focused on the efficacy of psychosocial treatment approaches to reducing the effects of inattention and hyperactivity in atypical populations other than in children and adolescents with ADHD alone. Given that there is now clear consensus that intellectual impairment does not necessarily imply global cognitive delay, alongside the reluctance of many professionals to rely solely on medication to alleviate ADHD symptoms, it is critical that research evaluates the effectiveness of behavior modification approaches that target disorder-specific profiles both in terms of short- and long-term outcomes. A key lesson from the ADHD literature is that the earlier the intervention, the more sustainable the improvement. We are only at the very beginning of this development in other disorders of attention.

Autism

There is now an emerging literature focused on behavioral modification programs that tap the core symptoms of the autism triad using random-ized and controlled designs (for an example, see McConachie & Diggle, 2007, for a systematic review of parent-implemented early intervention programs). As yet, this is still an evolving field with no substantive evi-dence to indicate whether successful or positive early interventions in young children with autism lead to greater long-term outcomes in later childhood or adulthood. In their recent review of comprehensive treat-ment approaches, Rogers and Vismara (2008) identify some factors that account for the wide variations in response to treatment, notably differ-ing research designs across studies, IQ, age at first treatment, duration of treatment, and family characteristics. To date, no one intervention study has focused on reducing the impact of ADHD symptoms in children with autism, but a number have reported an improvement in maladaptive behaviors that are likely to include inattentive behaviors. See Francis (2005), and Humphrey and Parkinson (2006), for critical analyses of cur-rent autism intervention methods.

Fragile X Syndrome

To date no studies have evaluated any psychosocial treatment approaches to reducing ADHD symptoms in children or adolescents with fragile X. However, there is now a considerable body of research that can inform future practices and treatment strategies. For example, parental stress and anxiety are unusually high in mothers of children with fragile X (it is currently unknown, although we suspect, that this high prevalence may be associated with the "carrier status" of many mothers) and have been shown to impact on the number and quality of parent–child interactions (see Wheeler, Hatton, Reichardt, & Bailey, 2007). Reducing stress levels by providing behavioral techniques that lead to reductions in problem behaviors, for example, those presented by severe inattention and hyper-activity, may reduce parental stress and thereby further benefit the child in a virtuous cycle of improvement.

Down Syndrome, Williams Syndrome, and 22q11 DS

As in the case of fragile X, there are currently no published studies that have assessed the efficacy of psychosocial treatments for reducing ADHD symptoms in children or adolescents with either Down syndrome, Williams syndrome, or 22q11 DS, although all three disorders present with high levels of inattention and hyperactive behaviors (see Chapter 9). However, the absence of any available data may well reflect the fact that knowledge about the ADHD profiles of these groups has only recently become available. Future studies need to take on board recent research updates about disorder-specific behavioral and cognitive profiles in order to identify avenues for intervention programs that might reduce the behavioral and cognitive manifestations of these disorders. Factors such as IQ, duration of treatment, other comorbidities in addition to ADHD, and family factors (such as parental stress) are likely to be critical mod-erator variables in treatment success.

Conclusions on Other Neurodevelopmental Disorders

The relative paucity of empirically based investigations of treatment effi-cacy in neurodevelopmental disorders that present with ADHD symp-toms at the behavioral and cognitive levels is striking, both in terms of psychopharmacology and psychosocial interventions. One potential avenue may be to *adapt* rather than duplicate intervention strategies from the ADHD literature, in order to take into account different levels of cognitive functioning and disorder-specific profiles. The wealth of data from the intervention studies with ADHD can then be used as a basis for programs appropriate to the strengths and weaknesses of each atypical population. In particular, we hope that much more intensive research will be conducted on behavioral intervention methods that explore the possibilities for these populations to develop their full potential and miti-gate the limitations on their abilities. Our brief overview of the current studies illustrates that the road ahead is not smooth, with many of the same difficulties encountered in the ADHD literature (e.g., inconsistency in research designs, small sample sizes) arising in studies of other neuro-developmental disorders.

Overall Conclusions

To date, the research on methods of treating attentional problems other than in ADHD is sparse and imprecise. We have stressed throughout this book that behavioral manifestations of inattention, impulsive behavior, and hyperactivity may conceal different patterns of underlying cognitive impairment. Consequently, when assessing the effect of any treatment, it is not enough to demonstrate a reduction in behavioral manifestations, but for adequate understanding of the effectiveness and applicability of the treatment we need to know which, if any, cognitive functions are affected. Some information of this type is available in the case of ADHD, though it is far from comprehensive and systematic, and virtually no evidence is available for whether a given improvement is specific to a particular weakness. Furthermore, such evidence is almost completely lacking in the case of the other disorders that we have considered.

In these latter cases, such information will be particularly important for psychosocial and educationally focused interventions. In view of our analysis of the different patterns of cognitive deficit that characterize different disorders, it is unlikely that the same treatment will be equally efficacious in all cases, and simply employing similar methods to those that have shown some results in the case of ADHD will be inadequate. In the case of treatments designed to improve educational performance, it will become vital to devise appropriate methods of targeting the specific deficits of each neurodevelopmental disorder, rather than to assume that more general approaches will suit all comers. For example, fragile X syndrome requires procedures specifically designed to improve response inhibition, while Williams syndrome should benefit by training in visual discrimination. Measures of the success of different methods clearly need not only to target general improvements in behavior or even in overall cognitive competence, but also to identify precisely which functions have been affected and which have not.

CHAPTER SUMMARY AND KEY FINDINGS

- There is now a considerable literature examining the positive impact of stimulants, namely MPH, and psychosocial treatments on reducing ADHD symptoms.
- Comparatively less research has focused on possible variations in the impact of these various approaches across developmental time from early childhood through to early adulthood, so trajectories of benefits and costs are as yet unknown.
- The extent to which inattentive behaviors and cognitive inattention benefit from similar treatment approaches in ADHD and other neurodevelopmental disorders remains a relatively unexplored area that is in urgent need of research innovation.

11

Conclusions and Pause for Thought

CHAPTER SUMMARY AND KEY FINDINGS
• Lessons and challenges in tracing typical and atypical *attention signatures*
• Can commonalities in inattentive behaviors across different neuro-developmental disorders arise from a common etiology?
• How important is development when predicting disorder-specific attention profiles?

Throughout this book our core goal has been to provide the reader with a current snapshot of the exciting and often innovative developments that have advanced our knowledge of the typical and atypical trajectories of attention at the brain, behavioral, and cognitive levels. Exponential advances across the disciplines of molecular genetics, neurobiology, brain imaging, and cognitive developmental neuroscience continue to move this field forward by creating a rich, fertile ground in which to tease apart, layer by layer, gene–brain–cognitive correlates. We still have much to learn. Indeed, one could argue, and we have, that a main contribution of research efforts over the past decade has been to demonstrate the implausibility of any assumption that postulates a simple one-to-one mapping of genes to different aspects of cognition, in our case, attention. When we add the role of development itself in shaping signature end states, then the reader can no doubt appreciate not only the complexity but also the excitement that comes with this challenge.

We summarize below our conclusions from the preceding chapters and then provide a review of the core strengths and weaknesses of this field to date. We focus on typical and atypical development, high-lighting findings from six neurodevelopmental disorders, all of which at first glance present with severe inattention behaviors: ADHD, autism, fragile X syndrome, Down syndrome, Williams syndrome, and 22q11 DS. We end our journey with consideration of some avenues for future research development that will hopefully bridge the current gaps in knowledge by providing much needed information to guide clinical treatments and educational interventions.

In Chapter 2 we set the scene with a description of how some attentional functions are exogenously controlled (automatic), such as alerting and orienting evoked by sudden-onset stimuli. In most everyday, complex situations, however, attention involves the operation of endogenous control systems: for example, excluding distraction (selection), maintaining focus and/or arousal (maintenance), and a number of executive functions such as setting up plans (from simple rules about stimulus–response linkages to complex strategies) and controlling behavioral sequences. Endogenously controlled plans can become automatic with practice. In our discussions, therefore, we consider three main aspects of attention as selection, maintenance, and control. Selection is basic to all attention, whether we refer to selection of target stimuli or selection of a specific course of action. Selection involves a pattern of excitation and inhibition of possible alternatives. Thus, in endogenously controlled attention, selection of a particular input or a specific course of action will always require the action of executive functions both to set up the appropriate pattern of brain activity and to control the required sequence of attentional switches and responses. We see attention, therefore, as a complex hierarchical system which coordinates the functioning of a complex of simpler processes to achieve effective behavior.

In Chapter 3 we focused on the rapid advances in our understanding of relations between genes and aspects of attentive behavior. This field of "cognitive genetics" is only emerging, but it is clear that simple relations between genes and specific attention functions rarely exist. Instead, such relations are rather complex and modulated by a variety of other factors that interact with multiple genes as well as environmental influences. Across all our neurodevelopmental disorders, the genetic factors influencing performance appear to have disparate origins, and as yet there is no clear consensus on specific "attention genes" and their trajectory across childhood and adolescence.

In Chapter 4 we focused on the major advances in brain imaging technologies and their potential to elucidate attention functions and pathways in both typically and atypically developing populations. Our knowledge of the functions of different areas of the brain is expanding rapidly with the variety of methods now available for viewing the active brain in vivo. While some attentional functions can be ascribed to particular areas of the brain, as exemplified in Posner's model of posterior (orienting), anterior (executive control), and arousal (alerting) systems, it is increasingly apparent that in any task, activity in several neural circuits is integrated to achieve a desired outcome. Traditional ideas of a modular system with different areas preset to carry out specified operations are giving way to a dynamic view of brain development in which organization is the result of competition and cooperation between different centers as they are bombarded with stimuli from the environment.

In Chapter 5 we introduced the reader to the wide range of attention rating scales and checklists that can serve as useful global measures of individual differences in inattentive behaviors. A main strength of these measures is their sensitivity in discriminating children with a diagnosis of ADHD alone from typically developing children. Conversely, their main weakness is that they appear to have limited utility in discriminating specific cognitive components of the attention weakness that are involved in atypical cognitive development. There is also a lack of normative data on age-related trajectories, especially for specific neurodevelopmental disorders.

In Chapter 6 we introduced the reader to the huge variety of cognitive tasks that claim to tap different attention functions and EFs, but many of these incorporate weaknesses or are unspecific in their demands. Few tasks have standardized data on disorders outside of ADHD yet are consistently used and reported upon across a wide range of neurodevelopmental disorders. A brief excursion through this vast literature will reveal how problematic these task measures can be in terms of producing floor and ceiling effects, when used with atypically developing populations. There is an urgent need to devise better tasks, or to use a range of tasks, in order to extract underlying attentional factors.

In Chapter 7 we began by identifying that the core change during development is from exogenous to endogenous control of attention. This involves the development of a range of executive functions, but the details of the developmental trajectories of these functions and differences between them are still a matter of intense debate. This rich data source, however, includes some stunningly innovative work, the findings of which can inform the design of similar studies in atypical development. The work of Adele Diamond and her colleagues (cited throughout this chapter) is one example of research excellence in this field.

In Chapter 8 we began our exploration of atypical attention development by examining the current theoretical explanations for ADHD. Here we demonstrated that no one current theory is entirely satisfactory. The most plausible theory is that ADHD is due to weak control systems (EFs) and in particular to weak inhibitory systems. However, performance on tasks that do not obviously involve inhibition is also impaired, suggesting that wider control functions are involved. Theories need to be refined to generate testable predictions. There is also a lack of longitudinal studies that trace the developmental trajectory of the disorder and its variants; there are some intriguing suggestions that this may be quite complex. Unfortunately, as yet there are no adequate explanations as to why attention weaknesses and hyperactivity, which have different cognitive correlates, co-occur so frequently. If these are indeed independent factors as is suggested by differences in their relation to a number of other variables, each should occur more frequently than their combination, but the reverse is the case, as in the classic ADHD syndrome. This puzzle is usually ignored, but it may be crucial to a full understanding of the condition.

In Chapter 9 we identified specific signatures that researchers are unraveling for the attentional weaknesses associated with some neuro-developmental disorders (fragile X and Williams syndrome) while others (Down syndrome) show more general problems. The picture is less clear for autism and 22q11 syndrome, but this may in part be due to sample composition and in particular to the variability in functioning levels that comprise different studies. Detecting sources of within-syndrome variations that may in some cases be related to different genetic variations is an important next step in understanding the pervasive inconsistencies across studies. A core goal of the next generation of research is to develop more sensitive experimental paradigms that can be used to test children and adolescents over a wide age and ability range and to enable disorder-specific differences and developmental trajectories to be explored.

In Chapter 10 we examined the application of these new research discoveries to the clinical and educational domains. There was a narrow range of research available, with the majority of all efficacy studies directed toward ADHD treatments and interventions. Although this finding is hardly surprising given the volume of research generated as a whole in the ADHD field, compared to other neurodevelopmental disorders, it does identify a gap that needs to be bridged. More and more children with developmental disorders are being prescribed medication to alleviate inattention symptoms, and therefore, clinical trials are urgently needed in order to evaluate the effectiveness and appropriateness of these treatment approaches. Perhaps one of the most disappointing findings was the paucity of current treatment studies that incorporate psychosocial approaches to reducing attention symptoms in disorders outside of ADHD. The continuing lack of treatment options needs to be urgently addressed in future research programs.

The Strengths

The strengths of findings from current research are plentiful, and we are seeing the dawn of a new age of collaborative research that will ultimately link specific clusters of genes to cognitive processes including attention and executive functions. The abundance of research in the field, both in terms of a well-established typical development literature and the emerging phenotypic literature bears testament to the importance of this work to the field of developmental cognitive neuroscience. We have seen how recent developments in genetics and neuroimaging techniques facilitate and enhance our understanding of attention at the behavioral and cognitive levels. These strengths are being accompanied by developments in experimental paradigms designed to test specific cognitive functions that move away from a reliance on traditional, standardized measures that in the past may have masked important but subtle differences in cognitive abilities across disorders. Instead, a new generation of attention paradigms has attempted to tease apart processing differences in terms of selection, maintenance, and control of attention, and to uncover the critical weaknesses that underlie overall cognitive performance.

The Weaknesses

Unfortunately, as with any relatively new research endeavor there are going to be inherent weaknesses that need to be acknowledged and addressed in order to advance the field. Indeed, there is no perfect research program and we are not flawless researchers. In both the typical and atypical literature, confounding factors are invariably related to methodology and design. Problems include small sample sizes, lack of appropriate comparison samples, and failure to include core demographic variables such as IQ, gender, chronological age, developmental level, ethnicity, and so forth. Reliance on standardized measures that tap more generalized cognitive abilities, rather than experimental paradigms that tap specific cognitive substrates, can make interpretation of findings difficult especially in children with atypical development. It is therefore critical that studies give careful consideration to all possible confounding or moderating variables that might influence the integrity of the findings and any subsequent interpretations.

The Critical Role of Development

One of the most persistent weaknesses in this field is the lack of a developmental framework to chart trajectories of attention functioning from infancy through to early adulthood. When age-related changes have been pursued, almost all studies to date have utilized cross-sectional designs, and although these findings provide important data, they are not truly developmental in nature and therefore cannot fully capture the subtle profiles that may distinguish attention pathways across different neurodevelopmental disorders. What is needed is a detailed profile over time in order to get the full picture of how the end state is reached, and preferably from the same group of children because development is not strongly preprogrammed and individual paths may differ. Thus, to assume that the adult end state had its beginnings in early development is likely to be erroneous. The theoretical and empirical work of Annette Karmiloff-Smith and her colleagues is an example of the pioneering research currently being undertaken in this field.

Final Thoughts and Future Avenues of Research

The huge strides in our understanding of the interplay between attention, genes, and developmental disorders were the driving force behind the writing of this book. We have described numerous examples of the innovative and seminal work that has shaped and will continue to shape this field over the next decade. Although the complexity of this relationship is acknowledged and new technologies have allowed us to explore hitherto uncharted territory that allows glimpses, if sometimes fleeting, of the combinations of gene–brain–behavior relationships, there still remain many challenges and unexplored issues.

We highlight below some of our thoughts on the most urgent future research directions.

First, one major goal of future research and analysis will be to address the problem of the imprecise and variable definitions of "attention" that pervade the literature. In line with current thinking, we have stressed the complexity of this concept and suggested that the most plausible approach is to regard attention as a complex hierarchy of processes, with control systems that organize subsidiary mechanisms in order to achieve current goals. However, the structure and components of this hierarchy need to be defined more precisely, particularly in regard to the different executive processes that select and link more basic components into a successful sequence of behavior.

This problem is particularly apparent in analysis of the wide range of tasks that have been claimed to test various aspects of attention. Too often a task is assumed to test, for example, selective attention, ignoring the complexity of the full range of functions required to carry it out (working memory, inhibition at both input and output, possibly switching, etc.). Such imprecision has implications for the interpretation of both performance and brain imaging data. Furthermore, the use of single tasks to measure a single hypothesized process should be largely abandoned in favor of extracting key underlying functions through factor analytic processes employing several tasks and measures.

Second, only with the development of superior investigatory tools will it become possible to develop precise criteria for ADHD, currently an amorphous syndrome that almost certainly covers several different sources of weakness. At present we rely entirely on behavioral criteria for discriminating different subcategories of this disorder, and, as we have frequently noted, similar behavior does not necessarily imply identity of underlying cognitive causes.

A third and related goal should be the design of better longitudinal tools that specifically enable tracking of the development of different critical processes throughout childhood, particularly in groups differing widely in overall ability. The current overreliance on cross-sectional designs needs to be complemented with approaches that can also capture subtle changes across discrete developmental time periods.

Fourthly, complex models will need to be developed for the relations between genes and cognitive processes, but these models will also have to include the role of development in producing the phenotypic outcomes we associate with a given disorder. Clearly, such models will also have to take into account the interactive effects of different genes, rather than simply the impact of single genes.

Fifthly, research needs to address how our increasing understanding of attentional processes can be applied more systematically to the design of therapeutic and remedial programs. It is essential that all treatment options undergo efficacy evaluation, but this gap is particularly evident in developmental disorders outside of ADHD. One avenue would be to transfer some of the important discoveries made in the ADHD field, notably in treatment evaluation of psychosocial interventions, and adapt them to atypical populations, taking into account individual differences in intellectual functioning and attention severity.

We do not expect these future research directions to be an easy task for the new generation of research scientists that will embark on such endeavors. Recording changes in human development across its complex pathways is not for the fainthearted, but the anticipated rewards and insights that will accrue from such an exploration will, we believe, be worth the journey.

CHAPTER SUMMARY AND KEY FINDINGS

- Commonalities in inattentive behaviors across differing neurodevelopmental disorders do not infer a common etiology or similar cognitive pathways.
- Close examination across multiple levels of analyses, genetic, brain, behavioral and cognitive, indicate disorder-specific start states with different attention trajectories throughout development.
- Development plays a crucial role in determining attention *signatures*, and so it is vital that research begins in infancy and journeys through childhood, adolescence, and into adulthood.
- Educational and clinical interventions need to recognize disorder-specific attention impairments and not to assume that strategies that work well in children with ADHD will have comparable effects in children whose attention problems may well arise from very different underlying genetic and neural mechanisms.

References

Abbeduto, L., Brady, N., & Kover, S. T. (2007). Language development and fragile X syndrome: Profiles, syndrome-specificity, and within-syndrome differences. *Mental Retardation and Developmental Disabilities Research Reviews, 13*(1), 36–46.

Abbeduto, L., Murphy, M. M., Cawthon, S. W., Richmond, E. K., Weissman, M. D., Karadottir, S., et al. (2003). Receptive language skills of adolescents and young adults with Down or fragile X syndrome. *American Journal on Mental Retardation, 108*(3), 149–160.

Abbeduto, L., Warren, S. F., & Conners, F. A. (2007). Language development in Down syndrome: From the prelinguistic period to the acquisition of literacy. *Mental Retardation and Developmental Disabilities Research Reviews, 13*(3), 247–261.

Abi-Dargham, A., Mawlawi, O., Lombardo, I., Gil, R., Martinez, D., Huang, Y.Y., et al. (2002). Prefrontal dopamine D-1 receptors and working memory in schizophrenia. *The Journal of Neuroscience, 22*(9), 3708–3719.

Abikoff, H. B., Vitiello, B., Riddle, M. A., Cunningham, C., Greenhill, L. L., Swanson, J. M., et al. (2007). Methylphenidate effects on functional outcomes in the Preschoolers with Attention-Deficit/Hyperactivity Disorder Treatment Study (PATS). *Journal of Child and Adolescent Psychopharmacology, 17*(5), 581–592.

Abrahams, B. S., & Geschwind, D. H. (2008). Advances in autism genetics: On the threshold of a new neurobiology. *Nature Reviews Genetics, 9*(5), 341–355.

Achenbach, T. M. (1991). *Manual for the Child Behavior Checklist/4-18 and 1991 Profile.* Burlington: University of Vermont, Department of Psychiatry.

Achenbach, T. M., & Rescorla, L. A. (2000). *Manual for the ASEBA Preschool Forms & Profiles.* Burlington, VT: University of Vermont, Research Center for Children, Youth, and Families.

Adler, S. A., & Orprecio, S. A. (2006). The eyes have it: Visual pop-out in infants and adults. *Developmental Science, 9*(2), 189–206.

Akshoomoff, N. (2002). Selective attention and active engagement in young children. *Developmental Neuropsychology, 22*(3), 625–642.

Alderson, R. M., Rapport, M. D., Sarver, D. E., & Kofler, M. J. (2008). ADHD and behavioral inhibition: A re-examination of the Stop-signal Task. *Journal of Abnormal Child Psychology, 36*(7), 989–998.

Allport, D. A. (1993). Attention and control: Have we been asking the wrong questions? A critical review of twenty-five years. In D. E. Meyer & S. Kornblum (Eds.), *Attention and performance XIV* (pp. 631–682). Cambridge, MA: MIT Press.

Allport, D. A., Antonis, B., & Reynolds, P. (1972). On the division of attention: A disproof of the single channel hypothesis. *Quarterly Journal of Experimental Psychology, 24*(2), 225–235.

Altafaj, X., Dierssen, M., Baamonde, C., Marti, E., Visa, J., Guimera, J., et al. (2001). Neurodevelopmental delay, motor abnormalities and cognitive deficits in transgenic mice overexpressing Dyrk1A (minibrain), a murine model of Down's syndrome. *Human Molecular Genetics, 10*(18), 1915–1923.

Altafaj, X., Ortiz-Abalia, J., Fernández, M., Potier, M.C., Laffaire, J., Andreu, N., et al (2008). Increased NR2A expression and prolonged decay of NMDA-induced calcium transient in cerebellum of TgDyrk1A mice, a mouse model of Down syndrome. *Neurobiology of Disease, 32*(3), 377–84.

Altshuler, D., Daly, M. J., & Lander, E. S. (2008). Genetic mapping in human disease. *Science, 322*(5903), 881–888.

Amano, S., Kezuka, E., & Yamamoto, A. (2004). Infant shifting attention from an adult's face to an adult's hand: A precursor of joint attention. *Infant Behavior and Development, 27*(1), 64–80.

American Psychiatric Association. (2000). *Diagnostic and statistical manual of mental disorders, 4th edition.* Washington, DC: American Psychiatric Association.

Amso, D., & Johnson, S. P. (2005). Selection and inhibition in infancy: Evidence from the spatial negative priming paradigm. *Cognition, 95*(2), B27–B36.

Amso, D., & Johnson, S. P. (2006). Learning by selection: Visual search and object perception in young infants. *Developmental Psychology, 42*(6), 1236–1245.

Anderson, V. A., Anderson, P., Northam, E., Jacobs, R., & Catroppa, C. (2001). Development of executive functions through late childhood and adolescence in an Australian sample. *Developmental Neuropsychology, 20*(1), 385–406.

Anderson, G. M., Gutknecht, L., Cohen, D. J., Brailly-Tabard, S., Cohen, J. H. M., Ferrari, P., et al. (2002). Serotonin transporter promoter variants in autism: Functional effects and relationship to platelet hyperserotonemia. *Molecular Psychiatry, 7*(8), 831–836.

Aneja, A., Fremont, W. P., Antshel, K. M., Faraone, S. V., AbdulSabur, N., Higgins, A. M., et al. (2007). Manic symptoms and behavioral dysregulation in youth with velocardiofacial syndrome (22q11.2 deletion syndrome). *Journal of Child and Adolescent Psychopharmacology, 17*(1), 105–114.

Anllo-Vento, L., Schoenfeld, M. A., & Hillyard, S. (2004). Cortical mechanisms of visual attention: Electrophysiological and neuroimaging studies. In M. I. Posner (Ed.), *Cognitive neuroscience of attention* (pp. 180–193). New York: Guilford Press.

Ansari, D., Donlan, C., & Karmiloff-Smith, A. (2007). Typical and atypical development of visual estimation abilities. *Cortex, 43*(6), 758–768.

Antshel, K. M., Aneja, A., Strunge, L., Peebles, J., Fremont, W. P., Stallone, K., et al. (2007). Autistic spectrum disorders in velo-cardio facial syndrome (22q11.2 deletion). *Journal of Autism and Developmental Disorders, 37*(9), 1776–1786.

Antshel, K. M., & Barkley, R. (2008). Psychosocial interventions in attention deficit hyperactivity disorder. *Child & Adolescent Psychiatric Clinics of North America, 17*(2), 421–437.

Antshel, K. M., Faraone, S. V., Fremont, W., Monuteaux, M. C., Kates, W. R., Doyle, A., et al. (2007). Comparing ADHD in velocardiofacial syndrome to idiopathic ADHD: A preliminary study. *Journal of Attention Disorders, 11*(1), 64–73.

Antshel, K. M., Fremont, W., & Kates, W. R. (2008). The neurocognitive phenotype in velo-cardio-facial syndrome: A developmental perspective. *Developmental Disabilities Research Reviews, 14*(1), 43–51.

Archibald, S. J., & Kerns, K. A. (1999). Identification and description of new tests of executive functioning in children. *Child Neuropsychology, 5*, 115–129.

Asato, M. R., Sweeney, J. A., & Luna, B. (2006). Cognitive processes in the development of TOL performance. *Neuropsychologia, 44*(12), 2259–2269.

Ashley-Koch, A. E., Jaworski, J., Ma, D. Q., Mei, H., Ritchie, M. D., Skaar, D. A., et al. (2007). Investigation of potential gene-gene interactions between APOE and RELN contributing to autism risk. *Psychiatric Genetics, 17*(4), 221–226.

Atkinson, J., Braddick, O., Anker, S., Curran, W., Andrew, R., Wattam-Bell, J., et al. (2003). Neurobiological models of visuospatial cognition in children with Williams syndrome: Measures of dorsal-stream and frontal function. *Developmental Neuropsychology, 23*(1–2), 139–172.

Atkinson, J., Braddick, O., Rose, F. E., Searcy, Y. M., Wattam-Bell, J., & Bellugi, U. (2006). Dorsal-stream motion processing deficits persist into adulthood in Williams syndrome. *Neuropsychologia, 44*(5), 828–833.

Aydin, M., Kabakus, N., Balci, T. A., & Ayar, A. (2007). Correlative study of the cognitive impairment, regional cerebral blood flow, and electroencephalogram abnormalities in children with Down's syndrome. *International Journal of Neuroscience, 117*(3), 327–336.

Aziz, M., Stathopulu, E., Callias, M., Taylor, C., Turk, J., Oostra, B., et al. (2003). Clinical features of boys with fragile X premutations and intermediate alleles. *American Journal of Medical Genetics Part B: Neuropsychiatric Genetics, 121B*(1), 119–127.

Azuma, R., Prinz, W., & Koch, I. (2004). Dual task slowing and the effects of cross-task compatibility. *Quarterly Journal of Experimental Psychology, Section A: Human Experimental Psychology, 57*, 693–713.

Baddeley, A. (2000). The episodic buffer: A new component of working memory? *Trends in Cognitive Sciences, 4*(11), 417–423.

Baddeley, A. D. (1986). *Working memory*. Oxford, UK: Oxford University Press.

Baddeley, A. D. (1993). Working memory or working attention? In A. D. Baddeley & L. Weiskrantz (Eds.), *Attention: Selection, awareness and control. A tribute to Donald Broadbent* (pp. 152–170). Oxford, UK: Oxford University Press.

Baddeley, A. D. (2001). Is working memory still working? *The American Psychologist, 56*(11), 851–864.

Baeyens, D., Roeyers, H., & Walle, J. V. (2006). Subtypes of Attention-Deficit/ Hyperactivity disorder (ADHD): Distinct or related disorders across measurement levels? *Child Psychiatry & Human Development, 36*(4), 403–417.

Bailey, D. B., Hatton, D. D., Skinner, M., & Mesibov, G. (2001). Autistic behavior, FMR1 protein, and developmental trajectories in young males with fragile X syndrome. *Journal of Autism and Developmental Disorders, 31*(2), 165–174.

Bailey, D. B., Raspa, M., Olmsted, M., & Holiday, D. B. (2008). Co-occurring conditions associated with FMR1 gene variations: Findings from a national parent survey. *American Journal of Medical Genetics Part A, 146A*(16), 2060–2069.

Bakker, S. C., van der Meulen, E. M., Oteman, N., Schelleman, H., Pearson, P. L., Buitelaar, J. K., et al. (2005). DAT1, DRD4, and DRD5 polymorphisms are not associated with ADHD in Dutch families. *American Journal of Medical Genetics Part B: Neuropsychiatric Genetics, 132B*(1), 50–52.

Banach, R., Thompson, A., Szatmari, P., Goldberg, J., Tuff, L., Zwaigenbaum, L., et al. (2008). Brief Report: Relationship between non-verbal IQ and gender in autism. *Autism Journal & Developmental Disorders, 39*(1), 188–193.

Banaschewski, T., & Brandeis, D. (2007). Annotation: What electrical brain activity tells us about brain function that other techniques cannot tell us—A child psychiatric perspective. *Journal of Child Psychology and Psychiatry, 48*(5), 415–435.

Band, G. P. H., van der Molen, M. W., & Logan, G. D. (2003). Horse-race model simulations studies of the stop signal procedure. *Acta Psychologica, 112*(2), 105–142.

Banerjee, T. D., Middleton, F., & Faraone, S. V. (2007). Environmental risk factors for attention-deficit hyperactivity disorder. *Acta Paediatrica, 96*(9), 1269–1274.

Baranek, G. T., Roberts, J. E., David, F. J., Sideris, J., Mirrett, P. L., Hatton, D. D., et al. (2008). Developmental trajectories and correlates of sensory processing in young boys with fragile X syndrome. *Physical & Occupational Therapy in Pediatrics, 28*(1), 79–98.

Barch, D., Carter, C., Braver, T., Sabb, F., MacDonald, A. E., Noll, D., et al. (2001). Selective deficits in prefrontal cortex function in medication-naive patients with schizophrenia. *Archives of General Psychiatry, 58*(3), 280–288.

Barkley, R. A. (1987). *Defiant children: A clinician's manual for parent training*. New York: Guilford Press.

Barkley, R. A. (1997). *ADHD and the nature of self-control*. New York: Guilford.

Barkley, R. A. (2004). Adolescents with attention-deficit/hyperactivity disorder: An overview of empirically based treatments. *Journal of Psychiatric Practice, 10*(1), 39–56.

Barkley, R. A., DuPaul, G. J., & McMurray, M. B. (1991). Attention deficit disorder with and without hyperactivity: Clinical response to three dose levels of methylphenidate. *Pediatrics, 87*(4), 519–531.

Barkley, R. A., Edwards, G., Laneri, M., Fletcher, K., & Metevia, L. (2001). The efficacy of problem-solving communication training alone, behavior management training alone, and their combination for parent-adolescent conflict in teenagers with ADHD and ODD. *Journal of Consulting and Clinical Psychology, 69*(6), 926–941.

Barkley, R. A., Guevremont, D. C., Anastopoulos, A. D., & Fletcher, K. E. (1992). A comparison of three family therapy programs for treating family conflicts in adolescents with attention-deficit hyperactivity disorder. *Journal of Consulting & Clinical Psychology, 60*(3), 450–462.

Barnett, J. H., Jones, P. B., Robbins, T. W., & Muller, U. (2007). Effects of the catechol-O-methyltransferase Val(158)Met polymorphism on executive function: A meta-analysis of the Wisconsin Card Sort Test in schizophrenia and healthy controls. *Molecular Psychiatry, 12*(5), 502–509.

Barnett, J. H., Scoriels, L., & Munafò, M. R. (2008). Meta-analysis of the cognitive effects of the catechol-O-methyltransferase gene Val158/108Met polymorphism. *Biological Psychiatry, 64*(2), 137–144.

Barnow, S., Schuckit, M., Smith, T., & Freyberger, H. J. (2006). Predictors of attention problems for the period from pre-teen to early teen years. *Psychopathology, 39*(5), 227–235.

Barr, C. L., Feng, Y., Wigg, K., Bloom, S., Roberts, W., Malone, M., et al. (2000). Identification of DNA variants in the SNAP-25 gene and linkage study of these polymorphisms and attention-deficit hyperactivity disorder. *Molecular Psychiatry, 5*(4), 405–409.

Barrouillet, P., Bernardin, S., & Camos, V. (2004). Time constraints and resource sharing in adults' working memory spans. *Journal of Experimental Psychology: General, 133*(1), 83–100.

Bassell, G. J., & Warren, S. T. (2008). Fragile X syndrome: Loss of local mRNA regulation alters synaptic development and function. *Neuron, 60*(2), 201–214.

Bawden, H. N., MacDonald, G. W., & Shea, S. (1997). Treatment of children with Williams syndrome with methylphenidate. *Journal of Child Neurology, 12*(4), 248–252.

Bearden, C. E., Jawad, A. F., Lynch, D. R., Sokol, S., Kanes, S. J., McDonald-McGinn, D. M., et al. (2004). Effects of a functional COMT polymorphism on prefrontal cognitive function in patients with 22q11.2 deletion syndrome. *American Journal of Psychiatry, 161*(9), 1700–1702.

Bearden, C. E., Woodin, M. F., Wang, P. P., Moss, E., McDonald-McGinn, D., Zackai, E., et al. (2001). The neurocognitive phenotype of the 22q11.2 deletion syndrome: Selective deficit in visual-spatial memory. *Journal of Clinical and Experimental Neuropsychology, 23*(4), 447–464.

Bechara, A., Damasio, A. R., Damasio, H., & Anderson, S. W. (1994). Insensitivity to future consequences following damage to human prefrontal cortex. *Cognition, 50*(1–3), 7–15.

Becker, A., Woerner, W., Hasselhorn, M., Banaschewski, T., & Rothenberger, A. (2004). Validation of the parent and teacher SDQ in a clinical sample. *European Child & Adolescent Psychiatry, 13*(Suppl. 2), II11–II16.

Bedard, A. C., Jain, U., Hogg-Johnson, S., & Tannock, R. (2007). Effects of methylphenidate on working memory components: Influence of measurement. *Journal of Child Psychology and Psychiatry and Allied Disciplines, 48*(9), 872–880.

Bedard, A. C., Nichols, S., Barbosa, J. A., Schachar, R., Logan, G. D., & Tannock, R. (2002). The development of selective inhibitory control across the life span. *Developmental Neuropsychology, 21*(1), 93–111.

Bedard, A. C., & Tannock, R. (2008). Anxiety, methylphenidate response, and working memory in children with ADHD. *Journal of Attention Disorders, 11*(5), 546–557.

Bellgrove, M. A., Domschke, K., Hawi, Z., Kirley, A., Mullins, C., Robertson, I. H., et al. (2005). The methionine allele of the COMT polymorphism impairs prefrontal cognition in children and adolescents with ADHD. *Experimental Brain Research, 163*(3), 352–60.

Bellgrove, M. A., Hawi, Z., Kirley, A., Gill, M., & Robertson, I. H. (2005). Dissecting the attention deficit hyperactivity disorder (ADHD) phenotype: Sustained attention, response variability and spatial attentional asymmetries in relation to dopamine transporter (DAT1) genotype. *Neuropsychologia, 43*(13), 1847–1857.

Bellgrove, M. A., Hawi, Z., Lowe, N., Kirley, A., Robertson, I. H., & Gill, M. (2005). DRD4 gene variants and sustained attention in attention deficit hyperactivity disorder (ADHD): Effects of associated alleles at the VNTR and-521 SNP. *American Journal of Medical Genetics Part B: Neuropsychiatric Genetics, 136B*(1), 81–86.

Bellgrove, M. A., Johnson, K. A., Barry, E., Mulligan, A., Hawi, Z., Gill, M., et al. (2009). Dopaminergic haplotype as a predictor of spatial inattention in children with attention-deficit/hyperactivity disorder. *Archives of General Psychiatry, 66*(10), 1135–42.

Bellgrove, M. A., & Mattingley, J. B. (2008). Molecular genetics of attention. *Molecular and Biophysical Mechanisms of Arousal, Alertness, and Attention, 1129*, 200–212.

Belser, R. C., & Sudhalter, V. (2001). Conversational characteristics of children with fragile X syndrome: Repetitive speech. *American Journal on Mental Retardation, 106*(1), 28–38.

Bennetto, L., & Pennington, B. F. (2002). Neuropsychology. In R. J. Hagerman & P. J. Hagerman (Eds.), *Fragile X syndrome: Diagnosis, treatment and research* (3rd ed.). Baltimore: The John Hopkins University Press.

Berger, A., Jones, L., Rothbart, M. K., & Posner, M. I. (2000). Computerized games to study the development of attention in childhood. *Behavior Research Methods, Instruments, and Computers, 32*(2), 297–303.

Berger, A., & Posner, M. I. (2000). Pathologies of brain attentional networks. *Neuroscience and Biobehavioral Reviews, 24*(1), 3–5.

Berlin, L., & Bohlin, G. (2002). Response inhibition, hyperactivity, and conduct problems among preschool children. *Journal of Clinical Child and Adolescent Psychology, 31*(2), 242–251.

Berman, T., Douglas, V. I., & Barr, R. G. (1999). Effects of methylphenidate on complex cognitive processing in attention-deficit hyperactivity disorder. *Journal of Abnormal Child Psychology, 108*(1), 90–105.

Berry-Kravis, E., & Potanos, K. (2004). Psychopharmacology in fragile X syndrome—Present and future. *Mental Retardation and Developmental Disabilities Research Review, 10*(1), 42–48.

Berry-Kravis, E., Sumis, A., Kim, O. K., Lara, R., & Wuu, J. (2008). Characterization of potential outcome measures for future clinical trials in fragile X syndrome. *Journal of Autism and Developmental Disorders, 38*(9), 1751–1757.

Bertin, E., & Bhatt, R. S. (2001). Dissociations between featural versus conjunction-based texture processing in infancy: analyses of three potential contributing factors. *Journal of Experimental Child Psychology, 78*, 291–311.

Berwid, O. G., Curko Kera, E. A. C., Marks, D. J., Santra, A., Bender, H. A., & Halperin, J. M. (2005). Sustained attention and response inhibition in young children at risk for Attention Deficit/Hyperactivity Disorder. *Journal of Child Psychology and Psychiatry, 46*(11), 1219–1229.

Bhatt, R. S., Rovee-Collier, C., & Weiner, S. (1994). Developmental changes in the interface between perception and memory retrieval. *Developmental Psychology, 30*(2), 151–162.

Bhatt, R. S., & Waters, S. E. (1998). Perception of three-dimensional cues in early infancy. *Journal of Experimental Child Psychology, 70*(3), 207–224.

Bhowmik, A. D., Dutta, S., Sinha, S., Chattopadhyay, A., & Mukhopadhyay, K. (2008). Lack of association between Down syndrome and polymorphisms in dopamine receptor D4 and serotonin transporter genes. *Neurochemical Research, 33*(7), 1286–1291.

Bialystok, E., & Martin, M. M. (2004). Attention and inhibition in bilingual children: Evidence from the dimensional change card sort task. *Developmental Science, 27*, 325–39.

Bianchi, L. (1922). *The mechanisms of the brain and the function of the frontal lobes.* Edinburgh, UK: Livingstone.

Biederman, J., Faraone, S. V., Keenan, K., Steingard, R. & Tsuang, M. T. (1991). Familial association between attention deficit disorder (ADD) and anxiety disorders. *American Journal of Psychiatry, 148*, 251–255.

Biederman, J., Faraone, S. V., Monuteaux, M. C., & Grossbard, J. R. (2004). How informative are parent reports of attention-deficit/hyperactivity disorder symptoms for assessing outcome in clinical trials of long-acting treatments? A pooled analysis of parents' and teachers' reports. *Pediatrics, 113*(6), 1667–1671.

Biederman, J., Gao, H. T., Rogers, A. K., & Spencer, T. J. (2006). Comparison of parent and teacher reports of attention-deficit/hyperactivity disorder symptoms from two placebo-controlled studies of atomoxetine in children. *Biological Psychiatry, 60*(10), 1106–1110.

Biederman, J., Monuteaux, M. C., Greene, R. W., Braaten, E., Doyle, A. E., & Faraone, S. V. (2001). Long-term stability of the child behavior checklist in a clinical sample of youth with attention deficit hyperactivity disorder. *Journal of Clinical Child Psychology, 30*(4), 492–502.

Biederman, J., Petty, C. R., Dolan, C., Hughes, S., Mick, E., Monuteaux, M. C., & Faraone, S. V. (2008). The long-term longitudinal course of oppositional defiant disorder and conduct disorder in ADHD boys: Findings from a controlled 10-year prospective longitudinal follow-up study. *Psychological Medicine, 38*(7), 1027–1036.

Biederman, J., Petty, C. R., Doyle, A. E., Spencer, T., Henderson, C. S., Marion, B., et al. (2008). Stability of executive function deficits in girls with ADHD: A prospective longitudinal follow-up study into adolescence. *Developmental Neuropsychology, 33*(1), 44–61.

Biederman, J., Petty, C. R., Fried, R., Doyle, A. E., Spencer, T., Seidman, L. J., et al. (2007). Stability of executive function deficits into young adult years: A prospective longitudinal follow-up study of grown up males with ADHD. *Acta Psychiatrica Scandinavica, 116*(2), 129–136.

Biederman, J., & Spencer, T. J. (2008). Psychopharmacological interventions. *Child and Adolescent Psychiatric Clinics of North America, 17*(2), 439–458.

Bilder, R. M., Volavka, J., Czobor, P., Malhotra, A. K., Kennedy, J. L., Ni, X., et al. (2002). Neurocognitive correlates of the COMT Val(158)Met polymorphism in chronic schizophrenia. *Biological Psychiatry, 52*(7), 701–707.

Bish, J. P., Ferrante, S. M., McDonald-McGinn, D., Zackai, E., & Simon, T. J. (2005). Maladaptive conflict monitoring as evidence for executive dysfunction in children with chromosome 22q11.2 deletion syndrome. *Developmental Science, 8*(1), 36–43.

Bitsakou, P., Psychogiou, L., Thompson, M., & Sonuga-Barke, E. J. (2008). Inhibitory deficits in attention-deficit/hyperactivity disorder are independent of basic processing efficiency and IQ. *Journal of Neural Transmission, 115*(2), 261–268.

Bitsakou, P., Psychogiou, L., Thompson, M., & Sonuga-Barke, E. J. (2009). Delay Aversion in Attention Deficit/Hyperactivity Disorder: An empirical investigation of the broader phenotype. *Neuropsychologia, 47*(2), 446–456.

Blakemore, S.-J. (2008). Development of the social brain during adolescence. *The Quarterly Journal of Experimental Psychology, 61*(1), 40–49.

Blakemore, S.-J., & Choudhury, S. (2006). Development of the adolescent brain: Implications for executive function and social cognition. *Journal of Child Psychology and Psychiatry, 47*(3), 296–312.

Blasi, G., Mattay, V. S., Bertolino, A., Elvevag, B., Callicott, J. H., Das, S., et al. (2005). Effect of catechol-O-methyltransferase Val(158)Met genotype on attentional control. *Journal of Neuroscience, 25*(20), 5038–5045.

Bontempo, D. E., Hofer, S. M., Mackinnon, A., Piccinin, A. M., Gray, K. M., Tonge, B. J., et al. (2008). Factor structure of the developmental behavior checklist using confirmatory factor analysis of polytomous items. *Journal of Applied Measurement, 9*(3), 265–280.

Booth, J. E., Carlson, C. L., & Tucker, D. M. (2007). Performance on a neurocognitive measure of alerting differentiates ADHD combined and inattentive subtypes: A preliminary report. *Archives of Clinical Neuropsychology, 22*(4), 423–432.

Booth, J. R., Burman, D. D., Meyer, J. R., Lei, Z., Trommer, B. L., Davenport, N. D., et al. (2005). Larger deficits in brain networks for response inhibition than for visual selective attention in attention deficit hyperactivity disorder (ADHD). *Journal of Child Psychology and Psychiatry, 46*(1), 94–111.

Botto, L. D., May, K., Fernhoff, P. M., Correa, A., Coleman, K., Rasmussen, S. A., et al. (2003). A population-based study of the 22q11.2 deletion: Phenotype, incidence, and contribution to major birth defects in the population. *Pediatrics, 112*(1), 101–107.

Botvinick, M. M. (2007). Conflict monitoring and decision making: Reconciling two perspectives on anterior cingulate function. *Cognitive, Affective, & Behavioral Neuroscience, 7*(4), 356–366.

Botvinick, M. M., Braver, T. S., Barch, D. M., Carter, C. S., & Cohen, J. D. (2001). Conflict monitoring and cognitive control. *Psychological Review, 108*, 624–652.

Botvinick, M. M., Cohen, J. D., & Carter, C. S. (2004). Conflict monitoring and anterior cingulate cortex: An update. *Trends in Cognitive Sciences, 8*(12), 539–546.

Boucher, J. (2003). Language development in autism. *International Journal of Pediatric Otorhinolaryngology, 67*(Suppl. 1), 159–163.

Bourdon, K. H., Goodman, R., Rae, D. S., Simpson, G., & Koretz, D. S. (2005). The strengths and difficulties questionnaire: US normative data and psychometric properties. *Journal of the American Academy of Child and Adolescent Psychiatry, 44*(6), 557–564.

Bouwknecht, J. A., Hijzen, T. H., van der Gugten, J., Mass, R. A. A., Hen, R., & Olivier, B. (2001). Absence of 5-HT1B receptors is associated with impaired impulse control in male 5-HT1B knockout mice. *Biological Psychiatry, 49*(7), 557–568.

Brahe, C., Bannetta, P., Serra, A., Opitz, J. M., & Arwert, F. (1986). The increased COMT activity in down-syndrome patients is not a consequence of dosage effect owing to location of the gene on chromosome-21: Further evidence. *American Journal of Medical Genetics, 24*(1), 203–204.

Braun, J. (1998). Divided attention: Narrowing the gap between brain and behavior. In R. Parasuraman (Ed.), *The attentive brain.* Cambridge, MA: MIT Press.

Brereton, A. V., Tonge, B. J., & Einfeld, S. L. (2006). Psychopathology in children and adolescents with autism compared to young people with intellectual disability. *Journal of Autism and Developmental Disorders, 36*(7), 863–870.

Brian, J. A., Tipper, S. P., Weaver, B., & Bryson, S. E. (2003). Inhibitory mechanisms in autism spectrum disorders: Typical selective inhibition of location versus facilitated perceptual processing. *Journal of Child Psychology and Psychiatry and Allied Disciplines, 44*(4), 552–560.

Britton, L. A., & Delay, E. R. (1989). Effects of noise on a simple visual attentional task. *Perceptual and Motor Skills, 68*(3), 875–878.

Broadbent, D. E. (1958). *Perception and communication.* London: Plenum Press.

Brock, J., & Jarrold, C. (2005). Serial order reconstruction in Down syndrome: Evidence for a selective deficit in verbal short-term memory. *Journal of child psychology and psychiatry, and allied disciplines, 46*(3), 304–16.

Brocki, K., Clerkin, S. M., Guise, K. G., Fan, J., & Fossella, J. A. (2009). Assessing the molecular genetics of the development of executive attention in children: Focus on genetic pathways related to the anterior cingulate cortex and dopamine. *Neuroscience 164*(1), 241–246. Retrieved from doi:10.1016/j.neuroscience.2009.01.029

Brocki, K. C., Nyberg, L., Thorell, L. B., & Bohlin, G. (2007). Early concurrent and longitudinal symptoms of ADHD and ODD: Relations to different types of inhibitory control and working memory. *Journal of Child Psychology and Psychiatry, 48*(10), 1033–1041.

Brophy, K., Hawi, Z., Kirley, A., Fitzgerald, M., & Gill, M. (2002). Synaptosomal-associated protein 25 (SNAP-25) and attention deficit hyperactivity disorder (ADHD): Evidence of linkage and association in the Irish population. *Molecular Psychiatry, 7*(8), 913–917.

Brown, J. H., Johnson, M. H., Paterson, S. J., Gilmore, R., Longhi, E., & Karmiloff-Smith, A. (2003). Spatial representation and attention in toddlers with Williams syndrome and Down syndrome. *Neuropsychologia, 41*(8), 1037–1046.

Bryan, J., Calvaresi, E., & Hughes, D. (2002). Short-Term Folate, Vitamin B-12 or Vitamin B-6 Supplementation Slightly Affects Memory Performance But Not Mood in Women of Various Ages. *Journal of Nutrition, 132*, 1345–1356.

Bryson, S. E., Zwaigenbaum, L., McDermott, C., Rombough, V., & Brian, J. (2008). The Autism Observation Scale for Infants: Scale development and reliability data. *Journal of Autism & Developmental Disorders, 38*(4), 731–738.

Buchmann, J., Gierow, W., Weber, S., Hoeppner, J., Klauer, T., Benecke, R., et al. (2007). Restoration of disturbed intracortical motor inhibition and facilitation in attention deficit hyperactivity disorder children by methylphenidate. *Biological Psychiatry, 62*(9), 963–969.

Bull, R., Espy, K. A., & Senn, T. E. (2004). A comparison of performance on the Towers of London and Hanoi in young children. *Journal of Child Psychology and Psychiatry, 45*(4), 743–754.

Bull, R., & Scerif, G. (2001). Executive functioning as a predictor of children's mathematics ability: Inhibition, switching, and working memory. *Developmental Neuropsychology, 19*(3), 273–293.

Bussing, R., Fernandez, M., Harwood, M., Hou, W., Garvan, C. W., Eyberg, S. M., et al. (2008). Parent and teacher SNAP-IV ratings of attention deficit hyperactivity disorder symptoms: Psychometric properties and normative ratings from a school district sample. *Assessment, 15*(3), 317–328.

Byrne, J. M., Bawden, H. N., DeWolfe, N. A., & Beattie, T. L. (1998). Clinical assessment of psychopharmacological treatment of preschoolers with ADHD. *Journal of Clinical and Experimental Neuropsychology, 20*(5), 613–627.

Callejas, A., Lupianez, J., & Tudela, P. (2004). The three attentional networks: On their independence and interactions. *Brain and Cognition, 54*(3), 225–227.

Campbell, L. E., Daly, E., Toal, F., Stevens, A., Azuma, R., Catani, M., et al. (2006). Brain and behaviour in children with 22q11.2 deletion syndrome: A volumetric and voxel-based morphometry MRI study. *Brain, 129*(5), 1218–1228.

Canfield, M. A., Honein, M. A., Yuskiv, N., Xing, J., Mai, C. T., Collins, J. S., et al. (2006). National estimates and race/ethnic-specific variation of selected birth defects in the United States, 1999–2001. *Birth Defects Research Part A: Clinical and Molecular Teratology, 76*(11), 747–756.

Carlson, S. A. (2005). Developmentally sensitive measures of executive function in preschool children. *Developmental Neuropsychology, 28*(2), 595–616.

Carlson, S. M., & Moses, L. J. (2001). Individual differences in inhibitory control of children's theory of mind. *Child Development, 72*(4), 1032–1053.

Carter, J. C., Capone, G. T., Gray, R. M., Cox, C. S., & Kaufmann, W. E. (2007). Autistic-spectrum disorders in Down syndrome: Further delineation and distinction from other behavioral abnormalities. *American Journal of Medical Genetics Part B: Neuropsychiatric Genetics, 144B*(1), 87–94.

Carver, A. C., Livesey, D. J., & Charles, M. (2001). Further manipulation of the stop signal task: Developmental changes in the ability to inhibit responding with longer stop signal delays. *International Journal of Neuroscience, 111*(1&2), 39–53.

Casey, B. J., Castellanos, F. X., Giedd, J. N., Marsh, W. L., Hamburger, S. D., Schubert, A. B., et al. (1997). Implication of right frontostriatal circuitry in response inhibition and attention-deficit/hyperactivity disorder. *Journal of the American Academy of Child and Adolescent Psychiatry, 36*(3), 374–383.

Casey, B. J., Davidson, M. C., Hara, Y., Thomas, K. M., Martinez, A., Galvan, A., et al. (2004). Early development of subcortical regions involved in non-cued attention switching. *Developmental Science, 7*(5), 534–542.

Casey, B. J., Galvan, A., & Hare, T. A. (2005). Changes in cerebral functional organization during cognitive development. *Current Opinion in Neurobiology, 15*(2), 239–244.

Casey, B. J., Getz, S., & Galvan, A. (2008). The adolescent brain. *Developmental Review, 28*(1), 62–77.

Casey, B. J., Jones, R. M., & Hare, T. A. (2008). The adolescent brain. *Annals of the New York Academy of Sciences, 1124*, 111–1126.

Casey B. J., Soliman F., Bath K. G., Glatt C. E. Imaging genetics and development: challenges and promises. *Hum Brain Mapping.* 2010 Jun;*31*(6):838–51.

Casey, B. J., Tottenham, N., Liston, C., & Durston, S. (2005). Imaging the developing brain: What have we learned about cognitive development? *Trends in Cognitive Sciences, 9*(3), 104–110.

Castellanos, F. X., Lee, P. P., Sharp, W., Jeffries, N. O., Greenstein, D. K., Clasen, L. S., et al. (2002). Developmental trajectories of brain volume abnormalities in children and adolescents with attention-deficit/hyperactivity disorder. *JAMA—The Journal of the American Medical Association, 288*(14), 1740–1748.

Castellanos, F. X., Sonuga-Barke, E. J. S., Milham, M. P., & Tannock, R. (2006). Characterizing cognition in ADHD: Beyond executive dysfunction. *Trends in Cognitive Sciences, 10*(3), 117–123.

Castellanos, F. X., & Tannock, R. (2002). Neuroscience of attention-deficit/hyperactivity disorder: The search for endophenotypes. *Nature Reviews Neuroscience, 3*(8), 617–628.

Castrén, M., Pääkkönen, A., Tarkka, I. M., Ryynänen, M., & Partanen, J. (2003). Augmentation of auditory N1 in children with fragile X syndrome. *Brain Topography, 15*(3), 165–171.

Cavanagh, P. (2004). Attention routines and the architecture of selection. In M. I. Posner (Ed.), *Cognitive neuroscience of attention* (pp. 13–28). New York: Guilford Press.

Cepeda, N. J., Kramer, A. F., & Gonzalez de Sather, J. C. M. (2001). Changes in executive control across the life-span: Examination of task switching performance. *Developmental Psychology, 37*, 715–730.

Cerella, J., & Hale, S. (1998). The rise and fall in information-processing rates over the life span. *Acta Psychologica (Amsterdam), 86*(2–3), 109–197.

Chambers, C. D., Garavan, H., & Bellgrove, M. A. (in press). Insights into the neural basis of response inhibition from cognitive and clinical neuroscience. *Neuroscience & Biobehavioral Reviews.*

Chandana, S. R., Behen, M. E., Juhasz, C., Muzik, O., Rothermel, R. D., Mangner, T. J., et al. (2005). Significance of abnormalities in developmental trajectory and asymmetry of cortical serotonin synthesis in autism. *International Journal of Developmental Neuroscience, 23*(2–3), 171–182.

Charman, T., Taylor, E., Drew, A., Cockerill, H., Brown, J. A., & Baird, G. (2005). Outcome at 7 years of children diagnosed with autism at age 2: Predictive validity of assessments conducted at 2 and 3 years of age and pattern of symptom change over time. *Journal of Child Psychology and Psychiatry, 46*(5), 500–513.

Chen, C. K., Chen, S. L., Mill, J., Huang, Y.S., Lin, S. K., Curran, S., et al. (2003). The dopamine transporter gene is associated with attention deficit hyperactivity disorder in a Taiwanese sample. *Molecular Psychiatry, 8*(4), 393–396.

Cherry, E. C. (1953). Some experiments on the recognition of speech with one and two ears. *Journal of the Acoustical Society of America, 25*(5), 975–979.

Cheuk, D. K. L., Li, S. Y. H., & Wong, V. (2006). No association between VNTR polymorphisms of dopamine transporter gene and attention deficit hyperactivity disorder in Chinese children. *American Journal of Medical Genetics Part B: Neuropsychiatric Genetics, 141B*(2), 123–125.

Cheuk, D. K. L., & Wong, V. (2006). Meta-analysis of association between a catechol-O-methyltransferase gene polymorphism and attention deficit hyperactivity disorder. *Behavior Genetics, 36*(5), 651–659.

Chiang, M., & Gau, S. S. (2008). Validation of attention-deficit-hyperactivity disorder subtypes among Taiwanese children using neuropsychological functioning. *Australia New Zealand Journal of Psychiatry, 42*(6), 526–535.

Christ, S. E., Holt, D. D., White, D. A., & Green, L. (2007). Inhibitory control in children with autism spectrum disorder. *Journal of Autism and Developmental Disorders, 37*(6), 1155–1165.

Chronis, A. M., Chacko, A., Fabiano, G. A., Wymbs, B. T., & Pelham, W. E., Jr. (2004). Enhancements to the behavioral parent training paradigm for families of children with ADHD: Review and future directions. *Clinical Child and Family Psychology Review, 7*(1), 1–27.

Chronis, A. M., Jones, H. A., & Raggi, V. L. (2006). Evidence-based psychosocial treatments for children and adolescents with attention-deficit/hyperactivity disorder. *Clinical Psychology Review, 26*(4), 486–502.

Clark, D., & Wilson, G. N. (2003). Behavioral assessment of children with Down syndrome using the Reiss psychopathology scale. *American Journal of Medical Genetics Part A, 118A*(3), 210–216.

Clarke, A. R., Tonge, B. J., Einfeld, S. L., & Mackinnon, A. (2003). Assessment of change with the Developmental Behaviour Checklist. *Journal of Intellectual Disability Research, 47*(3), 210–212.

Clifford, S., Dissanayake, C., Bui, Q. M., Huggins, R., Taylor, A. K., & Loesch, D. Z. (2007). Autism spectrum phenotype in males and females with fragile X full mutation and premutation. *Journal of Autism and Developmental Disorders, 37*(4), 738–747.

Coghill, D. R., Rhodes, S. M., & Matthews, K. (2007). The neuropsychological effects of chronic methylphenidate on drug-naive boys with attention-deficit/hyperactivity disorder. *Biological Psychiatry, 62*(9), 954–962.

Cohen, J. D., Aston-Jones, G., & Gilzenrat, M. S. (2004). A systems-level perspective on attention and cognitive control: Guided activation, adaptive gating, conflict monitoring and exploitation versus exploration. In M. I. Posner (Ed.), *Cognitive neuroscience of attention* (pp. 71–90). New York: Guilford.

Colby, C. L., & Goldberg, M. E. (1999). Space and attention in parietal cortex. *Annual Review of Neuroscience, 22*, 319–349.

Collett, B. R., Ohan, J. L., & Myers, K. M. (2003). Ten-year review of rating scales. VI: Scales assessing externalizing behaviors. *Journal of the American Academy of Child and Adolescent Psychiatry, 42*(10), 1143–1170.

Colombo, J. (2001). The development of visual attention in infancy. *Annual Review of Psychology, 52*, 337–367.

Colombo, J., Ryther, J. S., Frick, J. E., & Gifford, J. J. (1995). Visual pop-out in infants: Evidence for preattentive search in 3- and 4-month-olds. *Psychonomic Bulletin and Review, 2*(2), 266–268.

Conlin, J. A., Gathercole, S. E., & Adams, J. W. (2005). Children's working memory: Investigating performance limitations in complex span tasks. *Journal of Experimental Child Psychology, 90*(3), 303–317.

Conners, C. K. (1997). Conners' Rating Scales—Revised. North Tonawanda, NY: Multi-Health Systems, Canada.

Conners, C. K. (2007). K—CPT™ V.5. Conners' Kiddie Continuous Performance Test Version 5. Ottawa, Canada: Multi-Health Systems.

Conners, C. K. (2008). Conners 3rd Edition (Conners 3). North Tonawanda, NY: Multi-Health Systems.

Conners, C. K., & MHS Staff. (2007). Continuous Performance Test II (CPT II) Version 5.1 for Windows. Ottawa, Canada: Multi-Health Systems.

Conners, C. K., Sitarenios, G., Parker, J. D., & Epstein, J. N. (1998a). Revision and restandardization of the Conners Teacher Rating Scale (CTRS-R): Factor structure, reliability, and criterion validity. *Journal of Abnormal Child Psychology, 26*(4), 279–291.

Conners, C. K., Sitarenios, G., Parker, J. D., & Epstein, J. N. (1998b). The revised Conners' Parent Rating Scale (CPRS-R): Factor structure, reliability, and criterion validity. *Journal of Abnormal Child Psychology, 26*(4), 257–268.

Conroy, J., Meally, E., Kearney, G., Fitzgerald, M., Gill, M., & Gallagher, L. (2004). Serotonin transporter gene and autism: A haplotype analysis in an Irish autistic population. *Molecular Psychiatry, 9*(6), 587–593.

Cook, E. H., Stein, M. A., Krasowski, M. D., Cox, N. J., Olkon, D. M., Kieffer, J. E., et al. (1995). Association of attention-deficit disorder and the dopamine transporter gene. *American Journal of Human Genetics, 56*(4), 993–998.

Coppus, A., Evenhuis, H., Verberne, G. J., Visser, F., van Gool, P., Eikelenboom, P., et al. (2006). Dementia and mortality in persons with Down's syndrome. *Journal of Intellectual Disability Research, 50*(10), 768–777.

Corkum, P. V., & Siegel, L. S. (1993). Is the continuous performance task a valuable research tool for use with children with attention-deficit-hyperactivity disorder? *Journal of Child Psychology and Psychiatry and Allied Disciplines, 34*(7), 1217–1239.

Cormack, K. F. M., Brown, A. C., & Hastings, R. P. (2000). Behavioural and emotional difficulties in students attending schools for children and adolescents with severe intellectual disability. *Journal of Intellectual Disability Research, 44*(2), 124–129.

Cornish, K., & Bramble, D. (2002). Cri du chat syndrome: Genotype-phenotype correlations and recommendations for clinical management. *Developmental Medicine and Child Neurology, 44*(7), 494–497.

Cornish, K. M., Levitas, A., & Sudhalter, V. (2007). Fragile X syndrome: The journey from genes to behavior. In M. M. M. Mazzocco & J. L. Ross (Eds.), *Neurogenetic developmental disorders: Manifestation and identification in childhood* (pp. 73–103). Cambridge, MA: MIT Press.

Cornish, K., Munir, F., & Wilding, J. (1997). Is there a link between pattern of attention deficit and severity of behavioural problems in boys with fragile X syndrome? *Health and Education Professionals in Attention Deficit/Hyperactivity Disorder.* Hurstpierpoint, Sussex, UK: International Psychology Services.

Cornish, K., Munir, F., & Wilding, J. (2001). A neuropsychological and behavioural profile of attention deficits in fragile X syndrome. *Revista de Neurologia, 1,* S24–S29.

Cornish, K., Scerif, G., & Karmiloff-Smith, A. (2007). Tracing syndrome-specific trajectories of attention across the lifespan. *Cortex, 43*(6), 672–685.

Cornish, K., Sudhalter, V., & Turk, J. (2004). Attention and language in fragile X. *Mental Retardation and Developmental Disabilities Research Reviews, 10*(1), 11–16.

Cornish, K., Turk, J., & Hagerman, R. (2008). The fragile X continuum: New advances and perspectives. *Journal of Intellectual Disability Research, 52*(6), 469–482.

Cornish, K. M., Li, L., Kogan, C. S., Jacquemont, S., Turk, J., Dalton, A., et al. (2008). Age-dependent cognitive changes in carriers of the fragile X syndrome. *Cortex, 44*(6), 628–636.

Cornish, K. M., Manly, T., Savage, R., Swanson, J., Morisano, D., Butler, N., et al. (2005). Association of the dopamine transporter (DAT1) 10/10-repeat genotype with ADHD symptoms and response inhibition in a general population sample. *Molecular Psychiatry, 10*(7), 686–698.

Cornish, K. M., Munir, F., & Cross, G. (1998). The nature of the spatial deficit in young females with Fragile-X syndrome: A neuropsychological and molecular perspective. *Neuropsychologia, 36*(11), 1239–1246.

Cornish, K. M., Munir, F., & Cross, G. (1999). Spatial cognition in males with Fragile-X syndrome: Evidence for a neuropsychological phenotype. *Cortex, 35*(2), 263–271.

Cornish, K. M., Turk, J., & Levitas, A. (2007). Fragile X syndrome and autism: Common developmental pathways? *Current Pediatrics Reviews, 3*(1), 61–68.

Cornish, K. M., Turk, J., Wilding, J., Sudhalter, V., Munir, F., Kooy, F., et al. (2004). Annotation: Deconstructing the attention deficit in fragile X syndrome: A developmental neuropsychological approach. *Journal of Child Psychology and Psychiatry, 45*(6), 1042–1053.

Cornish, K. M., & Wilding, J. (2006). Does cognitive performance increase or decrease across development in Fragile X syndrome: Impact of verbal mental age and gender. In J. Mallard (Ed.), *Advances in fragile X research* (pp. 23–36). New York: Nova Publishing.

Cornish, K. M., Wilding, J., & Grant, C. (2006). Deconstructing working memory in developmental disorders of attention. In S. J. Pickering (Ed.), *Working memory and education* (pp. 157–188). Burlington, MA: Elsevier Science.

Cornish, K. M., Wilding, J. M., & Hollis, C. (2008). Visual search performance in children rated as good or poor attenders: The differential impact of DAT1 genotype, IQ and chronological age. *Neuropsychology, 22*(2), 217–225.

Corteen, R. S., & Wood, B. (1972). Autonomic responses to shock associated words in an unattended channel. *Journal of Experimental Psychology, 94*(3), 308–313.

Cragg, L., & Nation, K. (2007). Self-ordered pointing as a test of working memory in typically developing children. *Memory, 15*(5), 526–535.

Cragg, L., & Nation, K. (2008). Go or no-go? Developmental improvements in the efficiency of response inhibition in mid-childhood. *Developmental Science, 11*(6), 819–827.

Crone, E. A., Donohue, S. E., Honomichl, R., Wendelken, C., & Bunge, S. A. (2006). Brain regions mediating flexible rule use during development. *Journal of Neuroscience, 26*(43), 11239–11247.

Crone, E. A., Ridderinkhof, K. R., Worm, M., Somsen, R. J. M., & van der Molen, M. W. (2004). Switching between spatial stimulus-response mappings: A developmental study of cognitive flexibility. *Developmental Science, 7*(4), 443–455.

Crosbie, J., Pérusse, D., Barr, C. L., & Schachar, R. J. (2008). Validating psychiatric endophenotypes: Inhibitory control and attention deficit hyperactivity disorder. *Neuroscience and Biobehavioral Reviews, 32*(1), 40–55.

Curran, S., Mill, J., Sham, P., Rijsdijk, F., Marusic, K., Taylor, E., et al. (2001). QTL association analysis of the DRD4 Exon 3 VNTR polymorphism in a population sample of children screened with a parent rating scale for ADHD symptoms. *Behavior Genetics, 31*(5), 450–451.

Dally, K. (2006). The influence of phonological processing and inattentive behavior on reading acquisition. *Journal of Educational Psychology, 98*(2), 420–437.

Daly, G., Hawi, Z., Fitzgerald, M., & Gill, M. (1999). Mapping susceptibility loci in attention deficit hyperactivity disorder: Preferential transmission of parental alleles at DAT1, DBH and DRD5 to affected children. *Molecular Psychiatry, 4*(2), 192–196.

Daneman, M., & Carpenter, P. A. (1980). Individual differences in working memory and reading. *Journal of Verbal Learning and Verbal Behavior, 19*, 450–466.

Davidson, M. C., Amso, D., Anderson, L. C., & Diamond, A. (2006). Development of cognitive control and executive function from 4 to 13 years: Evidence from manipulation, inhibition, and task-switching. *Neuropsychologia, 44*(11), 2037–2078.

Davies, D. R., & Parasuraman, R. (1977). Cortical evoked potentials and vigilance: A decision theory analysis. In R. R. Mackie (Ed.), *Vigilance: Theory, operational performance and physiological correlates* (pp. 285–306). New York: Plenum Press.

Daviss, W. B. (2008). A review of co-morbid depression in pediatric ADHD: Etiology, phenomenology, and treatment. *Journal of Child and Adolescent Psychopharmacology, 18*(6), 565–567.

Deb, S., Dhaliwal, A. J., & Roy, M. (2008). The usefulness of Conners' Rating Scales-Revised in screening for Attention Deficit Hyperactivity Disorder in children with intellectual disabilities and borderline intelligence. *Journal of Intellectual Disability Research, 52*(2), 107–113.

Degirmenci, B., Miral, S., Kaya, G. C., Iyilikci, L., Arslan, G., Baykara, A., et al. (2008). Technetium-99m HMPAO brain SPECT in autistic children and their families. *Psychiatry Research: Neuroimaging, 162*(3), 236–243.

Delabar, J. M., Theophile, D., Rahmani, Z., Chettouh, Z., Blouin, J. L., Prieur, M., et al. (1993). Molecular mapping of twenty-four features of Down syndrome on chromosome 21. *European Journal of Human Genetics 1*, 114–124.

Delis, D. C., Kaplan, E., & Kramer, J. H. (2001). The Delis-Kaplan Executive Function System. San Antonio, TX: The Psychological Corporation.

Denckla, M. B. (1996). A theory and model of executive function: A neuropsychological perspective. In G. R. Lyon & N. A. Krasnegor (Eds.), *Attention, memory and executive function* (pp. 263–278). Baltimore: Paul Brookes.

Dereboy, C., Senol, S., Sener, S., & Dereboy, F. (2007). Validation of the Turkish versions of the short-form Conners' teacher and parent rating scales. *Turk Psikiyatri Dergisi, 18*(1), 48–58.

Derefinko, K. J., Adams, Z. W., Milich, R., Fillmore, M. T., Lorch, E. P., & Lynam, D. R. (2008). Response style differences in the inattentive and combined subtypes of attention-deficit/hyperactivity disorder. *Journal of Abnormal Child Psychology, 36*(5), 745–758.

Derks, E. M., Hudziak, J. J., Dolan, C. V., van Beijsterveldt, T., Verhulst, F. C., & Boomsma, D. I. (2008). Genetic and environmental influences on the relation between attention problems and attention deficit hyperactivity disorder. *Behavior Genetics, 38*, 11–23.

Desimone, R., & Duncan, J. (1995). Neural mechanisms of selective visual attention. *Annual Review of Neuroscience, 18*, 193–222.

de Sonneville, L. M. J. (1999). Amsterdam neuropsychological tasks: A computer-aided assessment program. In B. P. L. M. den Brinker, P. J. Beek, A. N. Brand, S. J. Maarse, & L. J. M. Mulder (Eds.), *Cognitive ergonomics, clinical assessment and computer-assisted learning: Computers in psychology* (Vol. 6, pp. 187–203). Lisse, The Netherlands: Swets.

D'Esposito, M., Ballard, D., Zarahn, E., & Aguirre, G. K. (2000). The role of prefrontal cortex in sensory memory and motor preparation: An event-related fMRI study. *NeuroImage, 11*(5), 400–408.

Deutsch, J. A., & Deutsch, D. (1963). Attention: Some theoretical considerations. *Psychological Review, 70*, 80–90.

Devito, E. E., Blackwell, A., Clark, L., Kent, L., Dezsery, A. M., Turner, D. C., et al. (2008). Methylphenidate improves response inhibition but not reflection-impulsivity in children with attention deficit hyperactivity disorder (ADHD). *Psychopharmacology (Berl)*.

Diamond, A. (1985). Development of the ability to use recall to guide action, as indicated by infants' performance on A-not-B. *Child Development, 56*(4), 868–883.

Diamond, A. (2007). Consequences of variations in genes that affect dopamine in prefrontal cortex. *Cerebral Cortex, 17*(Suppl.1), i161–i170.

Diamond, A., Barnett, W. S., Thomas, J., & Munro, S. (2007). Preschool program improves cognitive control. *Science, 318*(5855), 1387–1388.

Diamond, A., Briand, L., Fossella, J., & Gehlbach, L. (2004). Genetic and neuro-chemical modulation of prefrontal cognitive functions in children. *The American Journal of Psychiatry, 161*(1), 125–132.

Diamond, A., Carlson, S. M., & Beck, D. M. (2005). Preschool children's performance in task switching on the dimensional change card sort task: Separating the dimensions aids the ability to switch. *Developmental Neuropsychology, 28*(2), 689–729.

Diamond, A., Kirkham, N., & Amso, D. (2002). Conditions under which young children can hold two rules in mind and inhibit a prepotent response. *Developmental Psychology, 38*(3), 352–362.

Diaz-Asper, C. M., Goldberg, T. E., Kolachana, B. S., Straub, R. E., Egan, M. F., & Weinberger, D. R. (2008). Genetic variation in catechol-O-methyltransferase: Effects on working memory in schizophrenic patients, their siblings, and healthy controls. *Biological Psychiatry, 63*(1), 72–79.

Dietz, C., Swinkels, S. H. N., Buitelaar, J. K., van Daalen, E., & van Engeland, H. (2007). Stability and change of IQ scores in preschool children diagnosed with autistic spectrum disorder. *European Child & Adolescent Psychiatry, 16*(6), 405–410.

DiMaio, S., Grizenko, N., & Joober, R. (2003). Dopamine genes and attention-deficit hyperactivity disorder: A review. *Journal of Psychiatry & Neuroscience, 28*(1), 27–38.

Di Martino, A., Melis, G., Cianchetti, C., & Zuddas, A. (2004). Methylphenidate for pervasive developmental disorders: Safety and efficacy of acute single dose test and ongoing therapy: An open-pilot study. *Journal of Child and Adolescent Psychopharmacology, 14*(2), 207–218.

Diorio, J., & Meaney, M. J. (2007). Maternal programming of defensive responses through sustained effects on gene expression. *Journal of Psychiatry & Neuroscience, 32*(4), 275–284.

Dobbs, J., Doctoroff, G. L., Fisher, P., & Arnold, D. H. (2006). The association between preschool children's socio-emotional functioning and their mathematics skills. *Journal of Applied Developmental Psychology, 27*, 97–108.

Dockstader, C., Gaetz, W., Cheyne, D., Wang, F., Castellanos, F. X., & Tannock, R. (2008). MEG event-related desynchronization and synchronization deficits during basic somatosensory processing in individuals with ADHD. *Behavioral and Brain Functions, 4*, 8.

Donnai, D., & Karmiloff-Smith, A. (2000). Williams syndrome: From genotype through to the cognitive phenotype. *American Journal of Medical Genetics, 97*(2), 164–171.

Donnelly, N., Cave, K., Greenway, R., Hadwin, J. A., Stevenson, J., & Sonuga-Barke, E. (2007). Visual search in children and adults: Top-down and bottom-up mechanisms. *The Quarterly Journal of Experimental Psychology, 60*(1), 120–136.

Driver, J. (2001). A selective review of selective attention research from the past century. *British Journal of Psychology, 92*(1), 53–78.

Driver, J., & Baylis, G. C. (1998). Attention and visual object segmentation. In R. Parasuraman (Ed.), *The attentive brain* (pp. 299–325). Cambridge, MA: MIT Press.

Du, Y., Kou, J., & Coghill, D. (2008). The validity, reliability and normative scores of the parent, teacher and self report versions of the Strengths and Difficulties Questionnaire in China. *Child Adolescent Psychiatry and Mental Health, 2*(1), 8.

Duncan, J. (1980). The locus of interference in the perception of simultaneous stimuli. *Psychological Review, 87*(3), 272–300.

Duncan, J. (2004). EPS Mid-career award 2004: Brain mechanisms of attention. *Quarterly Journal of Experimental Psychology, 59*, 2–27.

Duncan, J., & Humphreys, G. W. (1989). Visual search and stimulus similarity. *Psychological Review, 96*, 443–458.

Duncan, J., Martens, S., & Ward, R. (1997). Restricted attentional capacity within but not between sensory modalities. *Nature, 387*(6635), 808–810.

Duncan, J., Williams, P., Nimmo-Smith, I., & Brown, I. (1993). The control of skilled behavior: Learning, intelligence and distraction. In E. Meyer & S. Kornblum (Eds.), *Attention and Performance* (Vol. 14, pp. 323–341). Cambridge, MA: MIT Press.

DuPaul, G.J., Power, T.J., Anastopoulos, AD.,& Reid, R. (1998). ADHD Rating Scale–IV: Checklists, Norms, and Clinical Interpretation. New York, NY: Guilford Press.

Durand, C. M., Betancur, C., Boeckers, T. M., Bockmann, J., Chaste, P., Fauchereau, F., et al. (2007). Mutations in the gene encoding the synaptic scaffolding protein SHANK3 are associated with autism spectrum disorders. *Nature Genetics, 39*(1), 25–27.

Durston, S., & Casey, B. J. (2006). What have we learned about cognitive development from neuroimaging? *Neuropsychologia, 44*(11), 2149–2157.

Durston, S., Davidson, M. C., Tottenham, N., Galvan, A., Spicer, J., Fossella, J. A., et al. (2006). A shift from diffuse to focal cortical activity with development. *Developmental Science, 9*(1), 1–8.

Dykens, E. M. (2007). Psychiatric and behavioral disorders in persons with Down syndrome. *Mental Retardation and Developmental Disabilities Research Reviews, 13*, 272–278.

Edgin, J. O., & Pennington, B. F. (2005). Spatial cognition in autism spectrum disorders: Superior, impaired, or just intact? *Journal of Autism and Developmental Disorders, 35*(6), 729–745.

Egan, M. F., Goldberg, T. E., Kolachana, B. S., Callicott, J. H., Mazzanti, C. M., Straub, R. E., et al. (2001). Effect of COMT Val(108/158) Met genotype on frontal lobe function and risk for schizophrenia. *Proceedings of the National Academy of Sciences of the United States of America, 98*(12), 6917–6922.

Einfeld, S. L., & Tonge, B. J. (1995). The developmental behavior checklist: The development and validation of an instrument to assess behavioral and emotional disturbance in children and adolescents with mental retardation. *Journal of Autism and Developmental Disorders, 25*(2), 81–104.

Einfeld, S. L., & Tonge, B. J. (1996a). Population prevalence of psychopathology in children and adolescents with intellectual disability: II. Epidemiological findings. *Journal of Intellectual Disability Research, 40*(2), 91–98.

Einfeld, S. L., & Tonge, B. J. (1996b). Population prevalence of psychopathology in children and adolescents with intellectual disability: I. Rationale and methods. *Journal of Intellectual Disability Research, 40*(2), 99–109.

Einfeld, S. L., & Tonge, B. J. (2002). *Manual for the Developmental Behaviour Checklist: Primary Carer Version (DBC-P) & Teacher Version (DBC-T)* 2nd ed. Clayton, Melbourne, Australia: Monash University Centre for Developmental Psychiatry and Psychology.

Einfeld, S. L., Tonge, B. J., & Florio, T. (1997). Behavioral and emotional disturbance in individuals with Williams syndrome. *American Journal on Mental Retardation, 102*(1), 45–53.

Einfeld, S. L., Tonge, B. J., & Rees, V. W. (2001). Longitudinal course of behavioral and emotional problems in Williams syndrome. *American Journal on Mental Retardation, 106*(1), 73–81.

Einfeld, S., Tonge, B., & Turner, G. (1999). Longitudinal course of behavioral and emotional problems in fragile X syndrome. *American Journal of Medical Genetics, 87*(5), 436–439.

Eisenberg, J., Mei-Tal, G., Steinberg, A., Tartakovsky, E., Zohar, A., Gritsenko, I., et al. (1999). Haplotype relative risk study of catechol-O-methyltransferase (COMT) and attention deficit hyperactivity disorder (ADHD): Association of the high-enzyme activity val allele with ADHD impulsive-hyperactive phenotype. *American Journal of Medical Genetics Part B: Neuropsychiatric Genetics, 88*(5), 497–502.

Eisenhower, A. S., Baker, B. L., & Blacher, J. (2005). Preschool children with intellectual disability: Syndrome specificity, behaviour problems, and maternal well-being. *Journal of Intellectual Disability Research, 49*(5), 657–671.

Elkins, I. J., McGue, M., & Iacono, W. G. (2007). Prospective effects of attention-deficit/hyperactivity disorder, conduct disorder, and sex on adolescent substance use and abuse. *Archives of General Psychiatry, 64*(10), 1145–1152.

Elsabbagh, M., Volein, A., Holmboe, K., Tucker, L., Csibra, G., Baron Cohen, S., et al. (2009). Neural correlates of eye gaze processing in the infant broader autism phenotype. *Journal of Child Psychology and Psychiatry, 50*(5), 637–642.

Engle, R. W., Tuholski, S. W., Laughlin, G. E., & Conway, A. R. A. (1999). Working memory, short-term memory and general fluid intelligence: A latent variable approach. *Journal of Experimental Psychology: General, 125*(3), 309–331.

Epstein, J. N., Conners, C. K., Hervey, A. S., Tonev, S. T., Arnold, L. E., Abikoff, H. B., et al. (2006). Assessing medication effects in the MTA study using neuropsychological outcomes. *Journal of Child Psychology and Psychiatry, 47*(5), 446–456.

Erford, B. T., & Hase, K. (2006). Reliability and validity of scores on the ACTeRS-2. *Measurement and Evaluation in Counseling and Development, 39*(2), 97–106.

Eriksen, B. A., & Eriksen, C. W. (1974). Effects of noise letters upon identification of a target letter in a nonsearch task. *Perception and Psychophysics, 16,* 143–149.

Espy, K. A., Kaufmann, P. M., McDiarmid, M. D., & Glisky, M. L. (1999). Executive functioning in preschool children: Performance on A-not-B and other delayed response format tasks. *Brain and cognition, 41,* 178–99.

Estes, A., Dawson, G., Sterling, L., & Munson, J. (2007). Level of intellectual functioning predicts patterns of associated symptoms in school children with autism spectrum disorder. *American Journal on Mental Retardation, 112*(6), 439–449.

Ewart, A. K., Morris, C. A., Atkinson, D., Jin, W., Sternes, K., Spallone, P., et al. (1993). Hemizygosity at the elastin locus in a developmental disorder, Williams-syndrome. *Nature Genetics, 5*(1), 11–16.

Fabiano, G. A., & Pelham, W. E. (2003). Improving the effectiveness of behavioral classroom interventions for attention-deficit/hyperactivity disorder: A case study. *Journal of Emotional and Behavioral Disorders, 11*(2), 122–128.

Fan, J., Flombaum, J. I., McCandliss, B. D., Thomas, K. M., & Posner, M. I. (2003). Cognitive and brain consequences of conflict. *NeuroImage, 18*(1), 42–57.

Fan, J., McCandliss, B. D., Sommer, T., Raz, A., & Posner, M. I. (2002). Testing the efficiency and independence of attentional networks. *Journal of Cognitive Neuroscience, 14*(3), 340–347.

Fan, J., & Posner, M. (2004). Human attentional networks. *Psychiatrische Praxis, 31,* S210–S214.

Fantz, R. L. (1964). Visual experience in infants: Decreased attention to familiar patterns relative to novel ones. *Science, 146,* 668–670.

Faraone, S. V., Biederman, J., & Mick, E. (2006). The age-dependent decline of attention deficit hyperactivity disorder: A meta-analysis of follow-up studies. *Psychological Medicine, 36*(2), 159–165.

Faraone, S. V., Perlis, R. H., Doyle, A. E., Smoller, J. W., Goralnick, J. J., Holmgren, M. A., et al. (2005). Molecular genetics of attention-deficit/hyperactivity disorder. *Biological Psychiatry, 57*(11), 1313–1323.

Faraone, S. V., & Wilens, T. E. (2007). Effect of stimulant medications for attention-deficit/hyperactivity disorder on later substance use and the potential for stimulant misuse, abuse, and diversion. *Journal of Clinical Psychiatry, 68*(Suppl. 11), 15–22.

Farre-Riba, A., & Narbona, J. (1997). Conners' rating scales in the assessment of attention deficit disorder with hyperactivity (ADD-H). A new validation and factor analysis in Spanish children. *Revista De Neurologia, 25*(138), 200–204.

Farroni, T., Csibra, G., Simion, F., & Johnson, M. (2002). Eye contact detection in humans from birth. *Proceedings of the National Academy of Science, 99*(14), 9602–9605.

Farroni, T., Johnson, M. H., & Csibra, G. (2004). Mechanisms of eye gaze perception during infancy. *Journal of Cognitive Neuroscience, 16*(8), 1320–1326.

Farroni, T., Massaccesi, S., Menon, E., & Johnson, M. H. (2007). Direct gaze modulates face recognition in young infants. *Cognition, 102*, 396–404.

Fatemi, S. H., Reutiman, T. J., Folsom, T. D., & Thuras, P. D. (2009). GABA(A) receptor downregulation in brains of subjects with autism. *Journal of Autism and Developmental Disorders, 39*(2), 223–230.

Fisch, G. S., Carpenter, N. J., Holden, J. J., Simensen, R., Howard-Peebles, P. N., Maddalena, A., et al. (1999). Longitudinal assessment of adaptive and maladaptive behaviors in fragile X males: Growth, development, and profiles. *American Journal of Medical Genetics, 83*(4), 257–263.

Fisch, G. S., Shapiro, L. R., Simensen, R., Schwartz, C. E., Fryns, J. P., Borghgraef, M., et al. (1992). Longitudinal changes in IQ among fragile X males: Clinical evidence of more than one mutation? *American Journal of Medical Genetics, 43*(1–2), 28–34.

Fisch, G. S., Simensen, R. J., & Schroer, R. J. (2002). Longitudinal changes in cognitive and adaptive behavior scores in children and adolescents with the fragile X mutation or autism. *Journal of Autism and Developmental Disorders, 32*(2), 107–114.

Fischer, M., Barkley, R. A., Smallish, L., & Fletcher, K. (2002). Young adult follow-up of hyperactive children: Self-reported psychiatric disorders, comorbidity, and the role of childhood conduct problems and teen CD. *Journal of Abnormal Child Psychology, 30*(5), 463–475.

Fisher, P. M., Munoz, K. E., & Hariri, A. R. (2008). Identification of neurogenetic pathways of risk for psychopathology. *American Journal of Medical Genetics Part C: Seminars in Medical Genetics, 148C*(2), 147–153.

Fisher, S. E., Francks, C., McCracken, J. T., McGough, J. J., Marlow, A. J., MacPhie, I. L., et al. (2002). A genomewide scan for loci involved in attention-deficit hyperactivity disorder. *American Journal of Human Genetics, 70*(5), 1183–1196.

Flanagan, T., Enns, J. T., Murphy, M. M., Russo, N., Abbeduto, L., Randolph, B., et al. (2007). Differences in visual orienting between persons with Down or fragile X syndrome. *Brain and Cognition, 65*(1), 128–134.

Fodor, J. (1983). *The Modularity of Mind*. MIT Press: Cambridge (Mass.).

Fombonne, E. (2009). Epidemiology of pervasive developmental disorders. *Pediatric Research, 65*(6), 591–598.

Forbes, G. B. (2001). A comparison of the Conner's Parent & Teacher Rating Scales, the ADD-H Comprehensive Teacher's Rating Scale, and the Child Behavior Checklist in the clinical diagnosis of ADHD. *Journal of Attention Disorders, 5*(1), 21–40.

Francis, K. (2005). Autism interventions: A critical update. *Developmental Medicine and Child Neurology, 47*(7), 493–499.

Frangiskakis, J. M., Ewart, A. K., Morris, C. A., Mervis, C. B., Bertrand, J., Robinson, B. F., et al. (1996). LIM-kinase1 hemizygosity implicated in impaired visuospatial constructive cognition. *Cell, 86*(1), 59–69.

Fray, P. J., Robbins, T. W., & Sahakian, B. J. (1996). Neuropsychological applications of CANTAB. *International Journal of Geriatric Psychiatry, 11*, 329–336.

Friedel, S., Saar, K., Sauer, S., Dempfle, A., Walitza, S., Renner, T., et al. (2007). Association and linkage of allelic variants of the dopamine transporter gene in ADHD. *Molecular Psychiatry, 12*(10), 923–933.

Friedman, N. P., Haberstick, B. C., Willcutt, E. G., Miyake, A., Young, S. E., Corley, R. P., et al. (2007). Greater attention problems during childhood predict poorer executive functioning in late adolescence. *Psychological Science, 18,* 893–900.

Friedman, N. P., & Miyake, A. (2004). The relations among inhibition and interference control functions: A latent-variable analysis. *Journal of Experimental Psychology: General, 133,* 101–135.

Frings, C., Feix, S., Rothig, U., Bruser, C., & Junge, M. (2007). Children do show negative priming: Further evidence for early development of an intact selective control mechanism. *Developmental Psychology, 43*(5), 1269–1273.

Fuchs, L. S., Compton, D. L., Fuchs, D., Paulsen, K., Bryant, J. D., & Hamlett, C. L. (2005). The prevention, identification, and cognitive determinants of math difficulty. *Journal of Educational Psychology, 97*(3), 493–513.

Funahashi, S. (2001). Neuronal mechanisms of executive control by the prefrontal cortex. *Neuroscience Research, 39*(2), 147–165.

Gage, N. M., Siegel, B., Callen, M., & Roberts, T. P. (2003). Cortical sound processing in children with autism disorder: An MEG investigation. *NeuroReport, 14*(16), 2047–2051.

Gainetdinov, R. R., Wetsel, W. C., Jones, S. R., Levin, E. D., Jaber, M., & Caron, M. G. (1999). Role of serotonin in the paradoxical calming effect of psychostimulants on hyperactivity. *Science, 283*(5400), 397–401.

Garber, K. B., Visootsak, J., & Warren, S. T. (2008). Fragile X syndrome. *European Journal of Human Genetics, 16*(6), 666–672.

Garon, N., Bryson, S. E., & Smith, I. M. (2008). Executive function in preschoolers: A review using an integrative framework. *Psychological Bulletin, 134*(1), 31–60.

Garretson, H. B., Fein, D., & Waterhouse, L. (1990). Sustained attention in children with autism. *Journal of Autism and Development Disorders, 20*(1), 101–114.

Garvey, M. A., & Mall, V. (2008). Transcranial magnetic stimulation in children. *Clinical Neurophysiology, 119*(5), 973–984.

Gathercole, S. (2008). Working memory in the classroom. *The Psychologist, 21*(5), 382–385.

Gathercole, S. E., & Pickering, S. J. (2000). Working memory deficits in children with low achievements in the national curriculum at 7 years of age. *British Journal Educational Psychology, 70*(2), 177–194.

Gaudino, E. A., Geisler, M. W., & Squires, N. K. (1995). Construct validity of the trail-making task: What makes Part B harder? *Journal of Clinical and Experimental Neuropsychology, 17*(4), 529–535.

Gauthier, J., Spiegelman, D., Piton, A., Lafrenière, R. G., Laurent, S., St-Onge, J., et al. (2009). Novel de novo SHANK3 mutation in autistic patients. *American Journal of Medical Genetics Part B: Neuropsychiatric Genetics, 150B*(3), 421–424.

Gerardi-Caulton, G. (2000). Sensitivity to spatial conflict and the development of self-regulation in children 24–36 months of age. *Developmental Science, 3/4,* 397–404.

Gerhardstein, P., & Rovee-Collier, C. (2002). The development of visual search in infants and very young children. *Journal of Experimental Child Psychology, 81*(2), 194–215.

Gerstadt, C. L., Hong, Y. J., & Diamond, A. (1994). The relationship between cognition and action: Performance of children 3 ½–7 years old on a Stroop-like day-night test. *Cognition, 53*(2), 129–153.

Geurts, H. M., Verté, S., Oosterlaan, J., Roeyers, H., & Sergeant, J. A. (2004). How specific are executive functioning deficits in attention deficit hyperactivity disorder and autism? *Journal of Child Psychology and Psychiatry, 45*(4), 836–854.

Geurts, H. M., Verté, S., Oosterlaan, J., Roeyers, H., & Sergeant, J. A. (2005). ADHD subtypes: Do they differ in their executive functioning profile? *Archives of clinical Neuropsychology, 20*(4), 457–477.

Giedd, J. N. (2008). The teen brain: Insights from neuroimaging. *The Journal of Adolescent Health, 42*(4), 335–343.

Gilbert, D. L., Sallee, F. R., Zhang, J., Lipps, T. D., & Wassermann, E. M. (2005). Transcranial magnetic stimulation-evoked cortical inhibition: A consistent marker of attention-deficit/hyperactivity disorder scores in Tourette syndrome. *Biological Psychiatry, 57*(12), 1597–1600.

Gilbert, D. L., Wang, Z. W., Sallee, F. R., Ridel, K. R., Merhar, S., Zhang, J., et al. (2006). Dopamine transporter genotype influences the physiological response to medication in ADHD. *Brain, 129*(8), 2038–2046.

Gilbert, S. J., Bird, G., Brindley, R., Frith, C. D., & Burgess, P. W. (2008). Atypical recruitment of medial prefrontal cortex in autism spectrum disorders: An fMRI study of two executive function tasks. *Neuropsychologia, 46*(9), 2281–2291.

Gillberg, C., & Rasmussen, P. (1994). Brief report: Four case histories and a literature review of Williams syndrome and autistic behavior. *Journal of Autism and Developmental Disorders, 24*(3), 381–393.

Gillis, J. J., Gilger, J. W., Pennington, B. F., DeFries, J. C. (1992). Attention deficit disorder in reading-disabled twins: Evidence for a genetic etiology. *Journal of Abnormal Child Psychology, 20*(3), 303–315.

Gioia, G. A., Espy, K. A. & Isquith, P. K. (2003). The Behavior Rating Inventory of Executive Function-Preschool version (BRIEF-P). Odessa, FL: Psychological Assessment Resources.

Gioia, G. A., Isquith, P. K., Guy, S. C., Kenworthy, L., & Baron, I. S. (2000). Test review: Behavior rating inventory of executive function. *Child Neuropsychology, 6*(3), 235–238.

Gioia, G. A., Isquith, P. K., Kenworthy, L., & Barton, R. M. (2002). Profiles of everyday executive function in acquired and developmental disorders. *Child Neuropsychology, 8*(2), 121–137.

Glaser, B., Debbane, M., Hinard, C., Morris, M. A., Dahoun, S. P., Antonarakis, S. E., et al. (2006). No evidence for an effect of COMT Val158Met genotype on executive function in patients with 22q11 deletion syndrome. *The American Journal of Psychiatry, 163*(3), 537–539.

Gliga, T., & Csibra, G. (2007). Seeing the face through the eyes: A developmental perspective on face expertise. *Progress in Brain Research, 164,* 323–339.

Goel, V., & Grafman, J. (1995). Are the frontal lobes implicated in "planning" functions? Interpreting data from the Tower of Hanoi. *Neuropsychologia, 33*(5), 623–642.

Gogtay, N., Giedd, J. N., Lusk, L., Hayashi, K. M., Greenstein, D., Vaituzis, A. C., et al. (2004). Dynamic mapping of human cortical development during childhood through early adulthood. *Proceedings of the National Academy of Sciences of the United States of America, 101*(21), 8174–8179.

Goldberg, M. C., Mostofsky, S. H., Cutting, L. E., Mahone, E. M., Astor, B. C., Denckla, M. B., et al. (2005). Subtle executive impairment in children with autism and children with ADHD. *Journal of Autism and Developmental Disorders, 35*(3), 279–293.

Goldberg, T. E., Egan, M. F., Gscheidle, T., Coppola, R., Weickert, T., Kolachana, B. S., et al. (2003). Executive subprocesses in working memory: Relationship to catechol-O-methyltransferase Val158Met genotype and schizophrenia. *Archives of General Psychiatry, 60*(9), 889–896.

Golden, J. C. (1978). *Stroop color and word test.* Chicago: Stoetling Company.

Goldman, D. Z., Shapiro, E. G., & Nelson, C. A. (2004). Measurement of vigilance in 2-year-old children. *Developmental Neuropsychology, 25*(4), 227–250.

Gomot, M., Belmonte, M. K., Bullmore, E. T., Bernard, F. A., Baron-Cohen, S. (2008). Brain hyper-reactivity to auditory novel targets in children with high-functioning autism. *Brain, 131*(9), 2479–88.

Goodale, M. A., & Milner, A. D. (1992). Separate visual pathways for the perception of action. *Trends in Neurosciences, 15*(1), 20–25.

Goodman, R. (1997). The Strengths and Difficulties Questionnaire: A Research Note. *Journal of Child Psychology and Psychiatry, 38,* 581–586.

Goodman, R. (2001). Psychometric properties of the strengths and difficulties questionnaire. *Journal of the American Academy of Child and Adolescent Psychiatry, 40*(11), 1337–1345.

Goodman R., & Scott S. (1999). Comparing the Strengths and Difficulties Questionnaire and the Child Behavior Checklist: Is small beautiful? *Journal of Abnormal Child Psychology, 27*(1), 17–24.

Goodship, J., Cross, I., & LiLing, J. (1998). A population study of chromosome 22q11 deletions in infancy. *Archives of Disease in Childhood, 79*(4), 348–351.

Gornick, M. C., Addington, A., Shaw, P., Bobb, A. J., Sharp, W., Greenstein, D., et al. (2007). Association of the dopamine receptor D4 (DRD4) gene 7-repeat allele with children with attention-deficit/hyperactivity disorder (ADHD): An update. *American Journal of Medical Genetics Part B: Neuropsychiatric Genetics, 144B*(3), 379–382.

Gothelf, D., Eliez, S., Thompson, T., Hinard, C., Penniman, L., Feinstein, C., et al. (2005). COMT genotype predicts longitudinal cognitive decline and psychosis in 22q11.2 deletion syndrome. *Nature Neuroscience, 8*(11), 1500–1502.

Gothelf, D., Hoeft, F., Hinard, C., Hallmayer, J. F., Stoecker, J. V., Antonarakis, S. E., et al. (2007). Abnormal cortical activation during response inhibition in 22q11.2 deletion syndrome. *Human Brain Mapping, 28*(6), 533–542.

Gothelf, D., Michaelovsky, E., Frisch, A., Zohar, A. H., Presburger, G., Burg, M., et al. (2007b). Association of the low-activity COMT (158) Met allele with ADHD and OCD in subjects with velocardiofacial syndrome. *The International Journal of Neuropsychopharmacology, 10*(3), 301–308.

Gothelf, D., Schaer, M., & Eliez, S. (2008). Genes, brain development and psychiatric phenotypes in velo-cardio-facial syndrome. *Developmental Disabilities Research Review, 14*(1), 59–68.

Grady, D. L., Harxhi, A., Smith, M., Flodman, P., Spence, M. A., Swanson, J. M., et al. (2005). Sequence variants of the DRD4 gene in autism: Further evidence that rare DRD4 7R haplotypes are ADHD specific. *American Journal of Medical Genetics Part B: Neuropsychiatric Genetics, 136B*(1), 33–35.

Graham, J. M., Rosner, B., Dykens, E., & Visootsak, J. (2005). Behavioral features of CHARGE syndrome (Hall-Hittner syndrome) comparison with Down syndrome, Prader-Willi syndrome, and Williams syndrome. *American Journal of Medical Genetics Part A, 133A*(3), 240–247.

Gray, K. M., & Tonge, B. J. (2005). Screening for autism in infants and preschool children with developmental delay. *Australian and New Zealand Journal of Psychiatry, 39*(5), 378–386.

Gray, K. M., Tonge, B. J., Sweeney, D. J., & Einfeld, S. L. (2008). Screening for autism in young children with developmental delay: An evaluation of the developmental behaviour checklist: Early screen. *Journal of Autism and Developmental Disorders, 38*(6), 1003–1010.

Gray, V., Karmiloff-Smith, A., Funnell, E., & Tassabehji, M. (2006). In-depth analysis of spatial cognition in Williams syndrome: A critical assessment of the role of the LIMK1 gene. *Neuropsychologia, 44*(5), 679–685.

Greber-Platzer, S., Fleischmann, C., Nussbaumer, C., Cairns, N., & Lubec, G. (2003). Increased RNA levels of the 25 kDa synaptosomal associated protein in brain samples of adult patients with Down Syndrome. *Neuroscience Letters, 336*(2), 77–80.

Greenberg, B. D., Tolliver, T. J., Huang, S. J., Li, Q., Bengel, D., & Murphy, D. L. (1999). Genetic variation in the serotonin transporter promoter region affects serotonin uptake in human blood platelets. *American Journal of Medical Genetics Part B: Neuropsychiatric Genetics, 88*(1), 83–87.

Greenhill, L., Kollins, S., Abikoff, H., McCracken, J., Riddle, M., Swanson, J., et al. (2006). Efficacy and safety of immediate-release methylphenidate treatment for preschoolers with ADHD. *Journal of the American Academy of Child and Adolescent Psychiatry, 45*(11), 1284–1293.

Greenhill, L. L., Posner, K., Vaughan, B. S., & Kratochvil, C. J. (2008). Attention deficit hyperactivity disorder in preschool children. *Child and Adolescent Psychiatric Clinics of North America, 17*(2), 347–366.

Greer, M. K., Brown, F. R., Pai, G. S., Choudry, S. H., & Klein, A. J. (1997). Cognitive, adaptive, and behavioral characteristics of Williams syndrome. *American Journal of Medical Genetics, 74*(5), 521–525.

Gregg, J. P., Lit, L., Baron, C. A., Hertz-Picciotto, I., Walker, W., Davis, R. A., et al. (2008). Gene expression changes in children with autism. *Genomics, 91*(1), 22–29.

Greimel, E., Herpertz-Dahlmann, B., Gunther, T., Vitt, C., & Konrad, K. (2008). Attentional functions in children and adolescents with attention-deficit/hyperactivity disorder with and without comorbid tic disorder. *Journal of Neural Transmission, 115*(2), 191–200.

Grice, S. J., de Haan, M., Halit, H., Johnson, M. H., Csibra, G., Grant, L., et al. (2003). ERP abnormalities of illusory contour perception in Williams Syndrome. *NeuroReport, 14*(14), 1773–1777.

Grigsby, J., Brega, A. G., Engle, K., Leehey, M. A., Hagerman, R. J., Tassone, F., et al. (2008). Cognitive profile of fragile X premutation carriers with and without fragile X-associated tremor/ataxia syndrome. *Neuropsychology, 22*(1), 48–60.

Groen, W. B., Zwiers, M. P., van der Gaag, R. J., & Buitelaar, J. K. (2008). The phenotype and neural correlates of language in autism: An integrative review. *Neuroscience & Biobehavoral Reviews, 32*(8), 1416–1425.

Groen, Y., Wijers, A. A., Mulder, L. J., Waggeveld, B., Minderaa, R., & Althaus, M. (2008). Error and feedback processing in children with ADHD and children with Autistic Spectrum Disorder: An EEG event-related potential study. *Clinical Neurophysiology, 119*(11), 2476–2493.

Grund, T., Lehmann, K., Bock, N., Rothenberger, A., & Teuchert-Noodt, G. (2006). Influence of methylphenidate on brain development – an update of recent animal experiments. *Behavioral and Brain Functions, 2*, 2.

Guitton, D., Buchtel, H. A., & Douglas, R. M. (1985). Frontal-lobe lesions in man cause difficulties in suppressing reflexive glances and in generating goal-directed saccades. *Experimental Brain Research, 58*(3), 455–472.

Gumenyuk, V., Korzyukov, O., Escera, C., Hamalainen, M., Huotilainen, M., Hayrinen, T., et al. (2005). Electrophysiological evidence of enhanced distractibility in ADHD children. *Neuroscience Letters, 374*(3), 212–217.

Gustavson, K. H., Floderus, Y., Jagell, S., Wetterberg, L., & Ross, S. B. (1982). Catechol-O-methyltransferase activity in erythrocytes in Down's syndrome: Family studies. *Clinical Genetics, 22*(1), 22–24.

Hagerman, P. J. (2008). The fragile X prevalence paradox. *Journal of Medical Genetics, 45*(8), 498–499.

Hagerman, R. J., Murphy, M. A., & Wittenberger, M. D. (1988). A controlled trial of stimulant medication in children with the fragile X syndrome. *American Journal of Medical Genetics, 30*(1–2), 377–392.

Hale, J. B., How, S. K., Dewitt, M. B., & Coury, D. L. (2001). Discriminant validity of the Conners' scales for ADHD subtypes. *Current Psychology, 20*(3), 231–249.

Halperin, J. M., Trampush, J. W., Miller, C. J., Marks, D. J., & Newcorn, J. H. (2008). Neuropsychological outcome in adolescents/young adults with childhood ADHD: Profiles of persisters, remitters and controls. *Journal of Child Psychology and Psychiatry, 49*(9), 958–966.

Hammond, P., Forster-Gibson, C., Chudley, A. E., Allanson, J. E., Hutton, T. J., Farrell, S. A., et al. (2008). Face-brain asymmetry in autism spectrum disorders. *Molecular Psychiatry, 13*(6), 614–623.

Hanisch, C., Konrad, K., Günther, T., & Herpertz-Dahlmann, B. (2004). Age-dependent neuropsychological deficits and effects of methylphenidate in children with attention-deficit/hyperactivity disorder: A comparison of pre- and grade-school children. *Journal of Neural Transmission, 111*(7), 865–868.

Hanisch, C., Radach, R., Holtkamp, K., Herpertz-Dahlmann, B., & Konrad, K. (2006). Oculomotor inhibition in children with and without attention-deficit hyperactivity disorder (ADHD). *Journal of Neural Transmission, 113*(5), 671–684.

Happé, F., Booth, R., Charlton, R., & Hughes, C. (2006). Executive function deficits in autism spectrum disorders and attention-deficit/hyperactivity disorder: Examining profiles across domains and ages. *Brain and Cognition, 61*(1), 25–39.

Happé, F., & Ronald, A. (2008). The 'fractionable autism triad': A review of evidence from behavioural, genetic, cognitive and neural research. *Neuropsychology Review, 18*(4), 287–304.

Harlow, J. (1868). Recovery after severe injury to the head. *Publications of the Massachusetts Medical Society, 2,* 327–336.

Hartley, S. L., Sikora, D. M., & McCoy, R. (2008). Prevalence and risk factors of maladaptive behaviour in young children with Autistic Disorder. *Journal of Intellectual Disability Research, 52*(10), 819–829.

Harvey, N., & Treisman, A. (1973). Switching attention from the ears to monitor tones. *Perception and Psychophysics, 14,* 51–59.

Hastings, R. P., Beck, A., Daley, D., & Hill, C. (2005). Symptoms of ADHD and their correlates in children with intellectual disabilities. *Research in Developmental Disabilities, 26*(5), 456–468.

Hastings, R. P., Brown, T., Mount, R. H., & Cormack, K. F. M. (2001). Exploration of psychometric properties of the developmental behavior checklist. *Journal of Autism and Developmental Disorders, 31*(4), 423–431.

Hatton, D. D., Hooper, S. R., Bailey, D. B., Skinner, M. L., Sullivan, K. M., & Wheeler, A. (2002). Problem behavior in boys with fragile X syndrome. *American Journal of Medical Genetics, 108*(2), 105–116.

Hattori, J., Ogino, T., Abiru, K., Nakano, K., Oka, M., & Ohtsuka, Y. (2006). Are pervasive developmental disorders and attention-deficit/hyperactivity disorder distinct disorders? *Brain & Development, 28*(6), 371–374.

Hattori, M., Fujiyama, A., Taylor, T. D., Watanabe, H., Yada, T., Park, H. S., et al. (2000). The DNA sequence of human chromosome 21 (Vol. 405, p. 311, 2000). *Nature, 407*(6800), 110–110.

Hawes, D. J., & Dadds, M. R. (2004). Australian data and psychometric properties of the Strengths and Difficulties Questionnaire. *Australian and New Zealand Journal of Psychiatry, 38*(8), 644–651.

Hawi, Z., Dring, M., Kirley, A., Foley, D., Kent, L., Craddock, N., et al. (2002). Serotonergic system and attention deficit hyperactivity disorder (ADHD): A potential susceptibility locus at the 5-HT1B receptor gene in 273 nuclear families from a multi-centre sample. *Molecular Psychiatry, 7*(7), 718–725.

Hawi, Z., Millar, N., Daly, G., Fitzgerald, M., & Gill, M. (2000). No association between catechol-O-methyltransferase (COMT) gene polymorphism and attention deficit hyperactivity disorder (ADHD) in an Irish sample. *American Journal of Medical Genetics Part B: Neuropsychiatric Genetics, 96*(3), 282–284.

Hay, D. A., Bennett, K. S., Levy, F., Sergeant, J., & Swanson, J. (2007). A twin study of attention-deficit/hyperactivity disorder dimensions rated by the Strengths and Weaknesses of ADHD-Symptoms and Normal-Behavior (SWAN) scale. *Biological Psychiatry, 61*(5), 700–705.

Hayes, L. (2007). Problem behaviours in early primary school children: Australian normative data using the Strengths and Difficulties Questionnaire. *Australian and New Zealand Journal of Psychiatry, 41*(3), 231–238.

Hazeltine, E., Ruthruff, E., & Remington, R. W. (2006). The role of input and output modality pairings in dual-task performance: Evidence for content-dependent central interference. *Cognitive Psychology, 52*(4), 291–345.

Heaton, R. K. (1981). *A manual for the Wisconsin Card Sorting Task*. Odessa, FL: Psychological Assessment Resources.

Heaton, S. C., Reader, S. K., Preston, A. S., Fennell, E. B., Puyana, O. E., Gill, N. et al. (2001). The Test of Everyday Attention for Children (TEA-Ch): Patterns of performance in children with ADHD and clinical controls. *Child Neuropsychology, 7*(4), 251–264.

Henik, A., & Salo, R. (2004). Schizophrenia and the Stroop effect. *Behavioral and Cognitive Neuroscience Reviews, 3*(1), 42–59.

Herrmann, C. S., & Knight, R. T. (2001). Mechanisms of human attention: Event-related potentials and oscillations. *Neuroscience and Biobehavioral Reviews, 25*(6), 465–476.

Hessl, D., Tassone, F., Cordeiro, L., Koldewyn, K., McCormick, C., Green, C., et al. (2008). Brief report: Aggression and stereotypic behavior in males with fragile X syndrome: Moderating secondary genes in a "single gene" disorder. *Journal of Autism and Developmental Disorders, 38*(1), 184–189.

Hinshaw, S. P. (2007). Moderators and mediators of treatment outcome for youth with ADHD: Understanding for whom and how interventions work. *Journal of Pediatric Psychology, 32*(6), 664–675.

Hitch, G. J., Towse, J. N., & Hutton, U. (2001). What limits children's working memory span? Theoretical accounts and applications for scholastic development. *Journal of Experimental Psychology: General, 130*, 184–198.

Hodapp, R. M. (2004). Behavioral phenotypes: Going beyond the two-group approach. In L. J. Glidden (Ed.), *International review of research in mental retardation* (Vol. 29, pp. 1–30). San Diego, CA: Academic Press.

Hodapp, R. M., Burack, J. A., & Zigler, E. (1998). Developmental approaches to mental retardation: A short introduction. In J. A. Burack, R. M. Hodapp, & E. Zigler (Eds.), *Handbook of mental retardation and development* (pp. 3–19). New York: Cambridge University Press.

Hoeft, F., Hernandez, A., Parthasarathy, S., Watson, C. L., Hall, S.S., & Reiss, A. L. (2007). Fronto-striatal dysfunction and potential compensatory mechanisms in male adolescents with fragile X syndrome. *Human Brain Mapping, 28*(6), 543–554.

Hoeft, F., Lightbody, A. A., Hazlett, H. C., Patnaik, S., Piven, J., & Reiss, A. L. (2008). Morphometric spatial patterns differentiating boys with fragile X syndrome, typically developing boys, and developmentally delayed boys aged 1 to 3 years. *Archives of General Psychiatry, 65*(9), 1087–1097.

Hollander, E., Kaplan, A., Cartwright, C., & Reichman, D. (2000). Venlafaxine in children, adolescents, and young adults with autism spectrum disorders: An open retrospective clinical report. *Journal of Child Neurology, 15*(2), 132–135.

Hollander, E., Phillips, A. T., & Yeh, C. C. (2003). Targeted treatments for symptom domains in child and adolescent autism. *Lancet, 362*(9385), 732–734.

Holmes, J., Gathercole, S. E., & Place, M. (2008). Working memory deficits can be overcome: Impacts of training and medication on working memory in children with ADHD. Manuscript submitted for publication.

Holmes, J., Payton, A., Barrett, J. H., Hever, T., Fitzpatrick, H., Trumper, A. L., et al. (2000). A family-based and case-control association study of the dopamine D4 receptor gene and dopamine transporter gene in attention deficit hyperactivity disorder. *Molecular Psychiatry, 5*(5), 523–530.

Holtmann, M., Bolte, S., & Poustka, F. (2007). Attention deficit hyperactivity disorder symptoms in pervasive developmental disorders: Association with autistic behavior domains and coexisting psychopathology. *Psychopathology, 40*(3), 172–177.

Hood, B. M., & Atkinson, J. (1993). Disengaging visual attention in the infant and adult. *Infant Behavior and Development, 16*(4), 405–422.

Hood, B. M., Willen, J. D., & Driver, J. (1998). Adult's eyes trigger shifts of visual attention in human infants. *Psychological Science, 9*(2), 131–134.

Hood, J., Baird, G., Rankin, P. M., & Isaacs, E. (2005). Immediate effects of methylphenidate on cognitive attention skills of children with attention-deficit-hyperactivity disorder. *Developmental Medicine and Child Neurology, 47*(6), 408–414.

Hoogenraad, C. C., Koekkoek, B., Akhmanova, A., Krugers, H., Dortland, B., Miedema, M., et al. (2002). Targeted mutation of Cyln2 in the Williams syndrome critical region links CLIP-115 haploin sufficiency to neurodevelopmental abnormalities in mice. *Nature Genetics, 32*(1), 116–127.

Hooper, S. R., Hatton, D., Sideris, J., Sullivan, K., Hammer, J., Schaaf, J., et al. (2008). Executive functions in young males with fragile X syndrome in comparison to mental age-matched with controls: Baseline findings from a longitudinal study. *Neuropsychology, 22*(1), 36–47.

Hooper, S. R., Poon, K. K., Marcus, L., & Fine, C. (2006). Neuropsychological characteristics of school-age children with high-functioning autism: Performance on the NEPSY. *Child Neuropsychology, 12*(4–5), 299–305.

Huang, C. H., & Santangelo, S. L. (2008). Autism and serotonin transporter gene polymorphisms: A systematic review and meta-analysis. *American Journal of Medical Genetics B: Neuropsychiatric Genetics, 147B*(6), 903–913.

Huang-Pollock, C. L., Nigg, J. T., & Halperin, J. M. (2006). Single dissociation findings of ADHD deficits in vigilance but not anterior or posterior attention systems. *Neuropsychology, 20*(4), 420–429.

Hughes, C., Russell, J., & Robbins, T. W. (1994). Evidence or executive dysfunction in autism. *Neuropsychologia, 32*(4), 477–492.

Huizinga, M., Dolan, C. V., & van der Molen, M. W. (2006). Age-related change in executive function: Developmental trends and a latent variable analysis. *Neuropsychologia, 44*(11), 2017–2036.

Humphrey, N., & Parkinson, G. (2006). Research on interventions for children and young people on the autistic spectrum: A critical perspective. *Journal of Research in Special Educational Needs, 6*(2), 75–86.

Huotilainen, M., Shestakova, A., & Hukki, J. (2008). Using magnetoencephalography in assessing auditory skills in infants and children. *International Journal of Psychophysiology, 68*(2), 123–129.

Huss, M., Poustka, F., Lehmkuhl, G., & Lehmkuhl, U. (2008). No increase in long-term risk for nicotine use disorders after treatment with methylphenidate in children with attention-deficit/hyperactivity disorder (ADHD): Evidence from a non-randomised retrospective study. *Journal of Neural Transmission, 115*(2), 335–339.

Iarocci, G., & Burack, J. A. (2004). Intact covert orienting to peripheral cues among children with autism. *Journal of Autism and Developmental Disorders, 34*(3), 257–264.

Inoue, Y., Inagaki, M., Gunji, A., Furushima, W., & Kaga, M. (2008). Response switching process in children with attention-deficit-hyperactivity disorder on the novel continuous performance test. *Developmental Medicine and Child Neurology, 50*(6), 462–466.

Iversen, S. D., & Iversen, L. L. (2007). Dopamine: 50 years in perspective. *Trends in Neurosciences, 30*(5), 188–193.

Jacobsen, C. F., Wolfe, J. B., & Jackson, T. A. (1935). An experimental analysis of the functions of the frontal association areas in primates. *Journal of Nervous and Mental Disease, 82*, 1–14.

Jansen, P. W., Duijff, S. N., Beemer, F. A., Vorstman, J. A. S., Klaassen, P. W. J., Morcus, M. E. J., et al. (2007). Behavioral problems in relation to intelligence in children with 22q11.2 deletion syndrome: A matched control study. *American Journal of Medical Genetics Part A, 143A*(6), 574–580.

Jarrold, C., Baddeley, A. D., & Hewes, A. K. (1999). Genetically dissociated components of working memory: Evidence from Down's and Williams syndrome. *Neuropsychologia, 37*(6), 637–651.

Jarrold, C., Baddeley, A. D., & Phillips, C. (2007). Long-term memory for verbal and visual information in Down syndrome and Williams syndrome: Performance on the Doors and People test. *Cortex, 43*(2), 233–247.

Jarrold, C., Gilchrist, I. D., & Bender, A. (2005). Embedded figures detection in autism and typical development: Preliminary evidence of a double dissociation in relationships with visual search. *Developmental Science, 8*(4), 344–351.

Jarrold, C., Thorn, A. S. C., & Stephens, E. (2009). The relationships among verbal short-term memory, phonological awareness, and new word learning: Evidence from typical development and Down syndrome. *Journal of experimental child psychology, 102*(2), 196–218.

Jensen, P. S., Arnold, L. E., Swanson, J. M., Vitiello, B., Abikoff, H. B., Greenhill, L. L., et al. (2007). Three-year follow-up of the NIMH MTA study. *Journal of the American Academy of Child and Adolescent Psychiatry, 46*(8), 989–1002.

Jeste, S. S., & Nelson, C. A. (2009). Event related potentials in the understanding of autism spectrum disorders: An analytical review. *Journal of Autism and Developmental Disorders, 39*(3), 495–510.

Johnson, J. A., & Zatorre, R. J. (2006). Neural substrates for dividing and focusing attention between simultaneous auditory and visual events. *NeuroImage, 31*(4), 1673–1681.

Johnson, K. A., Barry, E., Bellgrove, M. A., Cox, M., Kelly, S. P., Daibhis, A., et al. (2008). Dissociation in response to methylphenidate on response variability in a group of medication naive children with ADHD. *Neuropsychologia, 46*(5), 1532–1541.

Johnson, K. A., Kelly, S. P., Robertson, I. H., Barry, E., Mulligan, A., Daly, M., et al. (2008). Absence of the 7-repeat variant of the DRD4 VNTR is associated with drifting sustained attention in children with ADHD but not in controls. *American Journal of Medical Genetics Part B: Neuropsychiatric Genetics, 147B*(6), 927–937.

Johnson, K. A., Robertson, I. H., Kelly, S. P., Silk, T. J., Barry, E., Daibhis, A., et al. (2007). Dissociation in performance of children with ADHD and high-functioning autism on a task of sustained attention. *Neuropsychologia, 45*(10), 2234–2245.

Johnson, M. H. (1994). Visual attention and the control of eye-movements in early infancy. In C. Umilta & M. Moscovitch (Eds.), *Attention and performance XV* (pp. 291–310). Cambridge, MA: MIT Press.

Johnson, M. H. (1995). The inhibition of automatic saccades in early infancy. *Developmental Psychobiology, 28*(5), 281–291.

Johnson, M. H. (2001). Functional brain development in humans. *Nature Reviews Neuroscience, 2*, 475–483.

Johnson, M. H., Halit, H., Grice, S. J., & Karmiloff-Smith, A. (2002). Neuroimaging of typical and atypical development: A perspective from multiple levels of analysis. *Development and Psychopathology, 14*(3), 521–536.

Johnson, M. H., Posner, M. I., & Rothbart, M. K. (1991). Components of visual orienting in early infancy: Contingency learning, anticipatory looking, and disengaging. *Journal of Cognitive Neuroscience, 3*(4), 335–344.

Johnstone, S. J., Dimoska, A., Smith, J. L., Barry, R. J., Pleffer, C. B., Chiswick, D. et al. (2007). The development of stop-signal and Go/Nogo response inhibition in children aged 7–12: Performance and event-related potential indices. *International Journal of Psychophysiology, 63*(1), 25–38.

Jones, L. B., Rothbart, M. K., & Posner, M. I. (2003). Development of executive attention in preschool children. *Developmental Science, 6*(5), 498–504.

Joober, R., Grizenko, N., Sengupta, S., Ben Amor, L., Schmitz, N., Schwartz, G., et al. (2007). Dopamine transporter 3'-UTR VNTR genotype and ADHD: A pharmaco-behavioural genetic study with methylphenidate. *Neuropsychopharmacology, 32*(6), 1370–1376.

Joseph, R. M., McGrath, L. M., & Tager-Flusberg, H. (2005). Executive dysfunction and its relation to language ability in verbal school-age children with autism. *Developmental Neuropsychology, 27*(3), 361–378.

Joseph, R. M., Steele, S. D., Meyer, E., & Tager-Flusberg, H. (2005). Self-ordered pointing in children with autism: Failure to use verbal mediation in the service of working memory? *Neuropsychologia, 43*(10), 1400–1411.

Jucaite, A., Fernell, E., Halldin, C., Forssberg, H., & Farde, L. (2005). Reduced midbrain dopamine transporter binding in male adolescents with attention-deficit/hyperactivity disorder: Association between striatal dopamine markers and motor hyperactivity. *Biological Psychiatry, 57*(3), 229–238.

Kabakus, N., Aydin, M., Akin, H., Balci, T. A., Kurt, A., & Kekilli, E. (2006). Fragile X syndrome and cerebral perfusion abnormalities: Single-photon emission computed tomographic study. *Journal of Child Neurology, 21*(12), 1040–1046.

Kadesjo, C., Hagglof, B., Kadesjo, B., & Gillberg, C. (2003). Attention-deficit-hyperactivity disorder with and without oppositional defiant disorder in 3- to 7-year-old children. *Developmental Medicine and Child Neurology, 45*(10), 693–699.

Kaffman, A., & Meaney, M. J. (2007). Neurodevelopmental sequelae of postnatal maternal care in rodents: Clinical and research implications of molecular insights. *Journal of Child Psychology and Psychiatry, 48*(3–4), 224–244.

Kagan, J., Rosman, B. L., Day, D., Albert, J., & Phillips, W. (1964). Information processing in the child: Significance of analytic and reflective attitudes. *Psychological Monographs, 78*, 1–37.

Kahneman, D. (1973). *Attention and effort.* Englewood Cliffs, NJ: Prentice-Hall.

Kahneman, D., & Treisman, A. M. (1984). Changing views of attention and automaticity. In D. R. Davies & R. Parasuraman (Eds.), *Varieties of attention.* Orlando, FL: Academic Press.

Kaland, N., Smith, L., & Mortensen, E. L. (2008). Brief report: Cognitive flexibility and focused attention in children and adolescents with Asperger Syndrome or high- functioning autism as measured on the computerized version of the Wisconsin Card Sorting Test. *Journal of Autism and Developmental Disorders, 38*(6), 1161–1165.

Kaller, C. P., Rahm, B., Spreer, J., Mader, I., & Unterrainer, J. M. (2008). Thinking around the corner: The development of planning abilities. *Brain and Cognition, 67*(3), 360–370.

Kalff, A. C., Hendriksen, J. G. M., Kroes, M., Vles, J. S. H., Steyaert, J., Feron, F. J. M., et al. (2002). Neurocognitive performance of 5- and 6-year-old children who met criteria for attention deficit/hyperactivity disorder at 18 months follow-up: Results from a prospective population study. *Journal of Abnormal Child Psychology, 30*(6), 589–598.

Kana, R. K., Keller, T. A., Minshew, N. J., & Just, M. A. (2007). Inhibitory control in high-functioning autism: Decreased activation and underconnectivity in inhibition networks. *Biological Psychiatry, 62*(3), 198–206.

Kandel, E. R., Schwartz, J. H., & Jessell, T. M. (2008). *Principles of neural science* (5th ed.). New York: McGraw-Hill.

Kane, M. J., & Engle, R. W. (2000). Working-memory capacity, proactive interference, and divided attention: Limits on long-term memory retrieval. *Journal of Experimental Psychology-Learning Memory and Cognition, 26*(2), 336–358.

Kannass, K. N., & Colombo, J. (2007). The effects of continuous and intermittent distractors on cognitive performance and attention in preschoolers. *Journal of Cognition and Development, 8*(1), 63–77.

Kannass, K. N., Oakes, L. M., & Shaddy, D. J. (2006). A longitudinal investigation of the development of attention and distractibility. *Journal of Cognition and Development, 7*, 381–409.

Kanner, L. (1943). Autistic disturbances of affective contact. *Nervous Child, 2*, 217–250.

Karatekin, C. (2006). Improving antisaccade performance in adolescents with attention-deficit/hyperactivity disorder (ADHD). *Experimental Brain Research, 174*(2), 324–341.

Karatekin, C., & Asarnow, R. F. (1998). Components of visual search in childhood-onset schizophrenia and attention-deficit/hyperactivity disorder. *Journal of Abnormal Child Psychology, 26*(5), 367–380.

Karmiloff-Smith, A. (1998). Development itself is the key to understanding developmental disorders. *Trends in Cognitive Sciences, 2*(10), 389–398.

Karmiloff-Smith, A. (2007). Atypical epigenesis. *Developmental Science, 10*(1), 84–88.

Karmiloff-Smith, A. (2009). Nativism versus neuroconstructivism: Rethinking the study of developmental disorders. *Developmental Psychology, 45*(1), 56–63.

Karmiloff-Smith, A., Grant, J., Ewing, S., Carette, M. J., Metcalfe, K., Donnai, D., et al. (2003). Using case study comparisons to explore genotype-phenotype correlations in Williams-Beuren syndrome. *Journal of Medical Genetics, 40*(2), 136–140.

Karmiloff-Smith, A., Thomas, M., Annaz, D., Humphreys, K., Ewing, S., Brace, N., et al. (2004). Exploring the Williams syndrome face-processing debate: The importance of building developmental trajectories. *Journal of Child Psychology and Psychiatry, 45*(7), 1258–1274.

Karoum, F., Chrapusta, S. J., & Egan, M. F. (1994). 3-Methoxytyramine is the major metabolite of released dopamine in the rat frontal cortex: Reassessment of the effects of antipsychotics on the dynamics of dopamine release and metabolism in the frontal cortex, nucleus accumbens, and striatum by a simple two-pool model. *Journal of Neurochemistry, 63*(3), 972–979.

Karrer, J. H., Karrer, R., Bloom, D., Chaney, L., & Davis, R. (1998). Event-related brain potentials during an extended visual recognition memory task depict delayed development of cerebral inhibitory processes among 6-month-old infants with Down syndrome. *International Journal of Psychophysiology, 29*(2), 167–200.

Kastner, S., & Ungerleider, L. G. (2000). Mechanisms of visual attention in the human cortex. *Annual Review of Neuroscience, 23*, 315–341.

Kau, A. S. M., Tierney, E., Bukelis, I., Stump, M. H., Kates, W. R., Trescher, W. H., et al. (2004). Social behavior profile in young males with fragile X syndrome: Characteristics and specificity. *American Journal of Medical Genetics Part A, 126A*(1), 9–17.

Kebir, O., Tabbane, K., Sengupta, S., & Joober, R. (2009). Candidate genes and neuropsychological phenotypes in children with ADHD: Review of association studies. *Journal of Psychiatry and Neuroscience, 34*(2), 88–101.

Keenan, K., & Wakschlag, L. S. (2000). More than the terrible twos: The nature and severity of behavior problems in clinic-referred preschool children. *Journal of Abnormal Child Psychology, 28*(1), 33–46.

Kelly, A. M., Margulies, D. S., & Castellanos, F. X. (2007). Recent advances in structural and functional brain imaging studies of attention-deficit/hyperactivity disorder. *Current Psychiatry Reports, 9*(5), 401–407.

Kemp, S. L., Kirk, U., & Korkman, M. (2001). *Essentials of NEPSY assessment.* New York: Wiley.

Kent, L., Ivans, J., Paul, M., & Sharp, M. (1999). Comorbidity of autistic spectrum disorders in children with Down syndrome. *Developmental Medicine and Child Neurology, 41*(3), 153–158.

Kessler, R. C., Adler, L., Barkley, R., Biederman, J., Conners, C. K., Demler, O., et al. (2006). The prevalence and correlates of adult ADHD in the United States: Results from the National Comorbidity Survey Replication. *The American Journal of Psychiatry, 163*(4), 716–723.

Kieling, C., Goncalves, R. R. F., Tannock, R., & Castellanos, F. X. (2008). Neurobiology of attention deficit hyperactivity disorder. *Child and Adolescent Psychiatric Clinics of North America, 17*(2), 285–307.

Kieling, C., Roman, T., Doyle, A. E., Hutz, M. H., & Rohde, L. A. (2006). Association between DRD4 gene and performance of children with ADHD in a test of sustained attention. *Biological Psychiatry, 60*(10), 1163–1165.

Kiley-Brabeck, K., & Sobin, C. (2006). Social skills and executive function deficits in children with the 22q11 Deletion Syndrome. *Applied Neuropsychology, 13*(4), 258–268.

King, S., Griffin, S., Hodges, Z., Weatherly, H., Asseburg, C., Richardson, G., et al. (2006). A systematic review and economic model of the effectiveness and cost-effectiveness of methylphenidate, dexamfetamine and atomoxetine for the treatment of attention deficit hyperactivity disorder in children and adolescents. *Health Technology Assessment Database, 10*(23), iii–iv, xiii–146.

Kirk, J. W., Mazzocco, M. M. M., & Kover, S. T. (2005). Assessing executive dysfunction in girls with fragile X or Turner syndrome using the Contingency Naming Test (CNT). *Developmental Neuropsychology, 28*(3), 755–777.

Kirkham, N. Z., Cruess, L. M., & Diamond, A. (2003). Helping children apply their knowledge to their behavior on a bimension-switching task. *Developmental Science, 6*, 449–467.

Klasen, H., Woerner, W., Rothenberger, A., & Goodman, R. (2003). The German version of the Strengths and Difficulties Questionnaire (SDQ-Deu)—Overview over first validation and normative studies. *Praxis der Kinderpsychologie und Kinderpsychiatrie, 52*(7), 491–502.

Klein, C., Wendling, K., Huettner, P., Ruder, H., & Peper, M. (2006). Intra-subject variability in attention-deficit hyperactivity disorder. *Biological Psychiatry, 60*(10), 1088–1097.

Klein-Tasman, B. P., & Mervis, C. B. (2003). Distinctive personality characteristics of 8-, 9-, and 10-year-olds with Williams syndrome. *Developmental Neuropsychology, 23*(1–2), 269–290.

Klingberg, T., Fernell, E., Olesen, P. J., Johnson, M., Gustafsson, P., Dahlstrom, K., et al. (2005). Computerized training of working memory in children with ADHD—A randomized, controlled trial. *Journal of the American Academy of Child and Adolescent Psychiatry, 44*(2), 177–186.

Kloo, D., Perner, J., Kerschhuber, A., Dabernig, S., & Aichhorn, M. (2008). Sorting between dimensions: Conditions of cognitive flexibility in preschoolers. *Journal of Experimental Child Psychology, 100*(2), 115–134.

Koch, I. (2008). Mechanisms of interference in dual tasks. *Psychologische Rundschau, 59*(1), 24–32.

Kochanska, G., Murray, K., & Harlan, E. T. (2000). Effortful control in early childhood: Continuity and change, antecedents, and implications for social development. *Developmental Psychology, 36*(2), 220–232.

Kochanska, G., Murray, K., Jacques, T. Y., Koenig, A. L., & Van der Geest, K. A. (1996). Inhibitory control in young children and its role in emerging internalization. *Child Development, 67*(2), 490–507.

Koczat, D. L., Rogers, S. J., Pennington, B. F., & Ross, R. G. (2002). Eye movement abnormality suggestive of a spatial working memory deficit is present in parents of autistic probands. *Journal of Autism and Developmental Disorders, 32*(6), 513–518.

Koelega, H. S. (1995). Is the continuous performance task useful in research with ADHD children? Comments on a review. *Journal of Child Psychology and Psychiatry and Allied Disciplines, 36*(8), 1477–1485.

Kogan, C. S., Bertone, A., Cornish, K., Boutet, I., Der Kaloustian, V. M., Andermann, E., Faubert, J., & Chaudhuri, A. (2004). Integrative cortical dysfunction and pervasive motion perception deficit in fragile X syndrome. *Neurology, 63*, 1634–1639.

Kogan, C. S., Boutet, I., Cornish, K., Zangenehpour, S., Mullen, K. T., Holden, J. J. A., Der Kaloustian, V. M., Andermann, E. & Chaudhuri, A. (2004). Differential impact of the *FMR1* gene on visual processing in fragile X syndrome. *Brain, 127*(3), 591–601.

Kollins, S. H., Epstein, J. N., & Conners, C. K. (2004). Conners' Rating Scales-Revised. In M. E. Maruish (Ed.), *The use of psychological testing for treatment planning and outcomes assessment.* Mahwah, NJ: Lawrence Erlbaum Associates.

Konstantareas, M. M., & Stewart, K. (2006). Affect regulation and temperament in children with autism spectrum disorder. *Journal of Autism and Developmental Disorders, 36*(2), 143–154.

Korkman, M., Kemp, S. L., & Kirk, U. (2001). Effects of age on neurocognitive measures of children ages 5 to 12: A cross-sectional study on 800 children from the United States. *Developmental Neuropsychology, 20*(1), 331–354.

Korkman, M., Kirk, U., & Kemp, S. (1998). *NEPSY: A developmental neuropsychological assessment.* San Antonio, TX: The Psychological Corporation.

Korkman, M., Kirk, U., & Kemp, S. (2007). *NEPSY—Second Edition (NEPSY-II).* Cambridge, U.K.: Pearson.

Koski, L., & Petrides, M. (2001). Time-related changes in task performance after lesions restricted to the frontal cortex. *Neuropsychologia, 39*(3), 268–281.

Kotsoni, E., Byrd, D., & Casey, B. J. (2006). Special considerations for functional magnetic resonance imaging of pediatric populations. *Journal of Magnetic Resonance Imaging, 23*(6), 877–886.

Krause, K. H., Dresel, S. H., Krause, J., la Fougere, C., & Ackenheil, M. (2003). The dopamine transporter and neuroimaging in attention deficit hyperactivity disorder. *Neuroscience and Biobehavioral Reviews, 27*(7), 605–613.

Kringelbach, M. L. (2005). The human orbitofrontal cortex: Linking reward to hedonic experience. *Nature Reviews Neuroscience, 6*(9), 691–702.

Kumar, G., & Steer, R. A. (2003). Factorial validity of the Conners' Parent Rating Scale-Revised: Short Form with psychiatric outpatients. *Journal of Personality Assessment, 80*(3), 252–259.

Kumar, R. A., KaraMohamed, S., Sudi, J., Conrad, D. F., Brune, C., Badner, J. A., et al. (2008). Recurrent 16p11.2 microdeletions in autism. *Human Molecular Genetics, 17*(4), 628–638.

Kustanovich, V., Merriman, B., McGough, J., McCracken, J. T., Smalley, S. L., & Nelson, S. F. (2003). Biased paternal transmission of SNAP-25 risk alleles in attention-deficit hyperactivity disorder. *Molecular Psychiatry, 8*(3), 309–315.

Kylliäinen, A., Braeutigam, S., Hietanen, J. K., Swithenby, S. J., & Bailey, A. J. (2006). Face and gaze processing in normally developing children: A magnetoencephalographic study. *European Journal of Neuroscience, 23*(3), 801–810.

Kylliäinen, A., & Hietanen, J. K. (2004). Attention orienting by another's gaze direction in children with autism. *Journal of Child Psychology and Psychiatry, 45*(3), 435–444.

Lachiewicz, A. M., Dawson, D.V., & Spiridigliozzi, G. A. (2000). Physical characteristics of young boys with fragile X syndrome: Reasons for difficulties in making a diagnosis in young males. *American Journal of Medical Genetics, 92*(4), 229–236.

Lahey, B. B., Pelham, W. E., Chronis, A., Massetti, G., Kipp, H., Ehrhardt, A., et al. (2006). Predictive validity of ICD-10 hyperkinetic disorder relative to DSM-IV attention-deficit/hyperactivity disorder among younger children. *Journal of Child Psychology and Psychiatry, 47*(5), 472–479.

Lahey, B. B., Pelham, W. E., Loney, J., Lee, S. S., & Willcutt, E. (2005). Instability of the DSM-IV subtypes of ADHD from preschool through elementary school. *Archives of General Psychiatry, 62*(8), 896–902.

Laing, E., & Jarrold, C. (2007). Comprehension of spatial language in Williams syndrome: Evidence for impaired spatial representation of verbal descriptions. *Clinical Linguistics & Phonetics, 21*(9), 689–704.

Lajiness-O'Neill, R., Beaulieu, I., Asamoah, A., Titus, J. B., Bawle, E., Ahmad, S., et al. (2006). The neuropsychological phenotype of velocardiofacial syndrome (VCFS): Relationship to psychopathology. *Archives of Clinical Neuropsychology, 21*(2), 175–184.

Landa, R. J., & Goldberg, M. C. (2005). Language, social, and executive functions in high functioning autism: A continuum of performance. *Journal of Autism and Developmental Disorders, 35*(5), 557–573.

Landry, R., & Bryson, S. (2004). Impaired disengagement of attention in young children with autism. *Journal of Child Psychology and Psychiatry, 45*(6), 1115–1122.

Landstrom, U., Kjellberg, A. & Byström, M. (1995). Acceptable levels of tonal and broadband repetitive and continuous sounds during the performance of non-auditory tasks. *Perceptual & Motor Skills, 81*, 803–816.

Lanfranchi, S., Cornoldi, C., Drigo, S., & Vianello, R. (2008). Working memory in individuals with fragile X syndrome. *Child Neuropsychology, 15*(2), 105–119.

Langberg, J. M., Epstein, J. N., & Graham, A. J. (2008). Organizational skills interventions in the treatment of ADHD. *Expert Review of Neurotherapeutics, 8*(10), 1549–1561.

Langley, K., Payton, A., Hamshere, M. L., Pay, H. M., Lawson, D. C., Turic, D., et al. (2003). No evidence of association of two 5HT transporter gene polymorphisms and attention deficit hyperactivity disorder. *Psychiatric Genetics, 13*(2), 107–110.

Lansbergen, M. M., Kenemans, J. L., & van Engeland, H. (2007). Stroop interference and attention-deficit/hyperactivity disorder: A review and meta-analysis. *Neuropsychology, 21*(2), 251–262.

Lara, C., Fayyad, J., de Graaf, R., Kessler, R. C., Aguilar-Gaxiola, S., Angermeyer, M., et al. (2009). Childhood predictors of adult attention-deficit/hyperactivity disorder: Results from the World Health Organization World Mental Health Survey Initiative. *Biological Psychiatry, 65*(1), 46–54.

Laurie-Rose, C. (2005). Equating tasks and sustaining attention in children and adults: The methodological and theoretical utility of d' matching. *Perception and Psychophysics, 67*, 254–263.

Lavie, N. (2001). The role of capacity limits in selective attention: Behavioral evidence and implications for neural activity. In J. Braun & C. Koch (Eds.), *Visual attention and cortical circuits* (pp. 49–68). Cambridge, MA: MIT Press.

Lavie, N., Hirst, A., de Fockert, J. W., & Viding, E. (2004). Load theory of selective attention and cognitive control. *Journal of Experimental Psychology-General, 133*(3), 339–354.

Lawrence, N. S., Ross, T. J., Hoffmann, R., Garavan, H., & Stein, E. A. (2003). Multiple neuronal networks mediate sustained attention. *Journal of Cognitive Neuroscience, 15*(7), 1028–1038.

Lee, J. S., Kim, B. N., Kang, E., Lee, D. S., Kim, Y. K., Chung, J. K., et al. (2005). Regional cerebral blood flow in children with attention deficit hyperactivity disorder: Comparison before and after methylphenidate treatment. *Human Brain Mapping, 24*(3), 157–64.

Lee, P. S., Foss-Feig, J., Henderson, J. G., Kenworthy, L.E., Gilotty, L., Gaillard, W.D., et al. (2007). Atypical neural substrates of Embedded Figures Task performance in children with Autism Spectrum Disorders. *Neuroimage, 38*(1), 184–193.

Lehto, J. E., Juujärvi, P., Kooistra, L., & Pulkkinen, L. (2003). Dimensions of executive functioning: Evidence from children. *British Journal of Developmental Psychology, 21*(1), 59–80.

Lejeune, J., Gautier, M., & Turpin, R. (1959). Etude des chromosomes somatiques de neuf enfants mongoliens. *Comptes Rendus Hebd Seances Acad Sci, 248*(11), 1721–1722.

Lenroot, R. K., & Giedd, J. N. (2006). Brain development in children and adolescents: Insights from anatomical magnetic resonance imaging. *Neuroscience and Biobehavioral Reviews, 30*(6), 718–729.

Leon-Carrion, J., Garcia-Orza, J., & Perez Santamaria, F. J. (2004). Development of inhibitory component of the executive functions in children and adolescents. *International Journal of Neuroscience, 114*(10), 1291–1311.

Lepistö, T., Kujala, T., Vanhala, R., Alku, P., Huotilainen, M., & Näätänen, R. (2005). The discrimination of and orienting to speech and non-speech sounds in children with autism. *Brain Research, 1066*(1–2), 147–157.

Levy, F. (2007). What do dopamine transporter and catechol-o-methyltransferase tell us about attention deficit-hyperactivity disorder? Pharmacogenomic implications. *Australia New Zealand Journal of Psychiatry, 41*(1), 6–10.

Levy, F. (2008). Pharmacological and therapeutic directions in ADHD: Specificity in the PFC. *Behavioral and Brain Functions, 4,* 12.

Levy, F., Hay, D. A., McStephen, M., Wood, C., & Waldman, I. (1997). Attention-deficit hyperactivity disorder: A category or a continuum? Genetic analysis of a large-scale twin study. *Journal of the American Academy of Child and Adolescent Psychiatry, 36*(6), 737–744.

Lewandowski, K. E., Shashi, V., Berry, P. M., & Kwapil, T. R. (2007). Schizophrenic-like neurocognitive deficits in children and adolescents with 22q11 deletion syndrome. *American Journal of Medical Genetics Part B: Neuropsychiatric Genetics, 144B*(1), 27–36.

Lewis, J. L. (1970). Semantic processing of unattended messages using dichotic listening. *Journal of Experimental Psychology, 85*(2), 225–228.

Leyfer, O. T., Folstein, S. E., Bacalman, S., Davis, N. O., Dinh, E., Morgan, J., et al. (2006). Comorbid psychiatric disorders in children with autism: Interview development and rates of disorders. *Journal of Autism and Developmental Disorders, 36*(7), 849–861.

Leyfer, O. T., Woodruff-Borden, J., Klein-Tasman, B. P., Fricke, J. S., & Mervis, C. B. (2006). Prevalence of psychiatric disorders in 4 to 16–year-olds with Williams syndrome. *American Journal of Medical Genetics Part B: Neuropsychiatric Genetics, 141B*(6), 615–622.

Lezak, M. D. (2004). *Neuropsychological assessment* (4th ed.). Oxford, UK: Oxford University Press.

Li, D., Sham, P. C., Owen, M. J., & He, L. (2006). Meta-analysis shows significant association between dopamine system genes and attention deficit hyperactivity disorder (ADHD). *Human Molecular Genetics, 15*(14), 2276–2284.

Li, H. H., Roy, M., Kuscuoglu, U., Spencer, C. M., Halm, B., Harrison, K. C., et al. (2009). Induced chromosome deletions cause hypersociability and other features of Williams-Beuren syndrome in mice. *EMBO Molecular Medicine, 1*(1), 50–65.

Li, J., Wang, Y., Zhou, R., Zhang, H., Yang, L., Wang, B., et al. (2007). Association between polymorphisms in serotonin transporter gene and attention deficit hyperactivity disorder in Chinese Han subjects. *American Journal of Medical Genetics Part B: Neuropsychiatric Genetics, 144B*(1), 14–19.

Lightbody, A. A., Hall, S. S., & Reiss, A. L. (2006). Chronological age, but not FMRP levels, predicts neuropsychological performance in girls with fragile X syndrome. *American Journal of Medical Genetics Part B: Neuropsychiatric Genetics, 141B*(5), 468–472.

Lijffijt, M., Kenemans, J. L., ter Wal, A., Quik, E. H., Kemner, C., Westenberg, H., et al. (2006). Dose-related effect of methylphenidate on stopping and changing in children with attention-deficit/hyperactivity disorder. *European Psychiatry, 21*(8), 544–547.

Lijffijt, M., Kenemans, J. L., Verbaten, M. N., & van Engeland, H. (2005). A meta-analytic review of stopping performance in attention deficit hyperactivity disorder: Deficient inhibitory motor control? *Journal of Abnormal Psychology, 114*(2), 216–222.

Lobaugh, N. J., Cole, S., & Rovet, J. F. (1998). Visual search for features and conjunctions in development. *Canadian Journal of Experimental Psychology, 54*(4), 201–211.

Loe, I. M., & Feldman, H. M. (2007). Academic and educational outcomes of children with ADHD. *Ambulatory Pediatrics, 7*(1), 82–90.

Loesch, D. Z., Bui, Q. M., Dissanayake, C., Clifford, S., Gould, E., Bulhak-Paterson, D., et al. (2007). Molecular and cognitive predictors of the continuum of autistic behaviours in fragile X. *Neuroscience and Biobehavioral Reviews, 31*(3), 315–326.

Loesch, D. Z., Bui, Q. M., Grigsby, J., Butler, E., Epstein, J., Huggins, R. M., et al. (2003). Effect of the fragile X status categories and the fragile X mental retardation protein levels on executive functioning in males and females with fragile X. *Neuropsychology, 17*(4), 646–657.

Loesch, D. Z., Huggins, R. M., Bui, Q. M., Epstein, J. L., Taylor, A. K., & Hagerman, R. J. (2002). Effect of the deficits of fragile X mental retardation protein on cognitive status of fragile X males and females assessed by robust pedigree analysis. *Journal of Developmental and Behavioral Pediatrics, 23*(6), 416–423.

Loetscher, T., & Brugger, P. (2007). A disengagement deficit in representational space. *Neuropsychologia, 45*(6), 1299–1304.

Logan, G. D., & Cowan, W. B. (1984). On the ability to inhibit thought and action: A theory of an act of control. *Psychological Review, 91*(3), 295–327.

Loo, S. K., Humphrey, L. A., Tapio, T., Moilanen, I. K., McGough, J. J., McCracken, J. T., et al. (2007). Executive functioning among Finnish adolescents with attention-deficit/hyperactivity disorder. *Journal of the American Academy of Child and Adolescent Psychiatry, 46*(12), 1594–1604.

Loose, R., Kaufmann, C., Tucha, O., Auer, D. P., & Lange, K. W. (2006). Neural networks of response shifting: Influence of task speed and stimulus material. *Brain Research, 1090*(1), 146–155.

Lord, C., Rutter, M., DiLavore, P. C., & Risi, S. (1999). *Autism Diagnostic Observation Schedule-WPS (WPS ed.)*. Los Angeles: Western Psychological Services.

Lord, C., Rutter, M., & Le Couteur, A. (1994). Autism Diagnostic Interview–Revised: A revised version of a diagnostic interview for caregivers of individuals with possible pervasive developmental disorders. *Journal of Autism and Developmental Disorders, 24*(5), 659–685.

Losh, M., Sullivan, P. F., Trembath, D., & Piven, J. (2008). Current developments in the genetics of autism: From phenome to genome. *Journal of Neuropathology and Experimental Neurology, 67*(9), 829–837.

Lowe, N., Kirley, A., Hawi, Z., Sham, P., Wickham, H., Kratochvil, C. J., et al. (2004). Joint analysis of the DRD5 marker concludes association with attention-deficit/ hyperactivity disorder confined to the predominantly inattentive and combined subtypes. *American Journal of Human Genetics, 74*(2), 348–356.

Lowenthal, R., Paula, C. S., Schwartzman, J. S., Brunoni, D., & Mercadante, M. T. (2007). Prevalence of pervasive developmental disorder in Down's syndrome. *Journal of Autism and Developmental Disorders, 37*(7), 1394–1395.

Lubke, G. H., Muthen, B., Moilanen, I. K., McGough, J. J., Loo, S. K., Swanson, J. M., et al. (2007). Subtypes versus severity differences in attention-deficit/hyperactivity disorder in the northern Finnish birth cohort. *Journal of the American Academy of Child and Adolescent Psychiatry, 46*(12), 1584–1593.

Luciana, M. (2003). Practitioner review: Computerized assessment of neuropsychological function in children: Clinical and research applications of the Cambridge Neuropsychological Testing Automated Battery (CANTAB). *Journal of Child Psychology and Psychiatry and Allied Disciplines, 44*(5), 649–663.

Luciana, M., & Nelson, C. A. (2002). Assessment of neuropsychological function through use of the Cambridge Neuropsychological Testing Automated Battery: Performance in 4-to 12-year-old children. *Developmental Neuropsychology, 22*(3), 595–624.

Lui, M., & Tannock, R. (2007). Working memory and inattentive behaviour in a community sample of children. *Behavior and Brain Functions, 3.*

Luk, S. L., & Leung, P. W. L. (1989). Conners Teachers Rating Scale—A validity study in Hong Kong. *Journal of Child Psychology and Psychiatry and Allied Disciplines, 30*(5), 785–793.

Luna, B., Doll, S. K., Hegedus, S. J., Minshew, N. J., & Sweeney, J. A. (2007). Maturation of executive function in autism. *Biological Psychiatry, 61*(4), 474–481.

Luna, B., Garver, K. E., Urban, T. A., Lazar, N. A., & Sweeney, J. A. (2004). Maturation of cognitive processes from late childhood to adulthood. *Child Development, 75*(5), 1357–1372.

Luria, A. R. (1973). *The working brain: An introduction to neuropsychology.* New York: Basic Books.

Mabbott, D. J., Noseworthy, M., Bouffet, E., Laughlin, S., & Rockel, C. (2006). White matter growth as a mechanism of cognitive development in children. *NeuroImage, 33*(3), 936–946.

Mackay, D. G. (1973). Aspects of theory of comprehension, memory and attention. *Quarterly Journal of Experimental Psychology, 25,* 22–40.

Mackworth, N. H. (1948). The breakdown of vigilance during prolonged visual search. *Quarterly Journal of Experimental Psychology, 1,* 6–21.

MacLeod, C. M. (1991). Half a century of research on the Stroop effect: An integrative review. *Psychological Bulletin, 109*(2), 163–203.

Macmillan, M. (2008). Phineas Gage—Unravelling the myth. *Psychologist, 21*(9), 828–831.

Magnusson, P., Smari, J., Gretarsdottir, H., & Prandardottir, H. (1999). Attention-Deficit/Hyperactivity symptoms in Icelandic schoolchildren: Assessment with the Attention Deficit/Hyperactivity Rating Scale-IV. *Scandinavian Journal of Psychology, 40*(4), 301–306.

Maher, B. S., Marazita, M. L., Ferrell, R. E., & Vanyukov, M. M. (2002). Dopamine system genes and attention deficit hyperactivity disorder: A meta-analysis. *Psychiatric Genetics, 12*(4), 207–215.

Mahone, E. M. (2005). Measurement of attention and related functions in the preschool child. *Mental Retardation and Developmental Disabilities Research Reviews, 11*(3), 216–225.

Mahone, E. M., Cirino, P. T., Cutting, L. E., Cerrone, P. M., Hagelthorn, K. M., Hiemenz, J. R., et al. (2002). Validity of the behavior rating inventory of executive function in children with ADHD and/or Tourette syndrome. *Archives of clinical Neuropsychology, 17*(7), 643–662.

Mahone, E. M., & Hoffman, J. (2007). Behavior ratings of executive function among preschoolers with ADHD. *Clinical Neuropsychologist, 21*(4), 569–586.

Mahone, E. M., Powell, S. K., Loftis, C. W., Goldberg, M. C., Denckla, M. B., & Mostofsky, S. H. (2006). Motor persistence and inhibition in autism and ADHD. *Journal of the International Neuropsychological Society, 12*(5), 622–631.

Makeyev, A. V., Erdenechimeg, L., Mungunsukh, O., Roth, J. J., Enkhmandakh, B., Ruddle, F. H., et al. (2004). GTF2IRD2 is located in the Williams-Beuren syndrome critical region 7q11.23 and encodes a protein with two TFII-I-like helix-loop-helix repeats. *Proceedings of the National Academy of Sciences of the United States of America, 101*(30), 11052–11057.

Makris, N., Buka, S. L., Biederman, J., Papadimitriou, G. M., Hodge, S. M., Valera, E. M., et al. (2008). Attention and executive systems abnormalities in adults with childhood ADHD: A DT-MRI study of connections. *Cerebral Cortex, 18*(5), 1210–1220.

Mall, V., Berweck, S., Fietzek, U. M., Glocker, F. X., Oberhuber, U., Walther, M. et al. (2004). Low level of intracortical inhibition in children shown by transcranial magnetic stimulation. *Neuropediatrics, 35*(2), 120–125.

Malmo, R. B. (1942). Interference factors in delayed response in monkeys after removal of frontal lobes. *Journal of Neurophysiology, 5*(4), 295–308.

Mandy, W. P. L., & Skuse, D. H. (2008). Research review: What is the association between the social-communication element of autism and repetitive interests, behaviours and activities? *Journal of Child Psychology and Psychiatry, 49*(8), 795–808.

Manly, T., Anderson, V., Nimmo-Smith, I., Turner, A., Watson, P., & Robertson, I. H. (2001). The differential assessment of children's attention: The Test of Everyday Attention for Children (TEA-Ch), normative sample and ADHD performance. *Journal of Child Psychology and Psychiatry and Allied Disciplines, 42*(8), 1065–1081.

Manor, I., Kotler, M., Sever, Y., Eisenberg, J., Cohen, H., Ebstein, R. P., et al. (2000). Failure to replicate an association between the catechol-O-methyltransferase polymorphism and attention deficit hyperactivity disorder in a second, independently recruited Israeli cohort. *American Journal of Medical Genetics Part B: Neuropsychiatric Genetics, 96*(6), 858–860.

Manor, I., Tyano, S., Eisenberg, J., Bachner-Melman, R., Kotler, M., & Ebstein, R. P. (2002). The short DRD4 repeats confer risk to attention deficit hyperactivity disorder in a family-based design and impair performance on a continuous performance test (TOVA). *Molecular Psychiatry, 7*(7), 790–794.

Martel, M., Nikolas, M., & Nigg, J. T. (2007). Executive function in adolescents with ADHD. *Journal of the American Academy of Child and Adolescent Psychiatry, 46*(11), 1437–1444.

Martens, M. A., Wilson, S. J., & Reutens, D. C. (2008). Research review: Williams syndrome: A critical review of the cognitive, behavioral, and neuroanatomical phenotype. *Journal of Child Psychology and Psychiatry, 49*(6), 576–608.

Martin, J. P., & Bell, J. (1943). A pedigree of mental defect showing sex-linkage. *Journal of Neurology, Neurosurgery and Psychiatry, 6*, 154–156.

Martin, L. A., Goldowitz., & Mittleman, G. (2010). Repetitive behavior and increased activity in mice with Purkinje cell loss: A model for understanding the role of cerebellar pathology in autism. *European Journal of Neuroscience, 31*, 544–555.

Martinussen, R., Hayden, J., Hogg-Johnson, S., & Tannock, R. (2005). A meta-analysis of working memory impairments in children with attention-deficit/hyperactivity disorder. *Journal of the American Academy of Child and Adolescent Psychiatry, 44*(4), 377–384.

Martinussen, R., & Tannock, R. (2006). Working memory impairments in children with attention-deficit hyperactivity disorder with and without comorbid language learning disorders. *Journal of Clinical and Experimental Neuropsychology, 28*(7), 1073–1094.

Marzocchi, G. M., Capron, C., Di Pietro, M., Tauleria, E. D., Duyme, M., Frigerio, A., et al. (2004). The use of the Strengths and Difficulties Questionnaire (SDQ) in Southern European countries. *European Child and Adolescent Psychiatry, 13*(Suppl. 2), 40–46.

Mason, D. J., Humphreys, G. W., & Kent, L. (2004). Visual search, singleton capture, and the control of attentional set in ADHD. *Cognitive Neuropsychology, 21*(6), 661–687.

Mazzocco, M. M. (2001). Math learning disability and math LD subtypes: Evidence from studies of Turner syndrome, fragile X syndrome, and neurofibromatosis type 1. *Journal of Learning Disabilities, 34*(6), 520–533.

Mazzocco, M. M., Singh Bhatia, N., & Lesniak-Karpiak, K. (2006). Visuospatial skills and their association with math performance in girls with fragile X or Turner syndrome. *Child Neuropsychology, 12*(2), 87–110.

Mazzocco, M. M. M., & Kover, S. T. (2007). A longitudinal assessment of executive function skills and their association with math performance. *Child Neuropsychology, 13*(1), 18–45.

McAlonan, G. M., Cheung, V., Cheung, C., Suckling, J., Lam, G. Y., Tai, K. S., et al. (2005). Mapping the brain in autism. A voxel-based MRI study of volumetric differences and intercorrelations in autism. *Brain, 128*(2), 268–276.

McCauley, J. L., Olson, L. M., Delahanty, R., Amin, T., Nurmi, E. L., Organ, E. L., et al. (2004). A linkage disequilibrium map of the 1-Mb 15q12 GABA(A) receptor subunit cluster and association to autism. *American Journal of Medical Genetics Part B: Neuropsychiatric Genetics, 131B*(1), 51–59.

McConachie, H., & Diggle, T. (2007). Parent implemented early intervention for young children with autism spectrum disorder: A systematic review. *Journal of Evaluation in Clinical Practice, 13*(1), 120–129.

McGough, J. J., McBurnett, K., Bukstein, O., Wilens, T. E., Greenhill, L., Lerner, M., et al. (2006). Once-daily OROS (R) methylphenidate is safe and well tolerated in adolescents with attention-deficit/hyperactivity disorder. *Journal of Child and Adolescent Psychopharmacology, 16*(3), 351–356.

McGowan, P. O., Meaney, M. J., & Szyf, M. (2008). Diet and the epigenetic (re)programming of phenotypic differences in behavior. *Brain Research, 1237*, 12–24.

McInnes, A., Bedard, A. C., Hogg-Johnson, S., & Tannock, R. (2007). Preliminary evidence of beneficial effects of methylphenidate on listening comprehension in children with attention-deficit/hyperactivity disorder. *Journal of Child and Adolescent Psychopharmacology, 17*(1), 35–49.

McLeod, P. (1978). Does probe RT measure central processing demand. *Quarterly Journal of Experimental Psychology, 30*, 83–89.

Mehta, M. A., Manes, F. F., Magnolfi, G., Sahakian, B. J., & Robbins, T. W. (2004). Impaired set-shifting and dissociable effects on tests of spatial working memory following the dopamine D-2 receptor antagonist sulpiride in human volunteers. *Psychopharmacology, 176*(3–4), 331–342.

Mellor, D. (2004). Furthering the use of the Strengths and Difficulties Questionnaire: Reliability with younger child respondents. *Psychological Assessment, 16*(4), 396–401.

Menon, V., Leroux, J., White, C. D., & Reiss, A. L. (2004). Frontostriatal deficits in fragile X syndrome: Relation to FMR1 gene expression. *Proceedings of the National Academy of Sciences of the United States of America, 101*(10), 3615–3620.

Merigan, W. H., & Maunsell, J. H. R. (1993). How parallel are the primate visual pathways? *Annual Review of Neuroscience, 11*(1), 369–402.

Mervis, C. B., & Becerra, A. M. (2007). Language and communicative development in Williams syndrome. *Mental Retardation and Developmental Disabilities Research Reviews, 13*(1), 3–15.

Mervis, C. B., Morris, C. A., Klein-Tasman, B. P., Bertrand, J., Kwitny, S., Appelbaum, L. G., et al. (2003). Attentional characteristics of infants and toddlers with Williams Syndrome during triadic interactions. *Developmental Neuropsychology, 23*(1), 243–268.

Meyer-Lindenberg, A., Kohn, P. D., Kolachana, B., Kippenhan, S., McInerney-Leo, A., Nussbaum, R., et al. (2005). Midbrain dopamine and prefrontal function in humans: Interaction and modulation by COMT genotype. *Nature Neuroscience, 8*(5), 594–596.

Meyer-Lindenberg, A., Mervis, C. B., & Berman, K. F. (2006). Neural mechanisms in Williams syndrome: A unique window to genetic influences on cognition and behaviour. *Nature Reviews Neuroscience, 7*(5), 380–393.

Meyer-Lindenberg, A., Zink, C. F. (2007). Imaging genetics for neuropsychiatric disorders. *Child and Adolescent Psychiatric Clinics of North America, 16*(3), 581–97.

Michaelovsky, E., Gothelf, D., Korostishevsky, M., Frisch, A., Burg, M., Carmel, M., et al. (2008). Association between a common haplotype in the COMT gene region and psychiatric disorders in individuals with 22q11.2DS. *International Journal of Neuropsychopharmacology, 11*(3), 351–363.

Mick, E., & Faraone, S. V. (2008). Genetics of attention deficit hyperactivity disorder. *Child and Adolescent Psychiatric Clinics of North America, 17*(2), 261–284.

Mill, J., & Petronis, A. (2008). Pre- and peri-natal environmental risks for attention-deficit hyperactivity disorder: The role of epigenetic processes in mediating susceptibility. *The Journal of Child Psychology and Psychiatry, 49*(10), 1020–1030.

Mill, J., Richards, S., Knight, J., Curran, S., Taylor, E., & Asherson, P. (2004). Haplotype analysis of SNAP-25 suggests a role in the aetiology of ADHD. *Molecular Psychiatry, 9*(8), 801–810.

Mill, J., Xu, X., Ronald, A., Curran, S., Price, T., Knight, J., et al. (2005). Quantitative trait locus analysis of candidate gene alleles associated with attention deficit hyperactivity disorder (ADHD) in five genes: DRD4, DAT1, DRD5, SNAP-25, and 5HT1B. *American Journal of Medical Genetics Part B: Neuropsychiatric Genetics, 133B*(1), 68–73.

Miller, M. L., Fee, V. E., & Netterville, A. K. (2004). Psychometric properties of ADHD rating scales among children with mental retardation I: Reliability. *Research in Developmental Disabilities, 25*(5), 459–476.

Mills, S., Langley, K., Van den Bree, M., Street, E., Turic, D., Owen, M. J., et al. (2004). No evidence of association between Catechol-O-Methyltransferase (COMT) Val153Met genotype and performance on neuropsychological tasks in children with ADHD: A case-control study. *BMC Psychiatry, 4*, 15.

Milner, A. D., & Goodale, M. A. (1995). *The visual brain in action.* Oxford, UK: Oxford University Press.

Milner, B. (1963). Effects of different brain lesions on card sorting: Role of frontal lobes. *Archives of Neurology, 9*(1), 90–100.

Milner, B. (1964). Some effects of frontal lobotomy in man. In J. Warren & K. Akert (Eds.), *The frontal granular cortex and behavior.* New York: McGraw Hill.

Milner, B. (1971). Interhemispheric differences in the localisation of psychological processes in man. *Cortex, 27,* 272–277.

Mirsky, A. F. (1996). Disorders of attention: A neuropsychological perspective. In G. R. Lyon & N. A. Krasnegor (Eds.), *Attention, memory and executive function* (pp. 71–95). Baltimore: Paul Brookes.

Mitsis, E. M., McKay, K. E., Schulz, K. P., Newcorn, J. H., & Halperin, J. M. (2000). Parent-teacher concordance for DSM-IV attention-deficit/hyperactivity disorder in a clinic-referred sample. *Journal of the American Academy of Child and Adolescent Psychiatry, 39*(3), 308–313.

Miyake, A., Friedman, N. P., Emerson, M. J., Witzki, A. H., Howerter, A., & Wager, T. D. (2000). The unity and diversity of executive functions and their contributions to complex "frontal lobe" tasks: A latent variable analysis. *Cognitive Psychology, 41*(1), 49–100.

Mobbs, D., Eckert, M. A., Menon, V., Mills, D., Korenberg, J., Galaburda, A. M., et al. (2007). Reduced parietal and visual cortical activation during global processing in Williams syndrome. *Development of Medical Child Neurology, 49*(6), 433–438.

Mobbs, D., Eckert, M. A., Mills, D., Korenberg, J., Bellugi, U., Galaburda, A. M., et al. (2007). Frontostriatal dysfunction during response inhibition in Williams Syndrome. *Biological Psychiatry, 62*(3), 256–261.

Moessner, R., Marshall, C. R., Sutcliffe, J. S., Skaug, J., Pinto, D., Vincent, J., et al. (2007). Contribution of SHANK3 mutations to autism spectrum disorder. *American Journal of Human Genetics, 81*(6), 1289–1297.

Monk, C. S., McClure, E. B., Nelson, E. E., Zarahn, E., Bilder, R. M., Leibenluft, E., et al. (2003). Adolescent immaturity in attention-related brain engagement to emotional facial expressions. *NeuroImage, 20*(1), 420–428.

Monsell, S. (1996). Control of mental processes. In V. Bruce (Ed.), *Unsolved mysteries of the mind* (pp. 93–148). Hove, UK: Erlbaum.

Montfoort, I., Frens, M. A., Hooge, I. T. C., Lagers-van Haselen, G. C., & van der Geest, J. N. (2007). Visual search deficits in Williams-Beuren syndrome. *Neuropsychologia, 45*(5), 931–938.

Monuteaux, M. C., Mick, E., Faraone, S. B., & Biederman, J. (2010). The influence of sex on the course and psychiatric correlates of ADHD from childhood to adolescence: A longitudinal study. *Journal of Child Psychology and Psychiatry, 51(3)*, 233–241.

Morein-Zamir, S., Hommersen, P., Johnston, C., & Kingstone, A. (2008). Novel measures of response performance and inhibition in children with ADHD. *Journal of Abnormal Child Psychology, 36*(8), 1199–1210.

Morris, C. A., Demsey, S. A., Leonard, C. O., Dilts, C., & Blackburn, B. L. (1988). Natural history of Williams syndrome: Physical characteristics. *The Journal of Pediatrics, 113*(2), 318–326.

Morris, C. A., Mervis, C. B., Hobart, H. H., Gregg, R. G., Bertrand, J., Ensing, G. J., et al. (2003). GTF2I hemizygosity implicated in mental retardation in Williams syndrome: Genotype-phenotype analysis of five families with deletions in the Williams syndrome region. *American Journal of Medical Genetics Part A,* 123A(1), 45–59.

Moses, J. A. (2004). Test review—Comprehensive Trail Making Test (CTMT). *Archives of Clinical Neuropsychology, 19*(5), 703–708.

Moss, J., & Howlin, P. (2009). Autism spectrum disorders in genetic syndromes: Implications for diagnosis, intervention and understanding the wider autism spectrum disorder population. *Journal of Intellectual Disability Research, 53(10),* 852–873.

Moy, S. S., & Nadler, J. J. (2008). Advances in behavioral genetics: Mouse models of autism. *Molecular Psychiatry, 13*(1), 4–26.

MTA Cooperative Group. (1999). A 14-month randomized clinical trial of treatment strategies for attention-deficit/hyperactivity disorder. The MTA Cooperative Group. Multimodal Treatment Study of Children with ADHD. *Archives of General Psychiatry, 56*(12), 1073–1086.

MTA Cooperative Group. (2004). National Institute of Mental Health Multimodal Treatment Study of ADHD follow-up: Changes in effectiveness and growth after the end of treatment. *Pediatrics, 113*(4), 762–769.

Mulas, F., Capilla, A., Fernandez, S., Etchepareborda, M. C., Campo, P., Maestu, F., et al. (2006). Shifting-related brain magnetic activity in attention-deficit/ hyperactivity disorder. *Biological Psychiatry, 59*(4), 373–379.

Muller, U., Dick, A. S., Gela, K., Overton, W. F., & Zelazo, P. D. (2006). The role of negative priming in preschoolers' flexible rule use on the dimensional change card sort task. *Child Development, 77*(2), 395–412.

Munakata, Y., Casey, B. J., & Diamond, A. (2004). Developmental cognitive neuroscience: Progress and potential. *Trends in Cognitive Sciences, 8*(3), 122–128.

Munir, F., Cornish, K. M., & Wilding, J. (2000a). Nature of the working memory deficit in Fragile-X syndrome. *Brain and Cognition, 44*(3), 387–401.

Munir, F., Cornish, K. M., & Wilding, J. (2000b). A neuropsychological profile of attention deficits in young males with fragile X syndrome. *Neuropsychologia, 38*(9), 1261–1270.

Muris, P., Meesters, C., Eijkelenboom, A., & Vincken, M. (2004). The self-report version of the Strengths and Difficulties Questionnaire: Its psychometric properties in 8- to 13-year-old non-clinical children. *British Journal of Clinical Psychology, 43*(4), 437–448.

Muris, P., Meesters, C., & van den Berg, F. (2003). The Strengths and Difficulties Questionnaire (SDQ)—Further evidence for its reliability and validity in a community sample of Dutch children and adolescents. *European Child & Adolescent Psychiatry, 12*(1), 1–8.

Murphy, K. C. (2005). Annotation: Velo-cardio-facial syndrome. *Journal of Child Psychology and Psychiatry, 46*(6), 563–571.

Murphy, M. M., & Mazzocco, M. M. M. (2008a). Mathematics learning disabilities in girls with fragile X or Turner syndrome during late elementary school. *Journal of Learning Disabilities, 41*(1), 29–46.

Murphy, M. M., & Mazzocco, M. M. M. (2008b). Rote numeric skills may mask underlying mathematical disabilities in girls with fragile X syndrome. *Developmental Neuropsychology, 33*(3), 345–364.

Nakagawa, A., & Sukigara, M. (2007). Infant eye and head movements toward the side opposite the cue in the anti-saccade paradigm. *Behavioral and Brain Functions, 3*.

Nebel, K., Wiese, H., Stude, P., de Greiff, A., Diener, H. C., & Keidel, M. (2005). On the neural basis of focused and divided attention. *Cognitive Brain Research, 25*(3), 760–776.

Nelson, C. A., & McCleery, J. P. (2008). Use of event-related potentials in the study of typical and atypical development. *Journal of the American Academy of Child and Adolescent Psychiatry, 47*(11), 1252–61.

Newcorn, J. H., Halperin, J. M., Jensen, P. S., Abikoff, H. B., Arnold, L. E., Cantwell, D. P., et al. (2001). Symptom profiles in children with ADHD: Effects of comorbidity and gender. *Journal of the American Academy of Child and Adolescent Psychiatry, 40*(2), 137–146.

Nichelli, F., Scala, G., Vago, C., Riva, D., & Bulgheroni, S. (2005). Age-related trends in Stroop and Conflicting Motor Response Task findings. *Child Neuropsychology, 11*(5), 431–443.

Nickels, K., Katusic, S. K., Colligan, R. C., Weaver, A. L., Voigt, R. G., & Barbaresi, W. J. (2008). Stimulant medication treatment of target behaviors in children with autism: A population-based study. *Journal of Developmental and Behavioral Pediatrics, 29*(2), 75–81.

Nigg, J. T. (1999). The ADHD response-inhibition deficit as measured by the stop task: Replication with DSM-IV combined type, extension, and qualification. *Journal of Abnormal Child Psychology, 27*(5), 393–402.

Nigg, J. T. (2000). On inhibition/disinhibition in developmental psychopathology: Views from cognitive and personality psychology and a working inhibition taxonomy. *Psychological Bulletin, 126*(2), 200–246.

Nigg, J.T. (2001). Is ADHD a disinhibitory disorder? *Psychological Bulletin, 127(5)*, 571–598.

Nigg, J. T., Blaskey, L. G., Huang-Pollock, C. L., & Rappley, M. D. (2002). Neuropsychological executive functions and DSM-IV ADHD subtypes. *Journal of the American Academy of Child and Adolescent Psychiatry, 41*(1), 59–66.

Niklasson, L., Rasmussen, P., Óskarsdóttir, S., & Gillberg, C. (2001). Neuropsychiatric disorders in the 22q11 deletion syndrome. *Genetics in Medicine, 3*(1), 79–84.

Niklasson, L., Rasmussen, P., Óskarsdóttir, S., & Gillberg, C. (2002). Chromosome 22q11 deletion syndrome (CATCH 22): Neuropsychiatric and neuropsychological aspects. *Developmental Medicine and Child Neurology, 44*(1), 44–50.

Niklasson, L., Rasmussen, P., Óskarsdóttir, S., & Gillberg, C. (2005). Attention deficits in children with 22q.11 deletion syndrome. *Developmental Medicine and Child Neurology, 47*(12), 803–807.

Niklasson, L., Rasmussen, P., Óskarsdóttir, S., & Gillberg, C. (2009). Autism, ADHD, mental retardation and behavior problems in 100 individuals with 22q11 deletion syndrome. *Research Developmental Disabilities, 30*(4), 763–773.

Nobre, A. C., Rao, A. L., & Chelazzi, L. (2006). Selective attention to specific features within objects: Behavioral and electrophysiological evidence. *Journal of Cognitive Neuroscience, 18*(4), 539–561.

Norman, D. A. & Shallice, T. (1980). Attention to action: Willed and automatic control of behavior (Report No. 8006). San Diego: University of California, Center for Human Information Processing.

Norman, D. A., & Shallice, T. (1986). Attention to action: Willed and automatic control of behavior. In R. J. Davidson, G. E. Schwartz, & D. Shapiro (Eds.), *Consciousness and self-regulation* (pp. 1–18). New York: Plenum Press.

Oakes, L. M., Kannass, K. N., & Shaddy, D. J. (2002). Developmental changes in endogenous control of attention: The role of target familiarity on infant's distraction latency. *Child Development, 73*(6), 1644–1655.

Oakes, L. M., Ross-Sheehy, S., & Kannass, K. N. (2004). Attentional engagement in infancy: The interactive influence of attentional inertia and attentional state. *Infancy, 5*(2), 239–252.

Oakes, L. M., Tellinghuisen, D. J., & Tjebkes, T. L. (2000). Competition for infants' attention: The interactive influence of attentional state and stimulus characteristics. *Infancy, 1*(3), 347–361.

Oberauer, K., Lange, E., & Engle, R. W. (2004). Working memory capacity and resistance to interference. *Journal of Memory and Language, 51*(1), 80–96.

Ollendick, T. H., Jarrett, M. A., Grills-Taquechel, A. E., Hovey, L. D., & Wolff, J. C. (2008). Comorbidity as a predictor and moderator of treatment outcome in youth with anxiety, affective, attention deficit/hyperactivity disorder, and oppositional/conduct disorders. *Clinical Psychology Review, 28*(8), 1447–1471.

Olson, L. E., Richtsmeier, J. T., Leszl, J., & Reeves, R. H. (2004). A chromosome 21 critical region does not cause specific Down syndrome phenotypes. *Science, 306*(5696), 687–690.

Olson, L. E., Roper, R. J., Sengstaken, C. L., Peterson, E. A., Aquino, V., Galdzicki, Z., et al. (2007). Trisomy for the Down syndrome 'critical region' is necessary but not sufficient for brain phenotypes of trisomic mice. *Human Molecular Genetics, 16*(7), 774–782.

Oosterlaan, J., & Sergeant, J. A. (1998). Effects of reward and response cost on response inhibition in AD/HD, disruptive, anxious, and normal children. *Journal of Abnormal Child Psychology, 26*(3), 161–174.

O'Riordan, M. A. (2000). Superior modulation of activation levels of stimulus representations does not underlie superior discrimination in autism. *Cognition, 77*(2), 81–96.

O'Riordan, M. A. (2004). Superior visual search in adults with autism. *Austim, 8*(3), 229–248.

O'Riordan, M. A., Plaisted, K. C., Driver, J., & Baron-Cohen, S. (2001). Superior visual search in autism. *Journal of Experimental Psychology-Human Perception and Performance, 27*(3), 719–730.

Óskarsdóttir, S., Holmberg E., Fasth, A., & Strömland, K. (2008). Facial features in children with the 22q11 deletion syndrome. *Acta Paediatrica, 97*(8), 1113–1117.

Óskarsdóttir, S., Vujic, M., & Fasth, A. (2004). Incidence and prevalence of the 22q11 deletion syndrome: A population-based study in western Sweden. *Archives of Disease in Childhood, 89*(2), 148–151.

Ozonoff, S., Cook, I., Coon, H., Dawson, G., Joseph, R. M., Klin, A., et al. (2004). Performance on Cambridge Neuropsychological Test Automated Battery subtests sensitive to frontal lobe function in people with autistic disorder: Evidence from the Collaborative Programs of Excellence in Autism Network. *Journal of Autism and Developmental Disorders, 34*(2), 139–150.

Ozonoff, S., & Jensen, J. (1999). Brief report: Specific executive function profiles in three neurodevelopmental disorders. *Journal of autism and developmental disorders, 29*(2), 171–7.

Ozonoff, S., & Strayer, D. L. (1997). Inhibitory function in nonretarded children with autism. *Journal of Autism and Developmental Disorders, 27*(1), 59–77.

Pal, D. K., Chaudhury, G., Das, T., & Sengupta, S. (1999). Validation of a Bengali adaptation of the Conners' Parent Rating Scale (CPRS-48). *British Journal of Medical Psychology, 72*(4), 525–533.

Pandolfi, V., Magyar, C. I., & Dill, C. A. (2009). Confirmatory factor analysis of the Child Behavior Checklist 1.5-5 in a sample of children with autism spectrum disorders. *Journal of Autism & Developmental Disorders, 39*(7), 986–995.

Parasuraman, R., & Davies, D. R. (1976). Decision-theory analysis of response latencies in vigilance. *Journal of Experimental Psychology-Human Perception and Performance, 2*(4), 578–590.

Parasuraman, R., Warm, J. S., & See, J. E. (1998). Brain systems of vigilance. In R. Parasuraman (Ed.), *The attentive brain* (pp. 221–256). Cambridge, MA: MIT Press.

Paterson, S. J., & Schultz, R. T. (2007). Neurodevelopmental and behavioral issues in Williams syndrome. *Current Psychiatry Reports, 9*(2), 165–171.

Paterson, S. J., Girelli, L., Butterworth, B., & Karmiloff-Smith, A. (2006). Are numerical impairments syndrome specific? Evidence from Williams syndrome and Down's syndrome. *Journal of Child Psychology and Psychiatry, 47*(2), 190–204.

Paus, T. (2005). Mapping brain maturation and cognitive development during adolescence. *Trends in Cognitive Sciences, 9*(2), 60–68.

Paylor, R., McIlwain, K. L., McAninch, R., Nellis, A., Yuva-Paylor, L. A., Baldini, A., et al. (2001). Mice deleted for the DiGeorge/velocardiofacial syndrome region show abnormal sensorimotor gating and learning and memory impairments. *Human Molecular Genetics, 10*(23), 2645–2650.

Pearson, D. A., Santos, C. W., Casat, C. D., Lane, D. M., Jerger, S. W., Roache, J. D., et al. (2004). Treatment effects of methylphenidate on cognitive functioning in children with mental retardation and ADHD. *Journal of the American Academy of Child and Adolescent Psychiatry, 43*(6), 677–685.

Pearson, D. A., Santos, C. W., Roache, J. D., Casat, C. D., Loveland, K. A., Lachar, D., et al. (2003). Treatment effects of methylphenidate on behavioral adjustment in children with mental retardation and ADHD. *Journal of the American Academy of Child and Adolescent Psychiatry, 42*(2), 209–216.

Pelham, W. E., & Fabiano, G. A. (2008). Evidence-based psychosocial treatments for attention-deficit/hyperactivity disorder. *Journal of Clinical Child and Adolescent Psychology, 37*(1), 184–214.

Pelham, W. E., & Hoza, B. (1996). Comprehensive treatment for ADHD: A proposal for intensive summer treatment programs and outpatient follow-up. In E. Hibbs & P. S. Jensen (Eds.), *Psychosocial treatments for child and adolescent disorders: Empirically based approaches* (pp. 311–340). Washington, DC: American Psychological Press.

Pennington, B. F., Bennetto, L., McAleer, O., & Roberts, R. J. (1996). Executive functions and working memory. In G. R. Lyon & N. A. Krasnegor (Eds.), *Attention, memory and executive function* (pp. 327–348). Baltimore: Paul Brookes.

Pennington, B. F., McGrath, L. M., Rosenberg, J., Barnard, H., Smith, S. D., Willcutt, E. G., et al. (2009). Gene X environment interactions in reading disability and attention-deficit/hyperactivity disorder. *Developmental Psychology, 45*(1), 77–89.

Pennington, B. F., Moon, J., Edgin, J., Stedron, J., & Nadel, L. (2003). The neuropsychology of Down syndrome: Evidence for hippocampal dysfunction. *Child Development, 74*(1), 75–93.

Pennington, B. F., & Ozonoff, S. (1996). Executive functions and developmental psychopathology. *Journal of Child Psychology and Psychiatry and Allied Disciplines, 37*(1), 51–87.

Perner, J., & Lang, B. (2002). What causes 3-year-olds' difficulty on the Dimensional Change Card Sorting Task? *Infant and Child Development, 11*(2), 93–105.

Perren, S., Stadelmann, S., von Wyl, A., & von Klitzing, K. (2007). Pathways of behavioural and emotional symptoms in kindergarten children: What is the role of pro-social behaviour? *European Child & Adolescent Psychiatry, 16*(4), 209–214.

Persico, A. M., D'Agruma, L., Maiorano, N., Totaro, A., Militerni, R., Bravaccio, C., et al. (2001). Reelin gene alleles and haplotypes as a factor predisposing to autistic disorder. *Molecular Psychiatry, 6*(2), 150–159.

Peterson, B. S. (2003). Conceptual, methodological, and statistical challenges in brain imaging studies of developmentally based psychopathologies. *Development and Psychopathology, 15*(3), 811–832.

Petrides, M., Alivisatos, B., Evans, A. C., & Meyer, E. (1993). Dissociation of human mid-dorsolateral from posterior dorsolateral frontal-cortex in memory processing. *Proceedings of the National Academy of Sciences of the United States of America, 90*(3), 873–877.

Petrides, M., & Milner, B. (1982). Deficits on subject-ordered tasks after frontal- and temporal-lobe lesions in man. *Neuropsychologia, 20*(3), 249–262.

Pfiffner, L. J., Yee Mikami, A., Huang-Pollock, C., Easterlin, B., Zalecki, C., & McBurnett, K. (2007). A randomized, controlled trial of integrated home-school behavioral treatment for ADHD, predominantly inattentive type. *Journal of the American Academy of Child and Adolescent Psychiatry, 46*(8), 1041–1050.

Philofsky, A., Hepburn, S. L., Hayes, A., Hagerman, R., & Rogers, S. J. (2004). Linguistic and cognitive functioning and autism symptoms in young children with fragile X syndrome. *American Journal on Mental Retardation, 109*(3), 208–218.

Picton, T. W., & Taylor, M. J. (2007). Electrophysiological evaluation of human brain development. *Developmental Neuropsychology, 31*(3), 249–278.

Pinter, J. D., Eliez, S., Schmitt, J. E., Capone, G. T., & Reiss, A. L. (2001). Neuroanatomy of Down's syndrome: A high-resolution MRI study. *American Journal of Psychiatry, 158*(10), 1659–1665.

Pliszka, S. R. (1998). Comorbidity of attention-deficit/hyperactivity disorder with psychiatric disorder: An overview. *The Journal of clinical psychiatry, 59*, Supplement 7, 50–8.

Pober, B. R., Johnson, M., & Urban, Z. (2008). Mechanisms and treatment of cardiovascular disease in Williams-Beuren syndrome. *Journal of Clinical Investigation, 118*(5), 1606–1615.

Polderman, T. J. C., Derks, E. M., Hudziak, J. J., Verhulst, F. C., Posthuma, D., & Boomsma, D. I. (2007). Across the continuum of attention skills: A twin study of the SWAN ADHD rating scale. *Journal of Child Psychology and Psychiatry, 48*(11), 1080–1087.

Posey, D. J., Aman, M. G., McCracken, J. T., Scahill, L., Tierney, E., Arnold, L. E., et al. (2007). Positive effects of methylphenidate on inattention and hyperactivity in pervasive developmental disorders: An analysis of secondary measures. *Biological Psychiatry, 61*(4), 538–544.

Posner, M. I., & Boies, S. J. (1971). Components of attention. *Psychological Review, 78*(5), 391–408.

Posner, M. I., & Dehaene, S. (1994). Attentional networks. *Trends in Neurosciences, 17*(2), 75–79.

Posner, M. I., & Petersen, S. E. (1990). The attention system of the human brain. *Annual Review of Neuroscience, 13*, 25–42.

Posner, M. I., & Rothbart, M. K. (2005). Influencing brain networks: Implications for education. *Trends in Cognitive Sciences, 9*(3), 99–103.

Posner, M. I., & Rothbart, M. K. (2007a). *Educating the Human Brain.* Washington, DC: American Psychology Association.

Posner, M. I., & Rothbart. M. K. (2007b). Research on attention networks as a model for the integration of psychological science. *Annual review of psychology, 58*, 1–23.

Posner, M. I., Rothbart, M. K., & Sheese, B. E. (2007). The anterior cingulate gyrus and the mechanism of self-regulation. *Cognitive Affective & Behavioral Neuroscience, 7*(4), 391–395.

Posner, M. I., Walker, J. A., Friedrich, F. J., & Rafal, R. D. (1984). Effects of parietal injury on covert orienting of attention. *Journal of Neuroscience, 4*(7), 1863–1874.

Power, T. J., Blum, N. J., Jones, S. M., & Kaplan, P. E. (1997). Brief report: Response to methylphenidate in two children with Williams syndrome. *Journal of Autism and Developmental Disorders, 27*(1), 79–87.

Power, T. J., Costigan, T. E., Leff, S. S., Eiraldi, R. B., & Landau, S. (2001). Assessing ADHD across settings: Contributions of behavioral assessment to categorical decision making. *Journal of Clinical Child Psychology, 30*(3), 399–412.

Power, T. J., Doherty, B. J., Panichelli-Mindel, S. M., Karustis, J. L., Eiraldi, R. B., Anastopoulos, A. D., et al. (1998). The predictive validity of parent and teacher reports of ADHD symptoms. *Journal of Psychopathology and Behavioral Assessment, 20*(1), 57–81.

Prasad, S. E., Howley, S., & Murphy, K. C. (2008). Candidate genes and the behavioral phenotype in 22q11.2 deletion syndrome. *Developmental Disabilities Research Review, 14*(1), 26–34.

Preissler, M. A. (2008). Associative learning of pictures and words by low-functioning children with autism. *Autism, 12*(3), 231–248.

Pribram, K. H., & McGuinness, D. (1975). Arousal, activation, and effort in control of attention. *Psychological Review, 82*(2), 116–149.

Pritchard, V. E., & Neumann, E. (2004). Negative priming effects in children engaged in nonspatial tasks: Evidence for early development of an intact inhibitory mechanism. *Developmental Psychology, 40*(2), 191–203.

Putnam, S. P., Gartstein, M. A., & Rothbart, M. K. (2006). Measurement of fine-grained aspects of toddler temperament: The early childhood behavior questionnaire. *Infant Behavior and Development, 29*(3), 386–401.

Putnam, S. P., & Rothbart, M. K. (2006). Development of short and very short forms of the Children's Behavior Questionnaire. *Journal of Personality Assessment, 87*(1), 102–112.

Quist, J. F., Barr, C. L., Schachar, R., Roberts, W., Malone, M., Tannock, R., et al. (2003). The serotonin 5-HT1B receptor gene and attention deficit hyperactivity disorder. *Molecular Psychiatry, 8*(1), 98–102.

Rabiner, D., & Coie, J. D. (2000). Early attention problems and children's reading achievement: a longitudinal investigation. The Conduct Problems Prevention Research Group. *Journal of the American Academy of Child and Adolescent Psychiatry, 39*(7), 859–67.

Rabiner, D. L., Malone, P. S., & Conduct Problems Prevention Research Group (2004). The impact of tutoring on early reading achievement for children with and without attention problems. *Journal of Abnormal Child Psychology, 32*(3), 273–284.

Rachidi, M., & Lopes, C. (2008). Mental retardation and associated neurological dysfunctions in Down syndrome: A consequence of dysregulation in critical chromosome 21 genes and associated molecular pathways. *European Journal of Paediatric Neurology, 12*(3), 168–182.

Raye, C. L., Johnson, M. K., Mitchell, K. J., Greene, E. J., & Johnson, M. R. (2007). Refreshing: A minimal executive function. *Cortex, 43*(1), 135–145.

Rebok, G. W., Smith, C. B., Pascualvaca, D. M., Mirsky, A. F., Anthony, B. J., & Kellam, S. G. (1997). Developmental changes in attentional performance in urban children from eight to thirteen years. *Child Neuropsychology, 3*(1), 28–46.

Reddy, K. S. (2005). Cytogenetic abnormalities and fragile-X syndrome in autism spectrum disorder. *BMC Medical Genetics, 6*, 3.

Redick, T. S., & Engle, R. W. (2006). Working memory capacity and attention network test performance. *Applied Cognitive Psychology, 20*(5), 713–721.

Reed, M., Pien, D., & Rothbart, M.K. (*1984*). Inhibitory self-control in preschool children. *Merrill-Palmer Quarterly, 30*, 131–148.

Reimers, S., & Maylor, E. A. (2005). Task switching across the life span: Effects of age on general and specific switch costs. *Developmental Psychology, 41*(4), 661–671.

Reiss, A. L., Eckert, M. A., Rose, F. E., Karchemskiy, A., Kesler, S., Chang, M., et al. (2004). An experiment of nature: Brain anatomy parallels cognition and behavior in Williams syndrome. *Journal of Neuroscience, 24*(21), 5009–5015.

Reitan, R. M. (1958). Validity of the Trail Making Test as an indicator of organic brain damage. *Perceptual and Motor Skills, 8*, 271–276.

Renner, P., Klinger, L. G., & Klinger, M. (2006). Exogenous and endogenous attention orienting in autism spectrum disorders. *Child Neuropsychology, 12*, 361–382.

Rennie, D. A. C., Bull, R., & Diamond, A. (2004). Executive functioning in preschoolers: Reducing the inhibitory demands of the dimensional change card sort task. *Developmental Neuropsychology, 26*(1), 423–443.

Research Units on Pediatric Psychopharmacology Autism Network (2005). Randomized, controlled, crossover trial of methylphenidate in pervasive developmental disorders with hyperactivity. *Archives of General Psychiatry, 62*(11), 1266–1274.

Reynolds, C. R. (2002). *Comprehensive Trail Making Test (CTMT)* (Vol. 19). Austin, TX: PRO-ED, Inc.

Reynolds, J. H., & Chelazzi, L. (2004). Attentional modulation of visual processing. *Annual Review of Neuroscience, 27*, 611–647.

Rezazadeh, S., Wilding, J., & Cornish, K. (2010). Elucidating the relationship between behavioural ratings of attention and inhibitory permanence. Submitted for publication.

Rhodes, S. M., Coghill, D. R., & Matthews, K. (2004). "Methylphenidate restores visual memory, but not working memory function in attention deficit-hyperkinetic disorder". *Psychopharmacology (Berl), 175*, 319–330.

Rhodes, S. M., Coghill, D. R., & Matthews, K. (2006). Acute neuropsychological effects of methylphenidate in stimulant drug-naïve boys with ADHD II—broader executive and non-executive domains. *Journal of Child Psychology and Psychiatry, 47*(11), 1184–1194.

Riccio, C. A., Waldrop, J. J. M., Reynolds, C. R., & Lowe, P. (2001). Effects of stimulants on the continuous performance test (CPT): Implications for CPT use and interpretation. *Journal of Neuropsychiatry and Clinical Neurosciences, 13*(3), 326–335.

Ridderinkhof, K. R., & van der Molen, M. W. (1995). A psychophysiological analysis of developmental differences in the ability to resist interference. *Child Development, 66*(4), 1040–1056.

Rietveld, M. J. H., Hudziak, J. J., Bartels, M., van Beijsterveldt, C. E. M., & Boomsma, D. I. (2004). Heritability of attention problems in children: Longitudinal results from a study of twins, age 3 to 12. *Journal of Child Psychology and Psychiatry, 45*(3), 577–588.

Rinehart, N. J., Bradshaw, J. L., Moss, S. A., Brereton, A. V., & Tonge, B. J. (2008). Brief report: Inhibition of return in young people with autism and Asperger's disorder. *Autism, 12*(3), 249–260.

Robbins, T. W. (2007). Shifting and stopping: Fronto-striatal substrates, neurochemical modulation and clinical implications. *Philosophical Transactions of the Royal Society B: Biological Sciences, 362*(1481), 917–932.

Robbins, T. W., Milstein, J. E., & Dalley, J. W. (2004). Neuropharmacology of attention. In M. I. Posner (Ed.), *Cognitive neuroscience of attention* (pp. 283–293). New York: Guilford Press.

Robbins, T. W., & Roberts, A. C. (2007). Differential regulation of fronto-executive function by the monoamines and acetylcholine. *Cerebral Cortex, 17*(Suppl. 1), i151–i160.

Roberts, R. J., & Pennington, B. F. (1996). An interactive framework for examining prefrontal cognitive processes. *Developmental Neuropsychology, 12*(1), 105–126.

Roberts, J., Price, J., Barnes, E., Nelson, L., Burchinal, M., Hennon, E. A., et al. (2007). Receptive vocabulary, expressive vocabulary, and speech production of boys with fragile X syndrome in comparison to boys with Down syndrome. *American Journal on Mental Retardation, 112*(3), 177–193.

Roberts, J. E., Mankowski, J. B., Sideris, J., Goldman, B. D., Hatton, D. D., Mirrett, P. L., et al (2009). Trajectories and predictors of the development of very young boys with fragile X syndrome. *Journal of Pediatric Psychology, 34*(8), 827–836. Retrieved from

Roberts, J. E., Schaaf, J. M., Skinner, M., Wheeler, A., Hooper, S., Hatton, D. D., et al. (2005). Academic skills of boys with fragile X syndrome: Profiles and predictors. *American Journal on Mental Retardation, 110*(2), 107–120.

Robertson, I. H., Manly, T., Andrade, J., Baddeley, B. T., & Yiend, J. (1997). 'Oops!': Performance correlates of everyday attentional failures in traumatic brain injured and normal subjects. *Neuropsychologia, 35*(6), 747–758.

Robertson, I. H., Ward, A., Ridgeway, V., & Nimmo-Smith, I. (1994). *Test of Everyday Attention.* Bury St Edmunds, UK: Thames Valley Test Company.

Rock, I., & Gutman, D. (1981). The effect of inattention on form perception. *Journal of Experimental Psychology: Human Perception and Performance, 7*(2), 275–285.

Rogers, S. J., & Vismara, L. A. (2008). Evidence-based comprehensive treatments for early autism. *Journal of Clinical Child and Adolescent Psychology, 37*(1), 8–38.

Rogers, S. J., Wehner, E. A., & Hagerman, R. (2001). The behavioral phenotype in fragile X: Symptoms of autism in very young children with fragile X syndrome, idiopathic autism, and other developmental disorders. *Journal of Developmental and Behavioral Pediatrics, 22*(6), 409–417.

Rojas, D. C., Benkers, T. L., Rogers, S. J., Teale, P. D., Reite, M. L., & Hagerman, R. J. (2001). Auditory evoked magnetic fields in adults with fragile X syndrome. *NeuroReport, 12*(11), 2573–2576.

Roman, T., Schmitz, M., Polanczyk, G., Eizirik, M., Rohde, L. A., & Hutz, M. H. (2001). Attention-deficit hyperactivity disorder: A study of association with both the dopamine transporter gene and the dopamine D4 receptor gene. *American Journal of Medical Genetics Part B: Neuropsychiatric Genetics, 105*(5), 471–478.

Romine, C. B., Lee, D., Wolfe, M. E., Homack, S., George, C., & Riccio, C. A. (2004). Wisconsin Card Sorting Test with children: A meta-analytic study of sensitivity and specificity. *Archives of Clinical Neuropsychology, 19*(8), 1027–1041.

Rommelse, N. N., Altink, M. E., Fliers, E. A., Martin, N. C., Buschgens, C. J., Hartman, C. A., et al. (2009). Comorbid problems in ADHD: Degree of association, shared endophenotypes, and formation of distinct subtypes. Implications for a future DSM. *Journal of Abnormal Child Psychology, 37*(6), 793–804.

Ronald, A., Happé, F., Bolton, P., Butcher, L. M., Price, T. S., Wheelwright, S., et al. (2006). Genetic heterogeneity between the three components of the autism spectrum: A twin study. *Journal of the American Academy of Child & Adolescent Psychiatry, 45*(6), 691–699.

Rondan, C., Mancini, J., Livet, M. O., & Deruelle, C. (2003). Perceptual and visuo-constructive performance in children with Williams syndrome. *Cognition, Brain, Behavior, 7*, 149–156.

Rondan, C., Santos, A., Mancini, J., Livet, M. O., & Deruelle, C. (2008). Global and local processing in Williams syndrome: Drawing versus perceiving. *Child Neuropsychology, 14*(3), 237–248.

Ross, T. P., Hanouskova, E., Giarla, K., Calhoun, E., & Tucker, M. (2007). The reliability and validity of the self-ordered pointing task. *Archives of Clinical Neuropsychology, 22*(4), 449–458.

Rosvold, H. E., Mirsky, A. F., Sarason, I., Bransome, E. D., & Beck, L. H. (1956). A continuous performance-test of brain-damage. *Journal of Consulting Psychology, 20*(5), 343–350.

Rothbart, M. K., Ahadi, S. A., Hershey, K. L., & Fisher, P. (2001). Investigations of temperament at three to seven years: The children's behavior questionnaire. *Child Development, 72*(5), 1394–1408.

Rovee-Collier, C., Bhatt, R. S., & Chazin, S. (1996). Set size, novelty, and visual pop-out in infancy. *Journal of Experimental Psychology. Human Perception and Performance, 22*, 1178–1187.

Rovee-Collier, C., Hankins, E., & Bhatt, R. S. (1992). Textons, visual pop-out effects, and object recognition in infancy. *Journal of Experimental Psychology, 121*(4), 435–445.

Roy, A. L. (2001). Biochemistry and biology of the inducible multifunctional transcription factor TFII-I. *Gene, 274*(1–2), 1–13.

Ruchkin, V., Lorberg, B., Koposov, R., Schwab-Stone, M., & Sukhodolsky, D. G. (2008). ADHD symptoms and associated psychopathology in a community sample of adolescents from the European north of Russia. *Journal of Attention Disorders, 12*(1), 54–63.

Rucklidge, J. J. (2008). Gender differences in ADHD: Implications for psychosocial treatments. *Expert Review of Neurotherapeutics, 8*(4), 643–655.

Rueckert, L., & Grafman, J. (1996). Sustained attention deficits in patients with right frontal lesions. *Neuropsychologia, 34*(10), 953–963.

Rueda, M. R., Fan, J., McCandliss, B. D., Halparin, J. D., Gruber, D. B., Lercari, L. P., et al. (2004). Development of attentional networks in childhood. *Neuropsychologia, 42*(8), 1029–1040.

Rueda, M. R., Posner, M. I., & Rothbart, M. K. (2005). The development of executive attention: Contributions to the emergence of self-regulation. *Developmental Neuropsychology, 28*(2), 573–594.

Rueda, M. R., Posner, M. I., Rothbart, M. K., & Davis-Stober, C. P. (2004). Development of the time course for processing conflict: an event-related potentials study with 4 year olds and adults. *BMC Neuroscience, 5*, 39.

Ruff, H. A., & Capozzoli, M. (2003). Development of attention and distractibility in the first 4 years of life. *Developmental Psychology, 39*(5), 877–890.

Ruff, H. A., Capozzoli, M., Dubiner, K., & Parrinello, R. (1990). A measure of vigilance in infancy. *Infant Behavior and Development, 13*(1), 1–20.

Ruff, H. A., Capozzoli, M., & Saltarelli, L. M. (1996). Focussed visual attention and distractibility in 10-month-old infants. *Infant Behavior and Development, 19*(3), 281–293.

Ruff, H. A., Capozzoli, M., & Weissberg, R. (1998). Age, individuality, and context as factors in sustained visual attention during the preschool years. *Developmental Psychology, 34*(3), 454–464.

Ruff, H. A., & Lawson, K. R. (1990). Development of sustained, focused attention in young children during free play. *Developmental Psychology, 26*, 85–93.

Russell, J., Jarrold, C., & Hood, B. (1999). Two intact executive capacities in children with autism: Implications for the core executive dysfunctions in the disorder. *Journal of Autism and Developmental Disorders, 29*(2), 103–112.

Russo, N., Flanagan, T., Iarocci, G., Berringer, D., Zelazo, P. D., & Burack, J. A. (2007). Deconstructing executive deficits among persons with autism: Implications for cognitive neuroscience. *Brain and Cognition, 65*(1), 77–86.

Ruthruff, E., Van Selst, M., Johnston, J. C., & Remington, R. (2006). How does practice reduce dual-task interference: Integration, automatization, or just stage-shortening? *Psychological Research—Psychologische Forschung, 70*(2), 125–142.

Rutter, M. (2005). Incidence of autism spectrum disorders: Changes over time and their meaning. *Acta Paediatrica, 94*(1), 2–15.

Sagvolden, T., Johansen, E. B., Aase, H., & Russell, V. A. (2005). A dynamic developmental theory of attention-deficit/hyperactivity disorder (ADHD) predominantly hyperactive/impulsive and combined subtypes. *Behavior and Brain Sciences, 28*(3), 397–419.

Sanders, A. F. (1983). Towards a model of stress and human-performance. *Acta Psychologica, 53*(1), 61–97.

Santosh, P. J., Baird, G., Pityaratstian, N., Tavare, E., & Gringras, P. (2006). Impact of comorbid autism spectrum disorders on stimulant response in children with attention deficit hyperactivity disorder: A retrospective and prospective effectiveness study. *Child Care, Health and Development, 32*(5), 575–583.

Savage, R., Cornish, K., Manly, T., & Hollis, C. (2006). Cognitive processes in children's reading and attention: The role of working memory, divided attention, and response inhibition. *British Journal of Psychology, 97*(3), 365–385.

Scerif, G., Cornish, K., Wilding, J., Driver, J., & Karmiloff-Smith, A. (2004). Visual search in typically developing toddlers and toddlers with Fragile X or Williams syndrome. *Developmental Science, 7*(1), 116–130.

Scerif, G., Cornish, K., Wilding, J., Driver, J., & Karmiloff-Smith, A. (2007). Delineation of early attentional control difficulties in fragile X syndrome: Focus on neurocomputational changes. *Neuropsychologia, 45*(8), 1889–1898.

Scerif, G., Karmiloff-Smith, A., Campos, R., Elsabbagh, M., Driver, J., & Cornish, K. (2005). To look or not to look: Typical and atypical development of oculomotor control. *Journal of Cognitive Neuroscience, 17*(4), 591–604.

Schachar, R., Ickowicz, A., Crosbie, J., Donnelly, G. A., Reiz, J. L., Miceli, P. C., et al. (2008). Cognitive and behavioral effects of multilayer-release methylphenidate in the treatment of children with attention-deficit/hyperactivity disorder. *Journal of Child and Adolescent Psychopharmacology, 18*(1), 11–24.

Schaer, M., & Eliez, S. (2007). From genes to brain: Understanding brain development in neurogenetic disorders using neuroimaging techniques. *Child and Adolescent Psychiatric Clinics of North America, 16*(3), 557–579.

Schatz, D. B., & Rostain, A. L. (2006). ADHD with comorbid anxiety: A review of the current literature. *Journal of Attention Disorders, 10*(2), 141–149.

Scheres, A., Oosterlaan, J., Geurts, H., Morein-Zamir, S., Meiran, N., Schut, H., et al. (2004). Executive functioning in boys with ADHD: Primarily an inhibition deficit? *Archives of Clinical Neuropsychology, 19*(4), 569–594.

Scheres, A., Oosterlaan, J., Swanson, J., Morein-Zamir, S., Meiran, N., Schut, H., et al. (2003). The effect of methylphenidate on three forms of response inhibition in boys with AD/HD. *Journal of Abnormal Child Psychology, 31*(1), 105–120.

Schiller, P. H., Logothetis, N. K., & Charles, E. R. (1990). Role of colour-opponent and broad-band channels in vision. *Visual Neuroscience, 5*(1), 321–346.

Schneider, H., & Eisenberg, D. (2006). Who receives a diagnosis of attention-deficit/hyperactivity disorder in the United States elementary school population? *Pediatrics, 117*(4), e601–e609.

Schneider, W., & Chein, J. M. (2003). Controlled and automatic processing: Behavior, theory, and biological mechanisms. *Cognitive Science, 27*(3), 525–559.

Schneider, W., & Shiffrin, R. M. (1977). Controlled and automatic human information-processing. 1. Detection, search, and attention. *Psychological Review, 84*(1), 1–66.

Scholl, B. J. (2001). Objects and attention: The state of the art. *Cognition, 80*(1–2), 1–46.

Schulz, K. P., Fan, J., Tang, C. Y., Newcorn, J. H., Buchsbaum, M. S., Cheung, A. M., et al. (2004). Response inhibition in adolescents diagnosed with attention deficit hyperactivity disorder during childhood: An event-related FMRI study. *American Journal of Psychiatry, 161*(9), 1650–1657.

Schwartz, M. F., Reed, E. S., Montgomery, M., Palmer, C., & Mayer, N. H. (1991). The quantitative description of action disorganization after brain-damage—A case-study. *Cognitive Neuropsychology, 8*(5), 381–414.

Seeger, G., Schloss, P., & Schmidt, M. H. (2001). Functional polymorphism within the promotor of the serotonin transporter gene is associated with severe hyperkinetic disorders. *Molecular Psychiatry, 6*(2), 235–238.

Seidman, L. J., Biederman, J., Monuteaux, M. C., Valera, E., Doyle, A. E., & Faraone, S. V. (2005). Impact of gender and age on executive functioning: Do girls and boys with and without attention deficit hyperactivity disorder differ neuropsychologically in preteen and teenage years? *Developmental Neuropsychology, 27*(1), 79–105.

Senju, A., Tojo, Y., Dairoku, H., & Hasegawa, T. (2004). Reflexive orienting in response to eye gaze and an arrow in children with and without autism. *Journal of Child Psychology and Psychiatry, 45*(3), 445–458.

Senn, T. E., Espy, K. A., & Kaufmann, P. M. (2004). Using Path Analysis to Understand Executive Function Organization in Preschool Children. *Developmental Neuropsychology, 26*, 445–464.

Serajee, F. J., Zhong, H. L., & Huq, A. (2006). Association of reelin gene polymorphisms with autism. *Genomics, 87*(1), 75–83.

Serences, J. T., Schwarzbach, J., Courtney, S. M., Golay, X., & Yantis, S. (2004). Control of object-based attention in human cortex. *Cerebral Cortex, 14*(12), 1346–1357.

Sergeant, J. (2000). The cognitive-energetic model: An empirical approach to Attention-Deficit Hyperactivity Disorder. *Neuroscience and Biobehavioral Reviews, 24*(1), 7–12.

Sergeant, J. A. (2005). Modeling attention-deficit/hyperactivity disorder: A critical appraisal of the cognitive-energetic model. *Biological Psychiatry, 57*(11), 1248–1255.

Sergeant, J. A., Geurts, H., & Oosterlaan, J. (2002). How specific is a deficit of executive functioning for Attention-Deficit/Hyperactivity Disorder? *Behavioural Brain Research, 130*(1–2), 3–28.

Sergeant, J. A., Oosterlaan, J., & van der Meere, J. (1999). Information processing and energetic factors in Attention-Deficit/Hyperactivity Disorder. In H. C. Quay & A. E. Hogan (Eds.), *Handbook of disruptive behavior disorders* (pp. 75–104). New York: Kluwer Academic.

Sershen, H., Hashim, A., & Lajtha, A. (2000). Serotonin-mediated striatal dopamine release involves the dopamine uptake site and the serotonin receptor. *Brain Research Bulletin, 53*(3), 353–357.

Servera, M., & Cardo, E. (2007). ADHD Rating Scale-IV in a sample of Spanish schoolchildren: Normative data and internal consistency for teachers and parents. *Revista De Neurologia, 45*, 393–399.

Shaffer, L. H. (1975). Multiple attention in continuous verbal tasks. In P. M. A. Rabbitt & S. Dornic (Eds.), *Attention and performance* (Vol. 5, pp. 157–167). New York: Academic Press.

Shalev, L., Tsal, Y., & Mevorach, C. (2007). Computerized Progressive Attentional Training (CPAT) program: Effective direct intervention for children with ADHD. *Child Neuropsychology, 13*(4), 382–388.

Shallice, T., & Burgess, P. W. (1991). Deficits in strategy application following frontal lobe damage in man. *Brain, 114*, 727–741.

Shallice, T., Stuss, D. T., Alexander, M. P., Picton, T. W., & Derkzen, D. (2008). The multiple dimension of sustained attention. *Cortex, 44*(7), 794–805.

Shanahan, M. A., Pennington, B. R., & Willcutt, E. W. (2008). Do motivational incentives reduce the inhibition deficit in ADHD? *Developmental Neuropsychology, 33*(2), 137–159.

Shapley, R. (1990). Visual sensitivity and parallel retinocortical channels. *Annual Review of Psychology, 41*(1), 635–658.

Sharp, S. A., McQuillin A., & Gurling, H. M D. (2009). Genetics of attention-deficit hyperactivity disorder (ADHD). *Neuropharmacology, 57*, 590–600.

Shaw, P., Eckstrand, K., Sharp, W., Blumenthal, J., Lerch, J. P., Greenstein, D., et al. (2007). Attention-deficit/hyperactivity disorder is characterized by a delay in cortical maturation. *Proceedings of the National Academy of Sciences of the United States of America, 104*(49), 19649–19654.

Shaw, P., Kabani, N. J., Lerch, J. P., Eckstrand, K., Lenroot, R., Gogtay, N., et al. (2008). Neurodevelopmental trajectories of the human cerebral cortex. *Journal of Neuroscience, 28*(14), 3586–3594.

Sherman, S. L., Allen, E. G., Bean, L. H., & Freeman, S. B. (2007). Epidemiology of Down syndrome. *Mental Retardation and Developmental Disabilities Research Reviews, 13*(3), 221–227.

Shiffrin, R. M., & Schneider, W. (1977). Controlled and automatic human information-processing .2. Perceptual learning, automatic attending, and a general theory. *Psychological Review, 84*(2), 127–190.

Shilling, V. M., Chetwynd, A., & Rabbitt, P. M. A. (2002). Individual inconsistency across measures of inhibition: An investigation of the construct validity of inhibition in older adults. *Neuropsychologia, 40*(6), 605–619.

Shomstein, S., & Yantis, S. (2004). Control of attention shifts between vision and audition in human cortex. *Journal of Neuroscience, 24*(47), 10702–10706.

Shprintzen, R. J. (2008). Velo-cardio-facial syndrome: 30 years of study. *Developmental Disabilities Research Review, 14*(1), 3–10.

Shur-Fen Gau, S., Shang, C. Y., Lui, S. K., Lin, C. H., Swanson, J., Liu, Y. C., et al. (2008). Psychometric properties of the Chinese version of the Swanson, Nolan, and Pelham, version IV scale-parent form. *International Journal of Methods in Psychiatric Research, 17*(1), 35–44.

Shur-Fen Gau, S., Soong, W.-T., Chiu, Y.-N., & Tsai, W.-C. (2006). Psychometric properties of the Chinese version of the Conners' Parent and Teacher Rating Scales-Revised: Short form. *Journal of Attention Disorders, 9*(4), 648–659.

Simon, T. J. (2008). A new account of the neurocognitive foundations of impairments in space, time and number processing in children with chromosome 22q11.2 deletion syndrome. *Developmental Disabilities Research Review, 14*(1), 52–58.

Simon, T. J., Bearden, C. E., Mc-Ginn, D. M., & Zackai, E. (2005). Visuospatial and numerical cognitive deficits in children with chromosome 22q11.2 deletion syndrome. *Cortex, 41*(2), 145–155.

Simon, T. J., Bish, J. P., Bearden, C. E., Ding, L. J., Ferrante, S., Nguyen, V., et al. (2005). A multilevel analysis of cognitive dysfunction and psychopathology associated with chromosome 22q11.2 deletion syndrome in children. *Development and Psychopathology, 17*(3), 753–784.

Simon, T. J., Takarae, Y., DeBoer, T., McDonald-McGinn, D. M., Zackai, E. H., & Ross, J. L. (2008). Overlapping numerical cognition impairments in children with chromosome 22q11.2 deletion or Turner syndromes. *Neuropsychologia, 46*(1), 82–94.

Simonoff, E., Pickles, A., Wood, N., Gringras, P., & Chadwick, O. (2007). ADHD symptoms in children with mild intellectual disability. *Journal of the American Academy of Child and Adolescent Psychiatry, 46*(5), 591–600.

Simpson, A., & Riggs, K. J. (2005a). Factors responsible for performance on the day-night task: Response set or semantics? *Developmental Science, 8*(4), 360–371.

Simpson, A., & Riggs, K. J. (2005b). Inhibitory and working memory demands of the day-night task in children. *British Journal of Developmental Psychology, 23*(3), 471–486.

Simpson, A., & Riggs, K. J. (2006). Conditions under which children experience inhibitory difficulty with a "button-press" go/no-go task. *Journal of Experimental Child Psychology, 94*(1), 18–26.

Simpson, A., & Riggs, K. J. (2007). Under what conditions do young children have difficulty inhibiting manual actions. *Developmental Psychobiology, 43*(2), 417–428.

Simsek, M., Al-Sharbati, M., Al-Adawi, S., Ganguly, S. S., & Lawatia, K. (2005). Association of the risk allele of dopamine transporter gene (DAT1*10) in Omani male children with attention-deficit hyperactivity disorder. *Clinical Biochemistry, 38*(8), 739–742.

Sinzig, J., Morsch, D., Bruning, N., Schmidt, M. H., & Lehmkuhl, G. (2008). Inhibition, flexibility, working memory and planning in autism spectrum disorders with and without comorbid ADHD-symptoms. *Child and Adolescent Psychiatry and Mental Health, 2*(4), 1–44.

Skaar, D. A., Shao, Y., Haines, J. L., Stenger, J. E., Jaworski, J., Martin, E. R., et al. (2005). Analysis of the RELN gene as a genetic risk factor for autism. *Molecular Psychiatry, 10*(6), 563–571.

Skinner, M., Hooper, S., Hatton, D. D., Robert, J., Mirrett, P., Schaaf, J., et al. (2005). Mapping nonverbal IQ in young boys with fragile X syndrome. *American Journal of Medical Genetics Part A, 132A*(1), 25–32.

Slagter, H. A., Giesbrecht, B., Kok, A., Weissman, D. H., Kenemans, J. L., Woldorff, M. G., et al. (2007). fMRI evidence for both generalized and specialized components of attentional control. *Brain Research, 1177*(66), 90–102.

Smallwood, J., Fishman, D. J., & Schooler, J. W. (2007). Counting the cost of an absent mind: Mind wandering as an underrecognized influence on educational performance. *Psychonomic Bulletin & Review, 14*(2), 230–236.

Smidts, D., Jacobs, R., & Anderson, V. (2004). The Object Classification Task for Children (OCTC): A measure of concept generation and mental flexibility in early childhood. *Developmental Neuropsychology, 26*, 385–402.

Smith, A., Taylor, E., Warner-Rogers, J., & Rubia, K. (2002). Temporal Processing in Children with ADHD: Evidence for a pure time perception deficit in ADHD. *Journal of Child Psychology & Psychiatry, 43*(4), 529–542.

Smith, A. D., Gilchrist, I. D., Hood, B., Tassasbehji., M., & Karmiloff-Smith, A. (2009). Inefficient search of large scale space in Williams syndrome: Further insights on the role of LIMK1 deletion in deficits of spatial cognition. *Perception, 38*, 694–701.

Smoller, J. W., Biederman, J., Arbeitman, L., Doyle, A. E., Fagerness, J., Perlis, R. H., et al. (2006). Association between the 5HT1B receptor gene (HTR1B) and the inattentive subtype of ADHD. *Biological Psychiatry, 59*(5), 460–467.

Sobin, C., Kiley-Brabeck, K., Daniels, S., Blundell, M., Anyane-Yeboa, K., & Karayiorgou, M. (2004). Networks of attention in children with the 22q11 deletion syndrome. *Developmental Neuropsychology, 26*(2), 611–626.

Sobin, C., Kiley-Brabeck, K., Daniels, S., Khuri, J., Taylor, L., Blundell, M., et al. (2005). Neuropsychological characteristics of children with the 22q11 deletion syndrome: A descriptive analysis. *Child Neuropsychology, 11*(1), 39–53.

Sobin, C., Kiley-Brabeck, K., & Karayiorgou, M. (2005a). Associations between pre-pulse inhibition and executive visual attention in children with the 22q11 deletion syndrome. *Molecular Psychiatry, 10*(6), 553–562.

Sobin, C., Kiley-Brabeck, K., & Karayiorgou, M. (2005b). Lower prepulse inhibition in children with the 22q11 deletion syndrome. *American Journal of Psychiatry, 162*(6), 1090–1099.

Sokhadze, E. M., El-Baz, A., Baruth, J., Mathai, G., Sears, L., & Casanova, M. F. (2009). Effects of low frequency repetitive transcranial magnetic stimulation (rTMS) on gamma frequency oscillations and event-related potentials during processing of illusory figures in autism. *Journal of Autism and Developmental Disorders, 39*(4), 619–634.

Solanto, M. V., Abikoff, H., Sonuga-Barke, E., Schachar, R., Logan, G. D., Wigal, T., et al. (2001). The ecological validity of delay aversion and response inhibition as measures of impulsivity in AD/HD: A supplement to the NIMH multimodal treatment study of AD/HD. *Journal of Abnormal Child Psychology, 29*(3), 215–228.

Sollner, T., Whiteheart, S. W., Brunner, M., Erdjument-Bromage, H., Geromanos, S., Tempst, P., et al. (1993). SNAP receptors implicated in vesicle targeting and fusion. *Nature, 362*(6418), 318–324.

Solomon, M., Ozonoff, S. J., Cummings, N., & Carter, C. S. (2008). Cognitive control in autism spectrum disorders. *International Journal of Developmental Neuroscience, 26*(2), 239–247.

Somsen, R. J. M. (2007). The development of attention regulation in the Wisconsin Card Sorting Task. *Developmental Science, 10*(5), 664–680.

Sonuga-Barke, E. J. S. (1994). On dysfunction and function in psychological theories of childhood disorder. *Journal of Child Psychology and Psychiatry and Allied Disciplines, 35*(5), 801–815.

Sonuga-Barke, E. J. S. (2002). Interval length and time-use by children with AD/HD: A comparison of four models. *Journal of Abnormal Child Psychology, 30*(3), 257–264.

Sonuga-Barke, E. J. S. (2005). Causal models of attention-deficit/hyperactivity disorder: From common simple deficits to multiple developmental pathways. *Biological Psychiatry, 57*(11), 1231–1238.

Sonuga-Barke, E. J. S. & Halperin, J. M. (2010). Developmental phenotypes and causal pathways in attention deficit/hyperactivity disorder: potential targets for early interventions. *Journal of Child Psychology and Psychiatry, 51*(4), 368–389.

Sonuga-Barke, E. J. S., Houlberg, K., & Hall, M. (1994). When is "impulsiveness" not Impulsive? The case of hyperactive childrens cognitive-style. *Journal of Child Psychology and Psychiatry and Allied Disciplines, 35*(7), 1247–1253.

Sonuga-Barke, E. J. S., Taylor, E., & Heptinstall, E. (1992). Hyperactivity and delay a version. 2. The effect of self versus externally imposed stimulus-presentation periods on memory. *Journal of Child Psychology and Psychiatry and Allied Disciplines, 33*(2), 399–409.

Sowell, E. R., Thompson, P. M., Leonard, C. M., Welcome, S. E., Kan, E., & Toga, A. W. (2004). Longitudinal mapping of cortical thickness and brain growth in normal children. *Journal of Neuroscience, 24*(38), 8223–8231.

Spelke, E., Hirst, W., & Neisser, U. (1976). Skills of divided attention. *Cognition, 4*(3), 215–230.

Spencer, T. J. (2006). ADHD and comorbidity in childhood. *Journal of Clinical Psychiatry, 67*(Suppl. 8), 27–31.

Sperling, G. (1960). The information available in brief visual presentations. *Psychological Monographs, 74*(11), 1–29.

Spinrad, T. L., Eisenberg, N., Gaertner, B., Popp, T., Smith, C. L., Kupfer, A., et al. (2007). Relations of maternal Socialization and toddlers' effortful control to children's adjustment and social competence. *Developmental Psychology, 43*(5), 1170–1186.

Spira, E. G., & Fischel, J. E. (2005). The impact of preschool inattention, hyperactivity, and impulsivity on social and academic development: A review. *Journal of Child Psychology and Psychiatry, 46*(7), 755–773.

Stanfield, A. C., McIntosh, A. M., Spencer, M. D., Philip, R., Gaur, S., & Lawrie S. M. (2008). Towards a neuroanatomy of autism: A systematic review and meta-analysis of structural magnetic resonance imaging studies. *European Psychiatry, 23*(4), 289–299.

Steele, M., Weiss, M., Swanson, J. M., Wang, J., Prinzo, R. S., & Binder, C. E. (2006). A randomized, controlled, effectiveness trial of OROS-methylphenidate in ADHD. *Canadian Journal of Clinical Pharmacology, 13*(1), 50–62.

Steele, S. D., Minshew, N. J., Luna, B., & Sweeney, J. A. (2007). Spatial working memory deficits in autism. *Journal of Autism and Developmental Disorders, 37*(4), 605–612.

Stein, M. A., Waldman, I. D., Sarampote, C. S., Seymour, K. E., Robb, A. S., Conlon, C., et al. (2005). Dopamine transporter genotype and methylphenidate dose response in children with ADHD. *Neuropsychopharmacology, 30*(7), 1374–1382.

Steingard, R., Biederman, J., Doyle, A., & Sprich-Buckminster, S. (1992). Psychiatric comorbidity in attention-deficit disorder: Impact on the interpretation of child-behavior checklist results. *Journal of the American Academy of Child and Adolescent Psychiatry, 31*(3), 449–454.

Steinhausen, H. C., Drechsler, R., Foldenyi, M., Imhof, K., & Brandeis, D. (2003). Clinical course of attention-deficit/hyperactivity disorder from childhood toward early adolescence. *Journal of the American Academy of Child and Adolescent Psychiatry, 42*(9), 1085–1092.

Stevens, J., Quittner, A. L., & Abikoff, H. (1998). Factors influencing elementary school teachers' ratings of ADHD and ODD behaviors. *Journal of Clinical Child Psychology, 27*(4), 406–414.

Stevens, J., Quittner, A. L., Zuckerman, J. B., & Moore, S. (2002). Behavioral inhibition, self-regulation of motivation, and working memory in children with attention deficit hyperactivity disorder. *Developmental Neuropsychology, 21*(2), 117–139.

Stevens, M. C., Pearlson, G. D., & Kiehl, K. A. (2007). An fMRI auditory oddball study of combined-subtype attention deficit hyperactivity disorder. *American Journal of Psychiatry, 164*(11), 1737–1749.

Stevenson, J., Asherson, P., Hay, D., Levy, F., Swanson, J., Thapar, A., et al. (2005). Characterizing the ADHD phenotype for genetic studies. *Developmental Science, 8*(2), 115–121.

Steyaert, J. G., & De la Marche, W. (2008). What's new in autism? *European Journal of Pediatrics, 167*(10), 1091–1101.

Striano, T., Henning, A., & Stahl, D. (2005). Sensitivity to social contingencies between 1 and 3 months of age. *Developmental Science, 8*(6), 509–518.

Stroh, C. M. (1971). *Vigilance: The problem of sustained attention.* Oxford, UK: Pergamon.

Stromme, P., Bjornstad, P. G., & Ramstad, K. (2002). Prevalence estimation of Williams syndrome. *Journal of Child Neurology, 17*(4), 269–271.

Stroop, J. R. (1935). Studies of interference in serial verbal reactions. *Journal of Experimental Psychology, 18*, 643–662.

Styles, E. A. (1997). *The Psychology of Attention.* Hove, U.K.: Psychology Press.

Sullivan, J. R., & Riccio, C. A. (2007). Diagnostic group differences in Parent and Teacher Ratings on the BRIEF and Conners' Scales. *Journal of Attention Disorders, 11*(3), 398–406.

Sullivan, K., Hatton, D., Hammer, J., Sideris, J., Hooper, S., Ornstein, P., et al. (2006). ADHD symptoms in children with FXS. *American Journal of Medical Genetics Part A, 140A*(21), 2275–2288.

Sullivan, K., Hatton, D. D., Hammer, J., Sideris, J., Hooper, S., Ornstein, P. A., et al. (2007). Sustained attention and response inhibition in boys with fragile X syndrome: Measures of continuous performance. *American Journal of Medical Genetics Part B: Neuropsychiatric Genetics, 144B*(4), 517–532.

Sundaram, S. K., Chugani, H. T., & Chugani, D. C. (2005). Positron emission tomography methods with potential for increased understanding of mental retardation and developmental disabilities. *Mental Retardation and Developmental Disabilities Research Reviews, 11*(4), 325–330.

Suskauer, S. J., Simmonds, D. J., Fotedar, S., Blankner, J. G., Pekar, J. J., Denckla, M. B., et al. (2008). Functional magnetic resonance imaging evidence for abnormalities in response selection in attention deficit hyperactivity disorder: Differences in activation associated with response inhibition but not habitual motor response. *Journal of Cognitive Neuroscience, 20*(3), 478–493.

Swanson, J., Arnold, L. E., Kraemer, H., Hechtman, L., Molina, B., Hinshaw, S., et al. (2008). Evidence, interpretation, and qualification from multiple reports of long-term outcomes in the Multimodal Treatment Study of Children With ADHD (MTA): Part I: Executive summary. *Journal of Attentional Disorders, 12*(1), 4–14.

Swanson, J. M. (1992). *School-based assessments and interventions for ADD students.* Irvine, CA: KC Publishing.

Swanson, J. M., Kraemer, H. C., Hinshaw, S. P., Arnold, L. E., Conners, C. K., Abikoff, H. B., et al. (2001). Clinical relevance of the primary findings of the MTA: Success rates based on severity of ADHD and ODD symptoms at the end of treatment. *Journal of the American Academy of Child and Adolescent Psychiatry, 40*(2), 168–179.

Swanson, J. M., Schuck, S., Mann, M., Carlson, C., Hartman, K., Sergeant, J., et al. (2005). Categorical and dimensional definitions and evaluations of symptoms of ADHD: The SNAP and the SWAN ratings scales. Available at: http://www.adhd.net/SNAP_SWAN.pdf.

Swets, J. A., Tanner, W. P., Jr. & Birdsall, T. G. (1961). Decision processes in perception. *Psychological Review, 68*(5), 301–340.

Swick, D., & Knight, R. T. (1998). Cortical lesions and attention. In R. Parasuraman (Ed.), *The attentive brain* (pp. 143–162). Cambridge, MA: MIT Press.

Swillen, A., Vandeputte, L., Cracco, J., Maes, B., Ghesquiere, P., Devriendt, K., et al. (1999). Neuropsychological, learning and psychosocial profile of primary school aged children with the velo-cardio-facial syndrome (22q11 deletion): Evidence for a nonverbal learning disability? *Child Neuropsychology, 5*(4), 230–241.

Szyf, M., McGowan, P., & Meaney, M. J. (2008). The social environment and the epigenome. *Environmental and Molecular Mutagenesis, 49*(1), 46–60.

Taffe, J. R., Gray, K. M., Einfeld, S. L., Dekker, M. C., Koot, H. M., Emerson, E., et al. (2007). Short form of the Developmental Behaviour Checklist. *American Journal on Mental Retardation, 112*(1), 31–39.

Tahir, E., Curran, S., Yazgan, Y., Ozbay, F., Cirakoglu, B., & Asherson, P. J. (2000). No association between low- and high-activity catecholamine-methyl-transferase (COMT) and attention deficit hyperactivity disorder (ADHD) in a sample of Turkish children. *American Journal of Medical Genetics Part B: Neuropsychiatric Genetics, 96*(3), 285–288.

Tanji, J., & Hoshi, E. (2008). Role of the lateral prefrontal cortex in executive behavioral control. *Physiological Reviews, 88*(1), 37–57.

Tannock, R. (2003). Neuropsychology of attention disorders. In I. Rapin & S. Segalowitz (Eds.), *Handbook of neuropsychology* (2nd ed.): *Child neuropsychology* (Vol. 8, Pt. II, pp. 753–784). Amsterdam: Elsevier.

Tannock, R., Campbell, B., Seymour, P., Quellett, D., Soares, H., Wang, P., et al. (2009). Towards a biological understanding of ADHD and discovery of novel therapeutic approaches. In R. A. McArthur & F. Borsini (Eds.) *Animal and translational models for CNS drug discovery: Psychiatric disorders* (Vol. 1). London: Elsevier Press.

Tannock, R., & Martinussen, R. (2007). ABCs of ADHD (Ch. 4): Rethinking ADHD in the classroom. From http://research.aboutkidshealth.ca/teachadhd

Tannock, R., Schachar, R., & Logan, G. (1995). Methylphenidate and cognitive flexibility: Dissociated dose effects in hyperactive children. *Journal of Abnormal Child Psychology, 23*(2), 235–266.

Tassabehji, M., & Donnai, D. (2006). Williams-Beuren syndrome: More or less? Segmental duplications and deletions in the Williams-Beuren syndrome region provide new insights into language development. *European Journal of Human Genetics, 14*(5), 507–508.

Tassabehji, M., Hammond, P., Karmiloff-Smith, A., Thompson, P., Thorgeirsson, S. S., Durkin, M. E., et al. (2005). GTF2IRD1 in craniofacial development of humans and mice. *Science, 310*(5751), 1184–1187.

Tassabehji, M., Metcalfe, K., Karmiloff-Smith, A., Carette, M. J., Grant, J., Dennis, N., et al. (1999). Williams syndrome: Use of chromosomal microdeletions as a tool to dissect cognitive and physical phenotypes. *American Journal of Human Genetics, 64*(1), 118–125.

Tassone, F., Hagerman, R. J., Chamberlain, W. D., & Hagerman, P. J. (2000). Transcription of the FMR1 gene in individuals with fragile X syndrome. *American Journal of Medical Genetics Part C: Seminars in Medical Genetics, 97*(3), 195–203.

Tassone, F., Hagerman, R. J., Taylor, A. K., Gane, L. W., Godfrey, T. E., & Hagerman, P. J. (2000). Elevated levels of FMR1 mRNA in carrier males: A new mechanism of involvement in the fragile-X syndrome. *The American Journal of Human Genetics, 66*(1), 6–15.

Taylor, E., Schachar, R., & Hepstinall, E. (1993). *Manual for Parental Account of Childhood Symptoms Interview*. London: Unpublished manuscript.

Taylor, M. J., Sunohara, G. A., Khan, S. C., & Malone, M. A. (1997). Parallel and serial attentional processes in ADHD: ERP evidence. *Developmental Neuropsychology, 13*(4), 531–539.

Taylor, P. C., Walsh, V., & Eimer, M. (2008). Combining TMS and EEG to study cognitive function and cortico-cortico interactions. *Behavioral Brain Research, 191*(2), 141–147.

Tellinghuisen, D. J., Oakes, L. M., & Tjebkes, T. (1999). The influence of attentional state and stimulus characteristics on infant distractibility. *Cognitive Development, 14*(2), 199–213.

Thomas, M. S., Annaz, D., Ansari, D., Scerif, G., Jarrold, C., & Karmiloff-Smith, A. (2009). Using developmental trajectories to understand developmental disorders. *Journal of Speech, Language, and Hearing Research, 52*(2), 336–358.

Thorell, L. B. (2007). Do delay aversion and executive function deficits make distinct contributions to the functional impact of ADHD symptoms? A study of early academic skill deficits. *Journal of Child Psychology and Psychiatry, 48*(11), 1061–1070.

Tillman, C. M., Thorell, L. B., Brocki, K. C., & Bohlin, G. (2008). Motor response inhibition and execution in the stop-signal task: Development and relation to ADHD behaviors. *Child Neuropsychology, 14*(1), 42–59.

Tipper, S. P. (1985). The negative priming effect: Inhibitory priming by ignored objects. *Quarterly Journal of Experimental Psychology Section A: Human Experimental Psychology, 37*(4), 571–590.

Tipper, S. P. (2001). Does negative priming reflect inhibitory mechanisms? A review and integration of conflicting views. *Quarterly Journal of Experimental Psychology Section A: Human Experimental Psychology, 54*(2), 321–343.

Todd, R. D., Huang, H. Y., Smalley, S. L., Nelson, S. F., Willcutt, E. G., Pennington, B. F., et al. (2005). Collaborative analysis of DRD4 and DAT genotypes in population-defined ADHD subtypes. *The Journal of Child Psychology and Psychiatry, 46*(10), 1067–1073.

Tolhurst, D. J. (1975). Sustained transient channels in human vision. *Vision Research, 15*(10), 1511–1155.

Toplak, M. E., & Tannock, R. (2005). Time perception: Modality and duration effects in Attention-Deficit-Hyperactivity Disorder (ADHD). *Journal of Abnormal Child Psychology, 33*, 639-654.

Toplak, M. E., Bucciarelli, S. M., Jain, U., & Tannock, R. (2009). Executive functions: Performance-based measures and the behavior rating inventory of executive function (BRIEF) in adolescents with attention deficit/hyperactivity disorder (ADHD). *Child Neuropsychology, 15*(1), 53–72.

Toplak, M. E., Connors, L., Shuster, J., Knezevic, B., & Parks, S. (2008). Review of cognitive, cognitive-behavioral, and neural-based interventions for Attention-Deficit/Hyperactivity Disorder (ADHD). *Clinical Psychology Review, 28*(5), 801–823.

Torrioli, M. G., Vernacotola, S., Peruzzi, L., Tabolacci, E., Mila, M., Militerni, R., et al. (2008). A double-blind, parallel, multicenter comparison of L-acetylcarnitine with placebo on the attention deficit hyperactivity disorder in fragile X syndrome boys. *American Journal of Medical Genetics A, 146*(7), 803–812.

Townsend, J., Courchesne, E., Covington, J., Westerfield, M., Harris, N. S., Lyden, P., et al. (1999). Spatial attention deficits in patients with acquired or developmental cerebellar abnormality. *Journal of Neuroscience, 19*(13), 5632–5643.

Towse, J. N., & Hitch, G. J. (1995). Is there a relationship between task demand and storage space in tests of working-memory capacity? *Quarterly Journal of Experimental Psychology Section A: Human Experimental Psychology, 48*(1), 108–124.

Towse, J. N., Hitch, G. J., & Hutton, U. (1998). A reevaluation of working memory capacity in children. *Journal of Memory and Language, 39*(2), 195–217.

Towse, J. N., Hitch, G. J., & Hutton, U. (2000). On the interpretation of working memory span in adults. *Memory & Cognition, 28*(3), 341–348.

Treisman, A. M. (1964). Selective attention in man. *British Medical Bulletin, 20*(1), 12–16.

Treisman, A. M. (1971). Shifting attention between ears. *Quarterly Journal of Experimental Psychology, 23*(2), 157–167.

Treisman, A. M. (1991). Search, similarity, and integration of features between and within dimensions. *Journal of Experimental Psychology: Human Perception and Performance, 17*(3), 652–676.

Treisman, A. M., & Gelade, G. (1980). Feature-integration theory of attention. *Cognitive Psychology, 12*(1), 97–136.

Tremblay, H., & Rovira, K. (2007). Joint visual attention and social triangular engagement at 3 and 6 months. *Infant Behavior and Development, 30*(2), 366–379.

Trezise, K. L., Gray, K. M., & Sheppard, D. M. (2008). Attention and vigilance in the child with Down's syndrome. *Journal of Applied Research in Intellectual Disabilities, 21*, 502–508.

Trick, L. M., & Enns, J. T. (1998). Lifespan changes in attention: The visual search task. *Cognitive Development, 13*(3), 369–386.

Tripp, G., Schaughency, E. A., & Clarke, B. (2006). Parent and teacher rating scales in the evaluation of attention-deficit hyperactivity disorder: Contribution to diagnosis and differential diagnosis in clinically referred children. *Journal of Developmental and Behavioral Pediatrics, 27*(3), 209–218.

Tsiouris, J. A., & Brown, W. T. (2004). Neuropsychiatric symptoms of fragile X syndrome: Pathophysiology and pharmacotherapy. *CNS Drugs, 18*(11), 687–703.

Tsuchiya, E., Oki, J., Yahara, N., & Fujieda, K. (2005). Computerized version of the Wisconsin card sorting test in children with high-functioning autistic disorder or attention-deficit/hyperactivity disorder. *Brain & Development, 27*(3), 233–236.

Tucha, O., Prell, S., Mecklinger, L., Bormann-Kischkel, C., Kubber, S., Linder, M., et al. (2006). Effects of methylphenidate on multiple components of attention in children with attention deficit hyperactivity disorder. *Psychopharmacology (Berl), 185*(3), 315–326.

Turk, J. (1998). Fragile X syndrome and attentional deficits. *Journal of Applied Research in Intellectual Disabilities, 11*, 175–191.

Turner, M. S., Cipolotti, L., Yousry, T., & Shallice, T. (2007). Qualitatively different memory impairments across frontal lobe subgroups. *Neuropsychologia, 45*(7), 1540–1552.

Ullmann, R. K., Sleator, E. K., & Sprague, R. L. (1991). ADD-H Comprehensive Teacher's Rating Scale. Champaign, IL: Metritech, Inc.

Ullmann, R. K., Sleator, E. K., & Sprague, R. L. (2000). ADD-H Comprehensive Teacher's Rating Scale—Second Edition. Champaign, IL: Metritech, Inc.

Ungerleider, L. G., & Mishkin, M. (1982). Two cortical visual systems. In D. J. Ingle, M. A. Goodale, & R. J. W. Mansfield (Eds.), *Analysis of visual behavior* (pp. 549–586). Cambridge, MA: MIT Press.

Unsworth, N., & Engle, R. W. (2007). The nature of individual differences in working memory capacity: Active maintenance in primary memory and controlled search from secondary memory. *Psychological Review, 114*(1), 104–132.

Valdovinos, M. G. (2007). Brief review of current research in FXS: Implications for treatment with psychotropic medication. *Research in Developmental Disabilities, 28*(6), 539–545.

van den Wildenberg, W. P. M., & van der Molen, M. W. (2004). Developmental trends in simple and selective inhibition of compatible and incompatible responses. *Journal of Experimental Child Psychology, 87*(3), 201–220.

van der Geest, J. N., Kemner, C., Camfferman, G., Verbaten, M. N., & van Engeland, H. (2001). Eye movements, visual attention, and autism: A saccadic reaction time study using the gap and overlap paradigm. *Biological Psychiatry, 50*(8), 614–619.

van der Meere, J. J., Vreeling, H. J., & Sergeant, J. A. (1992). A motor presetting study in hyperactive, learning disabled and control children. *Journal of Child Psychology and Psychiatry, 33*(8), 1347–1354.

Van der Oord, S., Prins, P. J. M., Oosterlaan, J., & Emmelkamp, P. M. G. (2008). Efficacy of methylphenidate, psychosocial treatments and their combination in school-aged children with ADHD: A meta-analysis. *Clinical Psychology Review, 28*(5), 783–800.

van Hagen, J. M., van der Geest, J. N., van der Giessen, R. S., Lagers-van Haselen, G. C., Eussen, H. J. F. M. M., Gille, J. J. P., et al. (2007). Contribution of CYLN2 and GTF2IRD1 to neurological and cognitive symptoms in Williams Syndrome. *Neurobiology of Disease, 26*(1), 112–124.

van Mourik, R., Oosterlaan, J., Heslenfeld, D. J., Konig, C. E., & Sergeant, J. A. (2007). When distraction is not distracting: A behavioral and ERP study on distraction in ADHD. *Clinical Neurophysiology, 118*(8), 1855–1865.

van Mourik, R., Oosterlaan, J., & Sergeant, J. A. (2005). The Stroop revisited: A meta-analysis of interference control in AD/HD. *Journal of Child Psychology and Psychiatry, 46*(2), 150–165.

Vaughan Van Hecke, A., Mundy, P. C., Acra, C. F., Block, J. J., Delgado, C. E. F., Parlade, M. V., Meyer, J. A., Neal, A. R., & Pomares, Y. B. (2007). Infant Joint Attention, Temperament, and Social Competence in Preschool Children. *Child Development, 78*, 53–69.

Verté, S., Geurts, H. M., Roeyers, H., Oosterlaan, J., & Sergeant, J. A. (2005). Executive functioning in children with autism and Tourette syndrome. *Developmental Psychopathology, 17*(2), 415–445.

Veugelers, D., Post-Uiterweer, A., Sergeant, J. A., & Oosterlaan, J. (2009). Interference control in children with attention deficit/hyperactivity disorder. *Journal of Abnormal Child Psychology, 37*(2), 293–303.

Vicari, S., Bellucci, S., & Carlesimo, G. A. (2006). Evidence from two genetic syndromes for the independence of spatial and visual working memory. *Developmental Medicine and Child Neurology, 48*(2), 126–131.

Vicari, S., & Carlesimo, G. A. (2006). Short-term memory deficits are not uniform in Down and Williams syndromes. *Neuropsychology Review, 16*(2), 87–94.

Viding, E., Williamson, D. E., & Hariri, A. R. (2006). Developmental imaging genetics: Challenges and promises for translational research. *Development and Psychopathology, 18*(3), 877–892.

Virji-Babul, N., Cheung, T., Weeks, D., Herdman, A. T., & Cheyne, D. (2007). Magnetoencephalographic analysis of cortical activity in adults with and without Down syndrome. *Journal of Intellectual Disability Research, 51*(12), 982–987.

Virji-Babul, N., Moiseev, A., Cheung, T., Weeks, D., Cheyne, D., & Ribary, U. (2008). Changes in mu rhythm during action observation and execution in adults with Down syndrome: Implications for action representation. *Neuroscience Letters, 436*(2), 177–180.

Visootsak, J., & Sherman, S. (2007). Neuropsychiatric and behavioral aspects of trisomy 21. *Current Psychiatry Reports, 9*(2), 135–140.

Visu-Petra, L., Benga, O., Tincaş, I., & Miclea, M. (2007). Visual-spatial processing in children and adolescents with Down's syndrome: A computerized assessment of memory skills. *Journal of Intellectual Disability Research, 51*(12), 942–952.

Visser, S. N., Lesesne, C. A., & Perou, R. (2007). National estimates and factors associated with medication treatment for childhood attention-deficit/hyperactivity disorder. *Pediatrics, 119*(Suppl. 1), S99–S106.

Vitiello, B., Abikoff, H. B., Chuang, S. Z., Kollins, S. H., McCracken, J. T., Riddle, M. A., et al. (2007). Effectiveness of methylphenidate in the 10-month continuation phase of the Preschoolers with ADHD Treatment Study (PATS). *Journal of Child and Adolescent Psychopharmacology, 17*(5), 593–603.

Vlamings, P. H., Jonkman, L. M., Hoeksma, M. R., van Engeland, H., & Kemner, C. (2008). Reduced error monitoring in children with autism spectrum disorder: An ERP study. *European Journal of Neuroscience, 28*(2), 399–406.

Volkow, N. D., Wang, G. J., Newcorn, J., Telang, F., Solanto, M. V., Fowler, J. S., et al. (2007). Depressed dopamine activity in caudate and preliminary evidence of limbic involvement in adults with attention-deficit/hyperactivity disorder. *Archives of General Psychiatry, 64*(8), 932–940.

Vorstman, J. A. S., Morcus, M. E. J., Duijff, S. N., Klaassen, P. W. J., Heineman-de Boer, J. A., Beemer, F. A., et al. (2006). The 22q11.2 deletion in children: High rate of autistic disorders and early onset of psychotic symptoms. *Journal of the American Academy of Child and Adolescent Psychiatry, 45*(9), 1104–1113.

Vostanis, P. (2006). Strengths and Difficulties Questionnaire: Research and clinical applications. *Current Opinion in Psychiatry, 19*(4), 367–372.

Waber, D. P., De Moor, C., Forbes, P. W., Almli, C. R., Botteron, K. N., Leonard, G., et al. (2007). The NTH MRI study of normal brain development: Performance of a population based sample of healthy children aged 6 to 18 years on a neuropsychological battery. *Journal of the International Neuropsychological Society, 13*(5), 729–746.

Wahlstedt, C., Thorell, L. B., & Bohlin, G. (2008). ADHD symptoms and executive function impairment: Early predictors of later behavioral problems. *Developmental Neuropsychology, 33*(2), 160–178.

Waldman, I. D., & Gizer, I. R. (2006). The genetics of attention deficit hyperactivity disorder. *Clinical Psychology Review, 26*(4), 396–432.

Waldman, I. D., Rowe, D. C., Abramowitz, S., Kozel, T., Mohr, J. H., Sherman, L., et al. (1998). Association and linkage of the dopamine transporter gene and attention-deficit hyperactivity disorder in children: Heterogeneity owing to diagnostic subtype and severity. *American Journal of Human Genetics, 63,* 1767–1776.

Wang, P. P., Doherty, S., Rourke, S. B., & Bellugi, U. (1995). Unique profile of visuo-perceptual skills in a genetic syndrome. *Brain and Cognition, 29*(1), 54–65.

Warner-Rogers, J., Taylor, A., Taylor, E., & Sandberg, S. (2000). Inattentive behavior in childhood: Epidemiology and implications for development. *Journal of Learning Disabilities, 33*(6), 520–536.

Wassink, T. H., Piven, J., & Patil, S. R. (2001). Chromosomal abnormalities in a clinic sample of individuals with autistic disorder. *Psychiatric Genetics, 11*(2), 57–63.

Waterland, R. A., & Michels, K. B. (2007). Epigenetic epidemiology of the developmental origins hypothesis. *Annual Review of Nutrition, 27*, 363–388.

Weiss, L. A., Shen, Y., Korn, J. M., Arking, D. E., Miller, D. T., Fossdal, R., et al. (2008). Association between microdeletion and microduplication at 16p11.2 and autism. *The New England Journal of Medicine, 358*(7), 667–675.

Weissberg, R., Ruff, H. A., & Lawson, K. R. (1990). The usefulness of reaction time tasks in studying attention and organization of behavior in young children. *Journal of Developmental and Behavioral Pediatrics, 11*(2), 59–64.

Welsh, M. C., & Pennington, B. F. (1988). Assessing frontal lobe functioning in children: Views from developmental psychology. *Developmental Neuropsychology, 4*(3), 199–230.

Wheeler, A., Hatton, D., Reichardt, A., & Bailey, D. (2007). Correlates of maternal behaviours in mothers of children with fragile X syndrome. *Journal of Intellectual Disability Research, 51*(6), 447–462.

Whyte, J., Grieb-Neff, P., Gantz, C., & Polansky, M. (2006). Measuring sustained attention after traumatic brain injury: Differences in key findings from the sustained attention to response task (SART). *Neuropsychologia, 44*(10), 2007–2014.

Wigal, S., Swanson, J. M., Feifel, D., Sangal, R. B., Elia, J., Casat, C. D., et al. (2004). A double-blind, placebo-controlled trial of dexmethylphenidate hydrochloride and d,l-threo-methylphenidate hydrochloride in children with attention-deficit/hyperactivity disorder. *Journal of the American Academy of Child and Adolescent Psychiatry, 43*(11), 1406–1414.

Wigg, K. G., Takhar, A., Ickowicz, A., Tannock, R., Kennedy, J. L., Pathare, T., et al. (2006). Gene for the serotonin transporter and ADHD: No association with two functional polymorphisms. *American Journal of Medical Genetics Part B: Neuropsychiatric Genetics, 141B*(6), 566–570.

Wilding, J. (2003). Attentional difficulties in children: Weakness in executive function or problems in coping with difficult tasks? *British Journal of Psychology, 94*(4), 427–436.

Wilding, J. (2005). Is attention impaired in ADHD? *British Journal of Developmental Psychology, 23*(4), 487–505.

Wilding, J., & Burke, K. (2006). Attentional differences between groups of preschool children differentiated by teacher ratings of attention and hyperactivity. *British Journal of Developmental Psychology, 24*(2), 283–291.

Wilding, J., & Cornish, K. (2004). Efficiency of working memory and attention in children with Down syndrome. In J. Mallard (Ed.), *Focus on Down syndrome research* (pp. 147–162). New York: Nova.

Wilding, J., & Cornish, K. (2007). Independence of speed and accuracy in visual search: Evidence for separate mechanisms. *Child Neuropsychology, 13*(6), 510–521.

Wilding, J., Cornish, K., & Munir, F. (2002). Further delineation of the executive deficit in males with fragile-X syndrome. *Neuropsychologia, 40*(8), 1343–1349.

Wilding, J., Munir, F., & Cornish, K. (2001). The nature of attentional differences between groups of children differentiated by teacher ratings of attention and hyperactivity. *British Journal of Psychology, 92*(2), 357–371.

Wilding, J., Pankhania, P., & Williams, A. (2007). Effects of speed and accuracy instructions on performance in a visual search task by children with good or poor attention. *British Journal of Psychology, 98*(1), 127–139.

Wilens, T. E., Faraone, S. V., Biederman, J., & Gunawardene, S. (2003). Does stimulant therapy of attention-deficit/hyperactivity disorder beget later substance abuse? A meta-analytic review of the literature. *Pediatrics, 111*(1), 179–185.

Wilens, T. E., McBurnett, K., Bukstein, O., McGough, J., Greenhill, L., Lerner, M., et al. (2006). Multisite controlled study of OROS methylphenidate in the treatment of adolescents with attention-deficit/hyperactivity disorder. *Archives of Pediatric and Adolescent Medicine, 160*(1), 82–90.

Wilens, T. E., Spencer, T. J., & Biederman, J. (2002). A review of the pharmacotherapy of adults with attention-deficit/hyperactivity disorder. *Journal of Attentional Disorders, 5*(4), 189–202.

Wilke, M., Krägeloh-Mann, I., Holland, S. K. (2007). Global and local development of gray and white matter volume in normal children and adolescents. *Experimental Brain Research, 178*(3), 296–307.

Willemsen, R., Oostra, B. A., Bassell, G. J., & Dictenberg, J. (2004). The fragile X syndrome: From molecular genetics to neurobiology. *Mental Retardation and Developmental Disabilities Research Reviews, 10*(1), 60–67.

Williams, D. L., Goldstein, G., & Minshew, N. J. (2006). The profile of memory function in children with autism. *Neuropsychology, 20*(1), 21–29.

Wilson, T. W., Rojas, D. C., Reite, M. L., Teale, P. D., Rogers, S, J. (2007). Children and adolescents with autism exhibit reduced MEG steady-state gamma responses. *Biological Psychiatry, 62*(3), 192–197.

Wingen, M., Kuypers, K. P., van de Ven, V., Formisano, E., & Ramaekers, J. G. (2008). Sustained attention and serotonin: A pharmaco-fMRI study. *Human Psychopharmacology, 23*(3), 221–230.

Wodka, E. L., Mahone, E. M., Blankner, J. G., Larson, J. C. G., Fotedar, S., Denckla, M. B., et al. (2007). Evidence that response inhibition is a primary deficit in ADHD. *Journal of Clinical and Experimental Neuropsychology, 29*(4), 345–356.

Woerner, W., Becker, A., & Rothenberger, A. (2004). Normative data and scale properties of the German parent SDQ. *European Child & Adolescent Psychiatry, 13*(Suppl. 2), 3–10.

Woerner, W., Fleitlich-Bilyk, B., Martinussen, R., Fletcher, J., Cucchiaro, G., Dalgalarrondo, P., et al. (2004). The Strengths and Difficulties Questionnaire overseas: Evaluations and applications of the SDQ beyond Europe. *European Child & Adolescent Psychiatry, 13*(Suppl. 2), ii47–ii54.

Wolraich, M. L., Lambert, E. W., Bickman, L., Simmons, T., Doffing, M. A., & Worley, K. A. (2004). Assessing the impact of parent and teacher agreement on diagnosing attention-deficit hyperactivity disorder. *Journal of Developmental and Behavioral Pediatrics, 25*(1), 41–7.

Woodcock, K. A., Oliver, C., & Humphreys, G. W. (2009). Task-switching deficits and repetitive behavior in genetic neurodevelopment disorders: Data from children with Prader-Willi syndrome chromosome 15 q11-q13 deletion and boys with Fragile X syndrome. *Cognitive Neuropsychology, 26*(2), 172–194.

Woodin, M., Wang, P. P., Aleman, D., McDonald-McGinn, d., Zackai, E., & Moss, E. (2001). Neuropsychological profile of children and adolescents with the 22q11.2 deletion syndrome. *American Journal of Medical Genetics, 2001*; 101: 17–19;

Wright, I., Waterman, M., Prescott, H., & Murdoch-Eaton, D. (2003). A new Stroop-like measure of inhibitory function development: Typical Development Trends. *Journal of Child Psychology and Psychiatry, 44*(4), 561–575.

Wuhr, P. (2007). A Stroop effect for spatial orientation. *Journal of General Psychology, 134*(3), 285–294.

Yirmiya, N., Pilowsky, T., Nemanov, L., Arbelle, S., Feinsilver, T., Fried, I., et al. (2001). Evidence for an association with the serotonin transporter promoter region polymorphism and autism. *American Journal of Medical Genetics Part B: Neuropsychiatric Genetics, 105*(4), 381–386.

Youngwirth, S. D., Harvey, E. A., Gates, E. C., Hashim, R. L., & Friedman-Weieneth, J. L. (2007). Neuropsychological abilities of preschool-aged children who display hyperactivity and/or oppositional-defiant behavior problems. *Child Neuropsychology, 13*(5), 422–443.

Ypsilanti, A., & Grouios, G. (2008). Linguistic profile of individuals with Down syndrome: Comparing the linguistic performance of three developmental disorders. *Child Neuropsychology, 14*(2), 148–170.

Zahn, T. P., Kruesi, M. J. P., & Rapoport, J. L. (1991). Reaction-time indices of attention deficits in boys with disruptive behavior disorders. *Journal of Abnormal Child Psychology, 19*(2), 233–252.

Zelazo, P. D. (2006). The Dimensional Change Card Sort (DCCS): A method of assessing executive function in children. *Nature Protocols, 1*, 297–301.

Zelazo, P. D., Burack, J. A., Benedetto, E., & Frye, D. (1996). Theory of mind and rule use in individuals with Down's Syndrome: A test of the uniqueness and specificity claims. *Journal of Child Psychology and Psychiatry, 37*(4), 479–484.

Zelazo, P. D., & Frye, D. (1997). Cognitive complexity and control: A theory of the development of deliberate reasoning and intentional action. In M. I. Stamenov (Ed.), *Language structure, discourse and the access to consciousness* (pp. 112–153). Amsterdam: Benjamins.

Zelazo, P. D., Jacques, S., Burack, J. A., & Frye, D. (2002). The relation between theory of mind and rule use: Evidence from persons with autism-spectrum disorders. *Infant and Child Development, 11*(2), 171–195.

Zelazo, P. D., Müller, U., Frye, D., Marcovitch, S., Argitis, G., Boseovski, J., et al. (2003). The development of executive function in early childhood. *Monographs of the Society for Research in Child Development, 68*(3), vii–137.

Zhan, J. Y., Wilding, J., Cornish, K., Shao, J., Xie, C. H., Wang, Y., et al. (2010). Charting the developmental trajectories of attention and executive function in Chinese school-aged children. Manuscript submitted for publication.

Zhang, A., Shen, C. H., Ma, S. Y., Ke, Y., El Idrissi, A. (2009). Altered expression of Autism-associated genes in the brain of Fragile X mouse model. *Biochemical & Biophysical Research Communications, 379*(4), 920–923.

Zhang, S., Faries, D. E., Vowles, M., & Michelson, D. (2005). ADHD Rating Scale IV: Psychometric properties from a multinational study as a clinician-administered instrument. *International Journal of Methods in Psychiatric Research, 14*(4), 186–201.

Zhao, A. L., Su, L. Y., Zhang, Y. H., Tang, B. S., Luo, X. R., Huang, C. X., et al. (2005). Association analysis of serotonin transporter promoter gene polymorphism with ADHD and related symptomatology. *International Journal of Neuroscience, 115*(8), 1183–1191.

Zigman, W. B., & Lott, I. T. (2007). Alzheimer's disease in Down syndrome: Neurobiology and risk. *Mental Retardation and Developmental Disabilities Research Reviews, 13*(3), 237–246.

Zimmermann, P., & Fimm, B. (1993). Testbatterie zur Aufmerksam keitspruefung (TAP). Version 1.02. Freiburg, Germany: Psytest.

Zinkstok, J., & van Amelsvoort, T. (2005). Neuropsychological profile and neuroimaging in patients with 22q11.2 deletion syndrome: A review. *Child Neuropsychology, 11*(1), 21–37.

Zoroglu, S. S., Erdal, M. E., Alasehirli, B., Erdal, N., Sivasli, E., Tutkun, H., et al. (2002). Significance of serotonin transporter gene 5-HTTLPR and variable number of tandem repeat polymorphism in attention deficit hyperactivity disorder. *Neuropsychobiology, 45*(4), 176–181.

Zwaigenbaum, L., Bryson, S., Rogers, T., Roberts, W., Brian, J., & Szatmari, P. (2005). Behavioral manifestations of autism in the first year of life. *International Journal of Developmental Neuroscience, 23*(2–3), 143–152.

Index

Note: Page numbers followed by "*b*," "*f*," and "*t*" denote boxes, figures, and tables, respectively.